ECOLOGY OF
PREDATOR-PREY INTERACTIONS

ECOLOGY OF
PREDATOR-PREY INTERACTIONS

Edited by

Pedro Barbosa and Ignacio Castellanos

OXFORD
UNIVERSITY PRESS

2005

OXFORD

UNIVERSITY PRESS

Oxford University Press, Inc., publishes works that further
Oxford University's objective of excellence
in research, scholarship, and education.

Oxford   New York
Auckland   Cape Town   Dar es Salaam   Hong Kong   Karachi
Kuala Lumpur   Madrid   Melbourne   Mexico City   Nairobi
New Delhi   Shanghai   Taipei   Toronto

With offices in
Argentina   Austria   Brazil   Chile   Czech Republic   France   Greece
Guatemala   Hungary   Italy   Japan   Poland   Portugal   Singapore
South Korea   Switzerland   Thailand   Turkey   Ukraine   Vietnam

Published by Oxford University Press, Inc.
198 Madison Avenue, New York, New York 10016

www.oup.com

Oxford is a registered trademark of Oxford University Press

Library of Congress Cataloging-in-Publication Data
Ecology of predator-prey interactions / edited by Pedro Barbosa and Ignacio Castellanos.
p.   cm.
ISBN-13 978-0-19-517120-4
ISBN 0-19-517120-9
1. Predation (Biology)   2. Predatory animals—Ecology.   I. Barbosa, Pedro, 1944–
II. Castellanos, Ignacio.
QL758.E29 2004
591.5'3—dc22   2004014732

9 8 7 6 5 4 3 2 1

Printed in the United States of America
on acid-free paper

I (PB) dedicate this book to Quinto and all his future cousins. I think about you with joyful anticipation, even before you all arrive.

Preface

There are four major types of interactions that encompass the most critical and significant interactions among and between animals and plants. They are competition, mutualism, herbivore-plant interactions, and predator-prey interactions. Over the last several decades these types of interactions have come under severe scrutiny. The importance of, and in some cases the very definition of and assumptions about, competition, mutualism, and herbivore-plant interactions have changed dramatically. Only recently has the role of predation, as a force shaping the behavior, ecology, and evolution of vertebrate and invertebrate species, been exposed to a similarly intense scrutiny. For example, only recently have researchers experimentally shown that in numerous systems predators are important not only because they eat prey species, but also because they may trigger morphological, behavioral, and life-history changes in prey species.

These and other indirect effects may be as important to the dynamics of some predator and prey populations as the traditional direct impacts of predators. Recent research has demonstrated, for example, that the impact of predators on prey populations may depend on the nature of interactions among predator species: intraguild predation, which appears to be more common than previously assumed, may determine the outcome of many predator-prey interactions. Perhaps most fundamentally central to our understanding of the ecology of predator-prey interactions is the extensive debate over whether species assemblages and communities are defined or structured by natural enemies such as predators (top-down effects), or whether the resource base available to plants (bottom-up effects) ultimately determines the abundance of species in the community. All of these issues, and others, have triggered renewed interest in the ecology of predator-prey interactions.

Although several important books have been published on predator-prey interactions, none has taken a holistic and comprehensive view of the ecology of predator-prey interactions. The renaissance of interest in predator-prey interactions has engendered a new focus on all aspects of the ecology of such interactions. In particular, it is becoming increasingly clear that the ecological bases and consequences of predator-prey interactions are best understood through incorporation of data and perspectives from various related subdisciplines not typically considered. These perspectives provide different but critical and informative evolutionary, physiological, and behavioral insights. This book attempts to respond to the renaissance of interest in predation by incorporating into its chapters empirical and theoretical perspectives resulting from the latest research on predator-prey interactions. A second objective of the book is to include chapters that challenge the reader and, where appropriate, present hypotheses that stir interest, advance our understanding of the ecology of predator-prey interactions, and stimulate new research.

Acknowledgments

We appreciate and thank the many reviewers listed below and other anonymous reviewers for their very important contributions. We know that we speak for all the authors when we note that their comments and suggestions have had an immense impact on the quality of the chapters in this book. In alphabetical order, the reviewers are Peter Abrams, Fred Adler, Anurag Agrawal, Dale Bottrell, Culum Brown, Tim Caro, Moshe Coll, James T. Cronin, Sebastian Diehl, Jim Dietz, Larry Dill, Hugh Dingle, Lee Dyer, Micky Eubanks, William Fagan, Rosemary G. Gillespie, Geoff Gurr, Jason Harmon, Brian A. Hazlett, Jason Hoeksema, Robert D. Holt, L. B. Kats, Suzanne Koptur, Doug Landis, Steve Lima, Per Nystrom, Kenneth F. Raffa, Heiner Römer, Jay Rosenheim, Ann L. Rypstra, Maurice W. Sabelis, Os Schmitz, William Snyder, Kyle Summers, John Terborgh, Teja Tscharntke, Roy Van Driesche, Kevin Warburton, Eric J. Warrant, Earl Werner, David Wise, David E. Wooster, Jayne Yack, Ronald C. Ydenberg, Jochen Zeil, and Pat Zollner.

Contents

PART III   Population- and Community-Level Interactions

PART IV   Applied Consequences of Predator-Prey Interactions

Contributors

Peter A. Abrams
Department of Zoology
Ramsay Wright Zoological Labs
25 Harbord Street
University of Toronto
Toronto, ON M58 1A1, Canada
abrams@zoo.utoronto.ca

Pedro Barbosa
Department of Entomology
Plant Sciences Building
University of Maryland
College Park, MD 20742
pbarbosa@umd.edu

Alice Boyle
Department of Ecology and
    Evolutionary Biology
University of Arizona
Tucson, AZ 85721
alboyle@email.arizona.edu

Judith L. Bronstein
Department of Ecology and
    Evolutionary Biology
University of Arizona
Tucson, AZ 85721
judieb@u.arizona.edu

Grant E. Brown
Department of Biology
Concordia University
1455 Boulevard de Maisonneuve O.
Montreal, QC H3G 1M8, Canada
gbrown@alcor.concordia.ca

Astrid Caldas
Department of Entomology
Plant Sciences Building
University of Maryland
College Park, MD 20742
acaldas@umd.edu

Ignacio Castellanos
Centro de Investigaciones Biológicas
Universidad Autónoma del Estado de
    Hidalgo A.P. 69
Pachuca, Hidalgo, 42001, Mexico
ignacioe@uaeh.reduaeh.mx

Gary C. Chang
Department of Entomology
Washington State University
Pullman, WA 99164-6382
gchang@wsu.edu

Douglas P. Chivers
Department of Biology
University of Saskatchewan
112 Science Place
Saskatoon, SK S7N 5E2, Canada
Doug.Chivers@usask.ca

Thomas W. Cronin
Department of Biological Sciences
University of Maryland, Baltimore
    County
1000 Hilltop Circle
Baltimore, MD 21250
cronin@umbc.edu

Robert F. Denno
Department of Entomology
Plant Sciences Building
University of Maryland
College Park, MD 20742
rdenno@umd.edu

Joseph C. Dickens
U.S. Department of Agriculture
Agricultural Research Service
Chemicals Affecting Insect Behavior
    Laboratory
Building 007 BARC-West
10300 Baltimore Avenue
Beltsville, MD 20705
dickensj@ba.ars.usda.gov

Lee A. Dyer
Department of Ecology and
    Evolutionary Biology
310 Dinwiddie Hall
Tulane University
New Orleans, LA 70118
ldyer@tulane.edu

Micky D. Eubanks
Department of Entomology and Plant
    Pathology
Auburn University
327 Funchess Hall
Auburn, AL 36849
eubanmd@auburn.edu

Deborah L. Finke
Department of Entomology
Plant Sciences Building
University of Maryland
College Park, MD 20742
dfinke@umd.edu

John L. Gittleman
Department of Biology
Gilmer Hall
University of Virginia
Charlottesville, VA 22904-4328
JLGittleman@virginia.edu

Matthew E. Gompper
Department of Fisheries and Wildlife
    Sciences
302 Anheuser-Busch Natural
    Resources Building
University of Missouri
Columbia, MO 65211-7240
gompperm@missouri.edu

Matthew H. Greenstone
U.S. Department of Agriculture
Agricultural Research Service
Plant Sciences Institute
Insect Biocontrol Lab
Building 011A, Room 214
Beltsville Agricultural Research
    Center-W
Beltsville, MD 20705
greenstm@ba.ars.usda.gov

J. Nathaniel Holland
Rice University, MS 170
6100 South Main Street
Houston, TX 77005-1892
jholland@rice.edu

Gail A. Langellotto
Louis Calder Center
Biological Field Station
Fordham University
53 Whippoorwill Road, Box 887
Armonk, NY 10504
langellotto@fordham.edu

Steven L. Lima
Department of Life Sciences
Indiana State University
Terre Haute, IN 47809
lslima@scifac.indstate.edu

Robert Matlock
Department of Ecology and
    Evolutionary Biology
310 Dinwiddie Hall
Tulane University
New Orleans, LA 70118
rmatlock@tulane.edu

Joshua H. Ness
Department of Ecology and
    Evolutionary Biology
University of Arizona
Tucson, AZ 85721
jness@email.arizona.edu

Renée P. Prasad
Department of Entomology
Washington State University
Pullman, WA 99164-6382
prasad@wsu.edu

Rick A. Relyea
University of Pittsburgh
Department of Biological Sciences
101A Clapp Hall
4249 Fifth Avenue
Pittsburgh, PA 15260
relyea+@pitt.edu

Susan E. Riechert
Ecology and Evolutionary Biology
The University of Tennessee
569 Dabney Hall
Knoxville, TN 37996-1610
sriecher@utk.edu

Oswald J. Schmitz
School of Forestry and Environmental
    Studies
Yale University
370 Prospect Street
New Haven, CT 06511
oswald.schmitz@yale.edu

Andrew Sih
Department of Environmental Science
    and Policy
University of California
One Shields Avenue
Davis, CA 95616
asih@ucdavis.edu

William E. Snyder
Department of Entomology
Washington State University
Pullman, WA 99164-6382
wesnyder@wsu.edu

Todd D. Steury
Department of Life Sciences
Indiana State University
Terre Haute, IN 47809
lssteury@scifac.indstate.edu

John Stireman
Department of Ecology, Evolution,
    and Organismal Biology
353 Bessey Hall
Iowa State University
Ames, IA 50011-1020
stireman@iastate.edu

Jeffrey D. Triblehorn
Department of Biology
Georgia State University
Atlanta, GA 30303
biojdt@langate.gsu.edu

David D. Yager
Department of Psychology
University of Maryland
College Park, MD 20742
ddyager@umd.edu

PART I

THE NATURE OF PREDATION AND PREDATOR-PREY INTERACTIONS

What is a predator? The answer is relatively straightforward for many species: an animal that eats another animal. However, recent research has shown that for other species the answer is more complex and that predatory species may not be as diet-limited as previously assumed. An alternative trophic lifestyle, omnivory is a more widespread feeding habit than previously assumed and includes thousands of invertebrate and vertebrate species. Omnivores often have a mixture of morphological, physiological, and behavioral traits possessed by their strictly predaceous relatives, and other traits possessed by strictly herbivorous relatives. In chapter 1, Micky Eubanks provides some excellent examples to illustrate how being an omnivore has significant consequences for so-called predator-prey interactions. Similarly, he shows that this lifestyle may have a major influence on the functional and numerical responses of omnivores (or species typically viewed solely as predators) to prey and, consequently, on the impact of omnivores on prey populations.

The blend of lifestyles may indicate alternative and broader perspectives on traditional predator-prey interactions. Indeed, one such perspective is to view predator-prey interactions as consumer-resource interactions. This broader perspective would include the use of nutrients by a plant, as well as the consumption of a gazelle by a lion, as a consumer-resource interaction. In this vein, J. Nathaniel Holland and colleagues argue in chapter 2 that, when predator-prey interactions are viewed as consumer-resource interactions, they may encompass other types of interactions (including nearly all mutualisms) not traditionally considered as predator-prey interactions. They further argue that, although the outcome of consumer-resource interactions often is broadly viewed as a +/– interaction, it need not be viewed in such a limited fashion.

Regardless of whether one defines predator-prey interactions narrowly or broadly, what is clear is that at one point or another in the lives of most animals they are predators, prey, or both. Thus, predation risk assessment is central to the lives of most species. In chapter 3, Grant Brown and Douglas Chivers argue that predation risk avoidance can be optimized when prey rely on learning as an adaptive response to predation risk, rather than relying on fixed (i.e., innate) avoidance responses to potential predation threats. Because predation risk is ever present, assessing danger and acting on it (defensively) is essential for all organisms. However, actions taken in defense are costly. Thus, there is a continuous evolutionary process to maximize defense while minimizing cost. Brown and Chivers suggest that one way the latter can be achieved is through learning, a process that allows for continual modification of responses to predation risk. They demonstrate that learning allows for the continual reinforcement of ecologically relevant information.

In contrast, what defines a predatory lifestyle is the biological and ecological adaptations that facilitate finding prey. These adaptations are reflected in foraging paradigms shaped by environmental factors and a species' genetics. Studies have identified potential patterns of inheritance of foraging traits in predators, as well as genetic influences on morphological traits that affect the nature of attack, the ease of taking prey, discrimination of prey among diet specialists, resistance to prey defense mechanisms, and the search patterns used during foraging. Susan Riechert provides a series of excellent examples in chapter 4 demonstrating that genetic factors play a significant role in foraging behavior.

1

Predaceous Herbivores and Herbivorous Predators

The Biology of Omnivores and the Ecology of Omnivore-Prey Interactions

MICKY D. EUBANKS

Until recently, omnivory (i.e., feeding at multiple trophic levels) was thought to be rare in nature. It is now realized that omnivory is a widespread feeding habit and that there are thousands of invertebrate and vertebrate omnivores. We are, however, just beginning to understand the ecological consequences of feeding at multiple trophic levels for omnivore-prey interactions. This chapter focuses on the large number of animal species that consume both prey and plant food. The goal of the chapter is to highlight the life history traits of omnivores that affect their interactions with prey and to examine what we know about the effect of these traits on the ecology of omnivore-prey interactions.

Omnivores often have a mixture of morphological, physiological, and behavioral traits possessed by their strictly predaceous or strictly herbivorous relatives. This unique combination of traits strongly affects their functional and numerical responses to prey and, consequently, the impact of omnivores on prey populations. These effects, however, are just beginning to be studied by ecologists. Several heuristic and mathematical models have explored the ecology of omnivore-prey interactions. Many of these models predict that omnivores can have a greater suppressive effect on prey populations than strict predators because an omnivore's ability to feed on alternative food ultimately results in increased prey suppression. A relatively large body of anecdotal observations, sampling data, and several empirical studies support this prediction. The degree to which feeding on alternative food affects prey suppression depends on the relative effects of omnivory on per capita prey consumption, omnivore persistence during periods of low or no prey abundance, and the numerical response of omnivores to prey and plant food.

The effects of feeding on alternative food on per capita prey consumption and prey preferences of omnivores are complex and poorly understood. For example, the fitness of omnivores is often strongly affected by diet mixing, whereby the nutritional quality of one type of food (e.g., plant material) is altered by the ingestion of another type of food (e.g., prey). Consequently, the prey preferences of omnivorous animals can appear to be "suboptimal" in the sense that prey of less, "stand-alone" nutritional value may be more strongly preferred than simple optimal foraging theory might predict. Furthermore, the food preferences of omnivores are constrained by their evolutionary history and differences in prey behavior, size, toxicity or abundance. This makes understanding food preferences and per capita prey consumption of these animals very difficult, yet understanding these preferences is critical.

The effects of prey and plant consumption on the numerical response of omnivores appear to be less complicated. Omnivores, especially omnivorous insects, appear to track variation in high-quality plants more effectively than variation in prey abundance. Consequently, omnivores may have weaker numerical responses to prey than their strictly predaceous relatives and may have their strongest effects on prey populations when plant quality is high, even if per capita prey consumption is reduced under these conditions.

Omnivory may strongly affect the conservation biology of an omnivorous species or the likelihood of a successful invasion by an omnivore. Some of the invasive species with the strongest impacts on native ecosystems, for example, are extremely omnivorous with exceptionally broad diet breadth. Consequently, management practices and goals associated with omnivorous species may be different from those associated with strictly predaceous or herbivorous species.

There is still much to be learned about the ecology of omnivore-prey interactions. Understanding the effects of alternative food on per capita prey consumption and the dispersal, or numerical, response of omnivores is critical to understanding the ecological consequences of omnivory. Future studies focusing on these areas would add much to our understanding of the ecological consequences of this widespread feeding habit.

Although predicted to be rare in nature by classical food web theory (Pimm and Lawton 1977, Pimm and Lawton 1978, Yodzis 1984), feeding at multiple trophic levels is now recognized as a widespread feeding habit in animals (Sprules and Bowerman 1988, Polis et al. 1989, Vadas 1989, Diehl 1993, Whitman et al. 1994, Rosenheim et al. 1995, Alomar and Wiedenmann 1996, Coll and Guershon 2002, Cooper and Vitt 2002). Thousands of insect species, for example, consume both prey and plant food, including grasshoppers, earwigs, thrips, true bugs, beetles, and ants (Whitman et al. 1994, Coll and Guershon 2002). In addition, many other arthropod species representing taxa as diverse as mites, copepods, and crustaceans include both prey and plant food in their diets (Kleppel 1993, Momen and El-Saway 1993, McMurtry and Croft 1997, Buck et al. 2003). Many vertebrate animals are also omnivorous. Over half of all lizard species, for instance, are omnivorous (Cooper and Vitt 2002), and feeding on prey and plant food is a common way of making a living in small mammals, birds, and fish (Stoddart 1979, Moyle and Cech 1988, Vadas 1989, Gill 1994, Klasing 1999, Campos et al. 2001).

Although there is growing interest in omnivory (e.g., a recent *Ecology* special feature on omnivory, Agrawal 2003), the traits associated with omnivory and the ecological consequences of feeding at more than one trophic level have only recently been intensively studied. This chapter highlights some of the traits associated with feeding on both prey and plant food and explores the ecological consequences of such feeding for omnivore-prey interactions. Intraguild predation, a form of omnivory in which predators that share the same prey species consume each other, has recently received extensive attention (Rosenheim et al. 1993, Rosenheim et al. 1995, Denno and Fagan 2003, Rosenheim and Corbett 2003) and is covered in other chapters of this book (e.g., Denno et al., ch. 10 in this volume), so intraguild predation is not discussed in this chapter. Although this chapter focuses to a large extent on arthropods, broad generalizations concerning the ecological consequences of omnivory and the questions that need to be addressed in the future are applicable to almost all invertebrate and vertebrate omnivores.

Traits of Omnivores

Omnivores that feed on both plants and prey represent a unique blend of morphological, physiological, and behavioral adaptations found in their predaceous and herbivorous relatives (Cooper 2002, Cooper and Vitt 2002, Eubanks et al. 2003). Omnivorous insects in the order Heteroptera (true bugs), for example, possess digestive tracts and accessory salivary glands that are intermediate in length, size, and placement compared to those found in their herbivorous and predaceous relatives (Slater and Carayon 1963, Goodchild 1966, Boyd et al. 2002). In addition, many species of omnivorous bugs produce protein-digesting enzymes (proteinases and phospholipases) and plant-digesting enzymes (amylases and pectinases), whereas their strictly herbivorous and predaceous cousins produce only a subset (Baptist 1941, Goodchild 1966, Kahn and Ford 1967, Miles 1972, Varis et al. 1983, Cohen 1990, Cohen 1996, Schaefer and Panizzi 2000, Wheeler 2001, Boyd et al. 2002). Likewise, jaw dentition, relative length of intestine, and intestinal coiling patterns of omnivorous fish are typically intermediate between those of strictly predaceous and strictly herbivorous species (Moyle and Cech 1988, Kabasakal 2001). Omnivorous lizards have evolved the ability to chemically detect and discriminate both prey and plant food, whereas close relatives that are strictly predaceous or herbivorous usually perceive only one type of food (Cooper et al. 2000, Cooper and Habegger 2000, Cooper and Habegger 2001, Cooper 2002). Similar morphological and physiological patterns are found in many omnivorous groups, ranging from copepods, to rodents, to birds, and appear to be the norm (Kleppel 1993, Gill 1994, Klasing 1999, Romano et al. 1999, Campos et al. 2001).

Diet Mixing

The unique mixture of traits found in omnivores frequently results in a phenomenon that seems to strongly influence the biology of omnivores and omnivore-prey interactions. Diet mixing, in which the nutritional quality of one type of food (e.g., plant material) is enhanced by the ingestion of another type of food (e.g., prey), occurs in many omnivorous species and can strongly affect their fitness (Waldbauer and Friedman 1991, Singer and Bernays 2003). Diet mixing can result in synergistic effects. For example, the caloric value of plant food (such as duck weed) of omnivorous turtles is increased when ingested with prey (i.e., insect larvae) (Bjorndal 1991). Similar effects of mixed prey and plant diets have been observed in copepods (Kumar and Rao 1999a, Bonnet and Carlotti 2001), crabs (Buck et al. 2003), and many insects (Evans 2000, Coll and Guershon 2002, Hurka and Jarosik 2003, Patt et al. 2003).

Diet mixing can also allow animals to mix food items to meet the full range of their nutritional needs when feeding on only one type of food would not (Waldbauer and Friedman 1991). A recent review of the effects of mixed diets on omnivorous insects found that supplementing prey with plant food resulted in faster juvenile development in 36 out of 50 cases (Eubanks and Styrsky 2004). This seems especially true when the plant food contains high concentrations of nitrogen or when the plant food supplements low-quality prey. When fed nitrogen-rich corn earworm (*Helicoverpa zea*) eggs as prey, big-eyed bug nymphs (*Geocoris punctipes*) developed more rapidly when this diet was supplemented with lima bean pods, a high-quality, high-nitrogen plant food, than when they were provided with lima bean leaves, a lower quality, low-nitrogen plant food (Eubanks and Denno 1999). Likewise, when fed corn earworm eggs and pea aphids (*Acyrthosiphon pisum*), big-eyed bug nymphs increased survival by 25% when the nymphs were provided lima bean pods as a supplement, but by just 10% when they were provided lima bean leaves.

Consequences of Omnivory for Omnivore-Prey Interaction

Clearly, omnivory is a widespread feeding habit and almost certainly has pervasive effects on the ecology of omnivore-prey interactions. Surprisingly, the ecological consequences of omnivory have only recently attracted serious attention. However, over the last 25 years, a relatively large number of heuristic and mathematical models have explored the ecology of omnivore-prey interactions (e.g., Pimm and Lawton 1977, Pimm and Lawton 1978, Fagan 1997, Holt and Polis 1997, McCann and Hastings 1997, Holyoak and Sachdev 1998, McCann et al. 1998).

Many of these models predict that omnivores can have a greater suppressive effect on prey populations than strict predators (Crawley 1975, Pimm and Lawton 1977, Pimm and Lawton 1978, Holt 1984, Polis et al. 1989, Polis 1991, Holt and Lawton 1994, Holt and Polis 1997). Omnivores are predicted to have a greater impact on intermediate prey (e.g., herbivores) because of apparent competition over the shared resource (e.g., plants; Holt and Polis 1997, Diehl and Feissel 2000). This occurs because the shared resource (i.e., the plant) contributes to sustaining a higher predation pressure on the intermediate prey (herbivores) by sustaining a higher and less variable population density of the omnivore (Diehl and Feissel 2000). A growing body of anecdotal observations, sampling data, and experimental studies support this prediction. The ability to feed on both prey and plant food, for instance, is associated with strong top-down effects of omnivorous mites on herbivorous mites (Walde 1994, McMurtry and Croft 1997), strong effects of omnivorous fish on plankton (Lazzaro et al. 2003), dramatic reductions in prey densities by estuarine crabs (Jensen and Asplen 1998), and intense effects of omnivorous fish and shrimp on benthic communities of tropical streams (Pringle and Hamazaki 1998). Increased prey suppression of omnivores has also been demonstrated in laboratory microcosms inhabited by bacteria and ciliates (Diehl and Feissel 2001). The effects of omnivores on prey, however, can vary temporally and spatially (Agrawal et al. 1999, Agrawal and Klein 2000, Eubanks and Denno 2000a, Aldana et al. 2002, Geddes and Trexler 2003). Variation in the degree to which plant feeding affects prey suppression depends on several factors. These include variation in the effects of plant feeding on omnivore persistence at low or no prey abundance, the effects of plant feeding on omnivore dispersal and distribution, and the effects of plant feeding on per capita prey consumption and omnivores' food preferences. The following sections review what we know about these factors.

Effects of Plant Feeding on Omnivore Persistence at Low or No Prey Abundance

Some studies support the idea that plant feeding allows omnivores to survive periods without prey. Plant feeding allows omnivorous mites to survive periods of prey scarcity that kills their strictly predaceous relatives (Magalhaes and Bakker 2002). Similarly, pollen feeding by preying mantids allows early instar nymphs to survive extended periods without prey (Beckman and Hurd 2003). Studies of the omnivorous stinkbug, *Podisus maculiventris*, show that adults survive almost 60 days without prey when provided with plant material (Valicente and O'Neil 1995, Wiedenmann et al. 1996). A study of big-eyed bugs (Eubanks and Denno 1999) found that both the nymphs and adults were significantly more likely to survive prey-free periods when plant food was available (as opposed to a water control). Both previous prey and plant quality, however, affected survival during prey-free periods, and the relative nitrogen content of focal and alternative food appears to affect the benefits of omnivory for many other omnivorous arthropods, as well (Denno and Fagan 2003). Unfortunately, few studies have addressed this important issue, so it is difficult to make broad generalizations about the effects of plant feeding on omnivore persistence.

Effects of Plant Feeding on Omnivore Dispersal and Distribution

Many studies suggest that variation in plant food affects the dispersal and distribution of omnivores. Hoverflies (Diptera: Syrphidae) are life history omnivores that consume nectar and pollen as adults. Adult hoverflies are strongly attracted to flowering plants, as well as artificial honeydew, and are more likely to remain in areas with large numbers of flowering plants (Cowgill et al. 1993, Evans and Swallow 1993, Hickman et al. 1995, Hickman and Wratten 1996, Sutherland et al. 1999, Hickman et al. 2001). The distribution of omnivorous mites in apple orchards and other habitats is strongly affected by the distribution of pollen (Kennett et al. 1979, Addison et al. 2000). Minute pirate bugs are also attracted to flowering plants and artificial nectar (Kiman and Yeargan 1985, Read and Lampman 1989, Evans and Swallow 1993) and are frequently more abundant on or around flowering plants (Coll 1996,

Eubanks and Denno 1999). This is not surprising, given that many of these omnivorous mites and bugs can complete their development on a diet of pollen (Kiman and Yeargan 1985, Van Rijn and Tanigoshi 1999). The lady beetle *Coleomigilla maculata* is also attracted to pollen and is more abundant in areas with flowering plants (Coll and Bottrell 1991, Coll and Bottrell 1992, Cottrell and Yeargan 1998, Harmon et al. 2000).

Few studies, however, have actually documented dispersal associated with variation in plant quality. Eubanks and Denno (1999) found that marked big-eyed bugs were significantly more likely to remain on lima bean plants with pods, a preferred and highly nutritious plant food, than on plants without pods (100% recaptured on plants with pods after 4 hours, vs. 36% recaptured on plants without pods after 4 hours). Although more detailed mark-and-recapture studies would be useful, it is probably safe to conclude that variation in plant food, especially flowers, has a strong effect on the dispersal of most omnivorous insect predators.

Effects of Plant Feeding on Per Capita Prey Consumption and Omnivore Food Preferences

Several studies have attempted to quantify the effect of supplemental plant food on the per capita consumption of prey by omnivorous predators. A few studies have found that prey consumption increases when omnivores are provided with plant food. For example, ciliate consumption by copepods was unaffected by the presence of algae, so long as the abundance of algae remained constant (Gismervik and Andersen 1997), and moth-egg consumption by the mirid bug *Dicyphus hesperus* actually increased when its diet was supplemented with tomato leaves (Gillespie and McGregor 2000). Most studies, however, have found that the presence of plant food decreases per capita prey consumption. The presence of algae, for example, significantly reduces the consumption of prey by copepods (Kumar and Rao 1999b). Several experiments have found that the presence of pollen significantly reduces prey consumption by omnivorous mites (McMurtry and Scriven 1966a, McMurtry and Scriven 1966b, Wei and Walde 1997). Similarly, Cottrell and Yeargan (1998) found that the presence of pollen reduced moth-egg predation by the lady beetle *C. maculata* in cornfields. The authors concluded that pollen acted as a preferred, alternative food and significantly reduced per capita predation rates by *C. maculata*.

Eubanks and Denno (2000b) conducted a similar study of the effects of plant feeding on the consumption of prey by big-eyed bugs. They quantified the effects of plant-quality variation (the presence or absence of bean pods) on the functional response of big-eyed bugs to aphids and moth eggs in the laboratory and, in the field (using caged bean plants), on the consumption of prey by big-eyed bugs. They found that the presence of pods significantly reduced the consumption of both prey species by big-eyed bugs in laboratory experiments. Pods, therefore, had an indirect, positive effect on both prey species, because the presence of pods reduced the consumption of prey by big-eyed bugs. Likewise, big-eyed bugs significantly reduced the size of pea aphid populations on caged, podless lima bean plants in the field, but the suppression of aphid populations was significantly reduced when caged plants had pods. Because Eubanks and Denno frequently saw big-eyed bugs feeding on lima bean pods, they concluded that big-eyed bugs spent considerable time feeding on pods and be-

ecology of omnivore-prey interactions are, what is the effect of plant feeding on the per capita prey consumption of omnivores, what is the effect of plant feeding on the persistence of omnivores when prey are scarce, and what is the effect of plant feeding on the dispersal or numerical response of omnivores? Additional studies of the effect of plant feeding or mixed diets on omnivore biology (survival, development, etc.) may have relatively little to offer at this point. If understanding the functional and numerical response of omnivorous predators can predict the outcome of omnivore-prey interactions, then relatively simple experiments may provide a powerful means of disentangling this complex trophic interaction.

Acknowledgments

This chapter benefited greatly from suggestions by P. Barbosa, S. Diehl, J. Harmon, and an anonymous reviewer. This work was supported by a grant from the National Science Foundation (DEB-0074556) and by a grant from the Southern Region Integrated Pest Management program of the U.S. Department of Agriculture.

Literature Cited

Addison, J. A., Hardman, J. M., and Walde, S. J. 2000. Pollen availability for predaceous mites on apple: spatial and temporal heterogeneity. Exper. Appl. Acar. 24:1–18.

Agrawal, A. A. 2003. Why omnivory? Ecology 84:2521.

Agrawal, A. A., and Klein, C. C. 2000. What omnivores eat: direct effects of induced plant resistance on herbivores and indirect consequences for diet selection by omnivores. J. Anim. Ecol. 69:525–535.

Agrawal, A. A., Kobayashi, C., and Thaler, J. S. 1999. Influence of prey availability and induced host-plant resistance on omnivory by western flower thrips. Ecology 80:518–523.

Aldana, M., Pulgar, J. M., Ogalde, F., and Ojeda, F. P. 2002. Morphometric and parasitological evidence for ontogenetic and geographical dietary shifts in intertidal fishes. Bull. Mar. Sci. 70:55–74.

Alomar, O., and Wiedenmann, R. N. 1996. Zoophytophagous Heteroptera: Implications for Life History and Integrated Pest Management. Thomas Say Publications in Entomology: Proceedings. Lanham, Md.: Entomological Society of America.

Badalamenti, F., D'Anna, G., Pinnegar, J. K., and Polunin, N. V. C. 2002. Size-related trophodynamic changes in three target fish species recovering from intensive trawling. Mar. Biol. 141:561–570.

Baptist, B. A. 1941. The morphology and physiology of the salivary glands of Hemiptera-Heteroptera. Quart. J. Microsc. Sci. 83:91–139.

Beckman, N., and Hurd, L. E. 2003. Pollen feeding and fitness in praying mantids: the vegetarian side of a tritrophic predator. Environ. Entomol. 32:881–885.

Bjorndal, K. A. 1991. Diet mixing: nonadditive interactions of diet items in an omnivorous freshwater turtle. Ecology 72:1234–1241.

Bluthgen, N, Gebauier, G., and Fiedler, K. 2003. Disentangling a rainforest food web using stable isotopes: dietary diversity in a species-rich ant community. Oecologia 137:426–435.

Bonnet, D., and Carlotti, F. 2001. Development and egg production in Centropages typicus (Copepoda: Calanoida) fed different food types: a laboratory study. Mar. Ecol. Prog. Ser. 224:133–148.

Boyd, D. W., Cohen, A. C., and Alverson, D. R. 2002. Digestive enzymes and stylet morphology of Deraeocoris nebulosus (Hemiptera: Miridae), a predacious plant bug. Ann. Entomol. Soc. Am. 95:395–401.

Buck, T. L., Breed, G. A., Pennings, S. C., Chase, M. E., Zimmer, M., and Carefoot, T. H. 2003. Diet choice in an omnivorous salt-marsh crab: different food types, body size, and habitat complexity. J. Exp. Mar. Biol. Ecol. 292:103–116.

Campos, C., Ojeda, R., Monge, S., and Dacar, M. 2001. Utilization of food resources by small and medium-sized mammals in the Monte Desert biome, Argentina. Aust. Ecol. 26:142–149.

Cohen, A. C. 1990. Feeding adaptations of some predaceous Hemiptera. Ann. Entomol. Soc. Am. 83:1215–1223.

Cohen, A. C. 1996. Plant feeding by predatory Heteroptera: evolutionary and adaptational aspects of trophic switching. In: Zoophytophagous Heteroptera: Implications for Life History and Integrated Pest Management (Alomar, O., and Wiedenmann, R. N., eds.). Thomas Say Publications in Entomology: Proceedings. Lanham, Md.: Entomological Society of America; 1–17.

Coll, M. 1996. Feeding and ovipositing on plants by an omnivorous insect predator. Oecologia 105:214–220.

Coll, M., and Bottrell, D. G. 1991. Microhabitat and resource selection of the European corn borer (Lepidoptera: Pyralidae) and its natural enemies in Maryland field corn. Environ. Entomol. 20:526–533.

Coll, M., and Bottrell, D. G. 1992. Mortality of European corn borer larvae by enemies in different corn microhabitats. Biol. Contr. 2:95–103.

Coll, M., and Guershon, M. 2002. Omnivory in terrestrial arthropods: mixing plant and prey diets. Ann. Rev. Entomol. 47:267–297.

Cooper, W. E. 2002. Convergent evolution of plant chemical discrimination by omnivorous and herbivorous scleroglosan lizards. J. Zool. 257:53–66.

Cooper, W. E., Al-Johany, A. M., Vitt, L. J., and Habegger, J. J. 2000. Responses to chemical cues from animal and plant foods by actively foraging insectivorous and omnivorous scincine lizards. J. Exp. Zool. 287:327–339.

Cooper, W. E., and Habegger, J. J. 2000. Elevated tongue-flicking and biting by the insectivorous lygosomine skink Mabuya macularia to prey, but not plant chemicals. Ethol. Ecol. Evol. 12:175–186.

Cooper, W. E., and Habegger, J. J. 2001. Prey, but not plant chemical discrimination by the lizard Gerrhosaurus nigrolineatus. Afric. Zool. 36:55–62.

Cooper, W. E., and Vitt, L. J. 2002. Distribution, extent, and evolution of plant consumption by lizards. J. Zool. 257:487–517.

Cottrell, T. E., and Yeargan, K. V. 1998. Effect of pollen on Coleomegilla maculata (Coleoptera: Coccinellidae) population density, predation, and cannibalism in sweet corn. Environ. Entomol. 27:1402–1410.

Cowgill, S. E., Wratten, S. D., and Sotherton, N. W. 1993. The effect of weeds on the numbers of hoverfly (Diptera: Syrphidae) adults and the distribution and composition of their eggs in winter wheat. Ann. Appl. Biol. 123:499–515.

Crawley, M. J. 1975. The numerical response of insect predators to changes in prey density. J. Anim. Ecol. 44:877–892.

Denno, R. F., and Fagan, W. F. 2003. Might nitrogen limitation promote omnivory among carnivorous arthropods? Ecology 84:2522–2531.

Diehl, S. 1993. Relative consumer sizes and the strengths of direct and indirect interactions in omnivorous feeding relationships. Oikos 68:151–157.

Diehl, S., and Feissel, M. 2000. Effects of enrichment on three-level food chains with omnivory. Am. Nat. 155:200–218.

Diehl, S., and Feissel, M. 2001. Intraguild prey suffer from enrichment of their resources: a microcosm experiment with ciliates. Ecology 82:2977–2983.

Eubanks, M. D., and Denno, R. F. 1999. The ecological consequences of variation in plants and prey for an omnivorous insect. Ecology 80:1253–1266.

Eubanks, M.D., and Denno, R. F. 2000a. Health food versus fast food: the effects of prey quality and mobility on prey selection by a generalist predator and indirect interactions among prey species. Ecol. Entomol. 25:140–146.

Eubanks, M. D., and Denno, R. F. 2000b. Host plants mediate omnivore-herbivore interactions and influence prey suppression. Ecology 81:936–947.

Eubanks, M. D., and Styrsky, J. D. 2004. Predator response: the effects of plant feeding on the biology of "omnivorous" predators. In: Plant-Provided Food and Plant-Carnivore Mutualism (Wäckers, F., Van Rijn, P., and Bruin, J., eds.). New York: Cambridge University Press.

Eubanks, M. D., Styrsky, J. D., and Denno, R. F. 2003. The evolution of omnivory in Heteropteran insects. Ecology 84:2549–2556.

Evans, E. W. 2000. Egg production in response to combined alternative foods by the predator Coccinella transversalis. Entomol. Exp. Appl. 94:141–147.

Evans, E. W., and Swallow, J. G. 1993. Numerical responses of natural enemies to artificial honeydew in Utah alfalfa. Environ. Entomol. 22:1392–1401.

Fagan, W. F. 1997. Omnivory as a stabilizing feature of natural communities. Am. Nat. 150:554–568.

Geddes, P., and Trexler, J. C. 2003. Uncoupling of omnivore-mediated positive and negative effects on periphyton mats. Oecologia 136:585–595.

Gill, F. B. 1994. Ornithology. New York: Freeman.

Gillespie, D. R., and McGregor, R. R. 2000. The functions of plant feeding in the omnivorous predator Dicyphus hesperus: water places limits on predation. Ecol. Entomol. 25:380–386.

Gismervik, I., and Andersen, T. 1997. Prey switching by Acartia clausi: experimental evidence and implications of intraguild predation assessed by a model. Mar. Ecol. Prog. Ser. 157:247–259.

Goodchild, A. J. P. 1966. Evolution of the alimentary canal in the Hemiptera. Biol. Rev. Camb. Phil. Soc. 41:97–140.

Harmon, J. P., Ives, A. R., Losey, J. E., Olson, A. C., and Rauwald, K. S. 2000. Coleomegilla maculata (Coleoptera: Coccinellidae) predation on pea aphids promoted by proximity to dandelions. Oecologia 125:543–548.

Hickman, J. M., Lovei, G. L., and Wratten, S. D. 1995. Phenology and ecology of hoverflies (Diptera: Syrphidae) in New Zealand. Environ. Entomol. 24:595–600.

Hickman, J. M., and Wratten, S. D. 1996. Use of Phacelia tanacetifolia strips to enhance biological control of aphids by hoverfly larvae in cereal fields. J. Econ. Entomol. 89:832–840.

Hickman, J. M., Wratten, S. D., Jepson, P. C., and Frampton, C. M. 2001. Effect of hunger on yellow water trap catches of hoverfly (Diptera: Syrphidae) adults. Agric. For. Entomol. 3:35–40.

Holt, R. D. 1984. The ecological consequences of shared natural enemies. Ann. Rev. Ecol. Syst. 25:495–520.

Holt, R. D., and Lawton, J. H. 1994. Apparent competition and enemy-free space in insect host-parasitoid communities. Am. Nat. 142:623–645.

Holt, R. D., and Polis, G. A. 1997. A theoretical framework for intraguild predation. Am. Nat. 149:745–764.

Holway, D. A., Suarez, A. V., Tsutsui, N. D., and Case, T. J. 2002. The ecological causes and consequences of ant invasions. Ann. Rev. Ecol. Syst. 33:181–233.

Holyoak, M., and Sachdev, S. 1998. Omnivory and the stability of simple food webs. Oecologia 117:413–419.

Hurka, K., and Jarosik, V. 2003. Larval omnivory in *Amara aenea* (Coleoptera: Carabidae). Eur. J. Entomol. 100:329–335.

Janssen, A., Willemse, E., and Van der Hammen, T. 2003. Poor host plant quality causes omnivore to consume predator eggs. J. Anim. Ecol. 72:478–483.

Jensen, G. C., and Asplen, M. K. 1998. Omnivory in the diet of juvenile Dungeness crab, *Cancer magister* Dana. J. Exp. Mar. Biol. Ecol. 226:175–182.

Kabasakal, H. 2001. Description of the feeding morphology and the food habits of four sympatric labrids (Perciformes, Labridae) from south-eastern Aegean Sea, Turkey. Neth. J. Zool. 51:439–455.

Kahn, M. R., and Ford, J. B. 1967. The distribution and localization of digestive enzymes in the alimentary canal and salivary glands of the cotton stainer, *Dysdercus fasciatus*. J. Insect Physiol. 13:1619–1627.

Kennett, C. E., Flaherty, D. L., and Hoffmann, R. W. 1979. Effect of wind-borne pollens in the population dynamics of *Amblyseius hibisci* (Acari: Phytoseiidae). Entomophaga 24:83–98.

Kiman, Z. B., and Yeargan, K. V. 1985. Development and reproduction of the predator *Orius insidiosus* (Hemiptera: Anthocoridae) reared on diets of selected plant material and arthropod prey. Ann. Entomol. Soc. Am. 78:464–467.

Klasing, K. C. 1999. Avian gastrointestinal anatomy and physiology. Seminar Avian Exot. Pet Med. 8:42–50.

Kleppel, G. S. 1993. On the diet of calanoid copepods. Mar. Ecol. Prog. Ser. 99:183–195.

Kooi, B. W., Kuijper, L. D. J., Boer, M. P., and Kooijman, S. A. L. M. 2002. Numerical bifurcation analysis of a tri-trophic food web with omnivory. Math. Biosci. 177:201–228.

Kumar, R., and Rao, T. R. 1999a. Demographic responses of adult *Mesocyclops thermocyclopoides* (Copepoda, Cyclopida) to different plant and animal diets. Fresh. Biol. 42:487–501.

Kumar, R., and Rao, T. R. 1999b. Effect of algal food on animal prey consumption rates in the omnivorous copepod, *Mesocyclops thermocyclopoides*. Intern. Rev. Hydrobiol. 84:419–426.

Lazzaro, X., Bouvy, M., Ribeiro, R. A., Oliviera, V. S., Sales, L. T., Vasconcelos, A. R. M., and Mata, M. R. 2003. Do fish regulate phytoplankton in shallow eutrophic Northeast Brazilian reservoirs? Fresh. Biol. 48:649–668.

Magalhaes, S., and Bakker, F. M. 2002. Plant feeding by a predatory mite inhabiting cassava. Exp. Appl. Acarol. 27:27–37.

McCann, K. S., and Hastings, A. 1997. Re-evaluating the omnivory-stability relationship in food webs. Proc. R. Soc. Lond. B 264:1249–1254.

McCann, K. S., Hastings, A., and Strong, D. R. 1998. Trophic cascades and trophic trickles in pelagic food webs. Proc. R. Soc. Lond. B 265:205–209.

McGlynn, T. P. 1999. The worldwide transfer of ants: geographical distribution and ecological invasions. J. Biog. 26:535–548.

McMurtry, J. A., and Croft, B. A. 1997. Life-styles of Phytoseiid mites and their roles in biological control. Ann. Rev. Entomol. 42:291–321.

McMurtry, J. A., and Scriven, G. T. 1966a. The influence of pollen and prey density on the numbers of prey consumed by *Amblyseius hibisci* (Acarina: Phytoseiidae). Ann. Entomol. Soc. Am. 59:149–157.

McMurtry, J. A., and Scriven, G. T. 1966b. Studies on predator-prey interactions between *Amblyseius hibisci* and *Oligonychus punicae* (Arcarina: Phytoseiidae, Tetranychidae) under greenhouse conditions. Ann. Entomol. Soc. Am. 59:793–800.

Miles, P. W. 1972. The saliva of Hemiptera. Adv. Insect Physiol. 9:183–255.

Momen, F. M., and El-Saway, S. A. 1993. Biology and feeding behaviour of the predatory mite, *Amblyseius swirskii* (Acari, Phytoseiidae). Acarologia 34:199–204.

Moyle, P. B., and Cech, J. J., Jr. 1988. Fishes: An Introduction to Ichthyology. Englewood Cliffs, N.J.: Prentice-Hall.

Murdoch, W. W. 1969. Switching in general predators: experiments on predator and stability of prey populations. Ecol. Mon. 39:335–354.

Nystrom, P., Svensson, O., Lardner, B., Bronmark, C., and Graneli, W. 2001. The influence of multiple introduced predators on a littoral pond community. Ecology 82:1023–1039.

Patt, J. M., Wainright, S. C., Hamilton, G. C., Whittinghill, D., Bosley, K., Dietrick, J., and Lashomb, J. H. 2003. Assimilation of carbon and nitrogen from pollen and nectar by a predaceous larva and its effects on growth and development. Ecol. Entomol. 28:717–728.

Pimm, S. L., and Lawton, J. H. 1977. The number of trophic levels in ecological communities. Nature 268:329–331.

Pimm, S. L., and Lawton, J. H. 1978. On feeding on more than one trophic level. Nature 275:542–544.

Polis, G. A. 1991. Complex trophic interactions in deserts: an empirical critique of food web theory. Am. Nat. 138:123–155.

Polis, G. A., Myers, C. A., and Holt, R. D. 1989. The ecology and evolution of intraguild predation: potential competitors that eat each other. Ann. Rev. Ecol. Syst. 20:297–330.

Polis, G. A, and Strong, D. R. 1996. Food web complexity and community dynamics. Am. Nat. 147:813–846.

Pringle, C. M., and Hamazaki, T. 1998. The role of omnivory in a neotropical stream: separating diurnal and nocturnal effects. Ecology 79:269–280.

Read, C. D., and Lampman, R. L. 1989. Olfactory responses of *Orius insidiosus* (Hemiptera: Anthocoridae) to volatiles of corn silks. J. Chem. Ecol. 15:1109–1115.

Romano, P., Feletti, M., Mariottini, G. L., and Carli, A. 1999. Ecological and nutritional implications of the mandibular structure in the Antarctic calanoid copepod *Metridia gerlachei* Giesbrecht 1902: an ultrastructural study. Polar Biol. 22:7–12.

Rosenheim, J. A., and Corbett, A. 2003. Omnivory and the indeterminacy of predator function: can a knowledge of foraging behavior help? Ecology 84:2538–2548.

Rosenheim, J. A., Kaya, H. K., Ehler, L. E., Marois, J. J, and Jaffee, B. A. 1995. Intraguild predation among biological-control agents: theory and evidence. Biol. Contr. 5:303–335.

Rosenheim, J. A., Wilhoit, L. R., and Armer, C. A. 1993. Influence of intraguild predation among generalist insect predators on the suppression of an herbivore population. Oecologia 96:439–449.

Schaefer, C. W., and Panizzi, A. R. 2000. Heteroptera of Economic Importance. New York: CRC Press.

Schoener, T. W., Spiller, D. A., and Losos, J. B. 2002. Predation on a common *Anolis* lizard: can the food-web effects of a devastating predator be reversed? Ecol. Mon. 72:383–407.

Singer, M. S., and Bernays, E. A. 2003. Understanding omnivory needs a behavioral perspective. Ecology 84:2532–2537.

Slater, J. A., and Carayon, J. 1963. Ethiopean Lygaeidae IV: a new predatory lygaeid with a discussion of its biology and morphology (Hemiptera: Heteroptera). Proc. R. Entomol. Soc. Lond. A 38:1–11.

Sprules, W. G., and Bowerman, J. E. 1988. Omnivory and food chain length in zooplankton food webs. Ecology 69:418–426.

Stenroth, P., and Nystrom, P. 2003. Exotic crayfish in a brown water stream: effects on juvenile trout, invertebrates and algae. Fresh. Biol. 48:466–475.

Stoddart, D. M. 1979. Ecology of Small Mammals. Dordrecht, the Netherlands: Kluwer Academic.

Sutherland, J. P., Sullivan, M. S., and Poppy, G. M. 1999. The influence of floral character on the foraging behavior of the hoverfly, *Episyrphus balteatus*. 93:157–164.

Tsutsui, N. D., and Suarez, A. V. 2003. The colony structure and population biology of invasive ants. Cons. Biol. 17:48–58.

Usio, N., and Townsend, C. R. 2002. Functional significance of crayfish in stream food webs: roles of omnivory, substrate heterogeneity and sex. Oikos 98:512–522.

Vadas, R. L. 1989. The importance of omnivory and predator regulation of prey in freshwater fish assemblages of North America. Environ. Biol. Fish. 27:285–302.

Valicente, F. H., and O'Neil, R. J. 1995. Effects of host plants and feeding regimes on selected life history characteristics of *Podisus maculiventris* (Say) (Heteroptera: Pentatomidae). Biol. Con. 5:449–461.

Van Rijn, P. C. J., and Tanigoshi, L. K. 1999. Pollen as food for the predatory mites *Iphiseius degenerans* and *Neosiulus cucumeris* (Acari: Phytoseiidae): dietary range and life history. Exp. App. Acar. 23:785–802.

Varis, A. L., Laurema, S., and Miettinen, H. 1983. Variation of enzyme activities in the salivary glands of *Lygus rulipennis* (Hemiptera, Miridae). Ann. Entomol. Fenn. 49:1–10.

Vinson, S. B. 1997. Invasion of the red imported fire ant (Hymenoptera: Formicidae): spread, biology, and impact. Am. Entomol. 43:23–39.

Waldbauer, G. P., and Friedman, S. 1991. Self-selection of optimal diets by insects. Ann. Rev. Entomol. 36:43–63.

Walde, S. J. 1994. Immigration and the dynamics of a predator-prey interaction in biological control. J. Anim. Ecol. 63:337–346.

Wei, Q., and Walde, S. J. 1997. The functional response of *Typhlodromus pyri* to its prey, *Panonychus ulmi:* the effect of pollen. Exp. App. Acar. 21:677–684.

Weiser, L. A., and Stamp, N. A. 1998. Combined effects of allelochemicals, prey availability, and supplemental plant material on growth of a generalist insect predator. Entomol. Exp. Appl. 87:181–189.

Wheeler, A. G. 2001. Biology of the Plant Bugs (Heteroptera: Miridae). Ithaca, N.Y.: Cornell University Press.

Whitman, D. W., Blum, M. S., and Slansky, F., Jr. 1994. Carnivory in phytophagous insects. In: Functional Dynamics of Phytophagous Insects (Ananthakrishnan, T. N., ed.). Lebanon, N.H.: Science; 161–205.

Wiedenmann, R. N., Legaspi, J. C., and O'Neil, R. J. 1996. Impact of prey density and facultative plant feeding on the life history of the predator *Podisus maculiventris* (Heteroptera: Pentatomidae). In: Zoophytoophagous Heteroptera: Implications for Life History and Integrated Pest Management (Alomar, O., and Wiedenmann, R. N., eds.). Thomas Say Publications in Entomology. Lanham, Md.: Entomological Society of America; 57–93.

Williams, D. F. 1994. Exotic Ants: Biology, Impact, and Control of Introduced Species. Boulder, Colo.: Westview Press.

Woodward, G., and Hildrew, A. G. 2001. Invasion of a stream food web by a new top predator. J. Anim. Ecol. 70:273–288.

Yodzis, P. 1984. How rare is omnivory? Ecology 65:321–323.

2

Mutualisms as Consumer-Resource Interactions

J. NATHANIEL HOLLAND
JOSHUA H. NESS
ALICE BOYLE
JUDITH L. BRONSTEIN

Recognition that predator-prey interactions involve consumer-resource interactions has served as a conceptual foundation for mechanistic understanding of the influences of predation on patterns and processes in ecology and evolutionary biology. Conversely, although mutualism is increasingly recognized as important in nature, it is still often perceived as an eccentric case of interspecific interactions with little relevance to major patterns and processes in ecology. This perspective stems, in part, from the paucity of general principles that can unify mutualistic systems varying greatly in natural history. In this chapter, we develop one such principle, which is that mutualism, like predator-prey interactions, is a consumer-resource interaction.

Almost all mutualisms involve the transfer of energy and nutrients between individuals of two species; one species functions as a consumer and the other as a resource. Yet mutualism differs from predator-prey interactions in that the outcome of the consumer-resource interaction results in a net positive effect on per capita reproduction, survival of both populations, or both. We identify three ways in which consumer-resource interactions take place within mutualisms: two-way consumer-resource, one-way consumer-resource, and indirect interactions. We examine many different mutualistic systems, identifying which of these three means of consumer-resource interaction each exemplifies. Using case examples, we discuss why the exchange of resources and services leads to mutualism, rather than to predation or competition. We further discuss how consumer-resource mutualisms are often context-dependent, varying with the supply of an extrinsic resource or with the presence or abundance of a predator or parasite of one of the mutualists. The consumer-resource interaction between individuals of mutualists' populations may often generate the mechanism underlying the interaction's effects on population-level attributes, such as the growth or size of a population.

Interspecific interactions play key roles in the ecological and evolutionary dynamics of populations, as well as in the structure and dynamics of food webs, communities, and ecosystems (Jones and Lawton 1995, Polis and Winemiller 1996, Fox et al. 2001). Historically, predation and competition has received much more attention than mutualism. This bias is exemplified in the attention given to interspecific interactions in ecology textbooks. Over the past 30 years, the focus on predation has remained fairly constant, coverage of competition has decreased, and coverage of mutualism has increased (Figure 2.1). Nevertheless, mutualism currently represents only 12% of text pages devoted to interspecific interactions, while predation and competition each represent over 40%. Moreover, recent monographs on population dynamics thoroughly explore the roles of predation and competition, but barely address mutualism (Murdoch et al. 2003, Turchin 2003).

Despite this relatively poor coverage, mutualisms are becoming increasingly recognized as fundamental to the structure and function of biological systems worldwide. Examples of some key mutualists in habitats throughout the world include pollinators and seed dispersers in tropical forests, nitrogen-fixing bacteria in deserts and agroecosystems, mycorrhizal fungi in grasslands, lichens in tundras, corals in marine systems, and microbes in deep-sea vents. Influences of mutualism transcend multiple levels of biological organization, ranging from cells to populations, communities, and ecosystems. For example, mutualism may have been key to the origin of eukaryotic cells and to the radiation of angiosperms (Margulis 1975, Crepet 1983). Mutualism can be critical to the reproduction and survival of many plants and animals, and to the cycling of nutrients. Moreover, the ecosystem services that mutualists provide, such as C, N, and P cycles associated with plant-microbial systems, are leading them to be increasingly considered a conservation priority (Costanza et al. 1997, Nabhan and Buchmann 1997, Wall and Moore 1999).

Our understanding of mutualism has been hindered by the relative lack of general principles that generate predictions and syntheses across mutualistic systems that differ greatly in their natural history. This lack may partially explain why treatments of mutualism in textbooks and monographs are often short and focus on natural-history stories of particular systems, rather than on broad ecological and evolutionary concepts, which occur more commonly in discussions of predation and competition. Turchin (2003, p. 30) suggests that mutualisms are not necessarily unimportant, "but unlike trophic [or consumer-resource] interactions, mutualisms do not seem to be of universal importance." "Mutualism could be the most important interaction in some specific population systems," Turchin explains, "but all organisms are consumers of something, and most are also a resource to some other species." Turchin (2003) suggests that the attention of researchers is devoted to trophic and consumer-resource interactions, rather than to mutualism, because such interactions appear most important to population dynamics. Indeed, the consumer-resource dichotomy has been central to understanding predator-prey interactions and competition (MacArthur 1972, Abrams 1980, Tilman 1980, Tilman 1982, Murdoch et al. 2003, Turchin 2003).

The point has generally been missed, however, that mutualism also involves trophic interactions that reflect a consumer-resource dichotomy. In this chapter, we argue that consumer-resource interactions are central to nearly all mutualisms, and that this mechanism of interaction between individuals can potentially unify our understand-

Figure 2.1. Representation of interspecific interactions of predation, competition, and mutualism in college-level general ecology textbooks from 1971 through 1999. Values are percentages of pages covering interspecific interactions. The 1973 data come from Risch and Boucher (1976) and include 11 textbooks spanning 1971–1974, the mean date being 1973. The 1981 data come from Keddy (1990) and include 11 text-books spanning 1973–1986, the mean date being 1981. The 1995 data come from 10 textbooks spanning 1990–1999, the mean date being 1995.

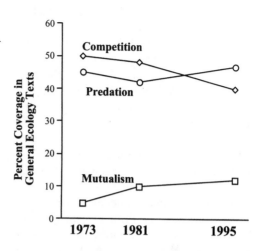

ing of mutualistic systems differing greatly in their natural history. Moreover, link-ing three interspecific interactions that fundamentally differ in outcome (mutualism, predation, and competition) offers both a challenge and a reward. By recognizing that consumer-resource interactions are central not only to predation and competi-tion, but also to mutualisms, we may be taking a step toward generalizing the influ-ences of interspecific interactions, and mutualism in particular, on population processes. We hope the consumer-resource approach will encourage the application of the large body of theoretical studies developed in the context of predation and competition to provide new insights into the study of mutualism. Below, we discuss the general applicability of a consumer-resource framework to interspecific interac-tions. We then show that mutualistic interspecific interactions can be characterized in one of three ways under a consumer-resource framework. We discuss how mutualisms are often context-dependent, varying with the supply of an extrinsic re-source, with the presence or abundance of a consumer of one of the mutualists, or with both. We conclude by discussing some of the advantages and limitations of treat-ing mutualisms as consumer-resource interactions.

Consumer-Resource Interactions

Interactions between populations occur when the actions, traits, or density of indi-viduals of one population result in a change in some attribute of another species' population (Abrams 1987). Population attributes may include per capita reproduc-tion or survival, population growth, population size or density, and mean character values of individuals. Population interactions can be classified in two general ways (Abrams 1987). First, interactions can be characterized according to effects or out-comes of the interaction between individuals. Thus, (+, –), (+, 0), (–, –), and (+, +), with +, 0, and – signs refer to positive, neutral, or negative effects, respectively, on the population attribute of interest. Second, interactions can be characterized according to the mechanism, that is, the ways in which individuals interact (e.g., a predator eating a prey, or an animal consuming nectar and pollinating a flower).

Consumer-resource interactions were primarily incorporated into the study of predation and competition in order to provide a mechanistic perspective on interactions among individuals and the resulting effects on population attributes (MacArthur and Levins 1967, MacArthur 1972, Abrams 1980, Tilman 1980, Tilman 1982). However, the consumer-resource interaction is currently used as often in an outcome-based context as in a mechanism-based context. Consumer-resource interactions have become almost synonymous with a +, - outcome of trophic interactions, including predator-prey, parasite-host, parasitoid-host, and herbivore-plant interactions. If consumer-resource interactions are classified based on +, – interaction outcomes, then by definition any interaction not resulting in a +, – outcome is not a consumer-resource interaction. Yet consumer-resource interactions are not limited to trophic interactions. They can include interactions between populations and abiotic resources, such as plant uptake of nutrients or detritivore use of decaying vegetation. Moreover, consumer-resource interactions do not always result in +, – outcomes. For example, two or more species using the same limited resource can lead to exploitative competition –, –. Competitors may include multiple predator species (consumers) using the same prey species (resource), or multiple plant species (consumers) using the same limited nutrient (resource). Similarly, as we will show, nearly all mutualisms +, + involve consumers and resources. Hence, consumer-resource interactions are a common feature of most interspecific interactions. The simple transfer of energy or nutrients between individuals of two populations does not necessarily mean the interaction outcome is +, –. Here, we suggest that the consumer-resource interaction is a general mechanism by which individuals of two different populations interact, often contributing to effects on population-level attributes.

Mutualism: Consumer-Resource Interactions

Mutualism has been characterized in terms of both the mechanism by which individuals interact and the effect or outcome on populations. The outcome definition of mutualism is simply that both species benefit. Nearly all mutualisms, however, involve both benefits and costs to both interacting species. Benefits result from acquiring a resource, often nutritional, or a service, often dispersal or protection, from a partner. Costs arise as a consequence of providing a resource or service to a partner. Both benefits and costs are implicitly understood to increase or decrease reproduction or survival (or both), or possibly some energetic currency, as these are the fundamental units for ecological and evolutionary processes (Brown 1995). Hence, the outcome of mutualism can be more precisely defined as net positive effects on per capita reproduction and/or survival of both interacting populations, with the understanding that interaction strengths are not simply (+, 0, –), but vary along a continuum (Paine 1980, Paine 1992). We must point out, however, that at this stage in the study of mutualism we do not really know how large the costs are of mutualism.

Mutualisms can also be characterized by how individuals interact. This characterization centers around the three most common benefits that mutualists provide one another: nutrition, transportation of gametes or progeny, and protection from natural enemies or the abiotic environment (Boucher et al. 1982). Although these characterizations have great utility for explaining mutualism, they are based on the perspective of only one of two interacting mutualists. For example, plant-pollinator interactions are classic examples of "transportation mutualisms," but this classification is obviously phytocentric, because the pollinator gains a food resource of nectar and/or pollen while the plant gains in reproduction and transportation of gametes. As we will show, nearly all mutualisms involve food, or nutrition, exchange in one or both directions.

We emphasize consumer-resource interactions in mutualisms as a way to describe the effects of the interaction on population-level attributes. We identify three ways in which consumer-resource interactions take place in mutualisms: two-way consumer-resource interactions, one-way consumer-resource interactions, and indirect interactions. We discuss each of these mechanisms of interaction and their relationships with the benefits and costs of mutualism. We define these mechanisms of interaction in Table 2.1, depict each of them graphically in Figure 2.2, and provide examples in Table 2.2.

Mutualisms with Two-Way Consumer-Resource Interactions

In one group of mutualistic interactions, individuals of each of the two species consume a resource provided by the other (Table 2.2; Figure 2.2A,B). We refer to these as *two-way consumer-resource mutualisms*. Familiar examples include lichens, corals, digestive symbioses, and plant-rhizobial interactions (Table 2.2; Figure 2.2A,B). Resources may be either produced by one mutualist and consumed by the other, or harvested from the environment by one mutualist and consumed by the other. In two-way consumer-resource mutualisms, each mutualist benefits from resources provided by its partner, and each mutualist incurs a cost of provisioning resources to its partner. Mutualistic outcomes occur when benefits of consuming the resource provided by each partner exceed costs of providing resources. Recently, Hoeksema and Schwartz (Schwartz and Hoeksema 1998, Hoeksema and Schwartz 2001, Hoeksema and Schwartz

Table 2.1. Three Means by Which Consumer-Resource Interactions Are Embedded within Mutualisms

---

Two-Way Consumer-Resource Mutualisms
Two species each consume a resource provided by the other. Resources may either be a product of one species, or be harvested by that species and then provided to the other. The interaction is mutualistic when benefits received by each species from the acquired resource exceed costs of providing a resource to the partner.

One-Way Consumer-Resource Mutualisms
Only one species consumes a resource provided or harvested by its partner; in return, the resource provider receives some service (dispersal or protection) from its partner, which directly benefits reproduction, survival, or both. The interaction is mutualistic when costs of producing or harvesting the resource do not exceed the benefits of receiving that service, and vice versa.

Indirect Mutualisms via a Third-Species Consumer or Resource
Two species, neither of which is a consumer or resource of the other, form an indirect mutualism via a third species that is a consumer or resource of one or both of the other two.

Two-Way Consumer-Resource Mutualisms.

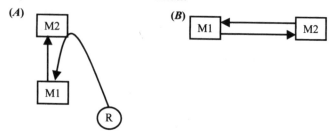

One-Way Consumer-Resource Mutualisms.

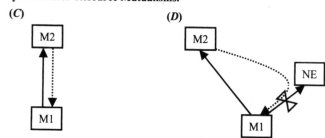

Indirect Mutualisms via a Third Species.

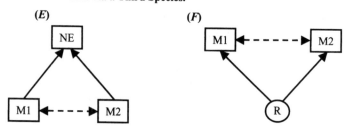

Figure 2.2. Graphical depictions (using topological approaches of food web diagrams) of the three means by which consumer-resource interactions are embedded within mutualisms: two-way consumer-resource (A,B), one-way consumer-resource (C,D), and indirect mutualistic interactions (E,F). Boxes with M1 and M2 represent mutualist species 1 and 2. A circle with R is a resource extrinsic to the pairwise interaction; the resource could be a third species or an abiotic nutrient. A box with NE is a natural enemy, such as predator or parasitoid. A solid, one-headed arrow is a consumer-resource interaction; the arrow points from the resource to the consumer and is the direction of energy flow or nutrient movement. A dotted arrow is a nontrophic, service (e.g., dispersal or protection). The arrow points from the mutualist providing the service to the mutualist receiving the service. A double-headed dashed arrow is a two-way indirect mutualistic interaction between M1 and M2 that arises via a third-party consumer or resource. In protection mutualisms (D), the service is provided by M2 by modifying the rate at which M1 interacts with NE. Joined triangles indicate rate modifer.

Table 2.2. Examples of Mutalistic Systems in Nature

Mutualism	Partners	Figure 2.2	Consumer
1. Two-Way Consumer-Resource Mutualisms			
Lichen	Fungi, algae	A	Fungi
			Algae
Coral	Coral, Zooxanthellae	A	Coral
			Zooxanthellae
Mycorrhizal	Plant, mycorrhizal fungi	A	Plants
			Mycorrhizae
Nitrogen fixation	Plant, rhizobial bacteria	A	Plants
			Rhizobium
Myrmecotrophy	Plants, ants	A	Plants
			Ants
Ant agriculture	Ants, fungi	A	Ants
			Fungi
Digestive symbioses	Aphid-bacteria	B	Aphids
			Bacteria
	Ruminant, bacteria/protozoa	B	Ruminants
			Bacteria/Protozoa
	Termite, protozoa	B	Termites
			Protozoa
2. One-Way Consumer-Resource Mutualisms			
Dispersal	Plant (pollen), animal	C	Animals
	Plant (pollen), pollinator/seed-eater	C	Pollinator progeny
	Fungus (spores), beetles	C	Beetle
	Plants (seed), vertebrates	C	Vertebrate
	Plants (seed), insects	C	Ants
Protection	Plants, ants	D	Ants
	Ant-Lycaenid caterpillar	D	Ants
	Ant-Homopteran	D	Ants
	Plants, fungal endophytes	D	Fungal endophytes
3. Indirect Mutualisms via a Third Species Consumer or Resource			
Cleaning	Cleaner-fish, client fish		Cleaner-fish
Müllerian mimicry	Two or more mimicking species	E	Predator of mimics
Mixed-species foraging groups	Two or more vertebrate species	E	Predators
Honey guide	Honey guide bird, honey badger	F	Bird, badger

Note: Figure 2.2 provides topological depictions of these examples. This table is intended to provide examples; it is not an exhaustive list.

2003) showed how what we identify as two-way consumer-resource mutualisms can be favored, given the comparative advantage of resource acquisition and trade. Although their theory and models are rooted in economics rather than in ecological consumer-resource theory, both approaches to mutualism involve similar principles.

The nitrogen-fixing interaction occurring between plants and rhizobial bacteria (Douglas 1994) is a two-way consumer-resource mutualism (Table 2.2; Figure 2.2A). Nitrogen is a limiting nutrient for plants in many ecosystems. Plants typically cannot convert atmospheric nitrogen to a chemical form that can be taken up by roots, although rhizobia can. Rhizobia form nodules on roots where they receive photo-synthates from plants; in exchange they provide plants with fixed nitrogen (Douglas 1994). In the broad sense, plants consume nitrogen that has been altered to a usable form by the bacteria, and bacteria consume photosynthates produced by plants. Hence, both plants and bacteria function as both consumers and resources. Plant-rhizobial mutualisms are in many ways similar to plant-mycorrhizal mutualisms, in which plants and fungi exchange photosynthates and nutrients, typically phosphorus or nitrogen. In this case, however, the nutrient is a harvested resource rather than a synthesized resource (Smith and Read 1997). Another example is the agricultural mutualism in-volving fungi and leaf-cutter ants (Currie 2001; Table 2.2; Figure 2.2A). Ants do not produce a resource consumed by fungi. Rather, they harvest a resource (leaf tissue), which they then provide to fungi. Fungi grow exclusively on the leaf tissue in ant nests, and, in turn, ants eat hyphae produced by fungi. Thus, as in plant-rhizobial interactions, both partners function as both consumers and resources.

Digestive symbioses also are examples of two-way consumer-resource interac-tions. Examples include interactions between termites and certain gut-inhabiting protozoa, and between ruminant mammals and certain bacteria and protozoa living within their rumen (Douglas 1994). Digestive symbionts, one partner in the mutual-ism, are provided with food as a result of the foraging of their hosts, the second part-ner (Douglas 1994). Yet hosts cannot utilize that food resource until it has been chemically altered by symbionts. Digestive symbioses differ from other two-way consumer-resource mutualisms in that both partners use a single resource that the host harvests from its environment. Because the host and symbiont utilize the same resource, they are depicted in Figure 2.2B as occupying the same trophic level. In contrast, other consumer-resource mutualisms involve mutualists on separate trophic levels (Figure 2.2A,C,D).

Mutualisms with One-Way Consumer-Resource Interactions

In *one-way consumer-resource mutualisms* (Table 2.2; Figure 2.2C,D), only one of the two interacting species consumes a resource produced by its partner. The consumer species benefits from the resource produced by the resource species, and producing the resource generally is a cost to the resource-providing species. Such interactions can result in a mutualistic outcome, because the consumer provides the resource species with a service, such as dispersal of the resource species' gametes or progeny (Table 2.2; Figure 2.2C) or protection from natural enemies or the abiotic environment (Table 2.2; Figure 2.2D). The interaction is mutualistic when benefits (resources) acquired by the consumer species outweigh costs of providing the service, *and* when benefits of the service to the resource species outweigh costs of producing resources.

Pollination mutualisms are one-way consumer-resource interactions in which the pollinator is the consumer and the plant is the resource. The pollinator benefits from receiving the resource of nectar or pollen (or both), while the resource-providing plant benefits from reproduction and gamete dispersal resulting from the consumer's ac-tions (Table 2.2; Figure 2.2C). The interaction is mutualistic when benefits of nectar consumption to the animal pollen vector outweigh costs of pollen transport *and* when benefits to plant reproduction and pollen dispersal outweigh costs of producing nectar. Seed dispersal by frugivorous vertebrates or ants is another example of a one-way consumer-resource mutualism involving plants and their visitors, but in this case dispersal increases propagule survival rather than reproduction of the visited plant. In these interactions, a food resource associated with the seeds (fleshy fruit or elaiosome) is produced by plants and consumed by dispersers.

Another kind of one-way consumer-resource mutualism is the interaction between lycaenid caterpillars and protective ants (Table 2.2; Figure 2.2D). Here, ants consume food secretions produced by lycaenid larvae, while larvae benefit when ants protect them by modifying the rate at which natural enemies interact with larvae (Pierce et al. 2002). This ant-lycaenid interaction is representative of many other protection mutualisms, including ant-plant protection, in which ants consume a food resource produced by plants and plants benefit from reduced herbivory resulting from ant protection (Heil and McKey 2003).

Indirect Mutualisms via an Intermediate Consumer or Resource

We have defined interactions between populations of two species as those in which the actions, traits, or density of individuals of one population cause a change in some attribute of another population. Such interactions are *direct* in nature. *Indirect* interactions between two populations differ in that changes in attributes of the two indirectly interacting species result from the actions, traits, or abundance of a third species or resource, this third party being a consumer or resource of one or both of the other two species (Holt 1977, Schoener 1993, Abrams et al. 1996, Werner and Peacor 2003).

Two common examples of indirect interactions are exploitative competition and apparent competition. Exploitative competition occurs between two species sharing a limiting resource. The resource could be another species or a limited nutrient. The two competitive species may not directly interact with one another, but each may have negative effects on the other's population density by reducing the abundance of the shared resource. Apparent competition occurs between two prey species and is mediated through a shared natural enemy (Holt 1977). The two prey species do not directly interact, but may have negative effects on each other's population density by contributing to their natural enemy's abundance (cf. Abrams et al. 1998). Similarly, indirect mutualisms can arise via a third species that may be either a consumer or resource of one or both mutualists (e.g., Boucher et al. 1982, Kawanabe et al. 1993, Wootton 1994, Abrams et al. 1998; Tables 2.1 and 2.2; Figure 2.2E,F).

Two species, neither of which is a consumer or resource of the other, can form an indirect mutualism via a third species that consumes one or both of them. Müllerian mimicry is one such example (Turner and Speed 1999). In these interactions, two species interact indirectly in the presence of a third species that is a predator of both (Table 2.2; Figure 2.2E), when the prey species share a color pattern that accurately advertises to the predator that each is toxic. The indirect interaction is mutualistic when predators learn the distasteful color pattern more quickly than when they interact with a population of only one of the prey species. The mimic species benefit by indirectly increasing each other's survival rates. A similar indirect mutualism is believed to occur between bird species feeding in mixed-species foraging groups (Hino 1998). Each species reduces per capita investment in vigilance for predators while increasing foraging intensity, thereby minimizing mortality risk while maximizing energy intake.

Cleaning mutualisms between fish (Poulin and Grutter 1996) also fit into this category of indirect interactions. Cleaner-fish consume ectoparasites of their client fish. Cleaners benefit by obtaining a food resource, while the client may benefit from reduced parasite loads. Even though close behavioral associations occur between cleaner-fish and client fish (Bshary and Grutter 2002), the cleaner-client interaction is an indirect mutualistic interaction mediated via ectoparasites. It differs from the previous two examples in that the mutualistic effect arises through a trophic cascade, exemplifying the adage that an enemy of one's enemy can be one's friend.

Indirect mutualistic interactions may occur when a third species is a shared resource, rather than a natural enemy of two species (Figure 2.2F). The mutualism between honeyguide birds and honey badgers (as well as humans) is one such example (Short and Horne 2002). Both birds and badgers use bee nests as food resources (wax and honey, respectively). However, honeyguides cannot gain access to the resource until the honey badger (or human) has first disrupted the nest and dislodged the bees. Honeyguides, in turn, have excellent spatial memories of local bee nests and through characteristic calling, flying, and perching behaviors are able to lead badgers and humans to those nests. Thus, birds and badgers benefit by sharing of a common resource, which neither could efficiently access without the other.

In these examples, indirect mutualistic effects are entirely mediated by a third-party consumer or resource (Figure 2.2E,F). Two-way and one-way consumer-resource mutualisms (Figure 2.2A–D) differ in that they involve direct consumer-resource interactions between two species (M1 and M2). As is shown in Figure 2.2A and D, however, even in direct mutualisms, a third-party resource (R) or natural enemy (NE) sometimes contributes to the mutualistic outcome of the interaction between M1 and M2. Indeed, as we will further discuss below, mutualisms are often context-dependent, varying with the presence or abundance of R and NE. Yet such context dependency differs from indirect interactions in which no direct consumer-resource relationship occurs between M1 and M2. For example, consider one-way consumer-resource mutualisms involving protection (Figure 2.2D). Effects of M1 on M2 are not mediated through NE. The M2 consumes a food resource produced by M1. Thus, populations interact directly, but the outcome of this interaction for M1 depends upon the service M2 provides by modifying the interaction between NE and M1. This example is similar to plant-mycorrhizal interactions, in which the mutualism arises in part through an extrinsic resource, R, rather than a natural enemy, provided to plants by fungi, but fungi directly consume photosynthates.

In many ways, as the examples we have offered demonstrate, our description of indirect mutualistic interactions is analogous to Connor's (1995) conceptual framework for mutualism of by-product benefits—the important difference being that we explicitly recognize that this type of mutualistic interaction is indirect, arising via a third species that is a consumer or resource of one or both of the indirectly interacting mutualistic species. Furthermore, the conceptual framework of benefits and costs arising via consumer-resource interactions is fundamentally different from that of

Connor (1995) in that the consumer-resource framework focuses on the ecological mechanism (or mechanisms) by which mutualistic interactions arise. In contrast, Connor's conceptual framework focuses on the origin and subsequent evolution of different types of benefits of mutualism. Below, we present examples of indirect mutualisms, many of which are the same examples of by-product mutualisms described by Connor (1995). We then contrast indirect mutualisms with direct consumer-resource mutualisms that also involve a resource or consumer extrinsic to the pairwise mutualistic interaction.

Context Dependency of Consumer-Resource Interactions

Effects of interspecific interactions on population attributes are rarely, if ever, static in space or time. Like predation and competition, strengths and outcomes of mutualistic interactions can vary with many factors, including life history traits, life stage or age, and density of interacting mutualists (Thompson 1988, Bronstein 1994a). Indeed, predatory lifestyles under appropriate circumstances may transition to omnivory (see Eubanks, ch. 1 in this volume). Here, we emphasize how mutualistic outcomes of consumer-resource interactions can depend on resources and predators (consumers) extrinsic to the pairwise mutualistic interaction (see also Bronstein and Barbosa 2002).

Two-way consumer-resource mutualisms can shift to commensalism or predation if M1 substantially reduces its production of resources for M2, or vice versa (Figure 2.2A,B). If ambient conditions are poor, such that M2 is able to provide little to no resources to M1, then M1 may either withdraw from the interaction, or reduce its reciprocal provision of resources to M2. In theory, any two-way consumer-resource interaction in which resource provision is potentially limited by environmental supply should be subject to this form of context dependency. Plant-mycorrhizal interactions are one well-studied example (Smith and Read 1997), because the supply of phosphorus provided to plants by mycorrhizae may vary with phosphorus availability in soil.

One-way consumer-resource mutualisms can be context-dependent, depending on the presence or abundance of natural enemies. In protection mutualisms (Figure 2.2D), if natural enemies are absent or rare, then the benefits of protection provided by M2 may not exceed the costs incurred by M2 in producing a resource for M1. In such cases, the interaction between M1 and M2 may shift to a commensal or parasitic consumer-resource interaction. However, M1 may also modify its production of resources for M2 in response to changes in its own requirement for protection. Examples include ant-defended plants that increase extrafloral nectar production in response to herbivory (Heil et al. 2001, Ness 2003) and ant-tended lycaenid caterpillars that offer secreted rewards at a greater rate after simulated predator attacks (Agrawal and Fordyce 2000), thereby attracting more mutualistic consumers. There is also evidence that resource-provisioning rates can increase in the presence of other, perhaps competing, reward-producers (Del-Claro and Oliviera 1993), or decrease in the absence of reward collection (Heil et al. 2000). Resource-supply rates, which are costs for M1 and benefits to M2, can thus be as varied in space and time as the natural enemies that necessitate protection.

Mutualisms also can have effects on consumer-resource dynamics of predator-prey interactions. When a mutualist is a prey item for a natural enemy and its protecting partner varies in space, time, or effectiveness, then the interaction between predator and prey (NE and M1) can similarly vary. For example, mutualist ant species can differ greatly in density, foraging pattern, aggressiveness, and effectiveness in deterring the natural enemies of their partner mutualist (Fraser et al. 2001). Thus, these characteristics mediate the interaction between natural enemy and prey (i.e., the mutualist).

The nature of consumer-resource interactions also can shift between being nominally predatory and mutualistic. For example, the likelihood that ants will consume individuals in a homopteran aggregation rather than protect them from natural enemies has been predicted to increase with the foraging distance required to collect the honeydew resource provided by Homoptera, the quality of rewards offered by Homoptera, and the relative need of those ant colonies for carbohydrates versus protein (Moya-Raygoza and Nault 2000, Cushman 1991, Fischer et al. 2001, Offenberg 2001). This context-dependent, qualitative change in consumer-resource interactions, from mutualism to predation, may even stabilize homopteran populations, given that their populations appear regulated in the presence of ants but explode or crash in their absence (Larsen et al. 2001). In assisting low-density populations and pruning high-density populations, the ants may be highly effective regulators of homopteran populations. Interactions typically interpreted as having (+, −) outcomes can also become mutualistic under some circumstances. For example, herbivory is typically classified as a (+, −) consumer-resource interaction. However, in some cases (e.g., large mammal grazers of African grasslands), some level of herbivory may increase biomass production and plant fitness (McNaughton 1979, de Mazancourt et al. 1998, de Mazancourt et al. 2001).

Discussion

Studies of mutualism have historically focused on details of interactions between particular partner species. The great diversity in natural history among mutualistic systems has obscured the many similarities among them and, consequently, has hampered recognition of the general influence mutualism has on patterns and processes in ecology. In the past two decades, however, the study of mutualism has begun to be unified under a number of organizing principles (Addicott 1984, Janzen 1985, Cushman and Beattie 1991, Bronstein 1994b, Schwartz and Hoeksema 1998, Bronstein 2001, Bronstein and Barbosa 2002, Holland et al. 2002). In this chapter, we continue this unification by proposing that, like predator-prey interactions, nearly all mutualisms have consumer-resource interactions embedded within them. Indeed, we have shown that mutualists almost always interact via one of the three types of consumer-resource interactions. These consumer-resource interactions between individuals are often the mechanism resulting in changes in population-level attributes, such as per capita reproduction, survival, population growth, population size or density, and mean trait values of individuals of a population. Furthermore, these three types of consumer-resource interaction explicitly incorporate the (reproductive and survival) services of dispersal and protection, as well as the benefits and costs of mutualism, which are some of the few general features of mutualisms. One advantage of examining the consumer-resource mechanism by which individuals of mutualistic populations interact is that it simplifies the diversity in natural histories of mutualists. Almost all of the many mutualistic systems occurring in nature appear to fit into only one of the three consumer-resource interactions (Tables 2.1 and 2.2), and nearly all mutualistic systems can be described by only one of a few topological diagrams of the consumer-resource interaction (Figure 2.2; Table 2.2). The very few exceptions include, for example, pollination mutualisms in which the pollinator does not consume a plant product but collects some substance to attract mates (e.g., some euglossine bees and orchids; Roubik 1989). By appreciating the ubiquity of consumer-resource interactions, ecologists can understand mutualisms as diverse as ant-fungus agriculture; plant-nitrogen fixing bacteria; and plant-mycorrhizal, coral, and lichen systems as reciprocal resource exchange mutualisms. Similarly, nearly all dispersal and protection mutualisms are one-way consumer-resource interactions, each with its own topological diagram that circumvents the past need to subdivide them into further categories based on interacting taxa and natural history.

Recognition that mutualisms have consumer-resource interactions embedded in them can lead to new thinking and perspectives in studies of mutualism. As an example, consider Turchin's (2003) conclusion based on a review of empirical studies of population dynamics. He suggested that trophic and consumer-resource interactions may be the most important mechanisms governing population dynamics. The rationale is that energy and nutrients obtained from food are the basic currency underlying many population processes. If Turchin's rationale is correct, then its extension to mutualism suggests that population dynamics of species that interact as consumers with their mutualist partners may be more influenced by the population changes of their partners than are species that receive services from their mutualists, such as dispersal or protection. The relative influence of one mutualistic population on another may be either enhanced or buffered, depending on which partners consume and which provide resources. One of the contributions that a consumer-resource framework may bring to the study of mutualism is a context in which to understand and frame our study of asymmetric influences of interacting species on one another. Further, mutualistic interactions may provide an opportunity to evaluate the importance ascribed to consumer-resource interactions, precisely because they include these interactions but are not limited to them.

We have argued that nearly all mutualistic interactions can fit into only one of three categories of consumer-resource mutualisms (Table 2.1), and into only one of six general interaction topologies (Figure 2.2). However, when a mutualistic system involves multiple forms of benefits and costs, other topologies can occur. This alternative will be particularly prevalent for mutualistic systems in which one of the interacting partners incurs benefits both through the acquisition of a resource and through a service such as dispersal or protection. One example may be the ant-fungus agriculture mutualism. We previously described this mutualism as a two-way consumer-resource interaction with the topology shown in Figure 2.2A. However, ants provide fungi not only with a resource (leaf tissues), but with dispersal; when ants leave their resident nests to establish new colonies, they take some fungi with them. This mutualism is a two-way consumer-resource interaction in either case, but the topology may be more accurately diagrammed by adding in Figure 2.2A a dotted arrow going from M2 (the ants) to M1 (the fungus) to represent the additional bene-fit of the ants' dispersing the fungus. As benefits and costs are more fully integrated into the consumer-resource approach to mutualistic interactions, more topologies may be needed to depict the full range of effects of M1 and M2 on one another.

Another concern with topological diagrams of consumer-resource mutualisms involves the ability to accurately discern trophic levels. In traditional food web dia-grams, solid arrows typically connect consumers with resources. The consumer is placed above the resource in the diagram, reflecting that the consumer occupies a higher trophic level. Representing two-way consumer-resource mutualisms in this way may be difficult. Because both mutualists provide one another with a resource, it can be argued that both mutualists are on the same trophic level (Figure 2.2B). Yet, in many two-way consumer-resource mutualisms, one partner is accurately depicted as occupying a trophic level above its partner (Figure 2.2A), as with bacte-ria or fungi that consume photosynthates from plant roots. On the other hand, hosts and endosymbionts involved in digestive symbioses are probably best depicted as being on the same trophic level, because they both utilize the same resource harvested from the environment by the host species.

Similar issues may arise when one is discerning whether a resource is intrinsic or extrinsic to a pairwise mutualistic interaction (e.g., R in Figure 2.2A). Some ecolo-gists may prefer that diagrams of digestive symbioses explicitly include an extrinsic resource, though not depicted in our Figure 2.2B, because the host harvests the re-source from the environment. However, one could argue that the extrinsic resource of Figure 2.2A be removed from the topological diagram, because the extrinsic re-source may be altered chemically before becoming available to M1. The questions whether to include the extrinsic resource in the topology and how to trophically de-pict mutualistic partners involved in two-way consumer-resource interaction are is-sues perhaps best left to the judgments of individual researchers.

Despite these few limitations, we believe the application of the consumer-resource paradigm to mutualism represents an advance in the study of mutualism. Further-more, we hope that the qualitative, verbal arguments we have presented here will encourage more theoretical and quantitative research on the population ecology of mutualistic systems. The remarkable diversity of natural histories that result in mutualisms make them some of the most fascinating interactions, yet this very di-versity has perhaps hampered an understanding of the general mechanics of inter-acting mutualistic populations. It is our hope that, by recognizing that almost all mutualisms involve consumer-resource interactions, the study of mutualism, like that of predation, will become a more unified and synthetic discipline.

Acknowledgments

We thank P. Barbosa, D. DeAngelis, D. Janos, B. McGill, M. Rosenzweig, and three anony-mous reviewers for discussion of this topic or commenting on a previous version. J.N.H. was supported by a National Parks Ecological Research Fellowship, a program funded by the National Park Foundation through a generous grant from the Andrew W. Mellon Founda-tion. J.H.N. was funded by the University of Arizona Center for Insect Science through the

National Institute of Health training grant no. 1k12GM00708. A.B. was funded by a Postgraduate Scholarship from the Natural Sciences and Engineering Research Council of Canada.

Literature Cited

Abrams, P. A. 1980. Consumer functional response and competition in consumer-resource systems. Theor. Pop. Biol. 17:80–102.

Abrams, P. A. 1987. On classifying interactions between populations. Oecologia 73:272–281.

Abrams, P. A., Holt, R. D., and Roth, J. D. 1998. Apparent competition or apparent mutualism? Shared predation when populations cycle. Ecology 79:201–212.

Abrams, P. A., Menge, B. A., Mittelbach, G. G., Spiller, D. A., and Yodzis, P. 1996. The role of indirect effects in food webs. In: Food Webs: Integration of Patterns and Dynamics (Polis, G. A., and Winemiller, K. O., eds.). New York: Chapman & Hall; 371–395.

Addicott, J. F. 1984. Mutualistic interactions in population and community processes. In: A New Ecology: Novel Approaches to Interactive Systems (Price, P. W., Slobodchikoff, C. N., and Gaud, B. S., eds.). New York: Wiley; 437–455.

Agrawal, A. A., and Fordyce, J. A. 2000. Induced indirect defense in a lycaenid-ant association: the regulation of a resource in a mutualism. Proc. R. Soc. Lond. B 267:1857–1861.

Boucher, D. H., James, S., and Keeler, K. H. 1982. The ecology of mutualism. Annu. Rev. Ecol. Syst. 13:315–47.

Bronstein, J. L. 1994a. Conditional outcomes in mutualistic interactions. Trends Ecol. Evol. 9:214–217.

Bronstein, J. L. 1994b. Our current understanding of mutualism. Q. Rev. Biol. 69:31–51.

Bronstein, J. L. 2001. Mutualisms. In: Evolutionary Ecology (Fox, C., Fairbairn, D., and Roff, D., eds.). New York: Oxford University Press; 315–330.

Bronstein, J. L., and Barbosa, P. 2002. Multi-trophic/multi-species mutualistic interactions: the role of non-mutualists in shaping and mediating mutualisms. In: Multitrophic Level Interactions (Hawkins, B., and Tscharntke, T., eds.). Cambridge: Cambridge University Press; 44–65.

Brown, J. H. 1995. Macroecology. Chicago, Ill.: University of Chicago Press.

Bshary, R., and Grutter, A. S. 2002. Asymmetric cheating opportunities and partner control in a cleaner fish mutualism. Anim. Behav. 63:547–555.

Connor, R. C. 1995. The benefits of mutualism: a conceptual framework. Biol. Rev. 70:427–457.

Costanza, R., d'Arge, R. C., de Groot, R., Farber, S., Grasso, M., Hannon, B., Limburg, K., O'Neill, R. V., Paruelo, J., Raskin, R. G., et al. 1997. The value of the world's ecosystem services and natural capital. Nature 387:253–261.

Crepet, W. L. 1983. The role of insect pollination in the evolution of angiosperms. In: Pollination Biology (Real, L., ed.). New York: Academic Press; 29–50.

Currie, C. R. 2001. A community of ants, fungi, and bacteria: a multilateral approach to studying symbiosis. Annu. Rev. Microbiol. 55:357–380.

Cushman, J. H. 1991. Host-plant mediation of insect mutualisms: variable outcomes in herbivore-ant interactions. Oikos 61:138–144.

Cushman, J. H., and Beattie, A. J. 1991. Mutualisms: assessing the benefits to hosts and visitors. Trends Ecol. Evol. 6:193–195.

Del-Claro, K., and Oliviera, P. S. 1993. Ant-Homoptera interaction: do alternative sugar resources distract tending ants? Oikos 68: 202–206.

de Mazancourt, C., Loreau, M., and Abbadie, L. 1998. Grazing optimization and nutrient cycling: when do herbivores enhance plant production? Ecology 79:2242–2252.

de Mazancourt, C., Loreau, M., and Dieckmann, U. 2001. Can the evolution of plant defense lead to plant-herbivore mutualism? Am. Nat. 158:109–123.

Douglas, A. E. 1994. Symbiotic Interactions. New York: Oxford University Press.

Fischer, M. K., Hoffman, K. H., and Völkl, M. 2001. Competition for mutualists in an ant-homopteran interaction mediated by hierarchies of ant attendance. Oikos 92:531–541.

Fox, C. W., Roff, D. A., and Fairbairn, D. J. 2001. Evolutionary Ecology. New York: Oxford University Press.

Fraser, A. M., Axén, A. H., and Pierce, N. E. 2001. Assessing the quality of different ant species as partners of a myrmecophilous butterfly. Oecologia 129:452–460.

Heil, M., Fiala, B., Baumann, B., and Linsenmair, K. E. 2000. Temporal, spatial, and biotic variations in extrafloral nectar secretion by Macaranaga tanarius. Func. Ecol. 14:749–757.

Heil, M., Koch, T., Hilbert, A., Fiala, B., Boland, W., and Linsenmair, K. E. 2001. Extrafloral nectar production of the ant-associated plant Macaranga tanarius is an induced, indirect, defensive response elicited by jasmonic acid. Proc. Natl. Acad. Sci. U.S.A. 98:1083–1088.

Heil, M., and McKey, D. 2003. Protective ant-plant interactions as model systems in ecological and evolutionary research. Annu. Rev. Ecol. Evol. Syst. 34:425–53.

Hino, T. 1998. Mutualistic and commensal organization of avian mixed-species foraging flocks in a forest of western Madagascar. J. Avian Biol. 29:17–24.

Hoeksema, J. D., and Schwartz, M. W. 2001. Modelling interspecific mutualisms as biological markets. In: Economics in Nature: Societal Dilemmas, Mate Choice, and Biological Markets (Noe, R., Van Hoff, J. A. R. A. M., and Mammerstein, P., eds.). Cambridge: Cambridge University Press; 173–183.

Hoeksema, J. D., and Schwartz, M. W. 2003. Expanding the comparative-advantage biological market models: contingency of mutualism on partners' resource requirements and acquisition trade-offs. Proc. R. Soc. Lond. B 270:913–919.

Holland, J. N., DeAngelis, D. L., and Bronstein, J. L. 2002. Population dynamics and mutualism: functional responses of benefits and costs. Am. Nat. 159:231–244.

Holt, R. D. 1977. Predation, apparent competition, and the structure of prey communities. Theor. Pop. Biol. 12:197–229.

Janzen, D. H. 1985. The natural history of mutualisms. In: The Biology of Mutualism (Boucher, D. H., ed.). New York: Oxford University Press; 40–99.

Jones, C. G., and Lawton, J. H. 1995. Linking Species and Ecosystems. New York: Chapman & Hall.

Kawanabe, H., Cohen, J. E., and Iwasaki, K. 1993. Mutualism and Community Organization: Behavioral, Theoretical and Food-Web Approaches. New York: Oxford University Press.

Keddy, P. 1990. Is mutualism really irrelevant to ecology? Bull. Ecol. Soc. Am. 71:101–102.

Larsen, K. J., Staehle, L. M., and Dotseth, E. J. 2001. Tending ants (Hymenoptera: Formicidae) regulate Dalbulus quinquenotatus (Homoptera: Cicadellidae) population dynamics. Environ. Entomol. 30:757–762.

MacArthur, R., and Levins, R. 1967. The limiting similarity, convergence, and divergence of coexisting species. Am. Nat. 101:377–385.

MacArthur, R. H. 1972. Geographical Ecology. New York: Harper & Row.

Margulis, L. 1975. Symbiotic theory of the origin of eukaryotic organelles. In: Symbiosis (Jenning, D. H., and Lee, D. L., eds.). Cambridge: Cambridge University Press; 21–38.

McNaughton, S. J. 1979. Grazing as an optimization process: grass-ungulate relationships in the Serengeti. Am. Nat. 113:691–703.

Moya-Raygoza, G., and Nault, L. R. 2000. Obligatory mutualism between Dalbulus quinquenotatus (Homoptera: Cicadellidae) and attendant ants. Ann. Entomol. Soc. Am. 93:929–940.

Murdoch, W. M., Briggs, C. J., and Nisbet, R. M. 2003. Consumer-Resource Dynamics. Princeton, N.J.: Princeton University Press.

Nabhan, G. P., and Buchmann, S. L. 1997. Services provided by pollinators. In: Nature's Services: Societal Dependence on Natural Ecosystems (Daily, G. C., ed.). Washington D.C.: Island Press; 133–150.

Ness, J. H. 2003. *Catalpa bignonioides* alters extrafloral nectar production after herbivory and attracts ant bodyguards. Oecologia 134:210–218.

Offenberg, J. 2001. Balancing between mutualism and exploitation: the symbiotic interaction between *Lasius* ants and aphids. Behav. Ecol. Sociobiol. 49:304–310.

Paine, R. T. 1980. Food webs: linkage, interaction strength, and community infrastructure. J. Anim. Ecol. 49:667–685.

Paine, R. T. 1992. Food-web analyses through field measurement of per capita interaction strength. Nature 355:73–75.

Pierce, N. E., Braby, M. F., Heath, A., Lohman, D. J., Mathew, J., Rand, D. B., and Travassos, M. A. 2002. The ecology and evolution of ant association in the Lycaenidae (Lepidoptera). Annu. Rev. Entomol. 47:733–771.

Polis, G. A., and Winemiller, K. O. 1996. Food Webs: Integration of Patterns and Dynamics. New York: Chapman & Hall.

Poulin, R., and Grutter, A. S. 1996. Cleaning symbioses: proximate and adaptive explanations. Bioscience 46:512–517.

Risch, S., and Boucher, D. 1976. What ecologists look for. Bull. Ecol. Soc. Am. 57:8–9.

Roubik, D. W. 1989. Ecology and Natural History of Tropical Bees. Cambridge: Cambridge University Press.

Schoener, T. W. 1993. On the relative importance of direct versus indirect effects in ecological communities. In: Mutualism and Community Organization: Behavioural, Theoretical, and Food-Web Approaches (Kawanabe, H., Cohen, J. E., and Iwasaki, K., eds.). New York: Oxford University Press; 365–411.

Schwartz, M. W., and Hoeksema, J. D. 1998. Specialization and resource trade: biological markets as a model of mutualisms. Ecology 79:1029–1038.

Short, L. L., and Horne, J. F. M. 2002. Family Indicatoridae (Honeyguides). In: Handbook of the Birds of the World (del Hoyo, J., Elliot, A., and Sargatal, J., eds.). Barcelona: Lynx Edicions; 274–295.

Smith, S. E., and Read, D. J. 1997. Mycorrhizal Symbiosis. New York: Academic Press.

Thompson, J. N. 1988. Variation in interspecific interactions. Annu. Rev. Ecol. Syst. 19:65–87.

Tilman, D. 1980. Resources: a graphical-mechanistic approach to competition and predation. Am. Nat. 116:362–393.

Tilman, D. 1982. Resource Competition and Community Structure. Princeton, N.J.: Princeton University Press.

Turchin, P. 2003. Complex Population Dynamics. Princeton, N.J.: Princeton University Press.

Turner, J. R. G., and Speed, M. P. 1999. How weird can mimicry get? Evol. Ecol. 13:807–827.

Wall, D. H., and Moore, J. C. 1999. Interactions underground—soil biodiversity, mutualism, and ecosystem processes. Bioscience 49:109–117.

Werner, E. E., and Peacor, S. D. 2003. A review of trait-mediated indirect interaction in ecological communities. Ecology 84:1083–1100.

Wootton, J. T. 1994. The nature and consequences of indirect effects in ecological communities. Annu. Rev. Ecol. Syst. 25:443–466.

3

Learning as an Adaptive Response to Predation

GRANT E. BROWN
DOUGLAS P. CHIVERS

An individual's response to local predation risk is shaped by the conflicting demands of predator avoidance and the benefits associated with a suite of fitness-related behaviors, such as foraging, mating, and territorial defense. In this chapter, we argue that this form of threat-sensitive trade-off can be optimized by prey that rely on learning as an adaptive response to predation, rather than fixed (i.e., innate) avoidance of all potential predation threats. A reliance on learning to recognize predation threats allows individuals increased flexibility to deal with variable predation pressures throughout their lives. In this review, we describe recent work examining (a) the selection for learning as an adaptive response, (b) mechanisms leading to the learned recognition of novel predators, (c) the enhancement of learning in intra- and interspecific groups, and (d) biological and ecological constraints that can inhibit the acquisition of information. Throughout this chapter we highlight a number of outstanding questions that need consideration from ecologists.

Predation is easily one of the most important and pervasive selection pressures faced by prey individuals over the course of their lives (Lima and Dill 1990, Kats and Dill 1998, Lima and Bednekoff 1999, Werner and Peacor 2003) and shapes their behavior, morphology, and life history traits. The ability to detect and avoid predation threats is therefore critical to an individual's lifetime fitness potential (Mirza and Chivers 2001). However, predator avoidance is likely to be costly (Werner and Anholdt 1996, Wisenden 2000, Golub and Brown 2003, Werner and Peacor 2003), because it tends to reduce time and energy available for a variety of other fitness-related activities, such as foraging, territorial defense, and mating. In addition, predation pressure may result in the use of suboptimal habitats (Gotceitas and Colgan 1987, Gotceitas and Brown 1993). Although not costly in terms of time, use of suboptimal habitats may result in an overall reduction in energy intake (Lima and Dill 1990). In addition, predator avoidance can lead to delayed ontogenetic niche shifts, resulting in decreased growth rates (Olson et al. 1995). Individuals should be at a selective advantage if they are capable of reliably assessing local predation risk and adjusting the intensity of their antipredator behavior to match their current risk (Chivers et al. 2001, Brown 2003).

Throughout their lives prey animals continually modify their responses to predation risk, and much of this change can be attributed to learning. Our goal in this chapter is to highlight the role learning (defined as individual behavior pattern changes based on experience) plays in allowing prey to cope with ever present predation threats. In particular, we focus on the way in which prey acquire information about the identity of predators and continually update their responses to predation threats. We also consider biological and ecological constraints on learning, constraints sometimes ignored by ecologists. For brevity, we restrict much of our discussion to learned recognition of predators, though learning is also of considerable importance in other aspects of predator-prey interactions—for example, in the determination of when and where it is safe to forage and reproduce. In keeping with our research strengths, we concentrate on aquatic systems for much of our review and synthesis. Nevertheless, the concepts that we highlight should be generally applicable to other systems, habitats, and predator-prey interactions.

Selection for Innate versus Learned Recognition Systems

The vast majority of organisms are prey to a diversity of predators. However, the form of predation and the degree of risk may change dramatically as prey individuals grow (i.e., size-dependent predation risk; Brönmark and Miner 1992, Relyea, ch. 9 in this volume) and shift habitat preferences with ontogeny (Werner and Gilliam 1984, Werner and Peacor 2003, Relyea, ch. 9). Prey also move between prey guilds (Olson et al. 1995, Olson 1996, Brown, LeBlanc, et al. 2001) or are subject to seasonal changes in biotic conditions, abiotic conditions, or both (Gilliam and Fraser 2001). Thus, form and frequency of predation acting upon prey may change year to year, season to season, day to day, or even moment to moment, as well as in space (i.e., across microhabitats). As a result of this variability in local predation risk, individuals adopting a strategy of simply avoiding all "potential" predators may not be making the most of the threat-sensitive trade-offs between predator avoidance and other fitness-related activities (Lima and Dill 1990). Rather, individuals that are able to "fine-tune" predator avoidance decisions (in a plastic fashion) in accordance with recent experience (i.e., learning) should be at a selective advantage. Individuals capable of altering their behavior patterns in accordance with learned information would be expected to have a greater degree of flexibility in their response to potential predation (Brown 2003). Thus, populations exposed to varying predation threats, over time, may be selected toward the use of a learned response, because it would allow individuals to optimize the threat-sensitive trade-offs between survival and other fitness-related benefits.

In populations characterized by high predation pressure or low predator diversity, we might predict low variability in predation risk; hence, prey may initially respond to novel predators with what might be considered an "innate" recognition of potential predators (Breden et al. 1987, Riechert, ch. 4 in this volume). Exposure to novel situations, including novel predators, may elicit an avoidance or antipredator response not because the cue is recognized as a predation threat, but simply because it is novel (neophobia). Such neophobia may be adaptive if predation pressure is very high, species diversity is low, or both situations exist (making it more likely that a larger individual is actually a predation threat).

Consider several experiments involving predator recognition in freshwater fish. Vilunen and Hirvonen (2003) showed that hatchery-reared Arctic charr (*Salvenlinus alpinus*), from a strain that was sympatric with predatory brown trout (*Salmo trutta*) and pikeperch (*Stizosedion lucioperca*), exhibited an elevated antipredator response to the odor of the two predators, without any prior experience. Similar results have been demonstrated for hatchery-reared Chinook salmon (*Oncorhynchus tshwaytscha*) exposed to the odor of the northern pikeminnow (*Ptchocheilis oregonensis;* Berijikian et al. 2003). Brown and Smith (1998) found, however, that hatchery-reared rainbow trout (*Oncorhynchus mykiss*) did not exhibit any change in behavior when exposed to the odor of northern pike (*Esox lucius*). Likewise, Mirza and Chivers (2000) found no response of brook charr (*Salvelinus fontinalis*) to cues of chain pickerel (*Esox niger*), suggesting no innate avoidance of predator odor. More telling are interpopulation studies. For example, fathead minnow eggs collected from a high-predator population and reared under laboratory conditions exhibited no innate response to chemical or visual cues of a predatory pike, but wild individuals of the same age and size did (Chivers and Smith 1994b). Thus, it appears that learning is necessary for predator recognition and that population-specific selection may operate to shape the degree to which individuals exhibit antipredator responses toward novel predator cues. In fact, population differences in responses to predators are widespread (e.g., spiders, Riechert and Hedrick 1990; salamanders, Ducey and Brodie 1991; and deermice, Hirsch and Bolles 1980).

Learned Recognition of Predators

It is evident that a wide variety of prey organisms do not possess innate recognition of potential predators. Rather, they must learn the identity of predators. As a way of illustrating some of the ways prey learn, we start by examining studies on the role of chemical cues in aquatic prey's learned recognition of predators. Next, we consider the social learning of predators, much of which is based primarily on visual displays. For brevity, the discussion omits the specific mechanisms responsible for the learning.

Chemical Cues and Learned Predator
Recognition by Aquatic Organisms

A variety of taxonomically diverse aquatic prey organisms rely on chemosensory information to assess local predation risk (Smith 1992, Chivers and Smith 1998, Brown 2003, Lima and Steury, ch. 8 in this volume). Responses occur to cues released by predators (e.g., predator odors) but also to chemical cues released by other prey (Kats and Dill 1998). Chemosensory cues can be released by prey before an attack (disturbance cues) or during an attack (damage-released chemical alarm cues). Disturbance cues have received relatively little study, but have been demonstrated in several species, including Iowa darters (*Etheostoma exile;* Wisenden et al. 1995), brook charr (*Salvelinus fontinalis;* Mirza and Chivers 2002) and red-legged frog tadpoles (*Rana aurora;* Kiesecker et al. 1999). Chemical cues released after mechanical damage to the skin, as would occur during an attack by a predator, have received considerable attention from chemical and behavioral ecologists. Such damage-released chemical alarm cues have been demonstrated in a wide range of aquatic organisms, including fish, amphibians, and invertebrates (Chivers and Smith 1998, Greenstone and Dickens, ch. 7 in this volume). If successful, a predator may become chemically labeled such that the chemical cues of the predator (i.e., its odor) contain detectable levels of prey alarm cues or some metabolite of its alarm cues (Chivers and Mirza 2001, Brown 2003). Such an array of reliable chemosensory information may provide significant opportunities for naïve prey to acquire ecologically relevant information regarding local predation risk.

Chemical alarm cues are often important in the facilitation of learning. The mechanism is relatively simple: when a predator-naïve prey individual detects a conspecific alarm cue at the same time as it detects a cue from a novel predator, the individual prey learns to recognize the predator as dangerous. Such learning is widely referred to as *conditioned predator recognition* (Chivers and Smith 1998). For example, prey fish like European minnows (*Phoxinus phoxinus*) and fathead minnows (*Pimpephales promelas*) acquire the visual and chemical recognition of potential predators amazingly quickly. Recognition is acquired after a single exposure to the predator cue paired with a conspecific chemical alarm cue (Magurran 1989, Chivers and Smith 1994a, Chivers and Smith 1994b). Similar associative learning has been shown for a wide range of aquatic prey, including planaria, (Wisenden and Millard 2001), snails (Rochette et al. 1998), crayfish (Hazlett and Schoolmaster 1998, Hazlett 2003), damselflies (Wisenden et al. 1997), amphibians (Woody and Mathis 1998), and fish (Yunker et al. 1999). Many species of aquatic prey that co-occur and share predators are members of the same prey guild. Chivers et al. (1995a) showed that acquired predator recognition could also result from direct experience with a predator cue and the alarm cues of a sympatric heterospecific individual that is a member of the same prey guild. This sort of flexibility in what is learned (i.e., recognized) allows individuals to exhibit predator avoidance responses only under appropriate contexts (Brown 2003), potentially optimizing the threat-sensitive trade-offs between predator avoidance and other fitness-related activities (such as foraging).

Recent studies have shown that learning can occur in the absence of an overt antipredator response by prey. For example, Brown and Smith (1996) have demonstrated that fathead minnows that had been food deprived for periods of 24 to 48 hours did not exhibit an overt response to the paired cue of conspecific alarm cues and the odor of a novel predator. However, after 4 days of feeding, all conditioned fish exhibited a significant learned response to the predator odor. In a similar study, Brown, Adrian, et al. (2001) exposed minnows to the paired cues of perch (*Perca flavenscens*) and hypoxanthine-3-*N*-oxide, the putative ostariophysan alarm "pheromone" (Pfeiffer et al. 1985, Brown et al. 2000, Brown et al. 2003) at concentrations ranging from 0.1 to 6.7 nM. During the initial conditioning trials, they reported no significant antipredator response by minnows exposed to hypoxanthine-3-*N*-oxide at concentrations below 0.4 nM. However, minnows exposed to hypoxanthine-3-*N*-oxide at concentrations as low as 0.1 nM (or 25% of the concentration required to elicit an overt behavioral response) exhibited a significant learned response when they were exposed to perch odor alone, demonstrating conditioned predator recognition in the absence of an overt antipredator response. These results illustrate that researchers must be extremely cautious when concluding that prey do not learn under a specific set of conditions.

Some prey animals also can learn to recognize the odor of unknown predators when they detect conspecific odors in the diet of their predators. In a landmark study, Mathis and Smith (1993) showed that predator-naïve fathead minnows exhibit an antipredator response to cues of minnow-fed pike but not to cues of pike fed another fish diet (swordtails, *Xiphophorus helleri*). In subsequent tests, the individuals that were initially exposed to pike fed minnows showed a response to pike fed swordtails. These results show that prey learned to recognize predators when the predator was labeled by the prey. Diet-dependent learning also has been demonstrated in damselflies (Chivers et al. 1996). However, additional studies are necessary to determine whether this type of learning is widespread.

Disturbance cues may allow for rapid transmission of alarm response within aggregations of aquatic prey. Mirza and Chivers (2002) have demonstrated that juvenile brook charr can learn to recognize a novel predator through the pairing of the predator cues with a disturbance cue (rather than the well-studied damage-released chemical alarm cue). Disturbance cues are thought to be urinary by-products, released by stressed prey (Hazlett 1990, Kieseker et al. 1999, Zulandt-Schneider and Moore 2000). As such, responses to heterospecific disturbance cues may be more plastic than the more specific response patterns found for damage-released alarm cues (Hazlett 1989, Hazlett 1990). We would predict, then, that individuals can learn to recognize novel predators according to the disturbance cues of unknown prey guild members. This prediction remains to be directly tested, however.

Social Learning of Predator Recognition

Social learning is defined as the acquisition of some biologically relevant information in the absence of direct experience (Mathis et al. 1996). For example, predator-naïve individuals may be alerted to potential predation threats as a result of their spatial associations with experienced conspecifics, heterospecifics, or both (Verheijen 1956,

Magurran and Higham 1988, Krause 1993, Mathis et al. 1996). As a result, these naïve prey can learn to recognize potential predation threats, even though they themselves have not directly interacted with the predator. The significance of this form of information transfer is that individuals responding to the presence of an experienced conspecific or heterospecific showing an "overt" antipredator response may learn to recognize risky cues without having to directly interact with the predator. As such, this learning mechanism may account for the rapid transmission of learned responses within groups or even populations (Mathis et al. 1996, Brown et al. 1997, Brown and Laland 2003). For example, Brown et al. (1997) stocked 39 juvenile northern pike into a 4 ha surface area pond containing a population of approximately 80,000 fathead minnows. Minnows sampled before stocking showed no recognition of either the chemical or visual cues of pike as a potential predator. However, minnows exhibited learned recognition of the chemical cues of pike within 2–4 days and visual cues of pike within 6–8 days poststocking, demonstrating very rapid transmission of predator recognition at the population level.

The presence of conspecifics and/or heterospecifics exhibiting an active antipredator response can elicit increased antipredator behavior in nearby conspecifics and heterospecifics (Smith 1989, Brown et al. 1999). Such social transmission between conspecifics and heterospecifics is likely to be based on a variety of sensory modalities, but most research has focused on visual alarm displays (the sight of an alarmed individual). For example, upon detection of a predation cue, the starry goby (*Asterropteryx semipunctatus*) engages in head-bobbing behavior, a conspicuous visual display that elicits increased antipredator behavior in nearby conspecifics (Smith 1989). Similarly, glowlight tetras (*Hemigrammus erthrozonus*) significantly increase their frequency and intensity of fin-flicking behavior upon detecting predation cues (Brown et al. 1999). Fin flicking is a conspicuous visual alarm signal that entails rapid flicking of the dorsal and caudal fins without change of the position of the fish in space. Nearby individuals will increase their antipredator behavior in response to a visually displaying conspecific. Tail flagging and stotting responses seen in cervids may likewise serve as visual alarm displays that evoke responses in nearby conspecifics (Caro 1986).

Several authors have shown that social transmission of antipredator responses can lead to social learning. Among birds, predator-naïve individuals learn to recognize predators by observing the mobbing responses of conspecifics. For example, this response occurs among jackdaws (*Corvus monedula*; Lorenz 1952), European blackbirds (*Turdus merula*; Curio et al. 1978), and zebra finches (*Taeniopygia guttata*; Vieth et al. 1980). Interestingly, Curio et al. (1978) demonstrated that the conditioned mobbing response of blackbirds was passed along a chain of six birds without any detectable decrease in response intensity.

Mathis et al. (1996) showed cultural learning in fish. Minnows naïve to pike could acquire recognition of the predator when they were paired with an experienced tutor exhibiting a response. Interestingly, the tutor did not have to be a conspecific. Minnows learn from observing conspecifics and sympatric brook stickleback. Griffin et al. (2001) demonstrated social learning when they showed that predator-naïve wallabies that observed a novel predator at the same time as they did an alarmed conspecific learned to recognize the unknown predator. The most interesting aspect of this work relates to the generalization of the learned responses. Wallabies that are trained to recognize foxes generalize the learning to other terrestrial carnivores, like cats, but not to nonpredators, like goats.

Social Enhancement of Learned Information

It is well established that many types of prey gain significant antipredator benefits from membership in social aggregations (Pitcher and Parrish 1993). Given that predator recognition can be socially and culturally transmitted, we would expect such information transfer to be enhanced within monospecies and mixed-species groups. Brown and Warburton (1999) demonstrated that groups of five rainbowfish (*Melanotaenia duboulayi*) learned to avoid a novel threat (a net moved toward the shoal at a constant speed) faster than fish in pairs. They reported no significant difference in latency to escape between large and small groups during the initial encounter, but after repeated exposure the larger group was significantly faster in finding an escape route (Brown and Warburton 1999). This result suggests that the increased amount of social information available to members of a larger group may enhance opportunities to acquire relevant information. Similarly, Kelley et al. (2003) demonstrated that Trinidadian guppies from low-predation streams were able to learn context-appropriate antipredator behaviors if they were paired with conspecifics from a high-predation stream, in the presence of a model predator. Again, this result suggests that there is ample opportunity to learn context-appropriate responses from group members. Social aggregations have also been shown to enhance learning of preferred mates, preferred prey, and foraging patterns (Swaney et al. 2001, Ward and Hart 2003).

A variety of additional factors may result in enhanced group learning, including dominance status, familiarity, and kinship. It is well established that many species are capable of long-term recognition of familiar individuals and/or kin (Colgan 1983, Brown and Brown 1996, Ward and Hart 2003). Chivers et al. (1995b) demonstrated that familiarity among shoalmates enhances antipredator responses in fathead minnows. Minnows preferentially aggregate with familiar, as opposed to unfamiliar, conspecifics (Brown and Smith 1994) and potentially gain significant antipredator benefits through enhanced antipredator responses to both chemical and visual predator cues (Chivers et al. 1995b). Similarly, rainbowfish show a strong preference for familiar rather than unfamiliar shoalmates, possibly enhancing antipredator benefits (Arnold 2000, Brown 2002). Familiar shoalmates may respond more intensely, for individuals may engage in coordinated behavior (Dugatkin and Alferi 1991, Chivers et al. 1995b, Dugatkin 1997, Griffiths 2003). Likewise, many animals are known to preferentially associate or defend territories near kin rather than nonkin (Brown and Brown 1996, Ward and Hart 2003). Juvenile convict cichlids (*Archocentrus nigrofaciatus*) respond more intensely to chemical alarm cues originating from full siblings than to those from unrelated conspecifics (Brown, G. E., and Mirza, R. S., unpublished data). An intriguing hypothesis is that learning may be enhanced because of the more intense response elicited by the alarm cues of kin. Although it remains to be directly tested, evidence suggests that conditioned predator recognition may be enhanced within aggregations of familiar or related individuals.

Social status is known to influence both the intensity of antipredator responses and the threat-sensitive trade-offs between predator avoidance and foraging benefits (Gotceitas and Godin 1991, Godin 1997). For example, socially dominant juvenile Atlantic salmon (*Salmo salar*) respond more intensely and for a longer time than subordinate group members (Gotceitas and Godin 1991). It is possible that by returning to normal activity sooner after exposure to a potential predator, subordinate individuals may increase their opportunities to acquire information regarding predator identities. This hypothesis, however, remains untested.

Individuals also differ in their degree of boldness, or risk-taking behavior, in the face of perceived predation risk (Wilson et al. 1994, Godin and Davis 1995, Fraser et al. 2001, Réale and Festa-Bianchet 2003, Sih et al. 2003). Such risk-taking strategies are known to enhance mating and foraging opportunities, but at a cost of increased predation risk (Godin and Davis 1995, Fraser et al. 2001). Bold individuals may have greater opportunity to acquire relevant information, especially during predator-inspection behavior (Godin and Davis 1995). Consequently, it is possible that bold individuals may be able to offset the elevated risk associated with increased risk-taking by enhancing learning of biologically relevant stimuli. Shy (risk-averse) individuals may benefit from associating with bold individuals, possibly gaining socially transmitted information. The relevance of "personality differences" in the learning of predator identities remains untested but is a potentially fruitful area of study.

Opportunities for Continued Reinforcement of Acquired Information

Previous studies demonstrate innate recognition of predator cues (Berjikian et al. 2003, Vilunen and Hirvonen 2003). Berijikian et al. (2003) note, however, that although juvenile Chinook salmon exhibit an innate avoidance of a novel predator odor, when this odor is paired with a Chinook salmon alarm cue, the response is significantly more intense. This result suggests that some species may show a neophobic response to a novel cue because the information may represent an increased risk. Through repeated exposures, individuals could, presumably, refine their response patterns and exhibit overt antipredator responses only when appropriate.

Most laboratory and field evaluations of conditioned predator recognition have involved the single exposure of naïve prey to the combination of alarm stimuli (chemical, visual, or both) and a predator cue. In such tests, prey are often able to retain this acquired recognition for biologically relevant time periods. For example, Brown and Smith (1998) demonstrated that juvenile rainbow trout retain the recognition of the chemical cues of a northern pike for at least 21 days. Mirza and Chivers (2000) showed that juvenile brook charr retain the ability to recognize predators and gain survival benefits, after exposure to chain pickerel, for at least 10 days following the initial exposure. Some prey may remember predator identities for considerably longer periods, more than one year, in the case of fathead minnows (Chivers and Smith 1994a). Few, if any, studies have directly examined the influence of repeated exposure of alarm stimuli and predator cues on learning.

Under natural conditions, prey are likely to be repeatedly exposed to both direct and indirect learning opportunities throughout their lives. Such repeated exposures may allow individuals to continually reinforce or refresh their ability to recognize biologically relevant cues. This is critically important for species that undergo onto-genetic niche shifts, such as juvenile centrarchids (Golub and Brown 2003, Marcus and Brown 2003). Juvenile largemouth bass, for example, shift from antipredator to foraging behavior in response to heterospecific alarm cues, typically within the first year of growth (Olson 1996). This shift is governed by threat-sensitive trade-offs between antipredator and foraging benefits (Werner and Gilliam 1984, Brown, Gershaneck, et al. 2002, Golub and Brown 2003). As an individual reaches the size at which it may include prey fish in its diet, responding to heterospecific or conspe-cific alarm cues (or both kinds of cues) with a foraging response may be reinforced (Golub and Brown 2003, Marcus and Brown 2003, Harvey and Brown 2004). At the same time, individuals are probably outgrowing common inshore predators. As a result, antipredator responses are less likely to be reinforced (Golub and Brown 2003). Though it has not been directly tested, a mechanism probably regulating such an ontogenetic shift is recent experience. This shift could be achieved if prey have a means of weighing recent experiences more heavily than past experiences.

Brian Hazlett has initiated a novel series of experiments designed to examine the influence of repeated exposure of prey to alarm stimuli and predator cues (Hazlett, B. A., personal communication). His research tests whether repeated learning events enhance or inhibit learning of multiple predators. Enhanced learning may occur through learning sets, whereby learning to recognize one predator enhances the ability to learn a second predator. Learning may be inhibited through interference, whereby learning to recognize one predator blocks or inhibits the recognition of a second. This is an exciting research program, because it attempts to move away from the one-predator-one-prey systems that have been the focus of most predator learning studies.

Upon encountering a potential predator, many prey fish exhibit predator inspec-tion behavior, during which individuals will approach the predator either alone or in groups (Dugatkin and Godin 1992, Pitcher 1992). One of the chief benefits associ-ated with this risky behavior pattern is thought to be the increased opportunity to learn to recognize novel predators (Brown and Godin 1999) and to assess local pre-dation risk (Brown and Schwarzbauer 2001, Brown and Zachar 2002, Brown 2003). Brown and Godin (1999) found that naïve inspecting prey could learn to recognize potential predators by the presence of conspecific alarm cues in the diet of a novel predator. Naïve glowlight tetras (*Hemigrammus erythrozonus*) were allowed to in-spect a predatory convict cichlid paired with the odor of a cichlid fed tetras or sword-tails. The inspectors exposed to the odor of the predator fed tetras altered their inspection behavior in threat-sensitive fashion. In addition, individuals that had in-spected a predator paired with the diet odor containing an alarm cue had learned to recognize the visual cues of the predator. In contrast, those that inspected the cichlid paired with the swordtail diet cues or those that did not inspect failed to acquire the visual recognition of the novel predator.

Inspection may allow individuals to continually update their information regard-ing local predation risk on the basis of both visual (Licht 1989, Murphy and Pitcher 1997) and chemical cues of the predator (Brown and Godin 1999, Smith and Belk

2001). Kelley and Magurran (2003) have recently shown that Trinidadian guppies from a high-predation population exhibited a more intense antipredator response than conspecifics from a low-predation-risk population. In addition, guppies from the high-risk population exhibited more intense inspection behavior. However, individuals from the same high-risk population, bred in the lab, were not different in either antipredator or inspection behavior from natural low-predation-risk population guppies. Brown and Dreier (2002) have shown that predator-naïve glowlight tetras exhibited threat-sensitive inspection patterns only in the presence of conspecific alarm cues in the diet of a predator. However, experienced tetras exhibited threat-sensitive inspection patterns, regardless of the predator odor present. Thus, it appears likely that predator inspection may serve as another mechanism allowing for the continual reinforcement of learned information.

Lima and Bednekoff (1999) recently proposed a model of risk allocation that predicts individuals will modify the intensity of antipredator and foraging behaviors according to recent experience with predation threats. Individuals faced with frequent or variable predation risk would, according to the prediction, respond less intensely during an encounter with a predator and forage more intensely during "safe" periods than individuals exposed to less frequent or more predictable predation threats. As such, this form of risk allocation may be seen as an example of recent experience shaping an individual's antipredator behavior. Several tests of the risk allocation hypothesis have recently appeared, providing partial support for the model (Hamilton and Heithaus 2001, Sih and McCarthy 2002, Van Buskirk et al. 2002, Pecor and Hazlett 2003). We predict that individuals exposed to frequent or variable predation threats have greater opportunities to learn recognition of novel predators. If so, then individuals may benefit by both optimizing threat-sensitive trade-offs (as predicted by Lima and Bednekoff 1999) and increasing opportunities to reinforce the learned avoidance of biologically relevant stimuli indicating potential predation risk.

Biological Constraints on Learning

There are a variety of biological constraints on learning that researchers need to consider when trying to understand the importance of learning in predator-prey interactions. Failing to respond appropriately to a predator the first time can be fatal, so prey can ill afford to make mistakes. Not surprisingly, learning occurs very rapidly, often with a single conditioning event (review Chivers and Smith 1998). This learning differs from learning in other contexts, where often mistakes can be tolerated. The neural capacity required for fast, efficient learning may place constraints on the degree of reliance on learning. A large-scale systematic study of the predator-learning ability of different taxa is in order. There are likely phylogenetic constraints that deserve consideration.

Although a considerable volume of literature has addressed the question of learned recognition of novel predators (Smith 1999, Brown 2003, Brown and Laland 2003), researchers have only recently addressed the question of how prey are able to avoid acquiring aversive responses toward biologically irrelevant stimuli (Wisenden and Harter 2001). Individuals should be expected to maximize potential fitness benefits by learning to respond only to biologically relevant stimuli (Wisenden and Harter 2001). However, learned aversions to nonrisky or irrelevant stimuli have been shown. For example, Magurran (1989) conditioned European minnows to show recognition of the sight of a nonpredatory goldfish (*Carassius auratus*). Similar findings have been reported by Chivers and Smith (1994a) and Yunker et al. (1999). Such learned responses would probably interfere with context-appropriate behavioral responses and, hence, represent a potential fitness cost to the individual. A number of mechanisms may, however, constrain the acquisition of such irrelevant associations.

Wisenden and Harter (2001) exposed predator-naïve fathead minnows to conspecific alarm cues and visual cues of artificial predator models, which varied in shape and movement. They reported that minnows would acquire an aversive response toward a moving (vs. a stationary) model, regardless of its shape. They argue that this behavior is adaptive, because predators, even ambush predators, must stalk and attack prey in order to capture them. As a result, movement may provide relatively reliable information regarding local predation risk (Murphy and Pitcher 1997). Thus, movement rather than shape appears to be the primary visual cue allowing prey to differentiate between relevant and irrelevant threats (Brown and Warburton 1997, Wisenden and Harter 2001).

The temporal associations of alarm cues and predator odors also may function as a constraint on the learning of novel, biologically relevant cues. Under natural conditions, the pairing of a relevant alarm cue and the chemical or visual cues of a predator may be temporally isolated, especially for prey typically found in large aggregations (Korpi and Wisenden 2001). However, most studies of acquired predator recognition have, typically, presented both alarm and predator cues simultaneously, or separated by a relatively short period of time. Korpi and Wisenden (2001) found that zebra danios (*Danio rerio*) are capable of acquiring the recognition of a novel predator even when the presentation of alarm cue and predator odor are separated by periods of up to 5 minutes. This result suggests that even after relatively long delays, learning of ecologically important associations are possible. Crayfish learning to avoid the odor of a novel predator exhibit a similar response (Hazlett 2003). There is likely to be some relatively short upper limit to this time-lag effect, because increasing the time between detection of stimuli may increase the likelihood of learning ecologically irrelevant associations (Korpi and Wisenden 2001).

Latent inhibition of a learned response may occur when an individual is repeatedly exposed to a neutral (predator) cue before experiencing the neutral cue paired with an alarm cue (Ferguson et al. 2001, Hazlett 2003). Acquistapace et al. (2003) have demonstrated that crayfish repeatedly exposed to the odor of a goldfish were unable to learn to avoid goldfish odor when it was paired with the odor of injured crayfish. It is not known whether repeated exposure to a novel cue before detection of a biologically relevant cue (alarm, disturbance, or predator diet cues) has a similar latent inhibition effect. Experiments are required to determine if there is such a relationship.

A third mechanism, learned irrelevance, may also account for the failure of individuals to make relevant associations between an alarm cue and the cue of a potential predator. If a known cue (i.e., an alarm cue) and an unknown cue (i.e., odor of a novel predator) are repeatedly encountered in random order, individual prey may fail to make any learned association between the two—a failure due to learned irrelevance. Hazlett (2003) exposed crayfish to the odor of a goldfish or a crayfish alarm cue in random order before pairing the two cues. He reported that even short-term random associations between neutral and recognized cues result in the inhibition or failure of individuals to learn the identity of a potential predator.

Ecological Constraints on Learning

Even though it is well documented that predation rate is influenced by ecological variables such as habitat type, we know surprisingly little about how ecological variables influence learning in the context of predator-prey interactions. Pollock and Chivers (2003) examined the influence of habitat complexity on the ability of fathead minnows to learn to recognize stickleback alarm cues. The two prey species co-occur and share many of the same predators; hence, cross-species responses to each other's alarm are likely adaptive (Pollock et al. 2003). Stickleback-naïve minnows were introduced into large artificial ponds that contained stickleback and predatory pike. Minnows introduced into ponds with a high amount of biotic structure (i.e., submergent vegetation) did not learn to recognize stickleback alarm cues after 8 days, but minnows introduced into ponds with little structure did. The high-complexity habitat may have reduced the possibility of minnows' observing stickleback being caught and hence may explain differences in learning. Minnows may also have reduced attention to predation cues when in relatively safe refuges. Moreover, learning could also be influenced if habitat differences alter predation rates or the spatial distribution of the two species. In a similar study, Pollock and Chivers (in press) showed that density of minnows and stickleback influenced the rate of learning.

Alarm cues and predator odors differ chemically (Brown et al. 2003), so they would likely have very different rates of transference within slow-moving or static water systems (i.e., ponds or small lakes lacking wind currents). Differences in dilution rates may result in temporal separation. Flowing waters, such as streams and rivers, present another set of potential constraints. An individual located downstream from a predation event may be exposed to one but not both types of cues, because of subtle differences in microcurrents, coupled with differential diffusion rates and differential breakdown rates of predator and prey cues (Hazlett 2003). Consequently, the mechanisms of temporal variability, latent inhibition, and learned irrelevance we have discussed may form significant barriers preventing individuals from acquiring biologically relevant information. Understanding how ecological variables influence learning in the context of predator-prey interactions is one of the most understudied topics in this field.

Recent studies have demonstrated that the ability of prey fish to detect and respond to conspecific and heterospecific alarm cues is significantly impaired by the presence of relatively low levels of anthropogenic pollutants, such as heavy metals (Scott et al. 2003) and acid rain (Brown et al. 2000). Under weakly acidic conditions (pH ~6.0), for example, fathead minnows are unable to detect conspecific alarm cues, because of a complete degradation of the alarm cue (Brown et al. 2000). Likewise, juvenile centrarchids and salmonids are impaired in their ability to detect these critically important cues under weakly acidic conditions (Leduc, Kelly, et al. 2004). Given the demonstrated survival benefits associated with responding to chemical alarm cues (Mirza and Chivers 2000, Mirza and Chivers 2002, Mirza and Chivers 2003a, Mirza and Chivers 2003b), an inability to detect such cues would likely have significant impacts on individual fitness and recruitment (Brown et al. 2000).

Such anthropogenic impacts may also significantly impair the learning of biologically relevant information, such as predator identities. Leduc, Ferrari, et al. (2004) have recently shown that juvenile rainbow trout are unable to learn the odor of a novel predator after acute exposure to relatively weak acidic conditions (pH 6.0). However, trout tested at pH 7.0 were able to learn this odor. The impairment of conditioned predator recognition appears to be due to the partial or complete degradation of the alarm cue itself, not to any chemical alteration of the predator odor (Leduc, Ferrari, et al. 2004). Additional studies are required to assess the long-term impact to individual survival and population recruitment of this phenomenon.

Survival Consequences of Learning

Despite the many studies that have demonstrated learning, few studies have directly tested the survival consequences of learning. This testing must be a priority if we are to gain a full understanding of mechanisms of, as well as constraints on, learning. In a series of recent studies, Mirza and Chivers (2000, 2002, 2003a) conditioned hatchery-reared brook charr to the odor of natural fish predators (chain pickerel and northern pike), and subsequent exposure to the predator cues elicited significant increases in antipredator behavior and a corresponding increase in survival. In a similar study, Gazdewich and Chivers (2002) conditioned minnows to recognize perch and then conducted survival trials. Interestingly, staged encounters revealed differential survival effects in different habitats.

Another important issue is the speed at which learning occurs and whether learning under field conditions is similar to that which occurs in laboratory conditions. Chivers and Smith (1995) and Brown et al. (1997) have demonstrated that entire populations of prey can rapidly acquire the visual and chemical recognition of novel predators within relatively short time frames. In fact, Brown et al. (1997) have shown that a population of fathead minnows learned (and responded to) the odor of a novel predator much faster than they did visual cues provided by the same predator. Asking the same question in habitats that vary (e.g., in turbidity) may provide an interesting comparison.

The studies in which fitness increases due to learning have been measured have been conducted in situations in which one prey is tested for a response to one predator. There is an obvious need to add complexity to these testing paradigms by simultaneously examining the effects of multiple prey and predators. Likewise, determining how predators counteract learned responses by the prey would be most useful.

Future Directions

In this chapter, we have attempted to highlight the critical role learning plays in allowing prey to respond to local predation threats. Recent studies have shown a remarkable degree of flexibility in the learned recognition of novel predators, which allows prey to acquire context-appropriate information. In turn, the flexible and contextual nature of these learned responses may allow prey to maximize threat-sensitive trade-offs between antipredator and other fitness-related activities, such as foraging, territorial defense, and mating. In addition, recent evidence demonstrates that learning allows for the continual reinforcement of ecologically relevant information. Learning thus appears to be an adaptive response to local predation pressure, rather than a simple mechanism to escape it. By relying on experience, prey can acquire only relevant and contextually appropriate responses, resulting in potentially selective advantages. Although we have focused on the acquisition of predator recognition, there are a variety of additional responses to predation threats in which learning is known to play a role; for example, habitat choice (Werner and Gilliam 1984, Werner and Peacor 2003), foraging (Reader et al. 2003, Warburton 2003), and diel activity patterns (Reebs 2002).

However, we do not yet have a complete picture of the role of learning. There are many unanswered questions. We have highlighted such questions throughout this chapter; however, general themes become apparent. Why do some populations exhibit neophobia, whereas others do not? Are there consistent ecological or population-specific differences that lead to the selection for or against innate versus learned responses? What is the role of multisensory learning for local risk assessment?

Much of the research discussed here focuses on the use of chemical and visual cues in the learning of the relevant information. However, aquatic and terrestrial organisms have at their disposal numerous other sensory modalities (i.e., auditory, mechanoreceptors, and electrical cues). Little work has been done on the potential synergistic effects of multiple sensory inputs. What is the influence of the local microhabitat on the detection of relevant information? For example, it is likely that microturbulence has an enormous influence on the degree of chemosensory information available to aquatic vertebrates, yet little is known regarding the behavior of such cues in dynamic flow environments. Such fine-scale habitat effects would likely have considerable effects on the amount of information received by prey at any given time. What are the long-term effects of anthropogenic pollutants or habitat degradation on learning? We know of only a single study (Leduc, Ferrari, et al. 2004) that directly addresses this question. Such a gap in our knowledge calls into question the ecological validity of habitat management and environmental monitoring programs.

To what degree do individual differences influence learning? Can we predict which classes of individuals are the most likely to learn in different contexts? We have highlighted the potential influence of social status, familiarity, kinship, and individual "personality" on learning. There are probably size, age, and sex differences that likewise influence learning, but these probable influences are largely speculative and require direct study.

One fundamental issue that should be addressed is the decay rate of learned information. Given that the vulnerability of an individual to a particular predator changes spatially and temporally, there must often be cases in which prey would be at a disadvantage if it remembered that a particular predator was a threat in the past. Experiments designed to understand the adaptive nature of memory loss may be particularly fruitful. Likewise, work in the field is essential, because much of what is known about the role of learning comes from laboratory studies. The importance of responding appropriately to predation risk is universal. It is a truism that being eaten dramatically reduces future fitness. Yet predation pressure is highly variable through time and space, so prey need to be able to respond to this fluctuation. Learning provides an adaptive means by which to do so.

Literature Cited

Acquistapace, P., Hazlett, B. A., and Gherardi, F. 2003. Unsuccessful predation and learning of predator cues by crayfish. J. Crustacean Biol. 23:364–370.

Arnold, K. E. 2000. Kin recognition in rainbowfish (*Melanotaenia eachamensis*): sex, sibs and shoaling. Behav. Ecol. Sociobiol. 48:385–391.

Berejikian, B. A., Tezak, E. P., and LaRae, A. L. 2003. Innate and enhanced predator recognition in hatchery-reared Chinook salmon. Environ. Biol. Fish. 67:241–251.

Breden, F., Scott, M., and Michel, E. 1987. Genetic differentiation for anti-predator behavior in the Trinidadian guppy, *Poecilia reticulata*. Anim. Behav. 35:618–620.

Brönmark, C., and Miner, J. G. 1992. Predator-induced phenotypic change in body morphology in crucian carp. Science 258:1348–1350.

Brown, C. 2002. Do female rainbowfish (*Melanotaenia* spp.) prefer to shoal with familiar individuals under predation pressure? J. Ethol. 20:89–94.

Brown, C., and Laland, K. 2003. Social learning in fishes: a review. Fish Fisher. 4:280–288.

Brown, C., and Warburton, K. 1997. Predator recognition and antipredator responses in the rainbow fish *Melanotaenia eachamensis*. Behav. Ecol. Sociobiol. 41:61–68.

Brown, C., and Warburton, K. 1999. Social mechanisms enhance escape responses in rainbowfish, *Melanotaenia duboulayi*. Environ. Biol. Fish. 56:455–459.

Brown, G. E. 2003. Learning about danger: chemical alarm cues and local risk assessment in prey fishes. Fish Fisher. 4:227–234.

Brown, G. E., Adrian, J. C., Jr., Lewis, M. G., and Tower, J. M. 2002. The effects of reduced pH on chemical alarm signalling in Ostariophysan fishes. Can. J. Fish. Aquat. Sci. 59:1331–1338.

Brown, G. E., Adrian, J. C., Jr., Naderi, N. T., Harvey, M. C., and Kelly, J. M. 2003. Nitrogen-oxides elicit antipredator responses in juvenile channel catfish, but not convict cichlids or rainbow trout: conservation of the ostariophysan alarm pheromone. J. Chem. Ecol. 29:1781–1796

Brown, G. E., Adrian, J. C., Jr., Patton, T., and Chivers, D. P. 2001. Fathead minnows learn to recognize predator odor when exposed to concentrations of artificial alarm pheromones below their behavioural response thresholds. Can. J. Zool. 79:2239–2245.

Brown, G. E., Adrian, J. C., Jr., Smyth, E., Leet, H., and Brennan, S. 2000. Ostariophysan alarm pheromones: laboratory and field tests of the functional significance of nitrogen oxides. J. Chem. Ecol. 26:139–154.

Brown, G. E., and Brown, J. A. 1996. Kin discrimination in salmonids. Rev. Fish Biol. Fish. 6:201–219.

Brown, G. E., Chivers, D. P., and Smith, R. J. F. 1997. Differential learning rates of chemical versus visual cues of a northern pike by fathead minnows in a natural habitat. Environ. Biol. Fish. 49:89–96.

Brown, G. E., and Dreier, V. M. 2002. Predator inspection behavior and attack cone avoidance in a characin fish: the effects of predator diet and prey experience. Anim. Behav. 63:1175–1181.

Brown, G. E., Gershaneck, D. L., Plata, D. L., and Golub, J. L. 2002. Ontogenetic changes in response to heterospecific alarm cues by juvenile largemouth bass are phenotypically plastic. Behavior 139:913–927.

Brown, G. E., and Godin, J.-G. J. 1999. Who dares, learns: chemical inspection behavior and acquired predator recognition in a characin fish. Anim. Behav. 57:475–481.

Brown, G. E., Godin, J.-G. J., and Pedersen, J. 1999. Fin flicking behavior: a visual anti-predator alarm signal in a characin fish (*Hemigrammus erythrozonus*). Anim. Behav. 59:469–476.

Brown, G. E., LeBlanc, V. J., and Porter, L. E. 2001. Ontogenetic changes in the response of largemouth bass (*Micropterus salmoides*, Centrarchidae, Perciformes) to heterospecific alarm pheromones. Ethology 107:401–414.

Brown, G. E., and Schwarzbauer, E. M. 2001. Chemical predator inspection and attack cone avoidance in a characin fish: the effects of predator diet. Behavior 138:727–739.

Brown, G. E., and Smith, R. J. F. 1994. Fathead minnows use chemical cues to discriminate shoalmates from unfamiliar conspecifics. J. Chem. Ecol. 20:3051–3061.

Brown, G. E., and Smith, R. J. F. 1996. Foraging trade-offs in fathead minnows (*Pimephales promelas*): acquired predator recognition in the absence of an alarm response. Ethology 102:776–785.

Brown, G. E., and Smith, R. J. F. 1998. Acquired predator recognition in juvenile rainbow trout (*Oncorhynchus mykiss*): conditioning hatchery-reared fish to recognize chemical cues of a predator. Can. J. Fish. Aquat. Sci. 55:611–617.

Brown, G. E., and Zachar, M. M. 2002. Chemical predator inspection in a characin fish (*Hemigrammus erythrozonus*, Characidae, Ostariophysi): the effects of mixed predator diets. Ethology 108:451–462.

Caro, T. M. 1986. The functions of stotting: a review of the hypotheses. Anim. Behav. 34:649–662.

Chivers, D. P., Brown, G. E., and Smith, R. J. F. 1995a. Acquired recognition of chemical stimuli from pike, *Esox lucius*, by brook sticklebacks, *Culaea inconstans* (Osteichthyes, Gasterosteidae). Ethology 99:234–242.

Chivers, D. P., Brown, G. E., and Smith, R. J. F. 1995b. Familiarity and shoal cohesion in fathead minnows (*Pimephales promelas*): implications for anti-predator behavior. Can. J. Zool. 73:955–960.

Chivers, D. P., and Mirza, R. S. 2001. Predator diet cues and the assessment of predation risk by aquatic vertebrates: a review and prospectus. In: Chemical Signals in Vertebrates, vol. 9 (Marchlewska-Koj, A., Lepri, J. J., and Müller-Schwarze, D., eds.). New York: Kluwer Academic; 277–284.

Chivers, D. P., Mirza, R. S., Bryer, P. J., and Kiesecker, J. M. 2001. Threat-sensitive predator avoidance by slimy sculpins: understanding the role of visual versus chemical information. Can. J. Zool. 79:867–873.

Chivers, D. P., and Smith, R. J. F. 1994a. Fathead minnows, *Pimephales promelas*, acquire predator recognition when alarm substance is associated with the sight of unfamiliar fish. Anim. Behav. 48:597–605.

Chivers, D. P., and Smith, R. J. F. 1994b. The role of experience and chemical alarm signaling in predator recognition by fathead minnows, *Pimephales promelas*. J. Fish Biol. 44:273–285.

Chivers, D. P., and Smith, R. J. F. 1995. Free-living fathead minnows rapidly learn to recognize pike as predators. J. Fish Biol. 46:949–954.

Chivers, D. P., and Smith, R. J. F. 1998. Chemical alarm signaling in aquatic predator-prey systems: a review and prospectus. Écoscience 5:338–352.

Chivers, D. P., Wisenden, B. D., and Smith, R. J. F. 1996. Damselfly larvae learn to recognize predators from chemical cues in the predator's diet. Anim. Behav. 52:315–320.

Colgan, P. W. 1983. Comparative Social Recognition. New York: Wiley.

Curio, E., Ernst, U., and Vieth, W. 1978. The adaptive significance of avian mobbing. II.

Cultural transmission of enemy recognition in blackbirds: effectiveness and some constraints. Z. Tierpsychol. 48:184–202.

Ducey, P. K., and Brodie, E. D., Jr. 1991. Evolution of antipredator behavior: individual and population variation in a neotropical salamander. Herpetologica 47:89–95.

Dugatkin, L. A. 1997. Cooperation among Animals: An Evolutionary Perspective. Oxford: Oxford University Press.

Dugatkin, L. A., and Alferi, M. 1991. Guppies and the tit-for-tat strategy: preference based on past interaction. Behav. Ecol. Sociobiol. 28:243–246.

Dugatkin, L. A., and Godin, J.-G. J. 1992. Prey approaching predators: a cost-benefit perspective. Ann. Zool. Fenn. 29:233–252.

Ferguson, H. J., Cobey, S., and Smith, B. H. 2001. Sensitivity to change in reward is heritable in the honeybee, *Apis mellifera*. Anim. Behav. 61:527–534.

Fraser, D. F., Gilliam, J. F., Daley, M. J., Le, A. N., and Skalski, G. T. 2001. Explaining leptokurtic movement distributions: intra population variation in boldness and exploration. Am. Nat. 158:124–135.

Gazdewich, K. J., and Chivers, D. P. 2002. Acquired predator recognition by fathead minnows: the influence of habitat characteristics on survival. J. Chem. Ecol. 28:439–445.

Gilliam, J. F., and Fraser, D. F. 2001. Movement in corridors: enhancement by predation threat, disturbance, and habitat structure. Ecology 82:124–135.

Godin, J.-G. J. 1997. Evading predators. In: Behavioural Ecology of Teleost Fishes (Godin, J.-G. J., ed.). Oxford: Oxford University Press; 191–226.

Godin, J.-G. J., and Davis, S. A. 1995. Who dares, benefits: predator approach behaviour in the guppy (*Poecilia reticulata*) deters predator pursuit. Proc. R. Soc. Lond. Ser. B Biol. Sci. 259:193–200.

Golub, J. L., and Brown, G. E. 2003. Are all signals the same? Ontogenetic change in the response to conspecific and heterospecific chemical alarm signals by juvenile green sunfish (*Lepomis cyanellus*). Behav. Ecol. Sociobiol. 54:113–118.

Gotceitas, V., and Brown, J. A. 1993. Substrate selection by juvenile Atlantic cod (*Gadus morhua*): effects of predation risk. Oecologia 93:31–37.

Gotceitas, V., and Colgan, P.W. 1987. Selection between densities of artificial vegetation by young bluegills avoiding predation. Trans. Am. Fish. Soc. 116:40–49.

Gotceitas, V., and Godin, J.-G. J. 1991. Foraging under the risk of predation in juvenile Atlantic salmon (*Salmo salar*): effects of social status and hunger. Behav. Ecol. Sociobiol. 29:255–261.

Griffin, A. S., Evans, C. S., and Blumstein, D. T. 2001. Learning specificity in acquired predator recognition. Anim. Behav. 62:577–589.

Griffiths, S. W. 2003. Learned recognition of conspecifics by fishes. Fish Fisher. 4:256–268.

Hamilton, I. M., and Heithaus, M. R. 2001. The effects of temporal variation in predation risk on anti-predator behavior: an empirical test using marine snails. Proc. R. Soc. Lond. Ser. B Biol. Sci. 268:2585–2588.

Harvey, M. C., and Brown, G. E. 2004. Dine or dash? Ontogenetic shifts in the response of yellow perch to conspecific alarm cues. Environ. Biol. Fish. 70:345–352.

Hazlett, B. A. 1989. Additional sources of disturbance pheromone affecting the crayfish *Orconectes virilis*. J. Chem. Ecol. 15:381–385.

Hazlett, B. A. 1990. Source and nature of disturbance-chemical system in crayfish. J. Chem. Ecol. 16:2263–2275.

Hazlett, B. A. 2003. Predator recognition and learned irrelevance in the crayfish *Orconectes virilis*. Ethology 109:765–780.

Hazlett, B. A., and Schoolmaster, D. R. 1998. Responses of cambarid crayfish to predator odor. J. Chem. Ecol. 24:1757–1770.

Hirsch, S. M., and Bolles, R. C. 1980. On the ability of prey to recognize predators. Z. Tierpsychol. 54:71–84.

Kats, L. B., and Dill, L. M. 1998. The scent of death: chemosensory assessment of predation risk by prey animals. Écoscience 5:361–394.

Kelley, J. L., Evans, J. P., Ramnarine, I. W., and Magurran, A. E. 2003. Back to school: can antipredator behavior in guppies be enhanced through social learning? Anim. Behav. 65:655–662.

Kelley, J. L., and Magurran, A. E. 2003. Effects of relaxed predation pressure on visual predator recognition in the guppy. Behav. Ecol. Sociobiol. 54:225–232.

Kiesecker, J. M., Chivers, D. P., Marco, A., Quilchano, C., Anderson, M. T., and Blaustein, A. R. 1999. Identification of a disturbance signal in larval red-legged frogs (*Rana aurora*). Anim. Behav. 57:1295–1300.

Korpi, N. L., and Wisenden, B. D. 2001. Learned recognition of novel predator odor by zebra danios, *Danio rerio*, following time-shifted presentation of alarm cue and predator odor. Environ. Biol. Fish. 61:205–211.

Krause, J. 1993. The effect of "Schreckstoff" on the shoaling behavior of a minnow: a test of Hamilton's selfish herd theory. Anim. Behav. 45:1019–1024.

Leduc, O. H. C., Ferrari, M. C. O., Kelly, J. M., and Brown, G. E. 2004. Learning to recognize novel predators under weakly acidic conditions: the effects of reduced pH on acquired predator recognition by juvenile rainbow trout (*Onchorhynchus mykiss*). Chemoecology 14:107–112.

Leduc, O. H. C., Kelly, J. M., and Brown, G. E. 2004. Detection of conspecific alarm cues by juvenile salmonids under neutral and weakly acidic conditions: laboratory and field tests. Oecologia 139:318–324.

Leduc, O. H. C., Noseworthy, M. K., Adrian, J. C., Jr., and Brown, G. E. 2003. Detection of conspecific and heterospecific alarm signals by juvenile pumpkinseed under weak acidic conditions. J. Fish Biol. 63:1331–1336.

Licht, T. 1989. Discriminating between hungry and satiated predators: the response of guppies (*Poecilia reticulata*) from high and low predation sites. Ethology 82:238–242.

Lima, S. L., and Bednekoff, P. A. 1999. Temporal variation in danger drives antipredator behavior: the predation risk allocation hypothesis. Am. Nat. 153:649–659.

Lima, S. L., and Dill, L. M. 1990. Behavioral decisions made under the risk of predation: a review and prospectus. Can. J. Zool. 68:619–640.

Lorenz, K. 1952. King Solomon's Ring. New York: Crowell.

Magurran, A. E. 1989. Acquired recognition of predator odor in the European minnow (*Phoxinus phoxinus*). Ethology 82:216–233.

Magurran, A. E., and Higham, A. 1988. Information transfer across fish shoals under predator threat. Ethology 78:153–158.

Marcus, J. P., and Brown, G. E. 2003. Response of pumpkinseed sunfish to conspecific chemical alarm cues: an interaction between ontogeny and stimulus concentration. Can. J. Zool. 81:1671–1677.

Mathis, A., Chivers, D. P., and Smith, R. J. F. 1996. Cultural transmission of predator recognition in fishes: intraspecific and interspecific learning. Anim. Behav. 51:185–201.

Mathis, A., and Smith, R. J. F. 1993. Fathead minnows (*Pimephales promelas*) learn to recognize pike (*Esox lucius*) as predators on the basis of chemical stimuli from minnows in the pike's diet. Anim. Behav. 46:645–656.

Mirza, R. S., and Chivers, D. P. 2000. Predator-recognition training enhances survival of brook trout: evidence from laboratory and field-enclosure studies. Can. J. Zool. 78:2198–2208.

Mirza, R. S., and Chivers, D. P. 2001. Do chemical alarm signals enhance survival of aquatic vertebrates? An analysis of the current research paradigm. In: Chemical Signals in

Vertebrates, vol. 9 (Marchlewska-Koj, A., Lepri, J. J., and Müller-Schwarze, D., eds.). New York: Kluwer Academic; 19–26.

Mirza, R. S., and Chivers, D. P. 2002. Behavioural responses to conspecific disturbance chemicals enhance survival of juvenile brook charr, *Salvelinus fontinalis*, during encounters with predators. Behavior 139:1099–1110.

Mirza, R. S., and Chivers, D. P. 2003a. Predator diet cues and the assessment of predation risk by juvenile brook charr: do diet cues enhance survival? Can. J. Zool. 81:126–132.

Mirza, R. S., and Chivers, D. P. 2003b. Response of juvenile rainbow trout to varying concentrations of chemical alarm cue: response thresholds and survival during encounters with predators. Can. J. Zool. 81:88–95.

Murphy, K. E., and Pitcher, T. J. 1997. Predator attack motivation influences the inspection behavior of European minnows. J. Fish Biol. 50:407–417.

Olson, M. H. 1996. Ontogenetic niche shifts in largemouth bass: variability and consequences for first-year growth. Ecology 77:179–190.

Olson, M. H., Mittelback, G. G., and Osenberg, C. W. 1995. Competition between predator and prey: resource-based mechanisms and implication for stage-structured dynamics. Ecology 76:1758–1771.

Pecor, K. W., and Hazlett, B. A. 2003. Frequency of encounter with risk and the tradeoff between pursuit and antipredator behaviours in crayfish: a test of the risk allocation hypothesis. Ethology 109:97–106.

Pfeiffer, W., Riegelbauer, G., Meier, G., and Scheibler, B. 1985. Effect of hypoxanthine-3-*N*-oxide and hypoxanthine-1-*N*-oxide on central nervous excitation of the black tetra, *Gymnocorymbus ternetzi* (Characidae, Ostariophysi, Pisces) indicated by dorsal light response. J. Chem. Ecol. 11:507–523.

Pitcher, T. J. 1992. Who dares, wins: the function and evolution of predator inspection behavior in shoaling fish. Neth. J. Zool. 42:371–391.

Pitcher, T. J., and Parrish, J. K. 1993. Functions of shoaling behavior in teleosts. In: Behavior of Teleost Fishes (Pitcher, T. J., ed.). New York: Chapman & Hall; 363–440.

Pollock, M. S., and Chivers, D. P. 2003. Does habitat complexity influence the ability of fathead minnows to learn to recognize heterospecific alarm cues? Can. J. Zool. 81:923–927.

Pollock, M. S., and Chivers, D. P. 2004. The effects of density on learned recognition of heterospecific alarm cues. Ethology 110:341–349.

Pollock, M. S., Chivers, D. P., Mirza, R. S., and Wisenden, B. D. 2003. Fathead minnows, *Pimephales promelas*, learn to recognize chemical alarm cues of introduced brook stickleback, *Culaea inconstans*. Environ. Biol. Fish. 66:313–319.

Reader, S. M., Kendal, J. R., and Laland, K. N. 2003. Social learning of foraging sites and escape routes in wild Trinidadian guppies. Anim. Behav. 66:729–739.

Réale, D., and Festa-Bianchet, M. 2003. Predator-induced natural selection on temperament in bighorn ewes. Anim. Behav. 65:463–470.

Reebs, S. G. 2002. Plasticity of diel and circadian activity rhythms in fishes. Rev. Fish Biol. Fish. 12:349–371.

Riechert, S. E., and Hedrick, A. V. 1990. Levels of predation and genetically based anti-predator behavior in the spider, *Agelenopsis aperta*. Anim. Behav. 40:679–687.

Rochette, R., Arsenault, D. J., Justome, B., and Himmelman, J. H. 1998. Chemically-mediated predator-recognition in a marine gastropod. Écoscience 5:353–360.

Scott, G. R., Sloman, K. A., Rouleau, C., and Wood, C. M. 2003. Cadmium disrupts behavioural and physiological responses to alarm substance in juvenile rainbow trout (*Oncorhynchus mykiss*). J. Exp. Biol. 206:1779–1790.

Sih, A., Kats, L. B., and Maurer, E. F. 2003. Behavioural correlations across situations and the evolution of antipredator behavior in a sunfish-salamander system. Anim. Behav. 65:29–44.

Sih, A., and McCarthy, T. M. 2002. Prey responses to pulses of risk and safety: testing the risk allocation hypothesis. Anim. Behav. 63:437–443.

Smith, M. E., and Belk, M. 2001. Risk assessment in western mosquitofish (*Gambusia affinis*): do multiple cues have additive effects? Behav. Ecol. Sociobiol. 51:101–107.

Smith, R. J. F. 1989. The response of *Asterropteryx semipunctatus* and *Gnatholepis anjerensis* (Pisces, Gobbidae) to chemical stimuli from injured conspecifics, an alarm response in gobies. Ethology 81:279–290.

Smith, R. J. F. 1992. Alarm signals in fishes. Rev. Fish Biol. Fish. 2:33–63.

Smith, R. J. F. 1999. What good is smelly stuff in the skin? Cross function and cross taxa effects in fish "alarm substances." In: Advances in Chemical Signals in Vertebrates (Johnston, R. E., Müller-Schwarze, D., and Sorensen, P. W., eds.). New York: Kluwer Academic.

Swaney, W., Kendal, J., Capon, H., Brown, C., and Laland, K. N. 2001. Familiarity facilitates social learning of foraging behavior in the guppy. Anim. Behav. 62:591–598.

Van Buskirk, J., Muller, C., Portmann, A., and Surbeck, M. 2002. A test of the risk allocation hypothesis: tadpole responses to temporal change in predation risk. Behav. Ecol. 13:526–530.

Verheijen, F. J. 1956. Transmission of a flight reaction amongst a school of fish and the underlying sensory mechanisms. Experientia 12:202–204.

Vieth, W., Curio, E., and Ernst, U. 1980. The adaptive significance of avian mobbing. III. Cultural transmission of enemy recognition in blackbirds: cross-species tutoring and properties of learning. Anim. Behav. 28:1217–1229.

Vilhunen, S., and Hirvonen, H. 2003. Innate antipredator response of Arctic charr (*Salvelinus alpinus*) depend on predator species and their diet. Behav. Ecol. Sociobiol. 55:1–10.

Warburton, K. 2003. Learning of foraging skills by fishes. Fish Fisher. 4:203–215.

Ward, A. J. W., and Hart, P. J. B. 2003. The effects of kin and familiarity on interactions between fish. Fish Fisher. 4:348–358.

Werner, E. E., and Anholdt, B. R. 1996. Predator-induced behavioral indirect effects: consequences to competitive interactions in anuran larvae. Ecology 77:157–169.

Werner, E. E., and Gilliam, J. F. 1984. The ontogenetic niche and species interactions in size-structured populations. Annu. Rev. Ecol. Syst. 15:393–425.

Werner, E. E., and Peacor, S. D. 2003. A review of trait-mediated indirect interactions in ecological communities. Ecology 84:1083–1100.

Wilson, D. S., Clark, A. B., Coleman, K., and Dearstyne, T. 1994. Shyness and boldness in humans and other animals. TREE 9:442–446.

Wisenden, B. D. 2000. Olfactory assessment of predation risk in the aquatic environment. Phil. Trans. R. Soc. Lond. Ser. B Biol. Sci. 355:1205–1208.

Wisenden, B. D., Chivers, D. P., and Smith, R. J. F. 1995. Early warning of risk in the predation sequence: a disturbance pheromone in Iowa darters (*Etheostoma exile*). J. Chem. Ecol. 21:1469–1480.

Wisenden, B. D., Chivers, D. P., and Smith, R. J. F. 1997. Learned recognition of predation risk by *Enallagma* damselfly larvae (Odonata, Zygoptera) on the basis of chemical cues. J. Chem. Ecol. 23:137–151.

Wisenden, B. D., and Harter, K. R. 2001. Motion, not shape, facilitates association of predation risk with novel objects by fathead minnows (*Pimephales promelas*). Ethology 107:357–364.

Wisenden, B. D., and Millard, M. C. 2001. Aquatic flatworms use chemical cues from injured conspecifics to assess predation risk and to associate risk with novel cues. Anim. Behav. 62:761–766.

Woody, D. R., and Mathis, A. 1998. Acquired recognition of chemical stimuli from an unfamiliar predator: associative learning by adult newts, *Notophthalmus viridescens*. Copeia 1998:1027–1031.

Yunker, W. K., Wein, D. E., and Wisenden, B. D. 1999. Conditioned alarm behavior in fathead minnows (*Pimephales promelas*) resulting from association of chemical alarm pheromone with a non-biological visual stimulus. J. Chem. Ecol. 25:2677–2686.

Zulandt Schneider, R. A., and Moore, P. A. 2000. Urine as a source of conspecific disturbance signals in the crayfish *Procambarus clarki*. J. Exp. Biol. 203:765–771.

4

Patterns of Inheritance of Foraging Traits in Predators

SUSAN E. RIECHERT

Optimal foraging theory predicts that animals will be opportunistic toward prey, taking advantage of complex and often changing prey availability. Yet a review of studies that have investigated potential patterns of inheritance of foraging traits in predators has identified genetic influences on morphological traits that affect the nature of attack, the ease of taking prey, discrimination of prey among diet specialists, resistance to prey defense mechanisms, and the search patterns used during foraging. Genes even determine whether individuals will exhibit solitary or aggregative feeding (in, e.g., nematode worms), as well as whether they adjust foraging effort in response to local prey availability. No evidence, however, has been found for genetic determination of patch staying time, which may simply reflect the fact that it has not been subjected to genetic study. In this chapter, the constraints and consequences of genetically controlled foraging decisions are considered, as is the extent to which patterns of inheritance of foraging traits are conserved throughout the animal kingdom. Some evidence is provided for the contention that phenotypic plastic responses are a first step toward increased foraging success gained with subsequent canalization of traits. More detailed studies are needed to test the generality of these findings.

Few studies have explicitly examined the genetic mechanisms underlying the foraging behavior of predators, and even fewer have considered the genetic mechanisms that underlie the coevolution of predators and prey. Genetic mechanisms may be overlooked to some extent when it comes to foraging behavior, because optimal foraging theory predicts that organisms should be opportunistic with respect to foraging to take advantage of complex and often changing prey environments.

However, genes underlie learning just as they do the more canalized behavioral and morphological phenotypes, and even developmental and experiential determinants of traits are influenced by allele variation and gene expression. Thus, while natural selection operates directly on complex phenotypes, it is ultimately the evolution of underlying components that shapes patterns of adaptation (Geoffeney et al. 2002). Genes are considered the principal determinants of all life processes, and foraging behavior is no exception. In this review I provide examples that illustrate the extent to which foraging behavior, and ultimately its consequences to respective predator and prey populations, are constrained by the genetic architecture of the predator.

In this chapter, I focus on the evolution of foraging traits of predators. Ideally, the study or the genetics of a system would consist of the following: (a) establishment of a genetic component of a foraging trait, (b) exploration of the mode of inheritance, (c) identification of the specific genes and pathways involved, and (d) increased precision in defining the phenotype (Wimer and Wimer 1985). Most genetic treatments addressing foraging traits of predators fall short of these goals. In this chapter, I have clustered studies by category of investigation in a progression from least to most insight provided into the pattern of inheritance. Included are (a) phylogenetic analyses and comparative studies completed on species radiations; (b) studies that examine variation among individuals, including conditioning and naïve-individual experiments; (c) population comparisons; and (d) molecular genetics and mutation studies. Sometimes more than one approach was taken in a given study. In these cases, the study is included under the heading that provided the most insight into potential genetic mechanisms.

Phylogenetic Analyses and Species Radiations

Large species proliferations have been observed in a number of lineages that have experienced isolation in some form (e.g., mountain tops, lakes, true islands). In these systems, there are frequently parallel adaptations in species groups on different "islands," which reflect the diversification of ecological roles through adaptation in response to selection. The fish family Cichlidae that has radiated in the Great Lakes of East Africa is tremendously diverse (approximately 1,100 species), with its initial proliferation due to isolation without diversification of ecological roles (nonadaptive radiation; Gittenberger 1991). There followed radiation attributed to the adaptability of the cichlid mouth, and even more recent diversification largely associated with male color, which reflects sexual selection (Stauffer et al. 1997, Albertson et al. 1999).

Mouth morphology in the cichlids is unique. Although many fish groups have teeth all over the mouth, in cichlids the bones bearing the teeth are located deep within the throat on a lower pharyngeal jaw. This arrangement permits the evolution of new uses for the jaws normally involved in grasping and masticating prey. The diversity of feeding modes in cichlids, therefore, includes such extremes as scale eating, algal scraping, snail crushing, detritus eating through sifting of the substrate, and eye biting. The asymmetrical mouths of seven scale-eating species of the genus *Perissodus* in Lake Tanganyika are of particular interest to our investigation of the evolution of foraging traits. Hori (1993) completed a field investigation of the adaptive significance of the asymmetrical mouth in the most common *Perissodus* species in the system, *P. microlepis*. To avoid detection by the prey, the cichlid approaches it from the rear and attacks obliquely from one side or the other. The species population has two phenotypes involving both a mouth asymmetry and an associated direction of attack. The morph with a left-skewed or left-handed mouth attacks the right flank of the prey, whereas the phenotype exhibiting a right-handed mouth attacks the left flank of its prey. The polymorphism exists because of frequency-dependent selection effected through the differential guarding response of the prey to that side of its body most vulnerable to *Perissodus* attack. As one morph increases in representation, it is encountered more and the prey fish adjust their vigilance to the side of the body most frequently attacked. The dominant phenotype's feeding success then decreases with corresponding losses in fecundity, thus favoring the other morph. This shift in phenotype success leads to an oscillation in the ratio of handedness that exhibits an amplitude of 0.15 and a periodicity of 5 years. The timing of the oscillation reflects a 2-year lag between reproduction and maturation of fry.

Hori (1993) used quantitative field observations to establish that the handedness in mouth morphology is a simple Mendelian trait. He noted that the asymmetrical mouth is already observed in *P. microlepis* fry, which are planktivores. Thus, phenotypic plasticity was eliminated as a possible explanation of morph shifts. Examination of the handedness of F_1 offspring of parents of known asymmetry suggests that handedness is a one-locus, two-allele system. There is a close correspondence between the observed ratios of left-handed to right-handed individuals and the expected ratios for particular crosses. Note, however, that the number of broods examined was small (i.e., three to five for different crosses). Also, this species is known to farm out its fry to other breeding pairs for brood care. To control parentage as much as possible, Hori (1993) eliminated from the counts fry that conspicuously differed in size from the majority of individuals in a given brood.

No data were presented on the degree to which direction of attack is learned or inherited, presumably through linkage to the gene controlling mouth "handedness." Laboratory breeding experiments are needed to both confirm the pattern of inheritance suggested for the mouth asymmetry and gain insight into the mechanisms underlying the observed phenotypic correlation between mouth handedness and direction of attack. Hori has not pursued the genetic mechanism further, but has instead extended the field investigations on dimorphism in lateral development of the mouth to other scale eaters and to other piscivorous fish in the Great Lakes of East Africa. A balanced polymorphism in laterality that is maintained by prey response to attack direction is widespread among the scale eaters and even other piscivorous fish in East Africa (Nakajima et al. 2004). Oscillations in the frequency of the respective morphs of 5 to 7 years were observed among the targeted species. Genetically determined laterality was also detected in all species tested (i.e., two cichlids and one goby). Right asymmetry was dominant over left asymmetry, and the dominant allele of the two-allele system was lethal in the homozygous condition.

While prey capture may be influenced by mouth morphology in many cichlid species, diet choice appears to be more labile. Liem and Summers (2000) found that experience with prey significantly influences the prey types that are included in the diet of the cichlid, *Petrotilapia tridentiger*. This species exhibited a repertoire of eight distinct capture patterns (motor patterns of the jaw muscles) toward various prey in the wild. Individuals brought into the lab lost all but two of these capture patterns within 21 days of conditioning to a single prey type. Similar effects were observed for four other cichlids. However, sample sizes were sufficiently low in these latter cases that it was not possible to demonstrate the effect statistically.

Manipulative feeding trials were also completed to test for the influence of prey type and availability on the prey consumption pattern of four species of cichlids. Two of the species classified as morphological specialists on the basis of the functional design of the mouth had broader diets than predicted, focusing on only the prey types they were adapted to capture at low prey abundances. This observation leads one to question the degree to which morphological and functional specializations are linked to narrow, or specialized, niches. Competition-based models of optimal foraging are based on the idea that prey specialization has evolved to avoid competition with the better competitors that remain prey generalists. From an isotope study of diet overlap in cichlids from Lake Malawi, Bootsma et al. (1996) found that, as in the lab experiments, food partitioning occurs primarily under low-prey conditions. One might conclude that, regardless of the morphological specializations observed in the cichlid jaw for particular feeding modes, cichlids are feeding opportunists when prey are abundant.

Inferences about potential genetic mechanisms underlying diet choice have also been drawn from phylogenetic studies involving groups that have not been subject to the degree of isolation observed in the cichlids. In a review of the literature on the larval provisioning habits of 132 species of Bembicine sand wasps (Hymenoptera, Sphecidae), Evans (2002) found that the majority of the species (76%) specialize on robust-bodied flies (suborder Brachycera) and that the remaining species were fly generalists, with two-thirds of these excluding flies altogether. Without imposing behavior on a morphological or molecular tree of the group, Evans assumed that specializing on flies is ancestral to generalist feeding. This assumption was based on the observed numerical bias toward fly specialization and on three additional observations. First, species that occasionally provision nests with prey other than flies do so when competition for larval food is high and fly abundance is low. Second, species that consistently take prey other than flies often occupy habitats where other prey types predominate (e.g., species inhabiting two arid habitats where *Periditis* bees and antlions, respectively, are the numerically prominent prey). And third, the remaining prey generalists co-occur with a fly specialist. Evans concluded that the radiation to other diets was necessitated by past shortages in dipterous prey, with poorer competitors shifting to alternative food sources. He also described the evolutionary steps involved (i.e., learning the new prey type was replaced by innate programming, over time). This conclusion parallels arguments presented by Tierney (1986) for shifts from neural plasticity to canalized behavior under natural selection.

Evan's (2002) scenario for a phylogenetic shift from prey specialization to generalist feeding is contrary to the basic tenet that species specialize to avoid competition. Because the study was limited in a number of respects (e.g., not grounded by a phylogeny that identified specialists as primitive and generalists as derived, and no explicit tests of the mode of change), it is not possible to test the validity of the results. Tauber et al. (1993) examined a phylogenetic shift in lacewings (genus *Chrysopa* of Neuroptera, Chrysopidae) that followed the expected trend from generalist to specialist feeder, and the researchers completed the appropriate tests to confirm this shift. Using two sister species of the genus *Chrysopa*, a feeding specialist on wooly alder aphid and a generalist feeder, they tested a three-step process for the development of a specialization: (a) establishment on novel prey, (b) evolutionary adaptation to the prey, and (c) reproductive isolation. This process had been previously identified for herbivore specialization on host plants. The generalist species, *Chrysopa quadripunctata*, is a widely distributed species throughout North America that exhibits considerable within- and among-site variability in its diet. A three-step test for the evolution of specialist feeding is as follows.

Step 1.   Some generalists developed adaptations to foraging on aphids after experiencing ant defense of the aphids. Like the aphid specialist *Chrysopa*, these individuals draped silk threads made by the aphids over their bodies as a camouflage against ant attacks.

Step 2.   Population variation occurred in the degree to which the camouflaging behavior was canalized in *C. quadripunctata*. Individuals from some populations showed repeatability in this behavior, suggesting that it was heritable in these populations.

Step 3.   Breeding-experiment results demonstrated that reproductive isolation could develop between populations that are generalist feeders and those that specialize on the aphids.

Studies of Individual Variation

Both steps 1 and 2 in the three-step test for the evolution of specialist feeding involved assays of individual performance within populations. The evolutionary progression is based on the assumption that inferences concerning the influence of genotype can be made from the results of tests for individual variation in phenotype. The recent interest in individual-based models of optimal foraging has led to studies that consider diet choice in individuals, as opposed to merely pooling individuals to obtain population means. For example, West (1986) completed a quantitative study of diet variation in the carnivorous marine snail (*Nucella emarginata*), which drills and then consumes other mollusks. Although individual snails were found to be consistent in the prey types they included in their diets, there was considerable dietary variation between individuals. West concluded that this pattern is evidence of trait polymorphism rather than of temporal or context flexibility in the diets of individuals. Because diet specialization was not correlated with the relative abundances of prey species in the vicinity of particular snails, one might further infer that genotype, as opposed to phenotypic plasticity, determined individual diet preference.

Studies completed on naïve individuals can better discriminate between experience and genotype influences than studies on the feeding habits of individuals that have had ample opportunity to sample the feeding environment. Mori and Tanaka (2001) examined innate preferences of the colubrid snake (*Leioheterodon madagascariensis*) by testing responses to chemical cues. Naïve hatchlings flicked their tongues significantly more frequently in response to chemicals from animal taxa that were part of their native diet than in response to those from taxa not included in the native diet. The authors concluded that this species has an innate disposition to attack particular prey species and that this preference is achieved through perception of chemical cues. Burghardt et al. (2000) obtained different results in neonate feeding trials completed on a restricted island population (Beaver Island) of the eastern garter snake, *Thanmnophis sirtalis*. Though *T. sirtalis* is a generalist forager that feeds on both fish and earthworms, this particular population feeds primarily on earthworms (85%) and is not known to forage on fish. Naïve neonates showed no differential tongue flicks and attack responses to fish and earthworm cues. The Beaver Island population studied by Burghardt et al. (2000) is only several thousand years removed from mainland populations, so one might not expect a loss in response to fish. Arnold (1980) suggests that a hundred thousand to hundreds of thousands of years would be required for large magnitude changes in chemoreceptive responses to prey to occur.

Burghardt et al. (2000) followed the prey cues test on neonates with a conditioning experiment in which all neonates were offered 12 feeding experiences with fish. Significant heritable differences (based on repeatability estimates) in snake responses to the two chemical cues were detected in the postexperience set of tests. Individual snakes apparently processed the experience with fish differently, suggesting that genotype determines how experience is used in modifying perception or sensory preference. Thus, heritable plasticity precedes the canalization of prey preferences, as predicted by Tierney (1986). Since heritable variation exists for the trait, we can also assume that many genes contribute to the phenotype (i.e., that the trait is polygenic).

In this Burghardt et al. (2000) study, repeatability estimates provided a rough estimate of the degree of heritable variation for learning to recruit a new prey into the diet. Marples and Brakefield (1995) completed a selection experiment that provides a more accurate estimate of the amount of heritable variation existing for this trait in quail, *Coturnix coturnix japanicus*. They selected two lines of quail for speed of response (low and high) to inclusion in the diet of a semipalatable prey (the two-spot beetle *Adalia punctata*). Significant differences were observed in recruitment speed between the two lines, suggesting that there is additive genetic variation for incorporating new prey into a diet. They concluded that the "recruitment rate" trait was a quantitative one involving many genes.

Population Comparisons

Population comparisons are useful in investigating genetic contributions, because species populations frequently exhibit ecotypic variation, reflecting adaptation to different selection pressures. I use Reeve and Sherman's (1993) definition of adaptation: in essence, "a phenotypic variant that results in the highest fitness among a specified set of variants in a given environment" (p. 1). Insight into the genetic architecture of traits can be gained through the following progression of tests: (a) quantification of trait value differences between populations, (b) completion of common-garden experiments, i.e., in which individuals are reared in the same selective environment, (c) completion of reciprocal crosses and tests for intermediate behavior in hybrids, and (d) completion of trait selection experiments or further genetic crosses in the event that intermediacy is not observed. This orderly progression of studies has been followed in a few cases described here.

Arnold's (1977, 1980, 1981a, 1981b) studies of snake feeding preferences are classic studies that have provided one of the first examples of parameter estimation of heritability and genetic correlation for a polygenic trait. Like Mori and Tanaka (2001), Arnold (1980) tested naïve newborn snakes for their chemoreceptive responses to different prey odors. His test species was the garter snake, *Thamnophis elegans*, and he exposed the neonates to 10 different prey odors. Unlike Mori and Tanaka, whose study was completed on individuals from a single population, Arnold tested individuals from different populations that specialized on particular prey. He studied a coastal population that exhibited terrestrial feeding on slugs, and an inland population consisting of aquatic feeders with a diet predominantly of fish and amphibians. Previous research had demonstrated that both populations were polymorphic for slug-eating tendency. In the coastal population the slug-eating phenotype predominated, and in the inland population the slug-refusing phenotype was dominant (Arnold 1977). Arnold (1980) calculated heritabilities and genetic correlations for each population by comparing the variation within and between sib groups (litters). Through completion of this common-garden experiment, he learned that the differences in slug feeding were congenital and developmentally stable.

In another study, Arnold (1981b) crossed the two population types by using two coastal and one inland population. The F_1 hybrids exhibited intermediate slug-eating (preference) scores, though there was a bias in the direction of the slug-refusal phenotype, suggesting at least partial dominance for slug refusal. Factor analyses applied to the genetic and phenotypic correlations between chemoreceptive responses to different prey species provided insight into the number of genes influencing feeding preferences (Arnold 1981a). At least three groups of genes appear to be involved in the prey preferences of *T. elegans*: one governing responses to salamanders and frogs, one to slugs, and one to a toxic salamander of the genus *Taricha*. Pleiotropic effects were evidenced, as well. For example, the fact that slugs and leeches exhibited a correlation of response of .89 indicates that a gene or group of genes affected the chemoreceptive response to a particular molecule shared between the two prey types. Arnold (1981a) concluded that genetic correlations of this type can place significant constraints on diet, in that selection for one species of prey could affect the snake's diet preferences toward many other species.

One of the species canalized in *T. elegans*' prey preference list is the toxic newt, *Taricha*. The genus *Thamnophis* seems to have had a long evolutionary history with these toxic newts, and the coevolutionary arms race of snake resistance to the newt toxin, tetrodotoxin (TTX), and corresponding increases in the toxicity of newt TTX has been well documented. For example, geographic variation in *Thamnophis sirtalis* resistance has been quantified through the study of 40 species populations (Brodie et al. 2002). Within populations, snake resistance to TTX shows individual variation that is heritable and nonplastic (Brodie and Brodie 1990, Riddenhour et al. 1999). Geoffeney et al. (2002) also have examined the physiological basis of the effect, identifying the expression of TTX-resistant sodium channels in skeletal muscle of *T. sirtalis*. Slower predator locomotion is a trade-off between increased resistance to

TTX (Brodie and Brodie 1999), which must influence the foraging behavior, and success of snakes in the high-resistance populations.

Studies of the desert spider *Agelenopsis aperta* provide another example of the insight that can be gained into the genetic influences on foraging behavior through the investigation of ecotypic variation. *Agelenopsis aperta* is a funnel-web-building spider that primarily occupies arid habitats in the western United States and Mexico. Time available for foraging is severely limited by high daytime and low nighttime temperatures, and prey availabilities are low, as well (Riechert and Tracy 1975). On the other hand, this species also occupies riparian habitats supported by spring-fed streams and rivers in the desert southwest. Insect availability in these "island" habitats is unlimited, and shading offered by the tree canopy permits extended day and night foraging (Riechert 1979). Field tests of spider attack rates toward different prey types were completed to test the optimal foraging hypothesis that, while individuals from food-limited habitats will exhibit a broad diet, individuals from habitats affording an abundance of prey will have a narrow diet of only the most profitable prey types (Riechert 1991). No diet breadth differences were observed between individuals from an arid habitat, desert grassland New Mexico (NM), and a riparian population, desert riparian Arizona (AZ). Both populations consisted of prey generalists that attempted capture of almost all prey encountered. Only beetles that had hard elytras and small dipterans, and wasps that were undetected on the web, were not included in the diet. Thus, the optimal foraging prediction concerning the influence of prey availability on diet breadth was not supported in this test system. The two populations did show markedly different capture attempt rates toward all prey encountered. Desert grassland spiders exhibited an average capture attempt rate of 78% across categories, and AZ desert riparian spiders exhibited an average of only 48% (Riechert 1991). I also observed that the highest attack responses by the spiders were toward prey that struggled vigorously on the web.

These results led to the hypothesis that arid and riparian populations of *A. aperta* have different latencies to attack prey (time elapsed between prey contact with the web, and spider approach and touch). Hedrick and Riechert (1989) tested attack latencies toward prey in spiders at their natural web sites in NM desert grassland and AZ desert riparian habitats, and in F_1 and F_2 lab-reared offspring of spiders collected in the respective habitats. The results of the common-garden experiment demonstrated that the between-population differences in attack rates exhibited toward prey (Riechert 1991) were not due to differences in hunger, experience, or maternal effects. Rather, population differences in attack rates reflected genetic differences in the latencies to attack prey. The F_1 and F_2 generation offspring of NM desert grassland spiders showed attack latencies shorter, by orders of magnitude, than those of AZ riparian spiders (Hedrick and Riechert 1989).

Possible genetic explanations for the population differences included (a) chance deviation in foraging behavior due to genetic drift, (b) the operation of natural selection operating on latencies to attack prey, and (c) selection acting on traits genetically correlated with attack latency. Electrophoretic studies completed on the two populations indicated that neither population was inbred (Riechert 1986) and that gene flow was extensive (Riechert 1993b). Thus, genetic drift was ruled out as a possible explanation of population divergence in the foraging behavior of *A. aperta*. Pleiotropy, though, is of particular interest, because it can create genetic correlations between the foraging phenotype and other fitness-linked behavioral traits observed in *A. aperta*. Phenotypic correlations were first observed between the size of an energy-based territory an individual spider demands and the level of aggression it exhibits in disputes with conspecifics over its web and associated territory (Riechert and Maynard Smith 1989). Riechert and Hedrick (1990) tested F_1 and F_2 lab-reared *A. aperta* from riparian populations for their response to predatory cues. Spider latency to return to a foraging mode, after retreat into the web funnel in response to a predatory cue, was found to covary with latency to respond to prey encountering the web sheet. Antipredatory behavior and foraging behavior thus showed significant phenotypic correlation. Further study (Riechert and Hedrick 1993) established that the phenotypes of the entire suite of fitness-linked characters, territory size, agonistic behavior, foraging behavior, and antipredator behavior were positively correlated. Riechert et al. (2001) added yet another fitness-linked trait to the suite, mating behavior. Genetic linkage can produce similar phenotypic correlations, though the results of the breeding experiments described below indicate that there are simply too many genes and different chromosomes involved for genetic linkage to explain the phenotypic correlations among traits that we have observed. Thus, pleiotropy is assumed.

Reciprocal crosses of individuals from NM desert grassland and AZ desert riparian populations produced F_1 hybrids that were not intermediate in behavioral scores, relative to the respective parental populations (Maynard Smith and Riechert 1984). Rather, individuals scored as more aggressive than either parental line. This result led to further breeding experiments that permitted the scoring of F_2 generation hybrids and of backcrosses to the respective parental lines for their aggressiveness (Riechert and Maynard Smith 1989). Level of aggressiveness is the trait that links all of the correlated behavioral phenotypes already listed. Aggressive spiders (arid-land phenotype characteristic of the NM desert grassland population) demanded larger territories, escalated to potentially injurious behavior patterns in territorial disputes, and attacked prey with a high frequency and short latency to attack. Aggressive individuals also did not respond to predatory cues with a retreat, or, if they did, they quickly returned to a foraging mode. Nonaggressive individuals (riparian phenotype, as in AZ riparian spiders) demanded smaller territories, if any at all, limited their behavior in contests largely to display, and showed long latencies to attack prey and thus low attack frequencies. They also quickly retreated in response to predatory cues and took a long time to return to a foraging position at the funnel entrance after the retreat (Riechert 1993a).

The genetic model of spider aggressiveness developed by Maynard Smith and Riechert (1984) is based on the antagonistic interaction of two gene complexes assumed to affect the production or threshold for the release of two antagonistic neurohormones, one governing some scalar as "tendency to attack" (aggression), and the other governing the scalar "tendency to flee" (fear). The aggression component is the major contributor to an individual's aggressiveness (Riechert et al. 2001). It is inherited on the sex chromosomes, and breeding-experiment results indicate that relatively few genes contribute to this component (Riechert and Maynard Smith 1989). The scalar "tendency to flee" (fear) is an autosomal, quantitative trait that appears to modulate level of aggression (Riechert et al. 2001). This model further predicts that directional dominance exists between the two populations of *A. aperta* with respect to the alleles individuals possess for aggression and fear. Because high aggression (A) is dominant over low aggression (a), and low fear is dominant (B) over high fear (b), the genotype of a desert grassland spider (arid-land phenotype) would be AA (high aggression), bb (high fear), whereas that of a desert riparian spider (desert riparian phenotype) would be aa (low aggression), BB (low fear). Although simplistic, this model adequately predicts the behavioral differences between desert grassland and desert riparian population types, as well as the behavior of F_1 hybrids between them (Maynard Smith and Riechert 1984, Riechert and Maynard Smith 1989, Riechert et al. 2001). The genotype of an F_1 hybrid would be Aa (high aggression), Bb (low fear). Directional dominance between fear and aggression produces hybrid females that are more aggressive than either parental type. However, hybrid males have different sex chromosomal contributions on the same autosomal background, because female *A. aperta* have two homologues of their sex chromosomes while males have only one (a contribution from the female parent). This means that a male *A. aperta* inherits its "tendency to attack" (aggression) component from its female parent.

As noted elsewhere, inheritance mechanisms may have associated trade-offs. The mixing of arid-land and riparian phenotypes that frequently occurs in riparian habitat patches leads to a predictable proportion of individuals (5%) that are either so aggressive that they attack every potential mate, or so fearful that they retreat from every potential mating (22%; Riechert et al. 2001). Breeding experiments and punnett square analyses of various genetic crosses show that both of these phenotypes are exhibited in the F_2 hybrids between riparian and arid-land phenotypes. The aggressive phenotype is represented also in backcrosses to an individual of arid-land origin, and the fearful type is represented in backcrosses to an individual of riparian origin. As gene exchange commonly occurs between arid habitats and riparian habitats (Riechert 1993b), directional dominance of the genes controlling fear and aggression lead to significant levels of gamete wastage.

Molecular Genetics and Mutation Studies

In recent years, molecular genetic techniques involving mapping of phenotypes onto underlying gene or protein arrays have provided major advances in the genetic analyses of complex behaviors, including foraging. *Caenorhabditis elegans*, a bacteria-feeding nematode with a body length of only 1 mm, is one of the three model systems to which high-density gene mapping techniques have been applied. It is the only predator. The nervous system of *C. elegans* is simple and well described, with 302 neurons of precisely known structure. Biologists have been successful in locating the genes associated with particular phenotypic traits by locating genetic markers on a high-density map of this species' genome.

One finding concerning foraging behavior in *C. elegans* is that a single nucleotide difference in a gene previously found to be associated with sensitivity to touch accounts for a difference in foraging behavior noted among natural populations (de Bono and Bargmann 1998, Sokolowski 2002). In some populations, individuals forage as solitary individuals when they encounter bacteria, and in others individuals aggregate and group forage. The thought is that the two forms represent alternative feeding strategies, each favored by different environmental conditions. The authors screened mutant strains for mutations affecting foraging strategy phenotype and inserted an artificially constructed gene that can induce solitary feeding in a wild aggregating population. Other work identified a null allele of the natural foraging strategy gene. This allele, which produces no gene product, created aggregating behavior that was more pronounced than that in native aggregating populations. This result, and species comparisons completed by Rogers et al. (2003), indicate that the aggregating foraging phenotype is ancestral to the solitary phenotype.

Through a combination of approaches, including the use of a fluorescent protein in the marking of relevant neurons, and ablation of implicated neuronal cells, members of the de Bono lab group determined where gene expression occurs (de Bono et al. 2002). The foraging strategy gene controls a set of anteriorly located neurons, one associated with the perception of mechanical and chemical cues, and the other with odors. Both neurons also detect pain and adverse conditions, such as overcrowding and the presence of chemical toxins. The two natural alleles of the foraging strategy gene both tend to cause dispersal in the presence of adverse stimuli. The chemical variant that produces aggregations is merely less sensitive to these stimuli than is the other variant. Another gene was found to control a signaling molecule composed of two types of neurons that regulate responses to signals. These neurons are located within the mesodermal wall of the body cavity and send antagonistic signals via the body fluid to the anterior neurons (Coates and de Bono 2002). Coates and de Bono think that communication between the anterior and mesodermal wall sets of neurons ensures that the aggregating behavior in the aggregating phenotype is induced only under conditions in which it offers a selective advantage (e.g., foraging on a dense patch of bacteria that may be producing toxic metabolic products).

Cylic GMP-dependent protein kinase (PKG) is another site-specific material that has been implicated for its regulatory effects on complex behaviors in animals, including insects and vertebrates. In *C. elegans*, the gene (*EGL-1*) that encodes PKG (Fujiwara et al. 2002) has been found to influence individual development and foraging behavior. Within individuals, low levels of the enzyme (PKG) produce roaming behavior (travel for long distances in search of prey, with infrequent turns), while high levels produce dwelling behavior (search for only short distances in short, jerky motions). Mutation studies demonstrate that PKG functions as a negative signaling molecule in the worm's ciliary sensory neurons (Fujiwara et al. 2002). The *for* gene in *Drosophila* and honey bees also encodes PKG, but the enzymes effect is the reverse of that noted in *C. elegans*. Allelic variation in the *for* locus determines whether foraging larval *Drosophila* will be either rover or sitter phenotypes (de Belle et al. 1989, Osborne et al. 1997), and gene expression of *amfor* (honey bee analog to *for*) initiates the shift from nursing behavior to foraging behavior in worker honey bees (Ben-Shahar et al. 2002). Higher levels of PKG in *Drosophila* produce the rover phenotype, and upregulation of the level of the enzyme in worker bees triggers the shift from nursing behavior to active forager behavior.

Although the specific effect of PKG on foraging in nematodes is opposite to its effect on insects, the similarities among the systems have been hypothesized to reflect conserved biological function and even a conserved neural mechanism. Fitzpatrick and

Sokolowski (2004) constructed PKG-protein phylogenies of 19 animal species, ranging from *Cnidaira* (jellyfish) to vertebrates. Three of the five different most parsimonious trees produced suggest that PKG is involved in the food-related behavior of a wide variety of animals, including vertebrates. The authors conclude that the role of cGMP-dependent protein kinase (PKG) is conserved in that it is involved in food-related behaviors throughout the animal kingdom. However, different modes of gene action have evolved, as evidenced in the reversal already noted. This fact may be related, in part, to the fact that two PKG-like proteins exist as a result of gene duplication early in the evolutionary sequence.

There is yet another entry to the *C. elegans* foraging story. This entry concerns the response of bacteria to predation by *C. elegans*. As already noted, one phenotype programmed by an allele of the foraging strategy gene, *NPR-1*, aggregates into feeding clumps in response to aversive stimuli, presumably released by their bacteria prey. Earlier I suggested that toxic by-products of metabolism might be the aversive stimuli cuing aggregation, given that bacteria are known to produce toxins, which form the basis of antibiotics against other bacteria. Darby et al. (2002) describe the production of another defense mechanism known for at least two Monerans. The two bacteria examined, *Yestis pestis* and *Yersinia pseudotuberculatus*, produce extracellular matrices called biofilms that obstruct nematode feeding by blocking the mouth. The biofilms are considered to be bacterial defense mechanisms against predation by invertebrates. Perhaps group foraging minimizes biofilm production or breaks up the matrix. Further studies are expected, because genetic analyses of prey and predator responses are feasible. So far, Darby (personal communication) has established that mutant *C. elegans* resistant to biofilm formation can be readily obtained with use of chemical mutagenesis.

Conclusions

That genetic factors play a significant role in foraging behavior is evident from the studies reviewed here. The examples I have provided clearly indicate that genes underlie foraging behavior just as they do other biological processes. Genetic influences have been identified for morphological traits that affect the nature of attack, ease of taking prey, discrimination of prey in diet specialists, resistance to prey defense mechanisms, and search patterns used during foraging. Genes even determine whether individuals will exhibit solitary or aggregative feeding (e.g., nematode worms). They also have been shown to adjust foraging effort to local prey availability environments. Patch staying time, a major parameter included in optimal foraging models, is a notable exception. Since foraging studies that emphasize patterns of inheritance are few in number compared to those that consider the adaptive significance of traits, it is possible that patch staying time has been overlooked as a potential subject for genetic study.

Genetic analyses applied to the foraging traits of predators provide insight into some important general questions, including the conservation of genetic mechanisms, the role of gene interactions in trait determination, and the degree to which trait canalization imposes constraints on evolution (thus favoring phenotypic plasticity). Combined gene and neural mapping of a nematode, *Drosophila*, and the mouse have led to recent molecular investigations into the extent to which genes and genetic mechanisms are conserved. Gene function has two meanings. It can be viewed from a proximate perspective, in the sense that it prescribes the production of a particular protein, or enzyme, or it can be viewed from an ultimate perspective, in the sense that it influences the fitness of the individual possessing it (i.e., has meaning to the organism). Few would argue against conserved gene function at the level of the protein, and there is evidence from other areas that the functional meaning of gene families may be conserved throughout the animal kingdom. Thus, the *Hox* gene family consists of homologous regulatory genes that are involved in caudal brain development in insects and mammals, indicating that the genetic mechanisms underlying pattern formation during the course of brain development are evolutionarily conserved (Hirth and Reichert 1999). Because *Hox* controls brain development both in insects that are protostomes and in mammals that are deuterostomes, a molecular blueprint for brain development must have existed in a common ancestor before the split of deuterostome (Ecinoderms, Chordates) and protostome (Annelids, Arthropods) lineages.

Molecular evidence available from gene mapping and mutation studies suggests that there might be a couple of conserved foraging equivalents to the *Hox* gene family. For instance, the *NPR* gene in nematode worms may well be a homolog of the *NPY* gene found in vertebrates (de Bono and Bargmann 1998, de Bono et al. 2002). Both genes are involved in the perception of and minimization of sensitivity to noxious stimuli. The *NPR* gene has been linked to foraging strategy exhibition in nematodes, but its potential role in vertebrate foraging has yet to be established. Genes programming production of the enzyme PKG may well be a second example of conservation of a foraging mechanism. The enzyme PKG is involved in programming foraging behavior in a wide variety of animals, ranging from the Cnidaria (hydra and their relatives) to mammals (Fitzpatrick and Sokolowski 2004). It at least is evolutionarily conserved. Although general association with foraging behavior is evidenced in both families, the specific mechanisms of gene control seem to vary with the particular system. Gene action is modified by other genes that may be acting antagonistically, acting in concert, or even determining when expression of the conserved gene occurs. Nevertheless, evidence for the conservation of specific genetic mechanisms was observed in some lineages. The best example comes from the monophyletic cichlid lineage, in which a simple two-allele Mendelian trait controlled mouth handedness in all of the species tested, even though they may have come from different isolated lake systems (Nakajima et al. 2004).

Most contemporary studies of animal behavior attempt to explain behavioral evolution by focusing on one functional category of behavior. We use the parameter we have measured and infer its contribution to the fitness of individuals. The approach implicitly assumes that patterns of behavior evolve independently of one another. However, this assumption is violated if behavior patterns are genetically correlated so that selection on one behavioral trait produces indirect selection on another. Because foraging behavior is complex, as are most behavioral traits linked to fitness, it is unlikely that such traits will be controlled by a single gene or gene complex (Hamer 2002). Genetic control of behavioral traits may involve one or more of the following mechanisms: many multiples of genes with an additive effect, gene interactions (e.g., epistasis; Grigorenko 2003), expression or regulatory differences between the same allelic forms of the genes (Robinson and Ben-Shahar 2002), and genes that simultaneously influence multiple traits (i.e., pleiotropy; Greenspan 2001). Each of these genetic phenomena has been identified in at least one of the systems reviewed here.

Genetic constraints are often cited as an explanation for the deviation of empirical estimates from model predictions in behavioral ecology. There may not be sufficient genetic variability for selection to operate. Traits that are found to have low heritabilities would be unlikely to respond to selection pressure. While heritability constrains the rate of evolution, epistasis, linkage, and pleiotropy all may constrain the direction of evolutionary responses, because modification of one trait has corresponding effects on other traits—for example, territory size and boldness toward predators in the spider, *A. aperta* (Riechert and Hedrick 1993), and feeding preferences for multiple species because their odors share some molecule in common in the snake, *T. elegans* (Arnold 1981a).

Patterns of trait inheritance also lead to evolutionary trade-offs. This finding is exemplified in the coevolutionary problem of snake adaptation to feeding on toxic newts. Snake resistance is associated with sodium channel function in muscle. A consequence of greater resistance is slower locomotory activity (Geoffeney et al. 2002). A trade-off to canalized level of aggressiveness in the spider *A. aperta* is the fact that, where ecotypes mix as a result of gene flow and the production of extreme phenotypes, significant levels of gamete wastage occur (Riechert et al. 2001).

This all brings us back to the question whether predators should be opportunistic with respect to foraging to take advantage of complex and often changing prey environments, or exhibit canalized decision processes and other traits to permit better utilization of particular prey. The trend within lineages at least seems to be toward canalized foraging traits. The model presented by Tierney (1986) seems a compelling one: as competition for limited prey is encountered, heritable plasticity evolves before the canalization of prey preferences and prey specialization.

In summary, we are in an exciting era in which new techniques and theoretical approaches can be brought to bear in testing many of the ideas expressed in this chapter. Chapter size limitations make it impossible to cover all of the potential research directions gene mechanistic studies could take with respect to foraging behavior and associated gene-based technologies that could be applied to biological control issues. The pursuit of these objectives promises to contribute much to our understanding of evolutionary processes as well. This is indeed a rich area for potential investigation.

Literature Cited

Albertson, R. C., Markert, J. A., Danley, P. D., and Koecher, T. D. 1999. Phylogeny of a rapidly evolving clade: the cichlid fishes of Lake Malawi, East Africa. Proc. Natl. Acad. Sci. 96:5107–5110.

Arnold, S. J. 1977. Polymorphism and geographic variation in the feeding behavior of the garter snake *Thamnophis elegans*. Science 197:676–678.

Arnold, S. J. 1980. The microevolution of feeding behavior. In: Foraging Behavior: Ecological, Ethological and Psychological Perspectives (Kamil, A., and Sargent, T., eds.). New York: Garland Press; 409–453.

Arnold, S. J. 1981a. Behavioral variation in natural populations. I. Phenotypic, genetic, and environmental correlations between chemoreceptive responses to prey in the garter snake, *Thamnophis elegans*. Evolution 35:489–509.

Arnold, S. J. 1981b. Behavioral variation in natural populations. II. The inheritance of a feeding response in crosses between geographic races of the garter snake, *Thamnophis elegans*. Evolution 35:510–515.

Ben-Shahar, Y., Robichon, A., Sokolowski, M. B., and Robinson, G. E. 2002. Influence of gene action across different time scales on behavior. Science 296:741–744.

Bootsma, H. A., Hecky, R. E., Hesslein, R. H., and Turner, G. F. 1996. Food partitioning among Lake Malawi nearshore fishes as revealed by stable isotope analyses. Ecology 77:1286–1289.

Brodie, E. D., III., and Brodie E. D., Jr. 1990. Tetrodoxin resistance in garter snakes: an evolutionary response of predators to dangerous prey. Evolution 44:651–659.

Brodie, E. D., III, and Brodie, E. D., Jr. 1999. The cost of exploiting poisonous prey: tradeoffs in a predator-prey arms race. Evolution 53:626–631.

Brodie, E. D., Jr., Ridenour, B. J., and Brodie, E. D., III. 2002. The evolutionary response of predators to dangerous prey: hotspots and coldspots in the geographic mosaic of coevolution between garter snakes and newts. Evolution 56:2067–2082.

Burghardt, G. M., Layne, D. G., and Konigsberg, L. 2000. The genetics of dietary experience in a restricted natural population. Psych. Sci. 11:69–72.

Coates, J. C., and de Bono, M. 2002. Antagonistic pathways in neurons exposed to body fluid regulate social feeding in *Caenorhabditis elegans*. Nature 419:925–929.

Darby, C. J., Hsu, W., Ghori, N., and Falkow, S. 2002. Plague bacteria biofilm blocks food intake. Nature 417:243–244.

de Belle, J. S., Hilliker, A. J., and Sokolowski, M. B. 1989. Genetic localization of foraging (*for*): a major gene for larval behavior in *Drosophila melanogaster*. Genetics 123:157–163.

de Bono, M., and Bargmann, C. I. 1998. Natural variation in a neuropeptide y receptor homolog modifies social behavior and food response in *C. elegans*. Cell 94:679–689.

de Bono, M., Tobin, D. M., Davis, M. W., Avery, L., and Bargmann, C. I. 2002. Social feeding in *Caenorhabditis elegans* is induced by neurons that detect aversive stimuli. Nature 419:899–903

Evans, H. E. 2002. A review of prey choice in Bembicine sand wasps (Hymenoptera: Sphecidae). Neotropic. Entomol. 31:1–11.

Fitzpatrick, M. J., and Sokolowski, M. B. 2004. In search of food: exploring the evolutionary link between cGmp-dependent protein kinase (PKG) and behavior. Integr. Comp. Biol. 44:28–36.

Fujiwara, M., Sengupta, P., and McIntire, S. L. 2002. Regulation of body size and behavioral state of *C. elegans* by sensory perception and the EGL-4 cGMP-dependent protein kinase. Neuron 36:1091–1102.

Geoffeney, S., Brodie, E. D., Jr., Ruben, P.C., and Brodie, E. D., III. 2002. Mechanisms of adaptation in a predator-prey arms race: TTX-resistant sodium channels. Science 297:1336–1339.

Gittenberger, E. 1991. What about non-adaptive radiation? Biol. J. Linn. Soc. 43:263–272.

Greenspan, R. J. 2001. The flexible genome. Nature Rev. 2:383–387.

Grigorenko, E. L. 2003. Epistasis and the genetics of complex traits. In: Behavioral Genetics in the Postgenomic Era (Plomin, R., DeFries, J. C., Craig, I. W., and McGuffin, P., eds.). Washington, D.C.: American Psychological Association; 247–266.

Hamer, D. 2002. Rethinking behavior genetics. Science 298:71–72.

Hedrick, A. V., and Riechert, S. E. 1989. Genetically-based variation between two spider populations in foraging behavior. Oecologia 80:533–539.

Hirth, F., and Reichert, H. 1999. Conserved genetic programs in insect and mammalian brain development. BioEssays 21:677–684.

Hori, M. 1993. Frequency-dependent natural selection in the handedness of scale-eating cichlid fish. Science 260:216–219.

Liem, K. F., and Summers, A. P. 2000. Integration of versatile functional design, population ecology, ontogeny and phylogeny. Neth. J. Zool. 50:245–259.

Marples, N. M., and Brakefield, P. M. 1995. Genetic variation for the rate of recruitment of novel insect prey into the diet of a bird. Biol. J. Linn. Soc. 55:17–27.

Maynard Smith, J., and Riechert, S. E. 1984. A conflicting tendency model of spider agonistic behaviour: hybrid-pure population line comparisons. Anim. Beh. 32:564–578.

Mori, A., and Tanaka, K. 2001. Preliminary observations on chemical preference, antipredator responses, and prey handling behavior of juvenile *Leioheterodon madagascariensis* (Colubridae). Curr. Herpet. 20:39–41.

Nakajima, M., Matsuda, H., and Hori, M. 2004. Persistence and fluctuation of lateral dimorphism in fishes. Am. Nat. 163:692–698.

Osborne, K. A., Robichon, A., Burgess, E., Butland, S., Shaw, R. A., Coulthard, A., Pereira, H. S., Greenspan, R. J., and Sokolowski, M. B. 1997. Natural behavior polymorphism due to a cGmP dependent protein kinase of *Drosophila*. Science 777:834–836.

Reeve, H. K., and Sherman, P. W. 1993. Adaptation and the goals of evolutionary research. Q. Rev. Biol. 68:1–32.

Riddenhour, B. J., Brodie, E. D., Jr., and Brodie, E. D., III. 1999. Repeated injections of TTX do not affect TTX resistance or growth in the garter snake *Thamnohis sirtalis*. Copeia 199:531–535.

Riechert, S. E. 1979. Games spiders play. II. Resource assessment strategies. Behav. Ecol. Sociobiol. 4:1–8.

Riechert, S. E. 1986. Between population variation in spider territorial behavior: hybrid-pure population line comparisons. In: Evolutionary Genetics of Invertebrate Behavior (Huettel, M., ed.). New York: Plenum Press; 33–42.

Riechert, S. E. 1991. Prey abundance versus diet breadth in a spider test system. Evol. Ecol. 5:327–338.

Riechert, S. E. 1993a. The evolution of behavioral phenotypes: lessons learned from divergent spider populations. Adv. Study Behav. 22:103–134.

Riechert, S. E. 1993b. Investigation of potential gene flow limitation of behavioral adaptation in an arid lands spider. Behav. Ecol. Sociobiol. 32:355–363.

Riechert, S. E., and Hedrick, A. V. 1990. Levels of predation and genetically based anti-predatory behavior in the spider, *Agelenopsis aperta*. Anim. Behav. 40:679–687.

Riechert, S. E., and Hedrick, A. V. 1993. A test for correlations among fitness-linked behavioural traits in the spider *Agelenopsis aperta* (Araneae, Agelenidae). Anim. Behav. 46:669–675.

Riechert, S. E., and Maynard Smith, J. 1989. Genetic analyses of two behavioural traits linked to individual fitness in the desert spider, *Agelenopsis aperta*. Anim. Behav. 37:624–637.

Riechert, S. E., Singer, F. D., and Jones, T. C. 2001. High gene flow levels lead to gamete wastage in a desert spider system. Genetica 112/113:297–319.

Riechert, S. E., and Tracy, C. R. 1975. Thermal balance and prey availability: bases for a model relating web-site characteristics to spider reproductive success. Ecology 56:265–284.

Robinson, G. E., and Ben-Shahar, Y. 2002. Social behavior and comparative genomics: new genes or new gene regulation? Genes, Brain and Behav. 1:197–203.

Rogers, C., Reale, V., Kim, K., Chatwin, H., Li, C., Evans, P., and de Bono, M. 2003. Inhibition of *Caenorhabditis elegans* social feeding by FMRFamide-related peptide activation of NPR-1. Nat. Neurosci. 6:1178–1185.

Sokolowski, M. B. 2002. Neurobiology: social eating for stress. Nature 419:893–894.

Stauffer, J. R., Jr., Bowers, N. J., Kellogg, K.A., and McKaye, K. R. 1997. A revision of the blue-black *Pseudotropheus* zebra (Teleostei: Cichlidae) complex from Lake Malawi, Africa, with a description of a new genus and ten new species. Proc. Acad. Natl. Sci., U.S.A. 148:189–230.

Tauber, M. J., Tauber, C. A., Ruberson, J. R., Milbrath, L. R., and Albuquerque, G. S. 1993. Evolution of prey specificity via three steps. Experientia 49:1113–1117.

Tierney, A. J. 1986. The evolution of learned and innate behavior contributions from genetics and neurobiology to a theory of behavioral evolution. Anim. Learn. Behav. 14:339–348.

West, L. 1986. Interindividual variation in prey selection by the snail *Nucella emarginata*. Ecology 67:798–809.

Wimer, R. E., and Wimer, C. C. 1985. Animal behavior genetics: a search for the biological foundations of behavior. Annu. Rev. Psych. 36:171–218.

PART II

SENSORY, PHYSIOLOGICAL, AND BEHAVIORAL PERSPECTIVES

Whether a species is a traditional predator (or an omnivore), or whether it responds to predation risk with fixed rather than learned behaviors, its success as a predator or a prey is determined by the capacity and efficiency of its sensory organs and the processing of information by the central nervous system (CNS). In chapter 5, Jeffrey Triblehorn and David Yager focus on the interactions between echolocating bats and flying insects to illustrate this important balancing act. Critical to their discussion is the recognition that not only is the CNS processing of sensory information important, but so are the mitigating and modulating influences of ecological factors.

This blending of sensory physiology and ecology provides a perspective that may be called *sensory ecology* (or, more specifically, in discussions of bats and insects, *acoustic ecology*; or *visual ecology* in Thomas Cronin's chapter). In essence, this subdiscipline focuses on the how the mandates of the ecology of a predator and its prey shape and influence information gathering, processing, and subsequent responses. For example, bats have evolved extraordinary capabilities for flight and echolocation that enable them to detect, track, and capture insect prey on the wing. However, the ecology of species entails interactions with the environment, potential mates, conspecifics, and so on. Thus, bats have to handle a large amount of incoming auditory information and must separate prey echoes from obstacles and other bats. For predators like bats there is no single combination of bat vocalization parameters that is optimal for all situations or that enable them to perceive all specifics of the environment. The same constraints plague the prey. That is, a hearing moth can fly toward a potential mate or evade capture by an attacking bat, but at any given moment it cannot do both. This fact suggests choice or compromise based on risk-benefit computation. In other words, it is the overall behavioral and ecological context that ultimately determines the requirements and adaptations for success of predator and prey, and the result is generally a compromise. Triblehorn and Yager propose that an understanding of ecological niche and habitat structure, combined, provides an effective way of classifying different bat echolocation behaviors. Ultimately, these interactions blend sensory physiology, ecology, and evolution.

In chapter 6, Cronin focuses on another sensory modality, vision. He provides a baseline of information on the visual adaptations of animals in order to demonstrate how vision functions in predator-prey interactions. More important, he makes a compelling argument for the contention that these adaptations shape the interplay of predators and prey, and thus the ecology of predator-prey interactions. As noted by Triblehorn and Yager, use of vision in predation risk assessment or prey detection is complicated by competing behavioral and ecological demands. Thus, vision gives a sense of location in space, helps monitor movement, and guides locomotion, as part of many crucial activities in the lives of animals; and an animal is required to perceive and discriminate among stimuli of many types. Visual systems must meet multiple needs. Most pertinent to the ecology of predator-prey interactions is that most predatory species are themselves prey, so they require the visual capacity to be effective predators but to simultaneously effectively detect their own predators.

Cronin describes the visual equivalent of Triblehorn and Yager's auditory scene analysis, in which bats perceptually group and segregate complex sounds, allowing them not only to discriminate between returning echoes, but to group them into different "chunks" (thus keeping track of objects within the environment over time). Cronin describes "visual natural scenes" that allow perceptual grouping and segregation of multiple elements of the environment. The issue of how predators and prey perceive and manage the multitude of cues from competitors, plants, predators, and prey is one that needs attention in many of the systems discussed in this book.

The role of chemical signals in the communication between invertebrate predators and their prey has been repeatedly discussed and reviewed by ecologists. However, Matthew Greenstone and Joseph Dickens blend ecological and physiological perspectives in their chapter 7 discussion of the role that various types of chemicals (e.g., kairomones, allomones, and pheromones) play in the communication between predator and prey, and thus in the outcome of their interactions. In particular, they provide excellent examples of predator responses to prey pheromones, a phenomenon observed in certain taxa but an interaction that has received less attention than predator responses to kairomones. Their focus on how predators eavesdrop on and exploit chemical signals mediating the interactions between other species is particularly interesting.

Greenstone and Dickens's sensory and physiological insights provide an understanding of the constraints and opportunities inherent in the use of chemical signals in communication. They convey the clear message that the outcome of predator-prey interactions is not solely dependent on the existence of chemical signals but also on the sensitivity of sensory structures of predators and prey. The constraints and opportunities provided by the sensory physiology of interacting predators and prey, and resulting fitness trade-offs, lead to the potential for significant evolutionary changes. In particular, they suggest that adaptations to facilitate the misappropriation of the chemical signals used by interacting species may provide opportunities for coevolution.

In chapter 8, Steven Lima and Todd Steury note that little attention has been directed toward understanding why some prey species do not respond to predators when it appears that they should. They explore the nature of information available to prey species regarding predation risk and the possibility that prey use certain rules of thumb in assessing risk. Although prey often escape death, they do not avoid being influenced by predators. Thus, the authors discuss how predators may "manage" the information available to prey.

The often implicit assumption of some research on predator-prey interactions is that, if an animal does not respond to an indicator of potential risk, then that piece of information must be unavailable to the animal in question (i.e., it is not perceived by the species). Thus, one might mistake the lack of a behavioral response for the lack of perceived information. In fact, the authors suggest that an animal's assessment of risk probably is not solely based on encounters with predators, but also may include a variety of other factors, not the least of which are experience and ecological variables.

Rick Relyea similarly expresses an interest in understanding the apparent lack of response by some prey to predators. He asks, what prevents prey from evolving or exhibiting predator-induced plasticity? In chapter 9, he discusses how phenotypic trade-offs, ontogeny, and phylogeny may constrain the evolution of antipredator responses. Relyea raises the issue of understanding induced antipredator behaviors in the context of multiple predator risks, because in the real world prey experience combinations of predators. He also explores the importance of the amount of resource available to prey in the trade-off between antipredator behaviors and other fitness-enhancing behaviors, such as foraging. The central question here is whether prey should exhibit stronger antipredator responses when there are plentiful resources.

Relyea continues, in a discussion of how prey traits and prey strategies change over ontogeny, by exploring how ontogeny might constrain predator-induced defenses. He highlights the importance of developmental windows in constraining the production of predator-induced defenses and proposes some interesting hypotheses that can be tested in future research. Finally, the author explores the question whether species possess a particular phenotype because of their evolutionary history, or because of the ecological conditions under which they currently exist.

5

Acoustic Interactions between Insects and Bats: A Model for the Interplay of Neural and Ecological Specializations

JEFFREY D. TRIBLEHORN
DAVID D. YAGER

Whether belonging to a predator or prey, an animal's central nervous system (CNS) and sensory systems have evolved to maximize its chances of survival. This evolution is rarely an optimization of structure and function, but rather a compromise, as demonstrated in one of the best-studied cases of the coevolution of CNS and sensory capabilities: echolocating bats and the flying insects they hunt at night. The bats echolocate using ultrasonic (frequencies over 20 kHz) sonar cries, and the nature of the cries and the way their echoes are processed in the CNS depend on the ecology of the particular species. The CNS processing of echolocation cries for swift and efficient prey capture must be balanced against the needs for obstacle avoidance, predator avoidance, and distinguishing their own pulse-echo pairs from those of nearby conspecifics. A partial solution may lie in *auditory scene analysis*, a CNS strategy of grouping and segregating complex sounds to facilitate their analysis.

Insects have evolved a range of clever strategies to evade echolocating bats, all of which require sensitive ultrasonic hearing. These evasive strategies are very successful, increasing the probability of surviving a bat attack by about 50%. Depending on the particular species, ultrasonic hearing involved the evolution of an entirely new sensory modality or the modification of a preexisting auditory system to include ultrasonic frequencies. In either case, the auditory system had to be linked to the CNS to produce these evasive responses. Acoustic ecology is important: ample evidence suggests that the range of hearing in an insect species is tailored evolutionarily to detect the cries of a multispecies assemblage of bats in its specific habitat. Thus, the CNS control of both predator and prey behavior involves a complex balance of sometimes competing ecological, behavioral, and social demands on the nervous system.

Interactions between predators and prey are a major driving force for evolution and adaptation in animals. In any single encounter, the prey has more at stake, because failure means death, whereas the predator misses only a feeding opportunity. However, changes that allow prey to effectively escape can deny predators too many meals. This possibility can pose problems for predators, forcing them to adapt to remain competitive. Improvements occur gradually over time, back and forth between predator and prey, forming an evolutionary "arms race" between the species (Dawkins and Krebs 1979). In some cases, the arms race reaches an end, with one side's improvements leading to the destruction of the other side, or with both sides reaching equilibrium. However, in most cases, the cycle of gradual improvements continues as species coevolve. Arms races can drive a variety of changes that improve a predator's ability to capture prey and a prey's ability to elude predators. These changes include morphological, biomechanical, and ecological changes, but can also involve nervous system modifications that lead to behavioral changes. The coevolution of insectivorous bats and the insects they hunt is a model system for studying how predator-prey interactions influence changes within the nervous system. Evolution of the profound ecological, anatomical, and neurophysiological specializations characteristic of hearing insects (i.e., ultrasound-sensitive audition, complex evasive maneuvers) appears to have been driven by bat predation. However, it is the case that predatory pressures also drove such specializations characteristic of extant bats (i.e., nocturnal flight, echolocation/sonar).

Neuroecology of Predator-Prey Interactions

Bat Echolocation, Ecology, and Evolution

According to all current theories (Fenton et al. 1995, Speakman 2001, Neuweiler 2003), bats evolved from small, diurnal, insectivorous animals that were driven into the night by predation pressure, probably from the rapidly diversifying bird taxa of the upper Paleocene (Fenton et al. 1994, Feduccia 1995, Fenton 1995, Rydell and Speakman 1995). Regardless of whether flight or echolocation came first, or whether they evolved together, nocturnal insectivores and their descendants, bats, became major predators of primitively nocturnal flying insects. Improvements in both flight and echolocation abilities over time have enabled bats to detect, track, and capture insect prey on the wing while simultaneously navigating obstacles—without requiring other sensory cues.

Echolocating bats probe their environment in a strobe-like manner by producing vocalizations (pulses) containing ultrasonic frequencies (i.e., those greater than 20 kHz) and then extracting information from echoes reflecting off objects (Simmons and Stein 1980, Moss and Sinha 2003). Returning echoes differ from the emitted vocalizations, depending on how the pulses reflect off objects. Bats extract information (such as object size and texture) from these differences (called acoustic glints). The time delay between the pulse and the returning echo provides the bat with object distance information (Hartridge 1945, Simmons 1973). Ultrasonic frequencies have short wavelengths that provide detailed information about small objects (such as insects), but these frequencies attenuate rapidly and limit a bat's detection range (Lawrence and Simmons 1982). Bats can compensate for this limitation by controlling various aspects of their echolocation emissions, including the rate of vocalization production, known as pulse repetition rate (PRR), as well as pulse duration, intensity, and frequency content. The PRR dictates the rate of information the bat receives, while the other three parameters affect the type of information each vocalization provides the bat.

The most widely studied aspect of bat echolocation behavior is the emission pattern produced during insect pursuit and capture (Simmons et al. 1979, Kick 1982, Surlykke and Moss 2000). This pattern is stereotypical across many bat species that catch insects on the wing. The pattern consists of three phases: search, approach, and the terminal buzz (Griffin et al. 1960). Insect detection occurs during the search phase and, once the insect is detected, the bat enters the approach phase by beginning its pursuit and closing in on the insect. As the bat attempts to capture the insect, it enters the terminal buzz phase. With each stage, PRRs increase (exceeding 100 pulses per second during the terminal buzz phase), while durations decrease. Bats may alter the frequency content and intensity of the vocalizations, as well. Figure 5.1 illustrates bat attack sequences for two species commonly used in laboratory echolocation experiments, *Eptesicus fuscus* (top) and *Rhinolophus ferrumequinum* (bottom).

With the exception of this generalized, stereotypical pattern of increasing PRR and decreasing duration during insect capture, there is no single combination of bat vocalization parameters optimal for all situations (Neuweiler 1984, Neuweiler 1989, Neuweiler 2003). Echolocation behavior must be dynamic to function in different situations (such as cruising, hunting, and obstacle avoidance), but also tailored to the habitat where a bat lives and hunts. Ecological niches serve as a useful basis for classifying different bat echolocation behaviors, because habitat partially dictates the echolocation vocalization bats emit and the hunting strategies they employ (Neuweiler 1984, Schnitzler and Kalko 2001). Schnitzler and Kalko (2001) defined the three main ecological niches within which bats operate as uncluttered space (i.e., open space away from vegetation and high above ground), background-cluttered space (i.e., along the edge of vegetation or within vegetation gaps), and highly cluttered space (i.e., within vegetation). Table 5.1 compares typical values for vocalization parameters characteristic of those emitted in the different ecological niches, and Figure 5.2 provides examples of actual vocalization structure from species that hunt within each of the niches. Together, they illustrate that environment and hunting strategy (i.e., capturing insects in the air vs. off a substrate) shape both the vocalizations and (by necessity) the underlying neural mechanisms for processing pulse-echo pairs.

Many insects fly at night and thus provide an abundant food source for bats during the time of day when their own predators are scarce, although, interestingly, bats are not immune to all predators (and may be subject to predation by owls, nocturnal raptors, and other bats; Fenton et al. 1994, Fenton 1995). Echolocation allows bats to exploit such a situation—a nontrivial issue, considering flight is energetically expensive (Rayner 1991, Speakman and Racey 1991). To appreciate the task facing bats, consider the following calculations from Hill and Smith (1984). Estimations indicate that bats typically eat between one-quarter to one-half their body weight nightly (e.g., 5–10 g of insects for a 20-g bat). In terms of feeding rates, the bat *Myotis*

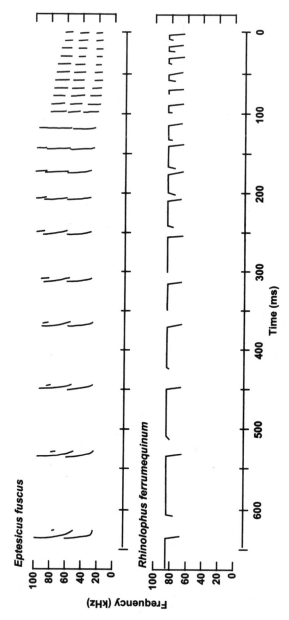

Figure 5.1. Time-frequency spectrograms of echolocation sequences emitted by a frequency-modulated (FM) bat (*Eptesicus fuscus*, top) and a constant-frequency–FM bat (*Rhinolophus ferrumequinum*, bottom) while pursuing and capturing an insect. *Eptesicus fuscus*'s vocalizations are frequency modulated (sweeping downward from high to low frequencies) and relatively short in duration. *Rhinolophus ferrumequinum*'s vocalizations are longer in duration and primarily contain a single frequency, with a very brief downward sweep at the end. In both examples, the vocalization durations decrease, and pulse repetition rate increases, as the bat pursues and captures the insect. (Adapted with permission from Simmons, J. A., Fenton, M. B., and O'Farrell, M. J. 1979. Echolocation and pursuit of prey by bats. Science 203:16–21. © 1979 AAAS.)

Table 5.1. Comparison of Echolocation Behaviors Used within Different Foraging Habitats

	Uncluttered	Background Cluttered	Highly Cluttered	Highly Cluttered
Hunting type	Aerial hawkers	Aerial hawkers	Aerial hawkers/hunt from perches	Capture insects off surfaces
General classification	None	FM bats Figure 5.1 (top)	CF-FM bats Figure 5.1 (bottom)	Gleaning bats
Echolocation pattern				
Vocalization bandwidth	Narrowband	Broadband, downward frequency modulated (FM) sweeps[a]	Dominated by narrow-band constant frequency (CF)[b]	Broadband
Frequency composition	Typically below 30 kHz	10–80 kHz but most energy 30–50 kHz	Typically over 60 kHz	
Signal duration	Medium (8–5 ms)	Short to medium, typically 3–10 ms	Long. 8–60 ms CF component; ends with brief FM component	Short, 1–3 ms
Intensity	High	High (100 dB SPL measured 10 cm away; Griffin 1958)[c]	High	Low ("whispering bats")
Doppler shift compensation?	No	No	Yes	No
Information from echoes	Good for detecting larger insects at greater distances but not discrimination; not good for navigation and obstacle avoidance in clutter (reason only used in uncluttered situations and bats change to background-cluttered echolocation signals when nearing clutter)	Target localization (range, elevation, azimuth); characteristics of prey (i.e., texture) and discrimination between prey and clutter echoes; E. fuscus can discriminate between two objects 12 mm apart (Simmons 1973)	Detecting and identifying targets in highly cluttered background, according to amplitude and frequency modulations in returning echoes caused by the beating wings of insects (CF component); some target localizations (i.e., range) due to brief FM component	Mostly for navigational purposes; short duration, low-intensityvocalizations reduce echoes from clutter; use passive hearing for prey detection, localization, and, capture
Examples	Molossidae (Molossus, Promops), Rhinopomatidae (Rhinopoma); Emballonuridae (Dielidurus, Peropteryx, Taphozous); Vespertilionidae (Lasiurus Nyctalusi)	Vespertilionidae (Eptesicus, Myotis, Pipistrellus); Mormoopidae (Mormoops, Pteronotus except P. parnellii); Emballonuridae (Saccopteryx)	Rhinolphidae; Hipposideridae; Mormooidae (Pteronotus parnellii)	Megadermatidae; Nycteridae; Phyllostomidae; Vespertilionidae

Source: Based primarily on Schnitzler and Kalko (2001).

Note: Compare table with vocalization examples in Figure 5.2.

[a] Many FM bats can operate either near clutter or within clutter, altering the structure of their calls to suit each situation. For example, Myotis myotis use longer duration (about 4 ms), narrowband vocalizations in uncluttered space, but emit shallow FM sweeps of shorter duration in background cluttered environments, and very short duration, steep FM sweeps in highly cluttered environments.

[b] CF-FM bat vocalizations are dominated by a long narrowband (CF) component (containing a specific frequency that varies slightly between individuals within a species) important for detecting fluttering insects. However, most vocalizations also contain a brief FM component (see Figure 5.2) at the end of the CF portion, which provides the bat with additional information for target localization that the CF component does not provide.

[c] SPL = sound pressure level.

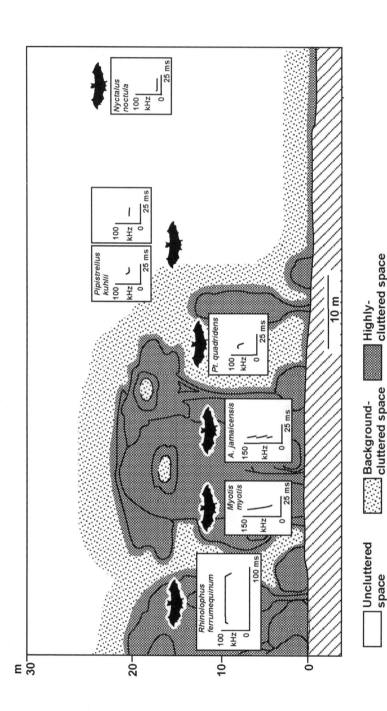

Figure 5.2. Bat vocalization structure varies with different foraging habitats, defined by the clutter situation (no clutter, background clutter, or heavy clutter). Vocalization examples come from species that hunt within each type of habitat. Bats hunting in uncluttered space (*Pipistrellus kuhlii* and *Nyctalus noctuis*) emit vocalizations with a narrow bandwidth. Bats hunting around vegetation (background clutter; *Myotis myotis* and *A. jamaicensis*) emit broad bandwidth, frequency-modulated vocalizations. Those echolocating deep within the vegetation (heavy clutter; *Rhinolophus ferrumequinum*) produce long vocalizations consisting primarily of a constant frequency portion, with a brief frequency-modulated portion at the end. (Adapted with permission from Schnitzler, H. U., and Kalko, E. K. V. 2001. Echolocation by insect-eating bats. BioScience 51:557–569. © 2001 American Institute of Biological Sciences. Reproduced with permission of the American Institute of Biological Sciences in the format Other Book via Copyright Clearance Center.)

lucifugus's average intake is about 1 g per hour. Given that the average prey size for this species is about 2 mg, each individual must detect and capture about 500 insects per hour. In the laboratory, *M. lucifugus* foraging in a room full of *Drosophila* (weighing about 2 mg) can capture up to 1,200 insects per hour (about one every 3 seconds). The ability of bats to meet these high intake requirements indicates that bats are very good at what they do.

Insect Hearing, Ecology, and Evolution

To date, ears have independently evolved in insects at least 20 times (Yager 1999, Yack and Hoy 2003). Insect ears appear in a variety of locations, including wings, legs, mouthparts, thorax, and abdomen (Yack and Fullard 1993, Minet and Surlykke 2003). Regardless of the location, and despite the independent evolutions, the majority of insect ears share the same basic functional anatomy. For sensitive hearing above 1 kHz, an ear typically requires three components: (a) a thin portion of cuticle forming a tympanum that vibrates in response to pressure-wave stimulation (similar to the vertebrate eardrum); (b) an air space behind the tympanum, allowing the tympanum to vibrate (although there are some notable exceptions, such as the fluid-filled lacewing ear; Miller 1970); and (c) a sensory tympanal organ, attached directly or indirectly to the tympanum, that converts the vibrating tympanum's mechanical energy into neural impulses. This tympanal organ is a modified chordotonal organ, a stretch receptor usually involved in proprioception (reviewed in Minet and Surlykke 2003). Discussion of specific insect ears is beyond our scope here, but reviews by Michelsen and Larsen (1985), Yager (1999), and Minet and Surlykke (2003) cover the topic in detail.

Nocturnal insectivores and their bat descendants have become the major nocturnal predators of primitively nocturnal insects like moths and have driven the evolution of ultrasound-triggered defense (Conner 1999, Yack and Fullard 2000). Bat predation is one of the major selective forces driving auditory evolution in insects; two others include intraspecific communication and parasitism (Hoy and Robert 1996, Conner 1999). The high predation rates of insectivorous bats impose a strong selective pressure on insects to adapt (Conner 1999, Fullard 1998, Rydell and Lancaster 2000, Rydell et al. 2002). In some insects, the ultrasound-sensitive auditory system involved the evolution of a new sensory modality, whereas other species (such as crickets and katydids) already possessed a functioning auditory system for intraspecific communication. These latter cases involved expanding the sensitivity range of the auditory system to include frequencies used by echolocating bats (Hoy 1992, Conner 1999, Miller and Surlykke 2001). The reverse situation has also occurred in certain moths, whereby ultrasonic hearing evolved initially for bat defense and later functioned in intraspecific communication (Fullard 1998, Sanderford and Conner 1990). Those insects that fly at night and possess ultrasound sensitivity include members of Orthoptera (e.g., locusts, grasshoppers, and katydids), Lepidoptera (e.g., several moth species and some butterflies), Mantodea (e.g., praying mantids), Coleoptera (e.g., tiger and scarab beetles), and Neuroptera (e.g., green lacewings). Insect species possessing ultrasound-sensitive auditory systems are in the minority, indicating that evolving an ear requires anatomical and neural precursors (Yager 1999) and sufficiently strong selective pressure.

Options for Prey Avoiding Bats

Insects that do not fly are obviously immune to attack by bats (other than gleaners). Some insect species are sexually dimorphic for flight. For instance, male gypsy moths (*Lymantria dispar*) fly and can hear, whereas females are flightless and their auditory system is secondarily reduced (Baker and Cardé 1978, Fullard 1988). A more complicated example occurs in praying mantids. Within mantid species, wing length is correlated with auditory abilities, and wing length and auditory ability can be sexually dimorphic (Yager 1990). In some mantid species, both males and females possess long wings and ultrasound-sensitive auditory systems with equivalent sensitivities. In others, males have long wings with typical sensitivity to ultrasound, whereas females have slightly reduced wings and hear ultrasound, though with reduced sensitivity compared to males. In a third group, only males fly, whereas females have severely reduced wings, do not fly, and are deaf.

Flying insects can exhibit other characteristics that protect them from bat predation (which may also reduce the selective pressure for evolving auditory defenses). In rare cases, nocturnal insects reduce their exposure to bat predation by becoming diurnal (and restricting the amount of time that they fly at night) or by flying only at dusk before bats emerge (Fullard et al. 1997, Fullard et al. 2000). Some of these "newly" diurnal insects possess a reduced, relatively insensitive ultrasound-sensitive auditory system. This reduced sensitivity may reflect degeneration due to the return to daytime activity and relaxation of the selective pressure (Fullard et al. 1997). A more common strategy some insects employ to avoid bat predation is to continue flying at night but to vary their seasonal activity so their activity does not coincide with bat activity (i.e., either just after or just before bats hibernate; Yack 1988, Svensson et al. 1999). Finally, a species' natural flight pattern can protect insects from bat predation, especially if the insect is fast and/or has a naturally erratic flight pattern. Insects flying close to vegetation may deter some bats, as well (e.g., *Hepialus humuli*; Rydell 1998), unless these insects are sympatric with bats that usually hunt insects near clutter.

Predator-Prey: Frequency-Modulated (FM) Bats and the Insects They Attack

The Bats: FM Emissions

Frequency-modulated (FM) bats emit vocalizations that sweep downward from high (around 80 kHz) to low (15–20 kHz) frequencies, although the exact frequency content can vary across species and within individual attack sequences (Simmons et al. 1979, Simmons and Kick 1983, Surlykke and Moss 2000). These FM signals contain multiple harmonics, with most of the energy falling within the 30–50 kHz range of the first harmonic. The 30–50 kHz range offers a good compromise, providing enough detailed information about targets, with a sufficient detection range (i.e., a 3- to 5-m maximum; Kick 1982). To attain a 5-m detection range, the vocalizations must be very loud (Table 5.1) to overcome the severe sound attenuation effects impinging on ultrasonic frequencies. Such intense sounds would deafen the bat (the *self-deafening* problem), preventing it from hearing the returning echoes, without protective measures. The most important protection comes from the bat's stiffening the middle ear bones just a few milliseconds (ms) before each vocalization and then relaxing them before the echoes return (Wever and Vernon 1961, Henson 1965, Suga and Jen 1975). The remarkable ability of FM bats to determine target range and discriminate objects (Table 5.1) comes from specializations within the auditory regions for processing signals in both time and frequency domains.

The Bats: Delay-Sensitive Neurons and Object Discrimination

Being auditory specialists, bats naturally have well-developed auditory areas in the brain. Many of these areas, including the inferior and superior colliculi (Dear and Suga 1995, Mittmann and Wenstrup 1995, Valentine and Moss 1997), the thalamus (Olsen and Suga 1991), and the auditory cortex (Suga and O'Neill 1979), contain specialized neurons sensitive to paired sound stimuli, such as the pulse-echo pairs of echolocating bats. Because the time interval between pulse-echo pairs has a direct relationship to the distance between the bat and objects within the environment, these delay-sensitive neurons are the neural mechanisms for range detection in echolocating bats. Recordings from delay-sensitive neurons show that each responds most consistently to a specific delay between the pulse-echo pair (known as the *best delay*, or BD) rather than to other delay intervals or single-pulse presentations (Figure 5.3). Because these neurons produce only a single action potential per stimulus, strong responses are defined in terms of consistency of the responses (how reliably it responds to each pulse-echo pair) and the consistency of the response latency. These BDs range from 2 to 28 ms (0.5–5 m; Dear, Fritz, et al. 1993, Dear, Simmons, et al. 1993).

Not only are there "populations" of delay-sensitive neurons with the same BDs, but also there exist subpopulations sensitive to different sound frequencies (Dear, Fritz, et al. 1993, Dear, Simmons, et al. 1993). Therefore, a single FM pulse-echo pair elicits a population neuronal response consisting of all delay-sensitive neurons with the same BDs sensitive to the frequencies contained within the emitted pulse and the returning echo. Thus, two objects equally distant but reflecting different echoes (i.e., echoes containing different frequencies) will elicit different neuronal responses and can be discriminated.

In the lab, bats and their delay-sensitive neurons process one pulse-echo pair at a time. However, in the natural situation, a bat may receive echoes from multiple sources within a small time window, with each source exciting different delay-sensitive neurons. Bats have to process a large amount of incoming auditory information and separate echoes from prey, important clutter (i.e., obstacles to avoid), and unimportant clutter. The problem becomes more complicated when they are hunting in the presence of other bats, because the bat has to distinguish echoes from its own vocalizations from those of the other bats (Obrist 1995, Moss and Surlykke 2001).

Although delay-sensitive neurons are an important neural mechanism, they alone cannot account for how bats process all of this information. Instead, it is more likely that an emergent cognitive, perceptual, property is derived from the initial processing

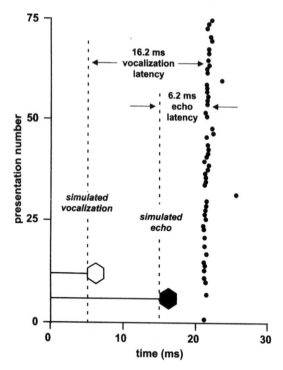

Figure 5.3. Delay-sensitive neurons as neural correlates for distance determination between a bat and objects in its environment, according to time delays between pulse emissions and returning echoes. This physiological recording result from a delay-sensitive neuron demonstrates the high selectivity for a precise time interval between two sounds (representing a pulse and returning echo). The stimulus consists of a simulated pulse followed by a simulated echo at some delay (in this case, 10 ms). The pulse-echo pair is repeated several times (75 times in this example). Dots indicate each spike the delay-sensitive neuron produced. The pulse-echo pair elicited only single-spike responses, occurring after most of the presentations with a very consistent latency (16.2 ms after the vocalization, 6.2 ms after the echo). Note that responses do not occur during the 10 ms delay between pulse and echo. Furthermore, responses do not occur to the pulse alone. (Adapted with permission from Dear, S. P., Simmons, J. A., and Fritz, J. 1993. A possible neuronal basis for representation of acoustic scenes in auditory cortex of the big brown bat. Nature 364:620–623. © 1993 Nature Publishing Group.)

of echolocation signals by such mechanisms. Perceptually, bats must discriminate between returning echoes and group them into different "chunks" to keep track of objects within the environment over time (Moss and Surlykke 2001). The perceptual grouping and segregation of complex sounds by the CNS is referred to as *auditory scene analysis* and has been described in humans (Bregman 1990). It is the auditory equivalent to the visual principles of perceptual grouping and segregation of Gestalt psychology (Bregman 1990, Surlykke and Moss 2001; see also Cronin's discussion of *natural scenes*, ch. 6 in this volume). However, all auditory animals, especially those coping with complex acoustic environments (e.g., vocalizing animals like birds and bats), are likely capable of auditory scene analysis. Recently, auditory scene analysis has been proposed as a driving force in the evolution of vertebrate hearing (Fay and Popper 2000).

The Bats: Auditory Scene Analysis

The ability of bats to perform auditory scene analysis has been studied by means of perceptual experiments, behavioral research in flying bats, and field observations. In a series of experiments, Moss and Surlykke (2001) demonstrated that bats likely perform auditory scene analysis and that control over their echolocation behavior is an important way this analysis is accomplished. Initial perceptual experiments using *E. fuscus* confirmed that these bats do monitor echo delay changes over time, in addition to single-echo delays, a necessary ability for auditory scene analysis (Moss and Surlykke 2001).

Behavioral and field observations revealed patterns in bat emissions that may aid auditory scene analysis. The most important strategy in the emissions pattern was revealed as *E. fuscus* captured mealworm targets either near clutter or in the open inside a large flight room (Moss and Surlykke 2001). As with the bat attack sequence described earlier (Figure 5.1), bats increased their PRR as they captured these targets. However, the increase was not continuous, because bats would emit several vocalizations at a stable PRR, followed by gaps. Moss and Surlykke (2001) labeled these stable emissions as sonar *strobe groups*. They hypothesized that stable strobe groups are a way bats facilitate the grouping and segregation of returning echoes through the emission pattern. The stable strobe groups may facilitate analysis of sonar scenes during capture attempts, because pulse-echo pairs occurring at constant intervals may be easier for the CNS to analyze. Ideally, these strobe groups elicit responses from the same delay-sensitive neurons and thus result in a stronger perceptual event (Moss and Surlykke 2001).

Finally, bat control over its emissions also can help it distinguish its own call from those of nearby conspecifics. Field recordings revealed that bats shift the frequencies of their emitted vocalizations slightly in the presence of nearby conspecifics but return to their normal frequencies once they are apart (Moss and Surlykke 2001). These frequency changes likely aid in segregating a bat's own echoes from those of conspecifics, serving as a "jamming" avoidance response.

The Insects: Tuning Specializations

Bat predation is a key factor determining the frequencies the insect can hear and, to some extent, how sensitive the system is to those frequencies generally referred to as the tuning of the insect's auditory system (Fullard 1998, Schul and Patterson 2003). Because most insectivorous bats (many of which use FM signals) produce sounds dominated by 20–50 kHz (Fullard 1988), these frequencies have the lowest thresholds in most ultrasound-sensitive insect auditory systems (Figure 5.4). Tuning in the ultrasound range is very broad across insects, especially when compared to the tuning of an auditory interneuron for intraspecific communication (thick line in Figure 5.4).

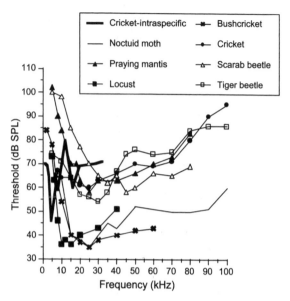

Figure 5.4. Auditory tuning curves from several different insects possessing ultrasound-sensitive auditory systems (lines with symbols). Although absolute sensitivities differ among insects (partially related to whether tuning was measured from afferents or primary inter-neurons), each insect's tuning demonstrates high sensitivity across a broad range of fre-quencies between 20 and 50 kHz. This region corresponds to the range of peak frequencies contained within the majority of insectivorous bat echolocation vocalizations. Broad tun-ing is characteristic of ultrasound-sensitive insect ears that mediate responses effective in evading bat predators. In contrast, many (though not all) interneurons sensitive to lower frequency intraspecific communication sounds are narrowly tuned (i.e., a cricket interneuron sensitive to lower frequency intraspecific calls, thick line, Huber 1978). Noctuid moth = A1 cell, *Barathra brassicae* (Roeder 1967); praying mantid = MR-501-T3, *Mantis religiosa* (Yager and Hoy 1989); locust = SN-5 neuron, *Locusta migratoria* (Römer et al. 1988); bushcricket = T-neuron, *Neoconocephalus ensiger* (Faure and Hoy 2000a); cricket = Int-1, *Teleogryllus oceanicus* (Moiseff and Hoy 1983); scarab beetle = afferent recordings, *Euetheola humilis* (Forrest et al. 1997); tiger beetle = afferent recordings, *Cicindela marutha* (Yager and Spangler 1995).

Several examples from research on moths illustrate that only the bats that feed on a particular insect, not all sympatric bats, influence the tuning of insect ears (Fullard 1998, Schoeman and Jacobs 2003). The top graph in Figure 5.5 displays a composite spec-trum of echolocation frequencies (solid line) employed by all bat species sympatric with notodontid moths on Isla Barro Colorado, Panama (a total of 37 species). Although the peak energy for the composite spectrum exceeds 80 dB sound pressure level (SPL) over a range of 40–130 kHz, the most sensitive region of the moth tuning curve over-laps with only a portion of this range (20–60 kHz). However, when the spectrum con-tains only those echolocation frequencies from bats feeding on these moths (and exerting selective pressure), the region of greater sensitivity in the moths overlaps with the peak energy of the composite spectrogram (Figure 5.5, bottom).

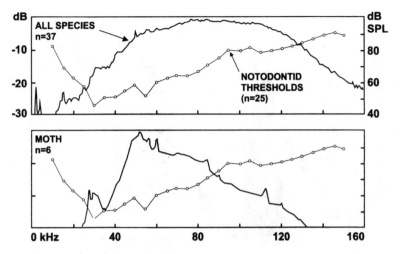

Figure 5.5. Tuning of ultrasound-sensitive hearing in insects as it corresponds to the frequencies contained in the vocalizations of bats that feed upon them (exerting the selective pressure) instead of all coexisting bats. The auditory tuning of notodontid moths on Isla Barro Colorado illustrates this. In the top figure, the composite frequency spectrum (line, no symbols) includes the echolocation calls from all bats (37 species) on Isla Barro Colorado. The most sensitive region of the moth auditory tuning (median audiogram of 25 sympatric notodontid moths) overlaps with only a portion of the bat emission frequency range containing the most energy. However, when the composite spectrum includes the echolocation calls from only the bats that heavily predate on moths (thus forming the actual selective pressure, 6 species), the region of greatest sensitivity in notodontid tuning overlaps with the frequency region containing the most energy (bottom). (Adapted with permission from Fullard, J. H. 1988. The tuning of moth ears. Experientia 44:423–428. © 1988 Birkhauser Verlag.)

For most insects using their auditory systems for bat avoidance, tuning drops off sharply below 20 kHz (Figure 5.4). However, this is not always the case. The Hawaiian island of Kauai contains only one bat species, *Lasiurus cinereus semotus* (Hawaiian hoary bat), a bat that feeds on lepidopterans. These bats not only echolocate (peak frequency around 29 kHz), but also emit audible social signals (around 10 kHz). The tuning of sympatric moths (e.g., the noctuid *Elydra nonagrica*, Figure 5.6) has two sensitive regions, one between 25 and 55 kHz (echolocation region) and one of even greater sensitivity between 5 and 15 kHz (the social region; Belwood and Fullard 1984, Fullard 1988). Noctuid moths from other regions, such as Canada and Africa, lack this low-frequency sensitivity. Although the change in tuning to include lower frequencies could be because of strong selection exerted by the hoary bat, there is little other experimental support for this conclusion. It is also unclear from the study whether lower frequencies elicit the same type of evasive responses as ultrasonic frequencies do.

Tuning is affected not only by bat predation, but also by ear biomechanics, forcing trade-offs between frequency range and overall sensitivity. To achieve the broader tuning advantageous for detecting a wide range of bats, there is some loss in sensitivity compared to the sensitivity in the narrower range of intraspecific sounds

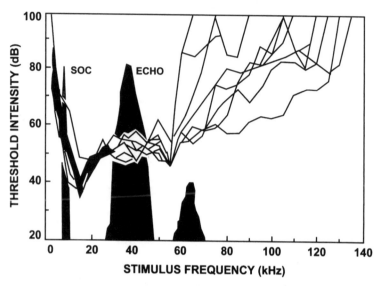

Figure 5.6. Another example how the high selective pressure of bat echolocation may influence the tuning of insect auditory systems. The bat *Lasiurus cinereus semotus* is the sole bat predator of the noctuid moth *Elydna nonagrica* on Kauai. These bats emit both high-frequency echolocation (ECHO) and lower frequency social (SOC) vocalizations. The tuning of these moths includes sensitivity to the mutually exclusive echolocation and social call frequencies. Noctuids from Africa and Nearctic regions lack this low-frequency sensitivity. (Adapted with permission from Fullard, J. H. 1988. The tuning of moth ears. Experientia 44:423–428. © 1988 Birkhauser Verlag.)

(Figure 5.4). Still, this trade-off between broader tuning and reduced sensitivity is not detrimental, because insects still detect the bats earlier than the bats detect the insects. Hearing ability comparisons of large and small moths also illustrate how ear biomechanics affect tuning and sensitivity (Surlykke and Filskov 1999, Norman and Jones 2000). As moth size increases, so does the size of the tympanum. A larger tympanum provides greater sensitivity at the best frequency, but the best frequency lower than in auditory systems of insects possessing smaller tympana. Surlykke Filskov (1999) noted that the proportional increase in tympanal size in larger was less than in nonauditory structures, indicating that there are tympana strictions that must be maintained to retain sensitivity to ultrasound.

The Insects: Evasive Responses

Each insect species typically exhibits a repertoire of ultrasound-' behaviors that vary with the level of danger posed by the bat. Ir level threat (i.e., a distant bat), many insects (i.e., moths, cri spond with gradual turns away from the stimulus (*negative p'* Popov and Shuvalov 1977, Moiseff et al. 1978, Robert 1ᶜ nondirectional gradual turns (Yager et al. 1990) because their auditory system appears unable to localize sound (Yager and Hoy 1989). Because insects can detect bats before bats detect them (Miller and Surlykke 2001), this response can be an effective strategy for avoiding trouble while expending little energy. If a bat manages to detect and "lock on" to a target (high-level threat as indicated at least by high-intensity ultrasound), insects perform more dramatic, and energetically costly, maneuvers to avoid capture. These maneuvers range from rapid spiral turns by moths, for example (Roeder 1967), or by butterflies (e.g., *Manataria maculate*; Rydell et al. 2003), and accelerating power dives, as exhibited by mantids (Yager et al. 1990), to passive falls, as exhibited by green lacewings and katydids (Miller and Olesen 1979, Libersat and Hoy 1991, Schulze and Schul 2001).

The effectiveness of these evasive maneuvers has been directly evaluated in only a few insects (e.g., moths, lacewings, and mantids). Results from these experiments do confirm that ultrasound-mediated maneuvers convey a significant advantage to hearing insects compared to their deafened counterparts: moths gain a 40% advantage (Roeder and Treat 1962), lacewings gain a 60% advantage (Miller and Olesen 1979), and mantids gain a 50% advantage (Triblehorn 2003). However, performing evasive behaviors can have drawbacks. Insects flying at night may be looking for a mate by following chemical and/or acoustic signals. This situation sets up an interesting conundrum for the insect. An insect risks losing track of the signal by performing an evasive response, but risks capture if it does not respond. Several recent studies (Acharya and McNeil 1998, Farris et al. 1998, Skals et al. 2003) have shown that the decision to pursue a nearby mate versus a response to avoid a nearby bat is a complex trade-off between perceived risk and benefit.

Not all insect auditory defenses involve flying maneuvers. Faure and Hoy (2000b) found ultrasound-elicited pauses in the singing behavior of katydids. This defense would be effective against bats that echolocate for navigation but rely on passive hearing for detecting and capturing insects (e.g., gleaning bats). Several insects, including arctiid moths (Dunning and Roeder 1965, Fullard and Fenton 1977), tiger beetles (Yager and Spangler 1997), and some butterflies (Möhl and Miller 1976), produce ultrasonic sounds (i.e., broadband "clicks") of their own in response to ultrasonic pulse trains. Arctiid moths produce their clicks by buckling two specialized ridged areas of cuticle on the thorax known as tymbal organs (Minet and Surlykke 2003). Tiger beetles employ a different method to produce their clicks: as part of a multicomponent response to ultrasound, tiger beetles move their elytra (the hard forewings that are not involved in flight) backwards so that they strike the beating hindwings (Yager and Spangler 1997). Each strike of the hindwings on the elytra produces an ultrasonic click. Dunning and Roeder (1965) first showed that bats avoid capturing and eating mealworm targets presented simultaneously with arctiid moth clicks. Even though clicking arctiids show little, if any, evasive behavior when attacked by an echolocating bat, the bat rarely actually contacts the moth (Acharya and Fenton 1992). However, it is unclear whether these clicks act as a form of acoustic aposematism (Dunning 1968, Dunning et al. 1992), as a startle response (Möhl and Miller 1976), or as an attempt to confuse or "jam" the bats (Fullard et al. 1979). Tougaard et al. (1998) demonstrated that click production altered responses in delay-sensitive neurons by either suppressing the response or altering the latency. Clicks affected responses only if they occurred within 3 ms before onset of the pulse-echo stimulus, which indicates that the moths would need to produce click trains to be effective.

The Insects: Sensorimotor Control of Evasive Responses

Regardless of whether ultrasound-sensitive hearing evolved as a new modality or through modification of a preexisting auditory system, insects had to incorporate this new stimulus into the CNS and link it to the motor system to produce these evasive responses. There is relatively little experimental data regarding the sensory-motor interface within the CNS that links ear stimulation with the production of evasive maneuvers. We know the most about the primary auditory interneurons receiving inputs from the auditory afferents. Moths possess several ultrasound-sensitive interneurons with different response properties (Roeder 1966, Boyan and Fullard 1986, Boyan et al. 1990). At least one ultrasound-sensitive interneuron has been identified (although there are likely others) in other insects exhibiting ultrasound-triggered evasive responses, including Int-1 in crickets (Casaday and Hoy 1977, Popov et al. 1978, Wohlers and Huber 1978), MR-501-T3 or simply 501-T3 in praying mantids (Yager and Hoy 1989), T-cell in katydids (Suga 1963, Faure and Hoy 2000a), and SN-5 in locusts (Römer et al. 1988). These interneurons typically have large axon diameters that yield high conduction velocities and short behavioral latency (30–230 ms, depending on species; see Faure and Hoy 2000b). In crickets, Int-1 ascends to the brain, where it connects to descending interneurons linked to flight behavior (Brodfuehrer and Hoy 1989, Brodfuehrer and Hoy 1990). Most of the other ultrasound-sensitive interneurons also travel toward the brain and may follow a similar circuit for initiating motor responses. For example, the brain is necessary for production of the evasive response in mantids, but not for flying (Cook, A. P., and Yager, D. D., unpublished data), and for normal click production in arctiid moths (Dawson and Fullard 1995).

There are only a few cases of direct evidence linking auditory neural responses to evasive behavior. In the case of crickets, the magnitude of the negative phonotactic turn is directly proportional to the firing rate of Int-1 (Nolen and Hoy 1984). Furthermore, Nolen and Hoy (1984) demonstrated that Int-1 is both necessary and sufficient to trigger evasive behavior, but only when the cricket is flying.

We now consider CF-FM bats that echolocate using signals dominated by a constant frequency (CF) signal. Like FM bats, CF-FM bats also possess delay-tuned neurons and likely perform auditory scene analysis. However, they also possess neural specializations that allow them to process the CF signal that FM bats lack.

Predator-Prey: CF-FM Bats and the Insects They Attack

The Bats: CF-FM Emissions

The CF portion dominates the CF-FM bats' vocalizations, but they also contain a brief FM component important for target range determination (Simmons et al. 1979; see Figure 5.1, bottom). In contrast to FM bat signals, the second harmonic of

CF-FM bat vocalizations contains the most energy, at frequencies over 50 kHz. Although these higher frequencies reduce the maximum detection range, these bats typically hunt in highly cluttered areas where detection distances would be shorter anyway (Neuweiler 1989, Neuweiler 1990). These frequencies also potentially make CF-FM bats less conspicuous to most auditory insects, so the bats are closer when insects detect them.

The CF component makes these signals suitable for hunting in a highly cluttered environment, because these signals are very sensitive for detecting fluttering, moving targets against nonmoving backgrounds (Schnitzler 1987, Schnitzler et al. 2003). When the CF signal reflects off objects moving toward or away from the bat (this includes flapping wings), the echo is shifted up or down in frequency (a Doppler shift).

The Bats: The Acoustic Fovea and Doppler-Shift Compensation

The narrow frequencies comprising the CF portion vary slightly across individual bats, and CF-FM bats are especially sensitive to the particular frequency in the CF portion of the call they emit (Moss and Sinha 2003, Smotherman and Metzner 2003). This sensitivity begins with specializations within the cochlea itself, which contains an overrepresentation of auditory receptors tuned to this frequency (referred to as the *acoustic fovea*; Suga and Jen 1976). The overrepresentation found in the acoustic fovea extends to all subsequent auditory brain regions processing the CF portion of the echolocation calls. Although CF-FM bats emit vocalizations using the frequency to which they are most sensitive, Doppler-shifted echoes move outside this sensitive region. To prevent Doppler-shifted echoes from falling outside the region of greatest sensitivity, bats adjust their emitted frequency so the returning echoes fall within the sensitive region (referred to as *Doppler-shift compensation*; Metzner et al. 2002). The long-duration CF component provides CF-FM bats with other information absent in the shorter FM signals. Their vocalizations reflect off fluttering insects for several wing beats. The moving wings introduce frequency and amplitude modulations in the returning echoes that CF-FM bats can use to discriminate between different insects (Schnitzler et al. 1983, Schnitzler 1987). By specializing in detecting moving prey, CF-FM bats incur problems detecting nonfluttering insects, a limitation insects could exploit, as will be discussed.

The echolocation strategy of CF-FM bats also solves other problems, such as self-deafening, pulse-echo overlap (when the echo returns before the end of the emission and prevents the bat from hearing the echo), and distinguishing their own pulse-echo pairs from those of conspecifics. The CF-FM bats have the same protective measures designed to prevent self-deafening as FM bats, but Doppler-shift compensation also helps prevent this self-deafening, because the emissions fall outside the acoustic fovea and the less intense echoes fall within it. Because the emissions and echoes fall within different frequency ranges and activate different regions of the auditory system, CF-FM bats avoid the pulse-echo overlap problem, even though the emissions and echoes overlap in time (Schuller 1986). The FM component also avoids the pulse-echo overlap problem in the same manner as in FM bats, because this portion has a short duration. The ability to distinguish between their own pulse-echo pairs and those of conspecifics may lie within specialized delay-sensitive neurons in the auditory cortex of CF-FM bats. Some of these neurons are sensitive to pulse-echo pairs of different harmonics (Suga and O'Neill 1979, Misawa and Suga 2001). For example, some neurons respond to combinations of first harmonic pulses of the CF component and the third harmonic echo of the CF component (CF1-CF3-sensitive neurons). There are CF1-CF2 neurons, as well. Although conspecifics may hear the loud CF2 component (containing the most energy), only the emitting bat can hear the low-energy first and third harmonics. As a result, only the CF1-CF3 and CF1-CF2 combination neurons will respond to these pulse-echo pairs, allowing the emitter to distinguish its own pulse-echo pairs.

The Insects: Tuning Specializations

The higher frequencies (>50 kHz) of CF-FM bat emissions lie outside the range of greatest sensitivity for many insects (i.e., 30–50 kHz), providing CF-FM bats an advantage over many ultrasound-sensitive insects. However, insect tuning could shift to include these higher frequencies, in a way similar to the tuning shift of the moths on Kauai to include the lower frequency bat social calls. This possibility would depend on CF-FM bats' exerting enough selection pressure, as well as on whether the biomechanics of the ear would allow it. Such may be the case in some mantid species (Triblehorn and Yager 2001). Although the majority of species are most sensitive within the 30–50 kHz range (called *narrowly tuned*, or NT, mantids; Figure 5.7), there is a subset of mantids (i.e., seven identified species) that are equally sensitive to frequencies over 50 kHz and between 30–50 kHz (and are thus called *broadly tuned*, or BT, mantids).

The Insects: Behavioral Specializations

The differences between NT and BT species extend beyond tuning to the evasive response (Triblehorn and Yager 2001). The NT species respond to ultrasonic pulse trains in tethered flight with a multicomponent response that involves foreleg extension, abdominal dorsiflexion, and changes in wing-beat phase and frequency (Yager and May 1990). However, BT species exhibit only a subset of these components in response to ultrasound, and the component set varies across presentations (Triblehorn and Yager 2001). When the same trains of ultrasonic pulses were played to BT mantids in tethered flight, they temporarily ceased flight in response to ultrasonic pulse trains, whereas NT species never do. Flight cessations eliminate the wing motion that CF-FM bats use as an important cue when they are hunting insects. Thus, the BT mantids "blend into the background," causing the bat to lose track of its target. Such a maneuver would likely not be effective against FM bats and is not included in NT evasive behavior. Although it is unknown what effect these variations would have on the production of the power dive response, which results from the full-component behavior, this variation in tethered flight responses may produce more unpredictable evasive behaviors in free flight. Because CF-FM bats detect insects at closer ranges, such unpredictability may provide greater protection from CF-FM bats.

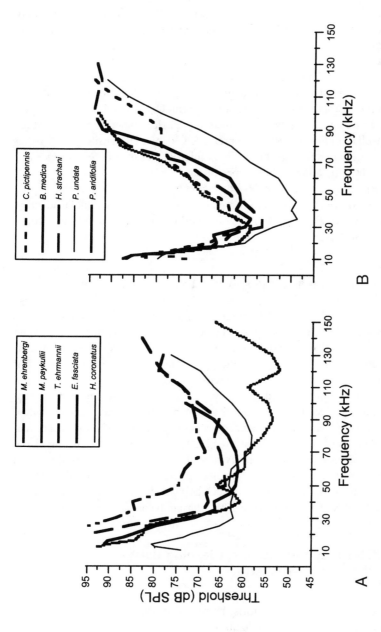

Figure 5.7. Insect auditory systems sensitive to frequencies higher than 50 kHz, possibly because of selective predation by CF-FM bats that echolocate at these higher frequencies. Although the majority of mantids are most sensitive within the 30–50 kHz range (termed *narrowly tuned*, or NT, mantids, bottom graph), some mantid species (termed *broadly tuned* mantids, top graph) exhibit sensitivity to higher frequencies. (Adapted with permission from Triblehorn, J. D., and Yager, D. D. 2001. Broad versus narrow auditory tuning and corresponding bat-evasive flight behavior in praying mantids. J. Zool. (Lond.) 254:27–40. © 2001 Cambridge University Press.)

Sudden unpredictable shifts in flight paths close to capture may prevent CF-FM bats from making the adjustments to successfully capture the mantid. Results from staged encounters do support the contention that the mantid high-frequency sensitivity of *Miomantis natalica*, a BT species, when compared to the deafness of *M. natalica*, provides protection from a CF-FM bat (*Rhinolophus clivosus* echolocating at 80–85 kHz; Cumming 1996).

It is not known what would cause CF-FM bats to exert more selective pressure on BT mantids. The BT mantids have been identified only from the Old World, which is also true of CF-FM bats, so there is geographic range overlap. However, NT mantids are found throughout the same parts of the world. One possibility is that differences in daily or seasonal activity cycles make BT species vulnerable to CF-FM bat predation, as well as to FM bats (thus the preserved 30–50 kHz sensitivity). Another possibility is that BT mantids inhabit different microniches than do NT species and thus are exposed to predation by both types of bats.

As in any arms race, as one side develops defenses for protection, the other side must adapt countermeasures to remain competitive. Insects are not alone in adapting to predatory influences. Bats can also exhibit modifications that may serve as countermeasures to insect defenses.

Bat Countermeasure Responses to Insect Defenses

Increasing flight speed decreases the amount of time between detection and capture (Fenton and Fullard 1979). This strategy can be a dangerous one in cluttered areas when bats will also have less time to avoid obstacles. Most strategies, however, involve altering echolocation vocalizations to make them less detectable by auditory insects. For example, the usefulness of high-frequency signals of CF-FM bats for echolocation in highly cluttered environments could also convey an advantage when the bats are capturing ultrasound-sensitive insects insensitive to such high frequencies. Some bats hunt and capture insects by using lower frequency echolocation calls. The European free-tailed bat, *Tadarida teniotis*, vocalizes at 11–12 kHz, which allows it to capture large lepidopteran and neuropteran prey with tympanate hearing sensitive to ultrasound (Rydell and Arlettaz 1994). Other bats, known as *gleaners*, use echolocation to navigate and locate prey but use other methods, such as passive hearing or vision, to detect insect movement (Hübner and Wiegrebe 2003) during the actual capture attempt.

Predator-Prey Interactions: Gleaning Bats and the Insects They Attack

The Bats: Gleaning Emissions, Hunting Strategies, and Acute Hearing Specializations

Gleaning bats emit short-duration (1–3 ms), very low intensity vocalizations (relative to aerial hawking bats) that can include either high frequencies (over 20 kHz; e.g., *Myotis evotis*, Faure et al. 1990, Faure and Barclay 1994; and *Myotis septentrionalis*, Faure et al. 1993), or low frequencies (below 20 kHz; e.g., *Euderma maculatum*, Fullard and Dawson 1997), depending on the species and their prey (Leippert et al. 2002). Many gleaners stop echolocating during attacks on detected insects, and others use only passive sound localization to hunt animals on the substrate (Bell 1982, Marimuthu and Neuweiler 1987; but see Schmidt et al. 2000). Some bats switch between aerial hawking and gleaning strategies, depending on the situation, and modify their echolocation emissions accordingly (e.g., *M. evotis*, Faure and Barclay 1994; and *M. septentrionalis*, Faure et al. 1993). Relatively little is known about peripheral and CNS specializations for gleaning and passive sound localization in bats (Guppy and Coles 1988). Some species have unusually large pinnae that would improve sound localization (Faure and Barclay 1994).

Insect Responses to Gleaning Bats

Several experiments have demonstrated that tympanate insects sensitive to ultra-sound (mostly moths) are unable to hear the vocalizations of gleaning bat species (e.g., *M. evotis*, Faure et al. 1990; *M. septentrionalis*, Faure et al. 1993; *Euderma maculatum*, Fullard and Dawson 1997), preventing auditory-mediated evasive responses. Insect defenses have, however, evolved specifically to address the threat of gleaning bats. Singing insects trying to attract mates make themselves obvious to bats by using passive sound localization (Bailey and Haythornwaite 1998). Their most common defense is to become acoustically "invisible" by song cessation when echolocating bats fly overhead (Spangler 1984, Belwood and Morris 1987, Faure and Hoy 2000b). Many insects also become immobile in order to eliminate the rustling that some bats use for passive sound localization of prey (Faure et al. 1990, Hübner and Wiegrebe 2003).

Conclusions and Future Directions

This chapter tells two primary stories about predators and their prey. The first concerns an evolutionary arms race. Insectivorous bats must consume an immense quantity of insects to survive and procreate. Nocturnal insects must evade the chiropteran onslaught for the same crucial reasons. Echolocation was countered by ultrasonic hearing and evasive behavior. The insectan countermeasures were answered by less intense echolocation signals at frequencies that are outside the insect's hearing range. Some insects compensated by extending their hearing range and sensitivity. A few bats have stopped using echolocation signals altogether during attacks on prey, and vulnerable insects thwart this strategy by becoming quiet and motionless. In the rare cases in which the "score" has been tallied (FM bats hunting hearing and deaf insects), the invertebrates are doing very well indeed against a larger, quicker, "smarter" vertebrate predator: hearing confers an increase in thwarted bat attacks from 20–40% to 70–90%. These studies probably underestimate bats' overall success, however, because they do not consider gleaning and highly plastic hunting strategies.

Underlying this arms race is the evolution of profound anatomical and neurological specializations. For bats, this evolution has meant refinements of the cochlea, including an acoustic fovea in some groups. The auditory system has elaborated and reorganized to evaluate minute echo delays and the equally minute spectral shifts introduced by the pinnae for three-dimensional localization and by Doppler shifts caused by flapping wings. For insects this evolution has taken the form of an entirely new sensory modality with the requisite CNS auditory neurons and synaptic regions. Neural processing circuitry of an auditory sensorimotor interface drives complex, multicomponent behaviors that ultimately save the insect's life.

The second story is one of compromise. Both predation and predator evasion are expensive and themselves dangerous. Some of the costs are metabolic, and these costs have been studied at length. Potentially more important, however, are the structural and behavioral trade-offs animals must make to be successful predators or long-lived prey. Bat echolocation signals and behavior must accommodate navigation, as well as the sometimes conflicting demands of optimal prey detection and tracking. Highly visual bats are poor echolocators; superb echolocators often have limited vision. This fact suggests a zero-sum situation involving CNS resources and overall anatomical options. A hearing moth can fly toward a potential mate or evade capture by an attacking bat, but at any given moment it cannot do both. This observation suggests choice or compromise based on risk-benefit computation. It is the overall behavioral and ecological context that ultimately determines the requirements and adaptations for success of predator and prey, and the result is generally a compromise.

Although there is certainly much to learn about CNS and behavioral adaptations of bats and moths as individual taxa, broader and more integrative questions of compromise driven by ecological, anatomical, physiological, and behavioral demands will be a key area of future research. For instance, there are currently no sensible measures of the complete cost of building and maintaining an insect auditory system. The zero-sum aspects must be considered and rarely have been (i.e., what is lost, as well as what is gained). We will need to know this information in order to understand why some flying, nocturnal insects listen to bats and others do not. That information must, in turn, be combined with a growing understanding of the overall ecology of bats, insects, and, most important, particular bat-insect pas de deux.

Literature Cited

Acharya, L., and Fenton, M. B. 1992. Echolocation behavior of vespertilionid bats (*Lasiurus cinereus* and *L. borealis*) attacking airborne targets including arctiid moths. Can. J. Zool. 70:1292–1298.

Acharya, L., and McNeil, J. N. 1998. Predation risk and mating behavior; the responses of moths to bat-like ultrasound. Behav. Ecol. 9:552–558.

Bailey, W. J., and Haythornwaite, S. 1998. Risks of calling by the field cricket *Teleogryllus oceanicus*; potential predation by Australian long-eared bats. J. Zool. (Lond.) 244:505–513.

Baker, T. C., and Cardé, R. T. 1978. Disruption of gypsy moth male sex pheromone behavior by high frequency sound. Nature 272:444–446.

Bell, G. P. 1982. Behavioral and ecological aspects of gleaning by a desert insectivorous bat, *Antrozous pallidus* (Chiroptera:Vespertilionidae). Behav. Ecol. Sociobiol. 10:217–223.

Belwood, J. J., and Fullard, J. H. 1984. Echolocation and foraging behaviour in the Hawaiian hoary bat, *Lasiurus cinereus semotus*. Can. J. Zool. 59:2113–2120.

Belwood, J. J., and Morris, G. K. 1987. Bat predation and its influence on calling behavior in Neotropical katydids. Science 238:64–67.

Boyan, G. S., and Fullard, J. H. 1986. Interneurones responding to sound in the tobacco bud-worm moth *Heliothis virescens* (Noctuidae): morphological and physiological characteristics. J. Comp. Physiol. A 158:391–404.

Boyan, G., Williams, L., and Fullard, J. 1990. Organization of the auditory pathways in the thoracic ganglia of noctuid moths. J. Comp. Neurol. 295:248–267.

Bregman, A. S. 1990. Auditory Scene Analysis. Cambridge, Mass.: MIT Press.

Brodfuehrer, P. D., and Hoy, R. R. 1989. Integration of ultrasound and flight inputs on descending neurons in the cricket brain. J. Exp. Biol. 145:157–171.

Brodfuehrer, P. D., and Hoy, R. R. 1990. Ultrasound sensitive neurons in the cricket brain. J. Comp. Physiol. A 166:651–662.

Casaday, G. B., and Hoy, R. R. 1977. Auditory neurons in the brain of the cricket *Teleogryllus oceanicus:* physiological and anatomical properties. J. Comp. Physiol. 121:1–13.

Conner, W. E. 1999. "Un chant d'appel amoureax": acoustic communication in moths. J. Exp. Biol. 202:1711–1723.

Cumming, G. S. 1996. Mantis movements by night and the interactions of sympatric bats and mantises. Can. J. Zool. 74:1771–1774.

Dawkins, R., and Krebs, J. R. 1979. Arms races between and within species. Proc. R. Soc. Lond. B 205:489–511.

Dawson, J. W., and Fullard, J. H. 1995. The neuroethology of sound production in tiger moths (Lepidoptera, Arctiidae). II. Location of the tymbal central pattern generator in *Cycnia tenera* Hübner. J. Comp. Physiol. A 176:541–549.

Dear, S. P., Fritz, J., Haresign, T., Ferragamo, M., and Simmons, J. A. 1993. Tonotopic and functional organizations in the auditory cortex of the big brown bat, *Eptesicus fuscus.* J. Neurophysiol. 70:1988–2009.

Dear, S. P., Simmons, J. A., and Fritz, J. 1993. A possible neuronal basis for representation of acoustic scenes in auditory cortex of the big brown bat. Nature 364:620–623.

Dear, S. P., and Suga, N. 1995. Delay-tuned neurons in the midbrain of the big brown bat. J. Neurophysiol. 73:1084–1100.

Dunning, D. C. 1968. Warning sounds of moths. Z. Tierpsychol. 25:129–138.

Dunning, D. C., Acharya, L., Merriman, C. B., and Dal Ferro, L. 1992. Interactions between bats and arctiid moths. Can. J. Zool. 70:2218–2223.

Dunning, D. C., and Roeder, K. D. 1965. Moth sounds and the insect-catching behavior of bats. Science 147:173–174.

Farris, H. E., Forrest, T. G., and Hoy, R. R. 1998. The effect of ultrasound on the attractiveness of acoustic mating signals. Physiol. Entomol. 23:322–328.

Faure, P. A., and Barclay, R. M. R. 1994. Substrate-gleaning versus aerial-hawking: plasticity in the foraging and echolocation behaviour of the long-eared bat, *Myotis evotis.* J. Comp. Physiol. A 174:651–660.

⸻, Fullard, J. H., and Barclay, R. M. R. 1990. The response of tympanate moths to ⸺cation calls of a substrate gleaning bat, *Myotis evotis.* J. Comp. Physiol. A

⸺ H., and Dawson, J. W. 1993. The gleaning attacks of the northern *is septentrionalis*, are relatively inaudible to moths. J. Exp. Biol.

⸺0a. Neuroethology of the katydid T-cell. I. Tuning and re-*Biol. 203:3225–3242.

⸺e sounds of silence: cessation of singing and song

pausing are ultrasound-induced acoustic startle behaviors in the katydid *Neoconocephalus ensiger* (Orthoptera; Tettigoniidae). J. Comp. Physiol. A 186:129–142.

Fay, R. R., and Popper, A. N. 2000. Evolution of hearing in vertebrates: the inner ears and processing. Hear. Res. 149:1–10.

Feduccia, A. 1995. Explosive evolution of tertiary birds and mammals. Science 267:637–638.

Fenton, M. B. 1995. Constraint and flexibility: bats as predators, bats as prey. In: Ecology, Evolution and Behavior of Bats (Racey, P.A., and Swift, S. M., eds.). Symposium 67 of the Zoological Society of London. Oxford: Oxford University Press; 277–290.

Fenton, M. B., Audet, D., Obrist, M. K., and Rydell, J. 1995. Signal strength, timing and self-deafening: the evolution of echolocation in bats. Paleobiology 21:229–242.

Fenton, M. B., and Fullard, J. H. 1979. The influence of moth hearing on bat echolocation strategies. J. Comp. Physiol. 132:77–86.

Fenton, M. B., Reutenbach, I. L., Smith, S. E., Swanepoel, C. M., Grosell, J., and Vanjaarsveld, J. 1994. Raptors and bats—threats and opportunities. Anim. Behav. 48:9–18.

Forrest, T. G., Read, M. P., Farris, H. E., and Hoy, R. R. 1997. A tympanal hearing organ in scarab beetles. J. Exp. Biol. 200:601–606.

Fullard, J. H. 1988. The tuning of moth ears. Experientia 44:423–428.

Fullard, J. H. 1998. The sensory coevolution of moths and bats. In: Comparative Hearing: Insects (Hoy, R. R., Popper, A. N., and Fay, R. R., eds.). New York: Springer-Verlag; 279–326.

Fullard, J. H., and Belwood, J. J. 1988. The echolocation assemblage: acoustic ensembles in a neotropical habitat. In: Animal Sonar, vol. 156 (Nachtigall, P. E., and Moore, P. W. B., eds.). NATO ASI Series, Series A: Life Sciences. New York: Plenum Press; 639–643.

Fullard, J. H., and Dawson, J. W. 1997. The echolocation calls of the spotted bat *Euderma maculatum* are relatively inaudible to moths. J. Exp. Biol. 200:129–137.

Fullard, J. H., Dawson, J. W., Otero, L. D., and Surlykke, A. 1997. Bat-deafness in day-flying moths (Lepidoptera, Notodontidae, Dioptinae). J. Comp. Physiol. A 181:477–483.

Fullard, J. H., and Fenton, M. B. 1977. Acoustic behavioral analyses of the sounds produced by some species of Nearctic Arctiidae (Lepidoptera). Can. J. Zool. 55:1213–1224.

Fullard, J. H., Fenton, M. B., and Simmons, J. A. 1979. Jamming bat echolocation: the clicks of arctiid moths. Can. J. Zool. 57:647–649.

Fullard, J. H., Otero, L. D., Orellana, A., and Surlykke, A. 2000. Auditory sensitivity and diel flight activity in neotropical Lepidoptera. Ann. Entomol. Soc. Am. 93:956–965.

Griffin, D. R. 1958. Listening in the Dark. New Haven, Conn.: Yale University Press.

Griffin, D. R., Webster, F. A., and Michael, C. R. 1960. The echolocation of flying insects by bats. Anim. Behav. 8:141–154.

Guppy, A., and Coles, R. B. 1988. Acoustical and neural aspects of hearing in the Australian bats, *Macroderma gigas* and *Nyctophilus gouldi*. J. Comp. Physiol. A 162:653–668.

Hartridge, H. 1945. Acoustic control in the flight of bats. Nature 156:490–494.

Henson, O. W., Jr. 1965. The activity and function of the middle ear muscles in echolocating bats. J. Physiol. (Lond.) 180:871–887.

Hill, J. E., and Smith, J. D. 1984. Bats: A Natural History. Austin: University of Texas Press.

Hoy, R. R. 1992. The evolution of hearing in insects as an adaptation to predation from bats. In: The Evolutionary Biology of Hearing (Webster, D. B., Fay, R. R., and Popper, A. N., eds.). New York: Springer-Verlag; 115–129.

Hoy, R. R., and Robert, D. 1996. Tympanal hearing in insects. Annu. Rev. Entomol. 41:433–450.

Huber, F. 1978. The insect nervous system and insect behaviour. Anim. Behav. 26:969–981.

Hübner, M., and Wiegrebe, L. 2003. The effect of temporal structure on rustling-sound detection in the gleaning bat, *Megaderma lyra*. J. Comp. Physiol. A 189:337–346.

Kick, S. A. 1982. Target detection by the echolocating bat, *Eptesicus fuscus*. J. Comp. Physiol. A 145:431–443.

Lawrence, B. D., and Simmons, J. A. 1982. Measurements of atmospheric attenuation at ultrasonic frequencies and the significance for echolocating bats. J. Acoust. Soc. Am. 71:585–590.

Leippert, D., Frank, E., Gabriel, P., Scheidermann, K. D., Von Stillfried, N., and Weeller, F. 2002. Prey-correlated spectral changes in echolocation sounds of the Indian false vampire bat, *Megaderma lyra*. Ethology 108:139–156.

Libersat, F., and Hoy, R. R. 1991. Ultrasonic startle behavior in bushcrickets (Orthoptera; Tettigonidae). J. Comp. Physiol. A 169:507–514.

Marimuthu, G., and Neuweiler, G. 1987. The use of acoustical cues for prey detection by the Indian false vampire bat, *Megaderma lyra*. J. Comp. Physiol. A 160:509–515.

Metzner, W., Zhang, S. Y., and Smotherman, M. S. 2002. Doppler-shift compensation behavior in horseshoe bats revisited: auditory feedback controls both a decrease and an increase in call frequency. J. Exp. Biol. 205:1607–1616.

Michelsen, A., and Larsen, O. N. 1985. Hearing and sound. In: Comprehensive Insect Physiology, Biochemistry, and Pharmacology (Kerkut, G., and Gilbert, L. I., eds.). Oxford: Pergamon Press; 495–556.

Miller, L. A. 1970. Structure of the green lacewing tympanal organ (*Crysopa carnea*, Neuroptera). J. Morphol. 131:359–382.

Miller, L. A., and Olesen, J. 1979. Avoidance behavior in green lacewings. I. Behavior of free flying green lacewings to hunting bats and ultrasound. J. Comp. Physiol. 131:113–120.

Miller, L. A., and Surlykke, A. 2001. How some insects detect and avoid being eaten by bats: tactics and countertactics of prey and predator. Bioscience 51:570–581.

Minet, J., and Surlykke, A. 2003. Auditory and sound producing organs. In: Lepidoptera, Moths and Butterflies, vol. 2: Morphology, Physiology and Development (Kristensen, N. P., ed.). New York: de Gruyter; 289–323.

Misawa, H., and Suga, N. 2001. Multiple combination-sensitive neurons in the auditory cortex of the mustached bat. Hear. Res. 151:15–29.

Mittmann, D. H., and Wenstrup, J. J. 1995. Combination-sensitive neurons in the inferior colliculus. Hear. Res. 90:185–191.

Möhl, B., and Miller, L. A. 1976. Ultrasonic clicks produced by the peacock butterfly: a possible bat-repellent mechanism. J. Exp. Biol. 64:639–644.

Moiseff, A., and Hoy, R. R. 1983. Sensitivity to ultrasound in an identified auditory interneuron in the cricket: a possible neural link to phonotactic behavior. J. Comp. Physiol. 152:155–167.

Moiseff, A., Pollock, G. S., and Hoy, R. R. 1978. Steering responses of flying crickets to sound and ultrasound: mate attraction and predator avoidance. Proc. Natl. Acad. Sci. U.S.A. 75:4052–4056.

Moss, C. F., and Sinha, S. R. 2003. Neurobiology of echolocation in bats. Curr. Opin. Neurobiol. 13:751–758.

Moss, C. F., and Surlykke, A. 2001. Auditory scene analysis by echolocation in bats. J. Acoust. Soc. Am. 110:2207–2226.

Neuweiler, G. 1984. Foraging, echolocation and auditory in bats. Naturwissenschaften 71:446–455.

Neuweiler, G. 1989. Foraging ecology and audition in echolocating bats. Trends Ecol. Evol. 6:160–166.

Neuweiler, G. 1990. Auditory adaptations for prey capture in echolocating bats. Physiol. Rev. 70:615–641.

Neuweiler, G. 2003. Evolutionary aspects of bat echolocation. J. Comp. Physiol. A 189:245–256.

Nolen, T. G., and Hoy, R. R. 1984. Initiation of behavior by single neurons: the role of behavioral context. Science 226:992–994.

Norman, A. P., and Jones, G. 2000. Size, peripheral auditory tuning and target strength in noctuid moths. Physiol. Entomol. 25:346–353.

Obrist, M. K. 1995. Flexible bat echolocation: the influence of individual, habitat and conspecifics on sonar signal design. Behav. Ecol. Sociobiol. 36:207–219.

Olsen, J. F., and Suga, N. 1991. Combination-sensitive neurons in the medial geniculate body of the mustached bat: encoding of the target range information. J. Neurophysiol. 65:1275–1296.

Popov, A. V., Markovich, A. M., and Andjan, A. S. 1978. Auditory interneurons in the prothoracic ganglion of the cricket, *Gryllus bimaculatus* deGeer. J. Comp. Physiol. 126:183–192.

Popov, A. V., and Shuvalov, V. F. 1977. Phonotactic behavior of crickets. J. Comp. Physiol. 119:111–126.

Rayner, J. M. V. 1991. The cost of being a bat. Nature 350:383–384.

Robert, D. 1989. The auditory behavior of flying locusts. J. Exp. Biol. 147:279–301.

Roeder, K. D. 1966. Acoustic sensitivity of the noctuid tympanic organ and its range for the cries of bats. J. Insect Physiol. 12:843–859.

Roeder, K. D. 1967. Nerve Cells and Insect Behavior. Cambridge, Mass.: Harvard University Press.

Roeder, K. D., and Treat, A. E. 1962. The acoustic detection of bats by moths. Proceedings of the 11th Entomological Congress 3:7–11.

Römer, H., Marquart, V., and Hardt, M. 1988. Organization of a sensory neuropile in the auditory pathway of two groups of Orthoptera. J. Comp. Neurol. 275:201–215.

Rydell, J. 1998. Bat defence in lekking ghost swifts (*Hepialus humuli*), a moth without ultrasonic hearing. Proc. R. Soc. Lond. B 265:1373–1376.

Rydell, J., and Arlettaz, R. 1994. Low-frequency echolocation enables the bat *Tadarida teniotis* to feed on tympanate insects. Proc. R. Soc. Lond. B 257:175–178.

Rydell, J., Kaerma, S., Hedelin, H., and Skals, N. 2003. Evasive response to ultrasound by the crepuscular butterfly *Manataria maculata*. Naturwissenschaften 90:80–83.

Rydell, J., and Lancaster, C. 2000. Flight and thermoregulation in moths were shaped by predation from bats. Oikos 88:13–18.

Rydell, J., McNeill, D. P., and Eklöf, J. 2002. Capture success of little brown bats (*Myotis lucifugus*) feeding on mosquitoes. J. Zool. (Lond.) 256:379–381.

Rydell, J., and Speakman, J. R. 1995. Evolution of nocturnality in bats—potential competitors and predators in their early history. Biol. J. Linn. Soc. 54:183–191.

Sanderford, M. V., and Conner, W. E. 1990. Courtship sounds of the polka-dot wasp moth, *Syntomeida epilais*. Natuwissen. 77:345–347.

Schmidt, S., Hanke, S., and Pillat, J. 2000. The role of echolocation in the hunting of terrestrial prey—new evidence for an underestimated strategy in the gleaning bat, *Megaderma lyra*. J. Comp. Physiol. A 186:975–988.

Schnitzler, H. U. 1987. Echoes of fluttering insects: information for echolocating bats. In: Recent Advances in the Study of Bats (Fenton, M. B., Racey, P. A., and Rayner, J. M. V., eds.). Cambridge: Cambridge University Press; 226–243.

Schnitzler, H. U., and Kalko, E. K. V. 2001. Echolocation by insect-eating bats. BioScience 51:557–569.

Schnitzler, H. U., Menne, D., Kober, R., and Heblich, K. 1983. The acoustical image of flut-tering insects in echolocating bats. In: Neuroethology and Behavioral Ecology (Huber, F., and Markl, H., eds.). Heidelberg, Germany: Springer-Verlag; 235–249.

Schnitzler, H. U., Moss, C. F., and Denzinger, A. 2003. From spatial orientation to food ac-quisition in echolocating bats. Trends Ecol. Evol. 18:386–394.

Schoeman, M. C., and Jacobs, D. S. 2003. Support for the allotonic frequency hypothesis in an insectivorous bat community. Oecologia 134:154–162.

Schul, J., and Patterson, A. C. 2003. What determines the tuning of hearing organs and the frequency of calls? A comparative study in the katydid genus *Neoconocephalus* (Ortho-ptera, Tettigoniidae). J. Exp. Biol. 206:141–152.

Schuller, G. 1986. Influence of echolocation pulse rate on Doppler-shift compensation con-trol system in the Greater Horseshoe Bat. J. Comp. Physiol. A 158:239–246.

Schulze, W., and Schul, J. 2001. Ultrasound avoidance behavior in the bushcricket *Tettigonia viridissima* (Orthoptera, Tettigoniidae). J. Exp. Biol. 204:733–740.

Simmons, J. A. 1973. The resolution of target range by echolocating bats. J. Acoust. Soc. Am. 54:157–173.

Simmons, J. A., Fenton, M. B., and O'Farrell, M. J. 1979. Echolocation and pursuit of prey by bats. Science 203:16–21.

Simmons, J. A., and Kick, S. A. 1983. Interception of flying insects by bats. In: Neuroethology and Behavioral Physiology (Huber, F., and Markl, H., eds.). Berlin: Springer-Verlag; 267–279.

Simmons, J. A., and Stein, R. A. 1980. Acoustic imaging in bat sonar: echolocation signals and the evolution of echolocation. J. Comp. Physiol. A 135:61–84.

Skals, N., Plepys, D., and Löfstedt, C. 2003. Foraging and mate-finding in the silver Y moth, *Autographa gamma* (Lepidoptera: Noctuidae) under the risk of predation. Oikos 102:351–357.

Smotherman, M., and Metzner, W. 2003. Fine control of call frequency by horseshoe bats. J. Comp. Physiol. A 189:435–446.

Spangler, H. G. 1984. Silence as a defense against predatory bats in two species of calling insects. Southwest. Nat. 29:481–488.

Speakman, J. R. 2001. The evolution of flight and echolocation in bats: another leap in the dark. Mammal. Rev. 31:111–130.

Speakman, J. R., and Racey, P. A. 1991. No cost of echolocation for bats in flight. Nature 350:421–423.

Suga, N. 1963. Central mechanisms of hearing and sound localization in Insects. J. Insect Physiol. 9:867–873.

Suga, N., and Jen, P. H. 1976. Disproportionate tonotopic representation for processing CF-FM signals in the mustache bat auditory cortex. Science 194:542–544.

Suga, N., and Jen, P. H. S. 1975. Peripheral control of acoustic signals in the auditory system of echolocating bats. J. Exp. Biol. 62:277–311.

Suga, N., and O'Neill, W. E. 1979. Neural axis representing target range in the auditory cor-tex of the mustached bat. Science 206:351–353.

Surlykke, A., and Filskov, M. 1999. Auditory relationships to size in noctuid moths: bigger is better. Naturwissenschaften 86:238–241.

Surlykke, A., and Moss, C. F. 2000. Echolocation behavior of big brown bats, *Eptesicus fuscus*, in the field and the laboratory. J. Acoust. Soc. Am. 108:2419–2429.

Svensson, M. G. E., Rydell, J., and Brown, R. 1999. Bat predation and flight timing of winter moths, *Epirrita* and *Operophtera* species (Lepidoptera, Geometridae). Oikos 84:193–198.

Tougaard, J., Casseday, J. H., and Covey, E. 1998. Arctiid moths and bat echolocation: broadband clicks interfere with neural responses to auditory stimuli in the nuclei of the lateral lemniscus of the big brown bat. J. Comp. Physiol. A 182:203–215.

Triblehorn, J. D. 2003. Multisensory integration in the ultrasound-triggered escape response of the praying mantis, *Parasphendale agrionina* (Ph.D. dissertation, University of Maryland, College Park).

Triblehorn, J. D., and Yager, D. D. 2001. Broad versus narrow auditory tuning and corresponding bat-evasive flight behavior in praying mantids. J. Zool. (Lond.) 254:27–40.

Valentine, D. E., and Moss, C. F. 1997. Spatially selective auditory responses in the superior colliculus of the echolocating bat. J. Neurosci. 17:1720–1733.

Wever, E. G., and Vernon, J. A. 1961. The protective mechanisms of the bat's ear. Ann. Otol. Rhinol. Laryngol. 70:1–13.

Wohlers, D. W., and Huber, F. 1978. Intracellular recording and staining of cricket auditory interneurons. J. Comp. Physiol. 127:11–28.

Yack, J. E. 1988. Seasonal partitioning of tympanate moths in relation to bat activity. Can. J. Zool. 66:753–755.

Yack, J. E., and Fullard, J. H. 1993. What is an insect ear? Entomol. Soc. Am. 86:677–682.

Yack, J. E., and Fullard, J. H. 2000. Ultrasonic hearing in nocturnal butterflies. Nature 403:265–266.

Yack, J. E., and Hoy, R. R. 2003. Hearing. In: Encyclopedia of Insects (Resh, V.H., and Cardé, R., eds.). San Diego: Academic Press.

Yager, D. D. 1990. Sexual dimorphism of auditory function and structure in praying mantises (Mantodea; Dictyoptera). J. Zool. (Lond.) 221:517–537.

Yager, D. D. 1999. Structure, development, and evolution of insect auditory systems. Microsc. Res. Tech. 47:380–400.

Yager, D. D., and Hoy, R. R. 1989. Audition in the praying mantis, *Mantis religiosa* L.: identification of an interneuron mediating ultrasonic hearing. J. Comp. Physiol. A 165:471–493.

Yager, D. D., and May, M. L. 1990. Ultrasound-triggered, flight-gated evasive maneuvers in the praying mantis *Parasphendale agrionina*. II. Tethered flight. J. Exp. Biol. 152:41–58.

Yager, D. D., May, M. L., and Fenton, M. B. 1990. Ultrasound-triggered, flight-gated evasive maneuvers in the praying mantis *Parasphendale agrionina*. I. Free flight. J. Exp. Biol. 152:17–39.

Yager, D. D., and Spangler, H. G. 1995. Characterization of auditory afferents in the tiger beetle, *Cicindela marutha* Dow. J. Comp. Physiol. A 176:587–600.

Yager, D. D., and Spangler, H. G. 1997. Behavioral response to ultrasound by the tiger beetle *Cicindela marutha* Dow combines aerodynamic changes and sound production. J. Exp. Biol. 200:649–659.

6

The Visual Ecology of Predator-Prey Interactions

THOMAS W. CRONIN

Vision is the sensory modality that normally provides the greatest awareness of events far removed from an animal's current location. Thus, it is often the most effective sense for early detection of predators or prey, as well as for enabling decisions about appropriate actions to take upon such detection. Indeed, it has been argued that the early appearance of good visual systems fueled the evolution of complex animals. Visual tasks of predators and prey require rather different specializations, reflecting the quite different role of each animal in the predator-prey relationship. Because the risk of overlooking prey involves slight relative costs to a predator, compared to the cost to a prey animal of overlooking the presence of a predator, prey visual systems tend to sample a broad spatial field relatively uniformly, searching for predators at virtually any location. Predators, instead, tend to focus on smaller visual fields, looking for concealed prey and making decisions about the potential for a successful attack. For both animals decisions must be made rapidly and often without a full analysis of the situation. The need for rapid decisions fosters the evolution not only of effective systems of information processing but also of systems of deceit that tap into weaknesses of other animals' mechanisms of visual processing. Some animals that would normally be prey have such effective defenses (e.g., stings or surface toxins) that their interests are best served by evolving clear signals of identity, such as bright patterns or distinctive postures, warning potential predators away. This chapter considers these aspects and others that demonstrate how vision functions in predator-prey interactions, not only discussing the visual adaptations of animals, but also considering how these adaptations shape the interplay of predators and prey.

In animals that are active in lighted environments, vision is almost always the sense responsible for rapid assessment of the surroundings. Consequently, it often plays the major role in maintaining vigilance in predators and prey alike. On the one hand, vision orients predatory attacks and, on the other, it enables prey escapes. It is therefore the critical sense in the great majority of predator-prey interactions. Vision permits rapid detection at a distance, enabling excellent spatial resolution and the capacity to sense passively (unlike sonar, for example; also see Triblehorn and Yager, ch. 5 in this volume). These benefits have fostered the evolution of vision even in environments where light levels are low and unreliable (e.g., on the forest floor at night or in the deep sea).

Even the simplest photoreceptor, a single photosensitive cell, can signal changes in light levels. Thus, it can mediate shadow reflexes, perhaps the earliest type of predator-avoidance behavior, by which a prey animal retreats in response to a sudden drop in illumination. The eyes of polychaete tubeworms and bivalve mollusks (e.g., ark clams, scallops, and giant clams), although relatively sophisticated compared to a single photoreceptive cell, are nothing but glorified shadow detectors (Land 1965, Nilsson 1994, Land 2003). When these animals sense a sudden dimming of light, they withdraw into their tubes or shells.

The visual systems of most of today's animals are highly evolved, complex, and generally associated with some of the most elaborate neural architecture and machinery found in living creatures. The sophistication and optical excellence of the human eye famously caused Charles Darwin himself to shudder when he considered how to account for its evolution. But Darwin realized that eyes like ours do arise from earlier, generally simpler designs, and recent work by Nilsson and Pelger (1994) finds that only 400,000 generations are required for an eye to evolve from a flat epithelial patch of photoreceptors to a fully competent lens eye with an optically perfect image (Land and Nilsson 2002).

Evolutionary advances, like those in the eyes of worms or clams, open new vistas for visual systems, beginning with the ability to ascertain the direction of simple stimuli. As visual organs and processing centers improve, visual systems gain the ability to organize the structure of visual fields, to detect and analyze motion, to see the forms of objects, and to segregate a scene into regions that vary in color or polarization. Each improvement makes vision more capable of alerting the nervous system to the presence, location, and identity of predators or prey. Parker (1998, 2000) maintains that the appearance of the first really competent eyes, about 550 million years ago, initiated the Cambrian evolutionary explosion. The event sparked a positive-feedback cycle in which prey were evolutionarily forced to become ever more efficient at detecting and escaping predators, who in turn were obligated to improve their own means of prey identification and predatory effectiveness. If Parker is correct, visual evolution was a key element in the appearance of increasingly complicated animal forms during the time that many body plans were tried and improved or rejected, leading ultimately to the biological designs we see today and shaping the behavior of modern animals (Land and Nilsson 2002).

Ecology and Evolution of Visual Systems

The great values of vision as a sensory modality (i.e., giving a sense of location in space, monitoring movement, and guiding locomotion) compel it to play many roles in the lives of animals. In particular, it provides awareness of the nature and distribution of objects in an animal's immediate environment, including those of special biological relevance—food, mates, home, and hiding places—as well as of the presence of predators or prey. Vision also receives biological signals: colors and patterns intended to affect the behavior of another individual (on occasion, even deceptive signals from potential predators or prey).

Thus, a visual system perceives and must generally discriminate among stimuli of many types. Typically, these stimuli originate from inanimate objects in the surroundings, from plants, or from other animals, such as mates, associates, and competitors, as well as predators and prey. The vistas that animals encounter in the wild are known as *natural scenes*, and visual systems have evolved in the contexts of such scenes. Natural scenes vary in their spatial structure (e.g., Burton and Moorhead 1987, Field 1987, Tolhurst et al. 1992, Ruderman 1994, Ruderman and Bialek 1994, van der Schaaf and van Hateren 1996), their temporal structure (Dong and Attick 1995, van Hateren and van der Schaaf 1996, Chiao, Osorio, et al. 2000), and their spectral structure (Burton and Moorhead 1987, Webster et al. 1996, Webster and Mollon 1997, Ruderman et al. 1998, Chiao, Cronin, et al. 2000, Chiao, Vorobyev, et al. 2000). Properties of a diversity of visual systems are demonstrably related to such scene properties (Field 1987, van Hateren 1993, Olshausen and Field 1996, Ruderman et al. 1998, Chiao, Vorobyev, et al. 2000).

Given that fundamental visual system properties are not shaped primarily by particular visual tasks, but rather by the statistical properties of the general views they take in, visual systems of animals in similar habitats will tend to function similarly. Obviously, this principle has its limits. Bees must find flowers, primates need fruit, and predators seek prey, so specializations that vary among species are built onto the robust underlying visual system structure. It is primarily these species-specific features that concern us in this chapter.

The study of how visual systems are specialized, both generally and for the special requirements of individuals and species, is called *visual ecology.* The best references in the field span most of a century, from Wall's *The Vertebrate Eye and Its Adaptive Radiation* (1942), through John Lythgoe's groundbreaking *The Ecology of Vision* (1979), to Land and Nilsson's new review, *Animal Eyes* (2002), and document a huge set of visual adaptations for particular lifestyles and environments. Not all of these adaptations relate to predator-prey interactions, of course, but here we focus on how visual function and design shape the biology, ecology, and behavior of predators and prey. To keep the subject manageable, I will select examples from a broad diversity of taxa and habitats, highlighting visual specializations that predators and prey use to enhance function in their respective roles and presenting some special features of predator and prey vision and visual behavior. I will then focus on one group of animals, the mantis shrimps, that exemplify how far a predatory animal can go toward developing visual specializations that disadvantage its prey. Finally, I will discuss how prey animals respond to the visual armament of their predators (and how predators respond to the gullibility of their prey), examining systems of camouflage, concealment, deception, and aposematism, and considering how animals ever manage to function and to communicate with each other without placing themselves in harm's way.

Visual Specializations of Predators and Prey

The sequences of events that occur, either when a predator first spots potential prey, or when a prey animal first recognizes the presence of a possible predator, are rather similar (Figure 6.1). Each player requires continuous vigilance to be aware of the other's proximity as early as possible. Unless the predator has a clear advantage, the prey will not be captured. The old proverb accounting for the escape of the rabbit from the fox explains this principle: "The fox is only running for his dinner, but the rabbit is running for his life."

Any prey species not extremely effective at predator evasion simply will not persist; prey species must escape, if all else is equal. The protracted, dramatic chases beloved of nature television shows are exceptional and generally involve only a very few vertebrate predator species with the stamina to prolong a pursuit after the initial attack: most predators have one fleeting chance at a particular prey item. Nevertheless, predator species survive, and to do so they must place themselves in a position of the best possible advantage from which to initiate an attack. Visual predators attempt to spot and identify prey early, often from a place of concealment. If possible, they then will launch a quick attack, again under visual control. In some predatory species, this feat first requires a stealthy, visually oriented stalk. The sequence of events is as indicated on the left side of Figure 6.1.

For prey, the situation is analogous but different in detail (right side of Figure 6.1). Predators can specialize in where they search for prey, but for most prey a predator can appear virtually anywhere. Thus, the detection system should be nearly omnidirectional and capable of extremely quick assessment of the level of danger. If necessary, the prey animal must initiate escape action instantaneously (the prey's counterpart to the predator's strike). More often, the prey may decide instead to minimize its visibility to the predator by changing posture, movement, or pattern, waiting for further action by the predator.

In this section of the chapter, I consider some features of eye design and performance that are particular to the requirements of predators and prey, focusing on each step of the detection-attack-evasion cycle. It is interesting, though perhaps not surprising, that much more attention has been paid to specializations in eyes of predators than to those of prey. Despite this bias, the work that is available permits some generalizing and, equally important, points the way toward critical issues that require

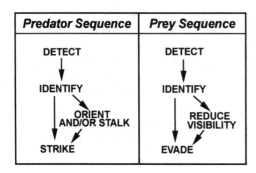

Figure 6.1. Sequences of perceptual and responsive events that occur during visual predation. Predators are schematized on the left; prey, on the right).

special research attention. Obviously, there is no clear distinction between predators and prey; one prey's predator is another predator's prey, but the asymmetries between the requirements of avoiding predators and obtaining prey do lead to asymmetric visual specializations for the different tasks.

Detection: Visual Field Specialization

Predators and prey sample visual space differently. Prey animals face possible threats that can emerge from many directions. They need eyes that provide very broad spatial coverage, commonly without bias except what is required for feeding or for intraspecific tasks (e.g., signal recognition or mate detection). Predators, on the other hand, can afford to specialize in their sampling of space. If this specialization sometimes causes them to overlook the presence of prey, the offsetting advantage is that it commonly allows them to visually outrange prey at the instant of detection. Even though predators are normally larger than their prey, they often can detect prey animals from locations and distances where they themselves are still invisible. Sometimes, they can even hunt from places and situations in which prey animals simply cannot detect them visually. Interestingly, some intraspecific behaviors replicate the predator-prey relationship. Males of some species stalk females by using tactics and visual designs not unlike those of visual predators, as will be illustrated later in this chapter. Animals that are predators as adults often pass through much more vulnerable developmental stages as infants or larvae: in frogs the eyes move from the lateralized, "prey" position in tadpoles to the more frontal, "predator" position in adults (Sivak and Warburg 1983).

Visual Fields: Predators

Binocular Vision

Most animals have two image-forming eyes, simple or compound, and either fixed in position or moveable. Although designs vary, simple (lens) eyes generally see up to as much as a hemisphere, whereas compound eyes can view nearly the full, spherical spatial field. Depending on eye orientation, the two visual fields may have little or no overlap, thus providing separate monocular fields; may nearly duplicate each other, providing a large binocular field; or, most commonly, may have a restricted binocular field nestled within two, more extensive, monocular ones. The extent of the binocular field varies in animals with moveable eyes, as well.

It is not difficult to determine the visual field of an eye, so many examples of visual field measurements are available. In vertebrates, some of the best available data are from birds, each eye of which provides nearly hemispherical coverage (Walls 1942, Martin 1984, Martin 1986, Martin and Katzir 1994, Martin and Katzir 1999, Martin and Prince 2001). Casual observation seems to suggest that many predatory birds have well-developed binocular vision. In reality, even owls have binocular fields encompassing less than 20% of the entire field coverage of both eyes, whereas in other predators, such as raptors, herons, seabirds, and penguins, the binocular field is typically 10% or less of the full field and can be eliminated by eye movements

(Martin 1984, Martin and Katzir 1999). In herons, for example visual fields are organized to provide a slightly overlapping view of prey almost vertically downwards from the head, permitting the bird to minimize its profile as seen by prey but still maintain excellent vision near its feet (Martin and Katzir 1994).

Binocularity, especially in predators, is often assumed to provide the advantage of depth measurement through binocular convergence, stereopsis (the formation of a sense of depth through neural comparison of disparities in the images seen by both eyes), or both. Eye movements in birds, however, are not well coordinated, and variations in binocular overlap make neural mechanisms of depth perception nearly impossible. Martin and Katzir (1999) suggest instead that, with the possible exception of owls, eye position and visual field sampling in birds is organized to provide appropriate optic flow (the movement of images over the retina during motion) for flight control, for target localization, and even for bill strikes. Binocularity would also provide redundancy, noise reduction, and the potential for symmetrical placement of images within the two eyes. As shall be described later, the predatory stoop of attacking raptors is actually controlled monocularly. Binocular vision, so highly regarded by us binocular primates, apparently plays no significant role in avian predation.

Many mammals, both predators (e.g., cats) and nonpredatory species (e.g., primates), do have binocular vision. Thus, there is no exclusive relationship between binocularity and predation: animals that require excellent depth perception at short distance (like nocturnally pouncing cats or tree-climbing primates) all benefit from stereopsis. Mammals that hunt large prey at extended ranges, or that have eyes designed for sensitivity rather than acuity, would be unlikely to use binocular mechanisms.

Invertebrate predators, on the other hand, often launch their ambushes without a preliminary stalk and capture prey at very short distances. For them, binocular disparity may be the ideal cue for distance measurement, particularly in insects, where the two eyes are geometrically fixed relative to each other. Praying mantids, for instance, determine the distance to a small target by using binocular disparity, which can be predictably disrupted with use of prisms to displace the apparent position of the image on one or both eyes (Rossel 1983). Larvae of tiger beetles and dragonflies also may use binocular cues to direct their attacks (Sherk 1977, Toh and Okamura 2001). Bulldog ants attack an approaching object viciously, snapping their mandibles to impale it, but they confuse size and distance. Large objects are attacked at much greater distances than small ones, which rules out stereopsis as the cuing mechanism (Via 1977). Instead, the stimulation of critical pairs of receptor sets initiates the snapping behavior.

Foveas and Acute Zones

Image-forming eyes rarely sample visual space uniformly. In lens eyes (the type found in vertebrates), receptor densities or the convergence of receptors onto higher order cells, such as retinal ganglion cells, can vary across the retina. Regions of high spatial acuity in these eyes are called *foveas*. In compound eyes, optical units (ommatidia) may crowd together to sample a small area of space, and here the region of greatest visual resolution is termed the *acute zone* (see Figure 6.2 for examples).

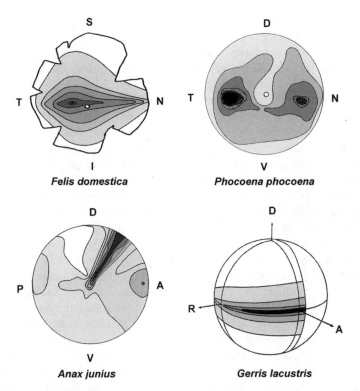

Figure 6.2. Visual fields (or retinal organization) of right eyes of vertebrate (top row) and invertebrate (bottom row) predators. Regions of increasing acuity, sampled by greater numbers of visual units (ganglion cells in the top row, ommatidia in the bottom row), are coded with increasing densities of fill, but do not represent absolute values of acuity across different panels of the figure. Empty circles near the centers of the visual fields of vertebrates represent the location of the optic disk, or blind spot, where the optic nerve exits the retina. Top left is retinal preparation of the house cat *Felix domestica*, to show ganglion cell densities, and thus visual acuity (modified from a figure in Land and Nilsson 2002). Note the presence of a spot-shaped fovea (black) and the suggestion of a horizontal visual band, or visual streak, of increased acuity. Top right is the visual field of the harbor porpoise, *Phocoena phocoena* (after Mass et al. 1986), mapped as ganglion cell densities, the two foveas. Bottom left is visual field densities of the dragonfly, *Anax junius* (after Sherk 1978). Note the strip-shaped upward-facing band of high acuity, the small anterior acute zone, and the suggestion of a weak posterior acute zone, as well. Bottom right shows projections of the visual fields of the waterstrider *Gerris lacustris* (after a drawing in Land and Nilsson 2002, based on data from Dahmen 1991). Note the extensive horizontal expanse of the zone of high acuity, representing the visual streak. Abbreviations: S, superior; I, inferior; T, temporal; N, nasal; D, dorsal; V, ventral; R, right; A, anterior; P, posterior.

Many vertebrates with laterally placed eyes have two foveas, one viewing the visual field directly ahead of the animal, and the other facing laterally. The areas of best vision are thus aimed at places of special interest: where food might be found or caught, and where other animals may approach. Double foveas exist in aquatic predators such as fish (Collin and Pettigrew 1988a, Collin and Pettigrew 1988b, Land 1990, Collin and Shand 2003) and marine mammals (Mass et al. 1986, Mass and Supin 1995; Figure 6.2). Even predatory fish of the deep-sea, living far below the reach of sunlight, have these double foveas. The presence of high-acuity eyes there seems paradoxical, but this is in fact the correct design for spotting tiny bioluminescent flashes indicating the presence of food (or a mate) deep in the gloom (Warrant 2000).

Among terrestrial vertebrate predators, double foveas are particularly common in birds (Meyer 1977, Land and Nilsson 2002), where one fovea is often devoted to detecting and localizing prey (Pettigrew et al. 1999, Tucker 2000, Collin and Shand 2003). The lateral foveas of raptors provide perhaps the most acute ("sharp") vision of any animal, but they face out to about 45 degrees to the side of the axis of the bird's head. Rather than either rotating the eye in its socket (an ability that is rather limited in birds), or turning the head to the side, raptors actually attack their prey in a spiral path, an approach that allows them constantly to fixate the target without sacrificing flight speed (Tucker 2000). Keeping the prey item in clear view permits a well-directed attack and instantly alerts the diving predator to evasive behavior on the part of the prey. Excellent vision, together with birds' need to spot small, mobile prey, has led to some of the most remarkable systems of head stabilization during body movement (even flight). Herons and other walking, predatory birds first thrust the head forward and then hold it steady through the rest of the step (Fujita 2003). Even when herons bounce up and down on a springy tree branch, their eyes are rigorously fixed in space (Martin and Katzir 1994). Flying terns track prey accurately on the ground beneath them by using head movements (Land 1999). All these systems of image stabilization must make prey motion in a static visual field particularly obvious.

Vertebrates are not the only predators with lens eyes. Many spiders have up to eight of them, which in wolf spiders (Lycosidae) and jumping spiders (Salticidae) can provide the most acute vision of any terrestrial invertebrate (Land 1985a). Spiders break up the overall visual field into regions of low- and high-acuity vision in a unique and parsimonious fashion. Each eye is aimed in a different direction, providing virtually full coverage of visual space above and around the spider (Land 1985a, Land 1985b). The principal eyes, the large pair that face to the front of the spider, orient predatory attacks and have by far the smallest visual fields and the best vision. The other eyes appear to be specialized for the detection of predators or prey in the spider's vicinity or for detecting movement (Duelli 1978). The extremely sharp vision of these cursorial hunters, coupled with other adaptations, such as moveable retinas and mirror reflectors behind the retina to brighten the image at night (Land 1985a, Land and Barth 1992), makes them outstanding prey spotters.

A few marine copepods (planktonic crustaceans) employ lens eyes to spot and track prey at ranges beyond which they themselves are visible to their quarry. In some of these tiny animals, the lenses are extremely complex, consisting of multiple elements with unusual optics (Land 1984a). The optical sophistication itself illustrates the selective pressure on predators, even minute ones, that fosters the evolution of superior vision.

Acute zones in compound eyes play the same roles as foveas in lens eyes. These are regions of the compound eye where optical axes of individual ommatidia (the units of a compound eye) are angled to sample space more densely than elsewhere (Horridge 1978). Acute zones are particularly common in animals (like predators) that fixate on a visual target. For example, each eye of a praying mantid has an acute zone used for fixation, tracking, and depth ranging (Rossel 1980). Among crustaceans, the best examples of acute zones are in the eyes of stomatopods (mantis shrimps), but some surprisingly small crustaceans, like cladocerans in the genus *Polyphemus* (Nilsson and Odselius 1982, Young 1988) and the mysid *Dioptromysis paucispinosa* (Nilsson and Modlin 1994), have compound eyes with acute zones. Here, the presence of the acute zone is clearly associated with predation, implying a role for detecting, fixating, and tracking other animals.

Acute zones need not sample only small spots in space; they can also have elongated or other geometrically odd visual fields. Unusual acute zones are found in some inhabitants of flat worlds, and one of the most interesting examples exists in the predatory waterstrider, *Gerris lacustris*, which lives on flat and smooth water surfaces (Dahmen 1991). This animal has paired compound eyes with binocular fields and an anterior acute zone centered in the binocular overlap. Each eye possesses a *visual streak*, a plane of ommatidia along the equator of the eye with very tightly packed optical axes providing excellent resolution at the level of the water surface (Figure 6.2). The benefit of such a design to an animal like *Gerris* is that almost everything it is likely to care about (i.e., prey, mates, and conspecifics) exists in just one plane, that of the water surface. Thus, an eye with best resolution at that level is unlikely to miss anything important to see. As we shall see later, visual streaks are found in prey animals of flat worlds, too.

Specializations for Vision Overhead

Predators will seize any opportunity they can to disadvantage their prey. A particularly interesting situation occurs in strongly biased light fields. For example, down to depths of perhaps 1000 m in the deep ocean, all illumination arriving from the sun is confined to the overhead light field. There is no primary production there, so life depends either on scavenging of material snowing down from above, or on predation. The region is a happy hunting ground for predators. Any prey above them that they can spot is completely at their mercy, because the prey is seen in silhouette, but the predator is completely invisible in the gloom of the depths. Of course, the predator itself assumes the same risk from other predators ever deeper. Hence, there is strong selection for enhancing vision to the greatest possible degree to see what is above, and for becoming as invisible as possible to confound what is below. Methods for concealment in open water are discussed later in the chapter; here we consider specialized visual designs to detect objects overhead.

Light is refracted as it enters water from air, confining light in the deep sea to an overhead patch 90 degrees or less across (Jerlov 1976). One way to maximize visual commitment for sighting prey within this patch of light is to discard all parts of the eye not aimed upward (Land 1990, Warrant et al. 2003). In a lens eye, the resulting fat and stubby organ is called a *tubular eye* (see Figure 6.3 for some examples). Photoreceptors in tubular eyes are large and tuned to the blue light of the deep sea (Douglas et al. 2003, Warrant et al. 2003). These eyes are extremely sensitive, permitting the visualization of very low contrast targets, such as transparent plankton or well-camouflaged fish. Most tubular eyes point dorsally, but some are parallel to the body axis of the fish (Figure 6.3B), suggesting that these species swim in a head-up posture, reducing their own silhouette and aligning their axis for a rapid vertical acceleration to snatch prey. Animals with tubular eyes either "float-and-wait" for prey, or slowly swim horizontally, trawling the overhead light field for food (Lythgoe 1979). The predatory squid (a species in the genus *Histioteuthis*) is an invertebrate with a tubular eye. One of its eyes has evolved into a perfect tubular eye, while the other has retained its wide-field optics (Land 1984b). The animal must swim on its side to use its asymmetrical eyes to advantage.

Some compound eyes are analogues of tubular eyes, having patches of crowded ommatidia that concentrate on downwelling light. Such eyes exist in predatory euphausiids and hyperiid amphipods (Land 1981, Land 1989, Land 1990, Land 2000). In the hyperiids, the crowding of axes causes a single point in space to be viewed by several ommatidia, a design that is superior for detecting tiny dark objects (Land 1989). *Phronima*, one species of hyperiid, has a hugely expanded optical region sampling the overhead light field and serving a tiny, compact retina (Figure 6.4A), while

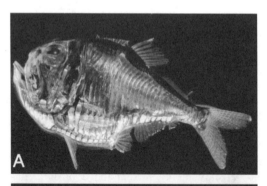

Figure 6.3. Examples of deep-sea fish with tubular eyes: (A) The hatchetfish, *Argyropelecus olfersi*; (B) *Stylephorus chordatus*. Note the bullet-shaped eyes of each species, which face upward in *A. olfersi* and forward in *S. chordatus*. Also note how the mouth of each species is adapted to take prey in the direction faced by the tubular eyes. *Argyropelecus olfersi* has shiny sides and ventral photophores (the ovoid structures) to hide its silhouette when seen from below. (Photographs courtesy of E. A. Widder. © 1999 Harbor Branch Oceanographic Institution.)

Figure 6.4. Adaptations of two species of hyperiid amphipods for life in the deep sea. Both species have transparent bodies (visualized by dark-field illumination in these photographs) and greatly enlarged compound eyes for sampling the overhead light field. (A) *Phronima lores*: The retinas are the small, dark condensed regions below the expanded upward-directed corneal regions. The dorsal expansions of the eye, resembling a human brain, are actually the optical units that intensively sample the overhead light field. Small extensions of the compound eyes face to the side and down for panoramic vision. (Photograph courtesy of T. N. Frank.) (B) *Streetsia* sp.: Note how the entire anterior end of the animal is a flattened compound eye for examining downwelling light. (Photograph courtesy of E. A. Widder. © 1999 Harbor Branch Oceanographic Institution.)

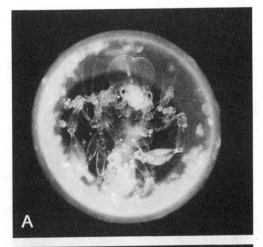

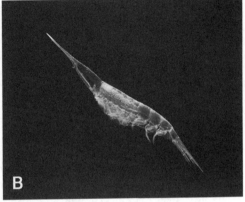

Streetsia has its entire "head" region formed into a long, flat pair of eyes (Figure 6.4B). An advantage that compound eyes have over lens eyes for deep-sea vision is that ommatidia outside the expanded upward-looking region are available for low-resolution vision to the side and below, picking up bioluminescence and providing some awareness of the approach of a predator. In Figure 6.4A the small spherical groups of ommatidia visible to the sides of the black, pigmented retinas form the second eye region.

At least one group of aerial "visual trawlers" can be found flying ceaseless beats over the surfaces of ponds. Some dragonflies, such as *Anax junius*, have a relatively weak forward-pointing acute zone, probably to handle flow-field monitoring and flight control, and a dorsal band of tightly packed ommatidia sampling a transverse strip of space about 30 degrees ahead of the vertical (Sherk 1978; Figure 6.2). As species with eyes like this fly forward, small objects above them are spotted against the sky and can be attacked from below.

Visual Field Specializations for Predation at the Air-Water Interface

A few aquatic species take prey that have fallen onto the water's surface. The paths that light takes to an eye from the part of the prey item extending up into the air, and from that protruding down into water, are different because of refraction of light from the aerial image as that image enters the water, separating the object into two apparent parts. Some animals, like the fish *Aplocheilus lineatus* (Land 1990) and the water bug *Notonecta glauca* (Schwind 1980) have two areas of acute vision, one aimed at the aquatic half of the image, and the other aimed at the aerial half! Possibly, the nervous system of each animal "stitches" the two halves back into a single image. Another aquatic predator that takes aerial prey is worth some thought. The archer fish, *Toxotes jaculatrix*, not only corrects its strike (squirting water onto a perched insect above the water) for the refractive displacement of the apparent position of an aerial object seen from under water, but even predicts where the prey animal will land on the water once it falls (Rossel et al. 2002). This is astonishing. The path the falling object takes through air is ballistic and displaced from its visualized path, but the fish is extremely accurate in making the appropriate corrections. Selection for predator efficiency must be ruthless to generate such precision in coping with complex optical geometry.

Visual Fields: Prey

In prey species, "specialization" usually means being prepared for the appearance of a predator anywhere. Visual field coverage of prey species therefore tends to be broad, often without high-resolution foveas or acute zones (although some parts of the visual field may be seem somewhat clearer than others). Vertebrate prey species commonly have their eyes placed almost opposite each other on the head (Walls 1942, Hughes 1977), giving nearly hemispheric coverage to each side and leaving virtually no blind area. As impressive examples, woodcocks (*Scolopax rusticola*) have nearly complete panoramic coverage (359 degrees) in the horizontal plane (Martin 1994), and shoveler ducks (*Anas clypeata*) see the entire hemisphere above the horizon and even down to 60 degrees below the horizon, except for the tiny patch where the bill protrudes (Guillemain et al. 2002). Both these bird species have binocular overlap to the rear. The shoveler is able to maintain vigilance even when the bill is submerged during feeding and rarely looks around between feeding bouts. Widgeons (*Anas penelope*), in contrast, have a narrow blind area to the rear and devote significantly more time to visual surveillance than the shovelers (Guillemain et al. 2002), a nice illustration of how visual field coverage and behavior can interact.

Mammalian prey species that live on plains or savannahs (e.g., horses, cattle, and rabbits) have retinas with visual streaks, visually sampling more densely along the line of the horizon (Walls 1942, Land and Nilsson 2002). The higher acuity in the area where predators are likely to appear, in grasses or behind bushes at ground level, provides wide-field vigilance. Similarly, fish that live on or near the bottom often have visual streaks (and laterally placed eyes) for the same reason (Collin and Pettigrew 1988a, Collin and Shand 2003).

Visual streaks in vertebrate eyes recall the specialized compound eyes of water-striders, so it should come as no surprise that flat-land prey with compound eyes have flat-land eyes. Fiddler crabs (species in the genus *Uca*) that inhabit mudflats and sand flats throughout the tropics have perky-looking eyes on long stalks that extend like periscopes above the carapace. The compound eyes themselves are vertically ovoid with nearly spherical fields of view and greatly increased vertical resolution along the equator of the eye (Zeil et al. 1986, Zeil et al. 1989, Land and Layne 1995a, Zeil and Al-Mutairi 1996, Zeil and Zanker 1997). Because the horizon is viewed with the best available resolution, any predator looming above it is instantly spotted. Other crabs of mangroves and rocky shores have very different eye designs, with eyes sitting far apart on short eyestalks that lack pronounced retinal specializations (Zeil et al. 1989). Such crabs lack specialized zones along the equatorial band, and, because they seem to walk, wander, and run in almost any direction, it appears that almost any part of the eye can lead. Fiddler crab eyes can be regarded as excellent predator detectors: the ommatidia looking upward from the horizon spot incoming birds, while the specialized band concentrates on terrestrial threats (Layne et al. 1997, Layne 1998).

This section on visual field design of prey would not be complete without a brief mention of prey with eyes specialized for predator-like behavior. Some winged insects, mainly dipteran flies, are the sexual predators of the invertebrate world, pouncing on unwary females from a position of advantage. This act is assisted by acute zones in the male compound eyes, which are aimed anteriorly and slightly above the horizontal. Their receptors drive neurons that control chasing behavior (Land and Eckert 1985, van Hateren et al. 1989, Gilbert and Strausfeld 1991, Burton and Laughlin 2003). These male eyes are specialized for detecting females as small, dark objects against the sky, an intraspecific example of "predation" based on visual asymmetry.

Detection: Contrast Enhancement

Many animals, whether predators or prey, use various optical tactics to remain inconspicuous. Predators, in particular, with their specializations for long-range vision, benefit from contrast-enhancing visual mechanisms to find some of these concealed prey. In environments where light is dim and spectrally limited, good contrast resolution is possible only by boosting sensitivity with visual sensitivities matched to the available light and by incorporating image-brightening optics (Land 1990, Douglas et al. 2003, Warrant et al. 2003,). In brighter visual worlds, contrast can be improved with spectral or polarizational adaptations.

A simple, but still widely unrecognized visual adaptation that improves prey capture in water is the use of ultraviolet (UV) vision to spot cryptic, transparent prey (Browman et al. 1994, McFarland and Loew 1994, Loew et al. 1996; reviewed in Losey et al. 1999). Many planktivorous fish and fish larvae increase the visibility of their tiny prey by using the bright background UV light of most natural waters as a backlight. The only available countermeasure is for prey to be as transparent as possible in the UV, an adaptation that is widespread (Johnsen and Widder 2001).

Visual systems of diurnal, predatory shallow-water fish maximize prey visibility by using cone photoreceptors (McFarland and Munz 1975). Predators that make high-speed attacks parallel to the surface, like mahi-mahi (*Coryphaena hippurus*), have cones that are very sensitive to the bright, broad-spectrum light reflected from a prey fish near the surface, and are insensitive to the bluer background light. Such receptors detect a very bright target against a dim background. If the attack is made from below, a receptor more sensitive to the background light is desirable, to sil-houette the prey. Thus, shallow-water marine predators commonly have two classes of cone photoreceptors (Lythgoe 1968, McFarland and Munz 1975), one for each type of attack. Another example of the use of two photoreceptor classes for vision in water is found in the deep sea. Some deep-sea fish produce red bioluminescence (Widder et al. 1984) and have a red-sensitive photoreceptor class in addition to the usual blue-sensitive one (O'Day and Fernandez 1974, Douglas et al. 1998). The red emission, it has been proposed, may illuminate prey at wavelengths visible only to the predator, permitting a deeply stealthy attack.

Color vision itself, though, does not seem to be particularly useful to predators or prey, at least not for spotting each other. For one thing, spectral properties of scenes, not of individual objects, seem to be more important in the evolution of general visual abilities like color vision. Camouflage systems based on spectral matching probably evolve very quickly, because a mismatch is lethal, negating any advantage that color itself may have as a prey-finding cue. Adding color-sensitive receptor classes to the retina takes up valuable space that could be used for sensi-tivity or acuity enhancement, compromising these critical aspects of vision (Wil-liams et al. 1991, Osorio et al. 1998). Color vision in mammalian predators, such as canids, is limited and may play only a minor role in prey capture (Neitz et al. 1989, Jacobs 1993, Jacobs et al. 1993). Passerine birds, the color-vision paragons among vertebrates, tend to have similar sets of cone receptors, whether their prey are subterranean worms, arboreal insects, or innocent seeds and fruits (Hart 2001), which implies that there is little adaptation to prey type. Kestrels apparently do use UV vision to spot urine marks left by voles, improving their ability to home in on prey (Viitala et al. 1995), but UV photosensitivity is widespread in small birds, so this enhancement seems more likely to be a fortuitous preadaptation than one evolved in the context of predation. In insects, an excellent comparative study of bees found little or no variation in their systems of color vision (Peitsch et al. 1992), and, indeed, lifestyle seems to have little influence on color vision throughout the Insecta (Briscoe and Chittka 2001). Thus, although color vision certainly is useful for many visual tasks, it rarely seems to play any special role in predator-prey in-teractions (with the important exception of aposematism, as will be discussed later in this chapter).

Polarization vision refers to the ability to see and analyze polarized light, and it gives at least one group of animals, the cephalopod mollusks (octopuses, squids, and cuttle-fish), a critical advantage in finding prey. Lythgoe and Hemmings (1967) noted that the ability to see polarized light in water could enhance the visibility of low-contrast objects. Shashar et al. (1998) showed that squids capture more transparent prey when the background is polarized. Cuttlefish use polarization vision to prey on shiny, sil-very fish (Shashar et al. 2000). Many marine invertebrates visually respond to polar-ized light, so polarization vision may be used by other marine predators. Polarization imaging is most useful when the object of interest reflects light with different polarization properties than the background, providing enhanced contrast. The use of this visual modality for predator evasion, however, has not yet been demonstrated.

Identification

Once a potential predator or prey animal has been detected, identification must be made reliably before appropriate action can be taken. This must almost always be done very quickly, so streamlined systems of identification play critical roles in most detection systems (see Curio 1993 for a review of some of these). Here, I will cover two sets of model identification systems used by vertebrate and invertebrate predators and prey.

Vertebrate Predator Model: The Common Toad

The first outstanding neuroethological description of visual predation by a vertebrate species was developed by Ewert, using the common toad, *Bufo bufo*. The brief discussion here is based on reviews published by Ewert (1980) and Ewert et al. (1983). The general features of objects that toads identify as prey can be determined behaviorally, because toads attack such objects by turning toward them, approaching if necessary, and lunging with an accompanying snap or flick of the tongue. To *Bufo*, the innate visual model of a prey item is a dark rectangle, smaller than a critical size, moving parallel to its long axis (e.g., a "worm"). The orientation of the rectangle is immaterial, but if the shape moves perpendicular to its axis, it becomes an "antiworm" (oddly, the toad itself is not the antiworm) and is ignored. Toad vision maintains size constancy. In other words, the change in apparent size with distance is compensated for, and therefore attacks are confined to objects in the correct size class and at the correct distance. Objects moving too fast or too slow are ignored. Thus, the best prey item is a smallish, dark, long object moving parallel to its long axis at a reasonable speed. Clearly, this is a very reduced version of prey, and, furthermore, toads change their behavior with experience (Ewert et al. 2001). Nevertheless, the basic concept of prey is genetically built into the visual system, and neurons acting as "prey detectors" can be identified in several regions of the central visual pathway. Toads are heterotherms, and on cool nights their retinas are chilly enough to reduce thermal noise sufficiently to permit visualization of model prey that would be invisible to a warm-blooded predator (Aho et al. 1993). These supposedly "primitive" vertebrates therefore have predatory advantages over nocturnal mammals.

Invertebrate Predator Model: The Praying Mantid

Ewert's work with toads inspired similar research into an invertebrate system, the praying mantid *Sphodromantis lineola*. Somewhat surprisingly, the fundamental criteria that define prey for a mantid are much like those for a toad (Prete 1999). Prey are best recognized when they have the right shape and size, are seen at about the right distance (which also produces a form of size constancy), and move parallel to their long axis at about the right speed. Interneurons in the ventral nerve cords of mantids respond to movement of prey-like objects in the correct part of the visual field, potentially coordinating predatory strikes (Gonka et al. 1999). One aspect of mantid predation behavior is odd and unexplained: males will attack much larger sham prey than females (Prete et al. 2002). The sexual disparity, whatever its source, illustrates the risk of generalizing freely about predatory behavior, even in relatively simple animals.

Vertebrate Prey Model: The Common Toad

Toads are intermediate-sized creatures, preying effectively on worms, insects, and tiny vertebrates while being vulnerable to attack from larger animals like snakes, birds, and mammals. An object passing rapidly overhead could be prey at short range or a predator further away, and little time is available to make the correct judgment. So, not surprisingly, just as for prey, *Bufo bufo* uses simple rules for identifying potential predators (Ewert 1980, Ewert 1983). Intermediate-sized square objects moving with the same velocity as small, attractive prey are ignored, but larger ones are avoided: instead of approaching, the toad turns away (Ewert 1980). In common with many predator-avoiding systems, objects seen in any direction that grow in size, or "loom," are powerful avoidance-generating stimuli (Ewert 1983). Nevertheless, the behavior is hardly stereotyped. Different prey configurations produce different means of defensive reaction. In response to snake-like objects, for instance, toads do not actively "avoid" the model but instead stand on tiptoe, puff up, and release aversive secretions (Ewert 1983).

Invertebrate Prey Model: Fiddler Crabs

Fiddler crabs, genus *Uca*, have already been introduced as examples of prey with eyes designed for vision in a flat world, sampling visual space near the horizon intensively. Fiddler crab eyes are on stalks, so the visual horizon extends in a plane from a height slightly above the bodies of conspecifics. This permits the fiddlers to live by a simple rule. If an object is in view *below* the horizon, it is nonthreatening and likely to be food, a competitor, or a potential mate. If an object is *above* the horizon, it is a predator. Even a passing vehicle, tens of meters away, will elicit an escape response if its profile breaks the horizon (Land and Layne 1995a, Land and Layne 1995b), and the likelihood of a response increases with the apparent (not actual) size of such an object (Layne 1998). Some species of crab escape by turning to place the potential predator to one side, and then running laterally away, crabwise (Land and Layne 1995b). Others head straight for their burrow (Zeil 1998). The simple predator identification rule used by crabs permits a rapid response, but elicits many false alarms. A butterfly passing at close range will empty a large beach area of crabs.

Stalking and Fleeing

Once identification has been made, the prey may decide to freeze, withdraw, or flee, and the predator might pounce or attempt to stalk the prey. These actions require no obvious visual specialization, although they certainly require visual sophistication to be successful. For instance, an araneophagic spider species (*Portia fimbriata*, a salticid) carefully observes its web-building prey (*Zosis geniculares*), approaching only when the prey spider's attention is diverted as it wraps its own prey already in the web (Jackson et al. 2002). The same predator often freezes when stalking other jumping spiders whenever they turn to face it, apparently to avoid detection by the prey's large principal eyes (Harland and Jackson 2000, Harland and Jackson 2002). These clever adaptations produce behavior much like that of a hunting mammal, and give the hunting spider an uncanny air of great presence of mind. Some predators manipulate visually oriented prey escape behavior. Painted redstarts (*Myrioborus pictus*) fan out their tail when hunting, panicking insect prey and driving them toward the bird's head where they are more likely to be caught. In a sense, the bird poses as a "backward" predator (Jablonski 2001), tapping into vulnerable features of visually directed escape behavior of its prey.

The Strike

Predators must know the precise distance of their prey to attack them with any assurance of success. Many visual cues are available (Collett and Harkness 1982), and those cues that serve predators best require no movements of the head or body that could alert prey. Thus, purely ocular mechanisms of depth perception are preferred, either binocular or monocular. Ewert (1980) noted that toad strikes at model prey are equally good, whether the toad views the prey monocularly, or binocularly. Using prisms and lenses, Collett (1977) demonstrated that toads use monocular cues (presumably accommodation, the focusing effort required to form a sharp image) when only one eye sees the target, but use binocular cues if both eyes are available. This is more flexible than the strictly binocular system of the praying mantid, already discussed (Rossel 1983), in which both eyes are required for the depth measurement; but, of course, compound eyes cannot accommodate.

Many other vertebrate predators use accommodation for depth measurement. Fire salamanders (*Salamandra salamandra*) shift the focus of their eyes to determine the range to prey (Werner and Himstedt 1984). More impressively, eyes of chameleons are supremely specialized for accommodative depth-ranging. The two eyes move independently, and there is no evidence of any binocular role in determining the distance to prey (Harkness 1977). They have telephoto optics (Ott and Schaeffel 1995), which magnify the retinal image and diminish depth of field, and therefore demand accurate focus. The independence of the eyes is perhaps a consequence of the excellence of the accommodative mechanism, which makes binocular vision superfluous. Astonishingly, a precisely analogous system of visual ranging exists in a small predatory fish, the sandlance (*Limnichthyes fasciatus*). It too has telephoto optics, independent eye movements, and an extreme ability to accommodate (Pettigrew et al. 1999). Chameleons and sandlances are ambush predators, and their spectacular eyes can take their strikes, whether by tongue or teeth, directly to their prey.

Vertebrate predators that are not well oriented to prey by vision may not use any visual depth cues. Bottlenose dolphins (*Tursiops truncatus*) find objects in murky water by using biosonar alone, and their eyes do not accommodate at all when fixating targets underwater (Cronin et al. 1998, Litwiler and Cronin 2001). On the other hand, another midwater predator, the cuttlefish, has excellent lens optics and good polarization vision, and it accommodates binocularly just before making a predatory attack by firing out the tentacles like biotorpedoes (Schaeffel et al. 1999). Cephalopods are the only invertebrates shown to focus their eyes at all, although retinal movements in some arthropods might achieve the same end. Almost incredibly, tiger beetle larvae (*Cincindela chinensis*) accurately estimate the distance to prey monocularly. They are disoriented in their attacks when fitted with larva-sized spectacles that displace the retinal image backward and focus a distant target, instead of a nearby one, on the retina (Toh and Okamura 2001). There is little likelihood that these simple lens eyes adjust focus, so Toh and Okamura (2001) suggest that the predatory leap is released only for objects at the distance that produces a sharply focused image. The routine success of monocular systems of prey localization leaves open the question whether any predator with lens eyes hunts prey by using strictly binocular vision.

A Case Study of Visual Predation: Stomatopod Crustacea

The stomatopod crustaceans, or mantis shrimps, are all marine predators. They are unique among the large modern Crustacea in that they launch strikes that smash or impale prey in ambush attacks, sometimes after brief chases (Caldwell and Dingle 1975). The attacks are visually controlled by some of the most specialized eyes ever to evolve. Stomatopods became evolutionarily separated from other Crustacea about 400 million years ago, and this deep divide from all other modern animals is no doubt partly responsible for the uniqueness of their visual systems.

Most stomatopod species live in shallow, tropical seas, occupying holes in the interstices of coral reefs or living in excavated burrows (Figure 6.5). From these homes, their odd eyes protrude to scan the environment for food, mates, and predators. Concealment inside the burrow, with only the antennae and eyes exposed, permits them to attack prey without warning. Mantis shrimp eyes are extremely mobile (Cronin 1986, Cronin et al. 1988, Cronin et al. 1991), operating independently (a disconcerting behavior to watch) and, like tiny radars, pointing at prey and other objects passing in front of their burrows. The mobile eyes and visual systems are responsible for finding, identifying, tracking, and finally directing accurate strikes at their prey.

Compound eyes of modern stomatopods have a triple organization, with separate dorsal and ventral masses of ommatidia flanking a middle band of several specialized ommatidial rows (Horridge 1978, Marshall et al. 1991a, Marshall et al. 1991b). Ommatidia in the dorsal and ventral regions are skewed toward the equatorial plane of the eye (Exner 1891, Horridge 1978, Marshall and Land 1993). Stomatopod eyes may be globular or bean-shaped (Figure 6.5). The overlap of visual axes is one specialization for predation: sets of units in the dorsal and ventral regions can simultaneously spot objects of interest, providing visual range by simple triangulation. This monocular range finder recalls the hardwired, binocular system of the praying mantid (Rossel 1983). Being so mobile and independent, compound eyes of mantis

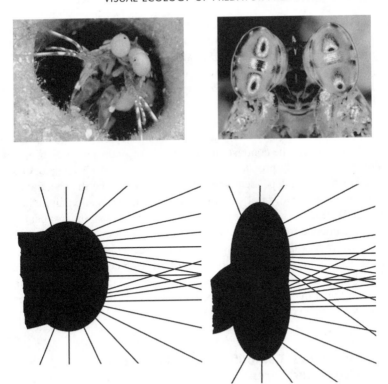

Figure 6.5. Stomatopod eyes and visual axes. Top right, *Haptosquilla trispinosa* in its burrow, showing how the globular eyes and antennules extend out to examine the surrounding waters. Top left, *Raoulserene komai*, to illustrate the triple eyes. Dark areas in each eye indicate the ommatidia facing the camera (because these ommatidia absorb almost all light in the camera's direction of view). Note how the triple organization is exhibited by the right eye, where there is a dark spot in the dorsal and ventral half, as well as in the midband of ommatidia. The left eye is tilted slightly upward, so ommatidia in the midband are directed above the camera's plane. The silhouettes below each photograph show the projected axes of ommatidia from the various eye regions, illustrating how the eye acts as a monocular range finder with given sets of ommatidia viewing objects at various distances from the eye, and with the axes of ommatidia in the midband projecting along this same viewing plane, providing color and polarization vision. (Photograph of *R. komai* courtesy of R. L. Caldwell.)

shrimps are analogous to the range-finding lens eyes of chameleons or sandlances. The mantis shrimp practice of dashing at prey and disabling it with an instantaneous strike makes accurate determination of distance a necessity.

Two receptor sets are sufficient for triangulation. What is the job description of the middle band? This part of the eye is specialized for contrast enhancement by spectral and polarizational analysis. It houses several UV spectral photoreceptor classes (Cronin, Marshall, Quinn, et al. 1994, Marshall and Oberwinkler 1999), eight narrow-band color receptor types (Cronin and Marshall 1989a, Cronin and Marshall

1989b), and a group of polarization receptors (Marshall et al. 1991a). The color classes incorporate anatomical features and colored filters that tune them to particular spectral regions (Cronin and Marshall 1989a, Cronin and Marshall 1989b, Cronin, Marshall, Caldwell, et al. 1994, Cronin et al. 2000). All these visual specializations search for contrast in more than a dozen spectral regions and at least a half-dozen orientations of polarizers.

Do mantis shrimps break the general rule that color vision rarely plays a role in predatory behavior? Probably not. They do recognize color (Marshall et al. 1996), but it seems likely that the contrast-enhancing role of midband photoreceptors produces color vision as a by-product. In all likelihood, the multiple receptor classes act as a series of mostly independent spectral bands (much like the spectral bands of artificial remote sensors), or at most as a series of dichromatic pairs (Cronin and Marshall 2004). Each compound eye of a mantis shrimp is an independent prey detector par excellence that determines prey range, provides spatial vision for prey identification, moves individually to track objects accurately (Cronin et al. 1988), and scans over them to register color and polarization contrast (Land et al. 1990, Cronin and Marshall 2001). Stomatopod vision can be viewed as a singular evolutionary experiment in designing an effective sensor system to direct prey capture in the difficult imaging conditions, under water.

Visual Ploys: Camouflage, Smoke Screens, Decoys, and Lures

Vision's major role in predator-prey interactions provides strong selective pressure on both predators and prey to develop means of being obscure, fostering the evolution of camouflage and cryptic coloration. Many species of prey animals bring other visual defenses into play, hiding behind smoke screens or leaving decoys behind them as they flee. On the other hand, some predators visually lure their prey into traps. In this section, we will consider the biology of some tactics that manipulate visual behavior to the advantage of predators or prey.

Camouflage: The Biology of Obscuration

Effective camouflage looks as much like the background as possible, at least to the target visual system. Of course, animals cannot always predict the background against which they will be viewed, which adds complexity, and theoretical interest, to the problem. Even in the featureless background of the open sea, the simplest situation in nature, brightness varies with direction of view, time of day, and depth, making it extremely difficult for an animal to be invisible in all situations. Three strategies, which can be combined in one individual, are possible: transparency, mirror reflectivity, and bioluminescent counterillumination (Nilsson 1997, Johnsen 2001, Johnsen 2002, Johnsen and Sosik 2003). Figure 6.3A shows a deep-sea fish that combines mirror reflectivity with counterillumination. All of the strategies have their limitations. Transparency means giving up hard parts, being at risk of exposure through visibility of food in the gut, and (in shallow water) assuming the risk of photodamage from UV radiation (Johnsen and Widder 2001). Mirror camouflage cannot be perfectly reflective at all angles (Denton 1970, Johnsen and Sosik 2003) and can be broken with use of polarization vision (Shashar et al. 2000). Bioluminescent counterillumination, the obliteration of a silhouette when seen from below by producing light to replace what is absorbed by the animal (Herring 1977), requires careful control of radiance angle, spectrum, and intensity (Warner et al. 1979, Young and Mencher 1980, Latz 1995). Nevertheless, even moderate reductions in contrast against background provide significant protection against visual predators hunting from below (Land 2000).

Camouflage against a patterned background is far more complicated. The colors, patterns, poses, adornments, and embellishments used in animal camouflage are astonishingly diverse (Cott 1940). Here, I discuss aspects of camouflage that relate to visual processes of predators or prey. Cuttlefish (*Sepia officinalis*) are masters of biological camouflage, producing dozens of patterns and being able to vanish against almost any background present in their world (Hanlon and Messenger 1988). Incredibly, even background colors are often well matched, even though cuttlefish are colorblind (Marshall and Messenger 1996), probably because the colors of underwater objects are often somewhat predictable. Cuttlefish, being both predators on small marine animals and prey to larger ones, such as fish, serve as an outstanding model of the flexible use of camouflage. The visual cues and neurobiological mechanisms they use to generate appropriate patterns are under intense investigation (Chiao and Hanlon 2001, Crook et al. 2002).

Some systems of camouflage use abnormal patterns, subtly co-opting visual processing mechanisms to enhance crypsis. Often, the spot patterns on animals have stronger contrast at their edges, which would seem to make them more visible by outlining each blotch. In reality, the contrast border ties into natural edge-enhancing neural processes, such as lateral inhibition, and increases the visual tendency to see the blotches as normal features of the background (Osorio and Srinivasan 1991). Biological patterns of cryptic coloration can never fully match a background selected at random. However, compromises between perfect and generalized matches apparently incur little penalty (Merilaita et al. 2001), perhaps because visual feature detectors, having evolved to cope with general properties of natural scenes, overlook local variations in pattern that are statistically "correct" for the scene in view. Visual systems are not point-by-point recorders of all scene details; rather, they capture particular statistical aspects of the scene.

A spectacular example of how perceptual limitations enable seemingly impossible camouflage matches to be made is seen in the crab spider, *Thomisus onustus* (Théry and Casas 2002). This species can mimic several flower colors, hiding in the flower to capture visiting bees while simultaneously confounding avian predators. The birds and the bees (and presumably the spiders, too) all have different visual systems, yet the match is good for all. With a limited selection of spider body pigments, it is simply impossible to match all host flowers perfectly. The spider can perform its multispecies, chameleonic matches because color-vision systems perceive many different physical spectra as having identical color. Thus, so long as the general reflectance spectrum of the spider is about right for the color receptors of birds and bees, it will vanish into the flower.

Another impressive camouflage technique, also based on variations in visual systems and exploiting the lighting properties of natural environments, is to use the same body colors to be either prominent or cryptic, depending on the situation. Reef fish are among the most colorful vertebrates, yet the reflectance spectra of many of their beautiful hues match the background when viewed at a distance, because water absorbs long-wavelength light (Marshall 2000a, Marshall 2000b, Marshall and Vorobyev 2003). In addition, the bright patterns seen from nearby (perhaps by a conspecific) blend into a single, dull, reef-like color when viewed from further away (perhaps by a predator). Here, the fish "has its cake and can eat it too," but the predator never even sees the crumbs. Bird plumage has similar versatility. Displayed at the right time and in the right place (for instance, on a tree branch in the sun), it is brilliant and unmistakable, but if the bird moves a few feet into the shade, it suddenly becomes cryptic (Endler and Théry 1996). The brightly colored fur of some primates may produce the same effect (Sumner and Mollon 2003). Trinidadian guppies, *Poecilia reticulata*, optimize their use of signal colors by courting conspecifics under favorable lighting conditions, and by varying the colors used among populations, depending on local lighting environment and the presence of particular predatory species (Endler 1991). It is possible for animals to be bright and conspicuous and to have impressive visual signals, so long as they display in appropriate circumstances. As Endler (1992, 1993, 2000) has pointed out, animals can choose when, where, and how they display to each other. Their choices permit conspecifics to see signals that are denied to predators.

Biological Subterfuges: Smoke Screens, Deterrents, Alarms, and Lures

Prey need not be defenseless victims waiting to be consumed. Even if unarmed, they may bring a diversity of protective mechanisms into play to protect themselves, sometimes going so far as to broadcast a biological request for help. Nor are predators obligated to wait for their victims to wander into range. They can increase the chance of a successful encounter by luring prey into attack range. In this section, I will briefly discuss a few of these biological tricks as they relate specifically to visual predation.

A simple ruse, found in many species, is to use body markings, such as false eyespots, to divert a predator's attack to a less vulnerable body region (Crook 1997). The diversion is even more effective when it is separated from the whole animal. Cephalopods are famous for jetting away in a cloud of ink when threatened. The black, sticky material is released into the mantle cavity and then fired out by the siphon. The ink may form a "pseudomorph," a predator distracter that momentarily may look like prey or a "smoke screen," as the ink is broadly dispersed in a cloud (Hanlon and Messenger 1996). Deep-sea squid, living in waters too dark for ink, instead produce a luminescent fog. Other deep-water animals, such as copepods or shrimps, emit a bright bioluminescent spew as a predator distracter or deterrent. In the void, the brilliant spot of light left behind is about all the predator has left to attack. The flashlight fish, *Photoblepharon palpebratus*, uses an imaginative tactic to confuse its predators. It flashes brilliantly, extinguishes the light, and immediately swims away rapidly. Morin et al. (1975) aptly term this behavior "blink-and-run," and it offers the predator a place to search where no fish exists. A final, particularly oblique use of bioluminescence is as a "burglar alarm." In the presence of an intermediate-level predator such as a shrimp or mysid, copepods or other planktonic prey flash desperately. The response can bring in higher-level predators such as fish or squid, forcing the copepod's enemy to vacate the area (Mensinger and Case 1992, Fleisher and Case 1995).

But bioluminescence shines both ways. Earlier, it was noted that dragon fish may use far-red luminescence as a "snooperscope" to illuminate prey (Widder et al. 1984). The sneaky flashlight fish is thought to use its light to attract photopositive plankton, calling in its prey at will (Morin et al. 1975). A recently discovered deep-sea octopus may do the same with its bioluminescent suckers (Johnsen et al. 1999). The technique could be widespread in planktivores. An escalation of this ruse is to use a bioluminescent organ as a prey mimic, attracting small predators to the presence of a large one. Lures like these are found in deep-sea fish such as anglerfish (Figure 6.6). Shallow-water anglerfish dispense with the bioluminescence but use lures that mimic small worms or injured fish. Terrestrial bioluminescent animals also may use their light as a predatory lure. Female fireflies of the genus *Photuris* mimic the flash patterns of male fireflies of other species, attracting them to their leafy bower and devouring them (Lloyd 1965, Lloyd 1975).

Spiderwebs act as lures in a different sense. Spiders in the genus *Argiope*, famous weavers of huge webs in open air, decorate the webs with UV-reflecting silk (Craig and Bernard 1990). The bright UV spot in the web may resemble a patch of open sky to a passing insect, attracting it into the web. The resident spiders often look very dark in the UV, concealing their presence. The same spiders add thick bands of silk called stabilimenta to the web, which makes them visible to birds flying the same routes as the insects; such features reveal the presence of the web and reduce bird impact damage (Eisner and Nowicki 1983). They also reflect UV, adding to the overall attractiveness of the web to insects, which do not resolve the strands until it is too late to avoid the web.

A recently described behavior of dragonflies also takes advantage of the poor spatial resolution of most insect prey. The Australian species *Hemianax papuensis* disappears through motion, mirroring the movements of its flying prey to remain in a constant position against the background (Mizutani et al. 2003). The shadowing is uncannily accurate and requires exceedingly rapid systems of motion analysis and flight control, but the end result is that the predator is literally invisible to the prey throughout its approach.

Prey That Advertise Their Presence: Aposematism

Prey are normally inconspicuous, but when an attack bears a significant risk to a predator, prey benefit from systems of visual display that warn predators away. *Aposematic* visual signals are those colors, patterns, or postures that signal to predators that prey are unpalatable, toxic, or dangerous. Aposematism introduces interesting

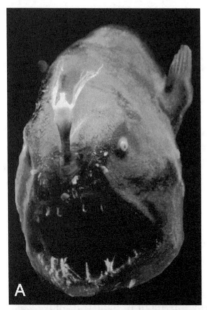

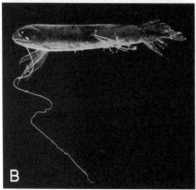

Figure 6.6. Bioluminescent lures in two species of deep-sea fish. (A) An unidentified species of anglerfish: the lure resembles a torch attached to the anterior skull between the eyes and just above the huge mouth. (B) *Bathophilus nigerrimus*: the lure is attached to the end of a long "fishing line" extending down from the lower jaw. Note also the photophore just below the eye, which emits blue light, possibly to attract smaller prey. (Photographs courtesy of E. A. Widder. © 1983, 1999 Harbor Branch Oceanographic Institution.)

evolutionary problems, such as the evolution of systems of mimicry and the maintenance of species-specific signals (to maintain species identity) independent of shared aposematic signals.

Color vision is usually unimportant in predator-prey relations, but in aposematic systems color can play a major role (Sillen-Tullberg 1985). The diversity of aposematic color systems, for instance, in the highly polymorphic populations of the poison-dart frog, *Dendrobates pumilio* (Summers et al. 2003), implies that predators may not be particularly concerned with the specific color of prey, but rather that they avoid colorful prey in general. Thus, the very visibility of colorful prey protects them (Guilford 1986). Many predators are neophobic (i.e., they avoid novel items); for instance, great tits (*Parus major*) are reluctant to attack conspicuous prey they have not encountered previously (Lindstrom et al. 2001). This generalized caution certainly assists in the origin and maintenance of aposematic signals, but an occasional attack must occur for any given system to persist. Baby birds may be born knowing the rule, "don't eat brightly colored frogs," but the rule requires an occasional test to maintain its genetic basis. Thus, aposematic prey, including dart frogs, monarch butterflies, bees, and pufferfish (among many other organisms) may confidently display their stunning colors and patterns to state that they are invulnerable, but they are never entirely free from the risk of an occasional attack. It is likely that predators attack conspicuous prey tentatively at best, reinforcing the impression that the prey are inedible while permitting most attacked prey to survive the halfhearted assault.

Freedom in the choice of signaling color occasionally produces extremely diverse morphotypes, even in single species. In *D. pumilio*, already mentioned, the diversity is thought to result from female choice (Summers et al. 1997), because female frogs discriminate among males on the basis of color (Summers et al. 1999). If so, the color signals must be both highly visible to predator (presumably avian and reptilian) visual systems and discriminable to conspecific visual systems. In fact, because male and female frogs face each other when meeting, females may discriminate by ventral coloration, whereas large predators see dorsal colors. Thus, different signaling systems may be directed at conspecifics and to predators, and this distinction may be rather common in aposematic species.

Conclusions

The fundamental asymmetry of the predator-prey relationship is reflected in many aspects of the visual systems of predatory and prey animals. Predators can specialize in how they locate and track prey, and their eyes are often adapted for high-quality vision in some directions of view, allowing them to detect prey at distances where they may still be visually out of range. Visual systems of prey are frequently designed to detect the presence of a predator at almost any location. Most predatory species are prey themselves for predators at higher trophic levels, so they require visual systems effective for predatory tasks but simultaneously capable of predator detection. Both players require alerting and response systems that are extremely rapid, forcing the evolution of streamlined means of detection and identification. Such streamlining permits the appearance of visual deceits and counterfeits that encourage predators to execute ineffective attacks or that lure prey into harm's way. Animals that are risky, or highly undesirable, as prey incorporate visual signaling systems that make them conspicuous and readily identified, these systems often being directed at different visual systems in several predatory species.

The intent of this review is to note general principles of how vision is adapted for, and how it shapes, predator-prey interactions. As can be seen, much has been learned concerning the ways designs and specializations of eyes and visual systems function to promote survival of both predators and prey. Some features of vision, including binocular vision and color vision, seem surprisingly unimportant in most species in this context. On the other hand, newly recognized contrast-enhancing tactics, including UV vision and polarization vision, are now recognized as significant in a number of species. Major current areas of inquiry include the functional design of systems of camouflage, the techniques animals use to produce conspicuous and effective intraspecific signals in the face of visual predation, and the unique problems facing aposematic animals that must maintain species identity while sending reliable, unambiguous signals to potential predators. The study of how vision functions in predator-prey interactions will continue to offer rich research rewards for the diligent investigator.

Literature Cited

Aho, A. C., Donner, K., Helenius, S., Larsen, L. O., and Rueter, T. 1993. Visual performance of the toad (*Bufo bufo*) at low light levels: retinal ganglion cell responses and prey-catching accuracy. J. Comp. Physiol. A 172:671–682.

Briscoe, A. D., and Chittka, L. 2001. The evolution of color vision in insects. Annu. Rev. Entomol. 46:471–510.

Browman, H. I., Novales-Flamarique, I., and Hawryshyn, C. W. 1994. Ultraviolet photoreception contributes to prey search behaviour in two species of zooplanktivorous fishes. J. Exp. Biol. 186:187–198.

Burton, B. G., and Laughlin, S. B. 2003. Neural images of pursuit targets in the photoreceptor arrays of male and female houseflies *Musca domestica.* J. Exp. Biol. 206:3963–3977.

Burton, G. J., and Moorhead, I. R. 1987. Color and spatial structure in natural scenes. Appl. Optics 26:157–170.

Caldwell, R. L., and Dingle, H. 1975. Ecology and evolution of agonistic behavior in stomatopods. Naturwissenschaften 62:214–222.

Chiao, C. C., Cronin, T. W., and Osorio, D. 2000. Color signals in natural scenes: characteristics of reflectance spectra and effects of natural illuminants. J. Opt. Soc. Am. A 17:218–224.

Chiao, C. C., and Hanlon, R. T. 2001. Cuttlefish camouflage: visual perception of size, contrast and number of white squares on artificial checkerboard substrata initiates disruptive coloration. J. Exp. Biol. 204:2119–2125.

Chiao, C. C., Osorio D., Vorobyev, M., and Cronin, T. W. 2000. Characterization of natural illuminants in forests and the use of digital video data to reconstruct illuminant spectra. J. Opt. Soc. Am. A 17:1713–1721.

Chiao, C. C., Vorobyev, M., Cronin, T. W., and Osorio, D. 2000. Spectral tuning of dichromats to natural scenes. Visi. Res. 40:3257–3271.

Collett, T. S. 1977. Stereopsis in toads. Nature 267:349–351.

Collett, T. S., and Harkness, L. I. K. 1982. Depth vision in animals. In: Analysis of Visual Behavior (Ingle, D. J., Goodale, M. A., and Mansfield, R. J. W., eds.). Cambridge, Mass.: MIT Press; 111–176.

Collin, S. P., and Pettigrew, J. D. 1988a. Retinal topography in reef teleosts. I. Some species with well-developed areae but poorly-developed streaks. Brain Behav. Evol. 31:269–282.

Collin, S. P., and Pettigrew, J. D. 1988b. Retinal topography in reef teleosts. II. Some species with prominent horizontal streaks and high density areae. Brain Behav. Evol. 31:283–295.

Collin, S. P., and Shand, J. 2003. Retinal sampling and the visual field in fishes. In: Sensory Processing in Aquatic Environments (Collin, S. P., and Marshall, N. J., eds.). Berlin: Springer-Verlag; 139–169.

Cott, H. B. 1940. Adaptive Coloration in Animals. London: Methuen.

Craig, C. L., and Bernard, G. D. 1990. Insect attraction to ultraviolet-reflecting spider webs and web decorations. Ecology 71:616–623.

Cronin, T. W. 1986. Optical design and evolutionary adaptation in crustacean compound eyes. J. Crust. Biol. 6:1–23.

Cronin, T. W., Fasick, J. I., and Howland, H. C. 1998. Video photoretinoscopy of the eyes of the small odontocetes *Tursiops truncatus*, *Phocoena phocoena*, and *Kogia breviceps*. Mar. Mam. Sci. 14:584–590.

Cronin, T. W., and Marshall, N. J. 1989a. Multiple spectral classes of photoreceptors in the retinas of gonodactyloid stomatopod crustaceans. J. Comp. Physiol. A 166:267–275.

Cronin, T. W., and Marshall, N. J. 1989b. A retina with at least ten spectral types of photo-receptors in a stomatopod crustacean. Nature 339:137–140.

Cronin, T. W., and Marshall, J. 2001. Parallel processing and image analysis in the eyes of mantis shrimps. Biol. Bull. 200:177–183.

Cronin, T. W., and Marshall, J. 2004. The visual world in mantis shrimps. In: Complex Worlds from Simple Nervous Systems (Prete, F., ed.). Cambridge, Mass.: MIT Press; 239–268.

Cronin, T. W., Marshall, N. J., and Caldwell, R. L. 2000. Spectral tuning and the visual ecol-ogy of mantis shrimps. Phil. Trans. R. Soc. Lond. B 355:1263–1267.

Cronin, T. W., Marshall, N. J., Caldwell, R. L., and Shashar, N. 1994. Specialization of reti-nal function in the compound eyes of mantis shrimps. Visi. Res. 34:2639–2656.

Cronin, T. W., Marshall, N. J., and Land, M. F. 1991. Optokinesis in gonodactyloid mantis shrimps (Crustacea; Stomatopoda; Gonodactylidae). J. Comp. Physiol. A 168:233–240.

Cronin, T. W., Marshall, N. J., Quinn, C. A., and King, C. A. 1994. Ultraviolet photorecep-tion in mantis shrimp. Vis. Res. 34:1443–1452.

Cronin, T. W., Nair, J. N., Doyle, R. D., and Caldwell, R. L. 1988. Visual tracking of rapidly moving targets by stomatopod crustaceans. J. Exp. Biol. 138:155–179.

Crook, A. C. 1997. Determinants of the physiological colour patterns of juvenile parrotfish, *Chlorurus sordidus*. Anim. Behav. 53:1251–1261.

Crook, A. C., Baddeley, R., and Osorio, D. 2002. Identifying the structure in cuttlefish visual signals. Phil. Trans. R. Soc. Lond. B 357:1617–1624.

Curio, E. 1993. Proximate and developmental aspects of antipredator behavior. In: Advances in the Study of Behavior, vol. 22 (Slater, P. J. B., Rosenblatt, J. S., Snowdon, C. T., and Milinski, M., eds.). New York: Academic Press; 135–238.

Dahmen, H. 1991. Eye specialization in waterstriders: an adaptation to life in a flat world. J. Comp. Physiol. A 169:623–632.

Denton, E. J. 1970. On the organization of reflecting structures in some marine animals. Phil. Trans. R. Soc. Lond. B 258:285–313.

Dong, D. W., and Attick, J. J. 1995. Statistics of natural time-varying images. Network 6:345–358.

Douglas, R. H., Hunt, D. M., and Bowmaker, J. K. 2003. Spectral sensitivity tuning in the deep sea. In: Sensory Processing in Aquatic Environments (Collin, S. P., and Marshall, N. J., eds.). Berlin: Springer-Verlag; 323–342.

Douglas, R. H., Partridge, J. C., Dulai, K., Hunt, D., Mullineaux, C. W., Tauber, A. Y., and Hynninen, P. H. 1998. Dragon fish see using chlorophyll. Nature 393:423–424.

Duelli, P. 1978. Movement detection in the posterolateral eyes of jumping spiders (*Evarcha arcuata*, Salticidae). J. Comp. Physiol. 124:15–26.

Eisner, T., and Nowicki, S. 1983. Spider web protection through visual advertisement: role of the stabilimentum. Science 219:185–187.

Endler, J. A. 1991. Variation in the appearance of guppy color patterns to guppies and their predators under different visual conditions. Vis. Res. 31:587–608.

Endler, J. A. 1992. Signals, signal conditions, and the direction of evolution. Am. Nat. 139:S125–S153.

Endler, J. A. 1993. Some general comments on the evolution and design of animal commu-nication systems. Phil. Trans. R. Soc. Lond. B 340:215–225.

Endler, J. A. 2000. Evolutionary implications of the interaction between animal signals and the environment. In: Animal Signals: Signalling and Signal Design in Animal Communication (Espmark, Y., Amundsen, T., and Rosenqvist, G., eds.). Trondheim, Norway: Tapir Academic Press; 11–46.

Endler, J. A., and Théry, M. 1996. Interacting effects of lek placement, display behaviour, ambient light, and color patterns in three neotropical forest-dwelling birds. Am. Nat. 148:421–452.

Ewert, J.-P. 1980. Neuroethology. Berlin: Springer-Verlag.

Ewert, J.-P., Burghagen, H., and Shrg-Pfeiffer, E. 1983. Neuroethological analysis of the innate releasing mechanism for prey-catching behavior in toads. In: Advances in Vertebrate Neuroethology (Ewert, J.-P., Capranica, R. R., and Ingle, D. J., eds.). New York: Plenum Press; 413–475.

Ewert, J.-P., Buxbaum-Conradi, H., Dreisvogt, F., Glagow, M., Merkel-Harff C., Röttgen A., Schürg-Pfeiffer, E., and Schwippert, W. W. 2001. Neural modulation of visuomotor functions underlying prey-catching behaviour in anurans: perception, attention, motor performance, and learning. Comp. Biochem. Physiol. A 128:417–461.

Exner, S. 1891. Die Physiologie der Facettirten Augen von Krebsen und Insecten. Leipzig: Deuticke.

Field, D. J. 1987. Relations between the statistics of natural images and the response properties of cortical cells. J. Opt. Soc. Am. A 11:3011–3019.

Fleisher, K. J., and Case, J. F. 1995. Cephalopod predation facilitated by dinoflagellate luminescence. Biol. Bull. 189:263–271.

Fujita, M. 2003. Head bobbing and the body movement of little egrets (*Egretta garzetta*) during walking. J. Comp. Physiol. A 189:53–58.

Gilbert, C., and Strausfeld, N. J. 1991. The functional organization of male-specific visual neurons in flies. J. Comp. Physiol. A 169:395–411.

Gonka, M. D., Laurie, T. J., and Prete, F. H. 1999. Responses of movement-sensitive visual interneurons to prey-like stimuli in the praying mantis *Sphodromantis lineola* (Burmeister). Brain Behav. Evol. 54:243–262.

Guilford, T. 1986. How do "warning colors" work? Conspicuousness may reduce recognition errors in experienced predators. Anim. Behav. 34:286–288.

Guillemain, M., Martin, G. R., and Fritz, H. 2002. Feeding methods, visual fields and vigilance in dabbling ducks (Anatidae). Funct. Ecol. 16:522–529.

Hanlon, R. T., and Messenger, J. B. 1988. Adaptive coloration in young cuttlefish (*Sepia officinalis* L.): the morphology and development of body patterns and their relation to behaviour. Phil. Trans. R. Soc. Lond. B 320:437–487.

Hanlon, R. T., and Messenger, J. B. 1996. Cephalopod Behaviour. Cambridge: Cambridge University Press.

Harkness, L. 1977. Chameleons use accommodation cues to judge distance. Nature 267:346–349.

Harland, D. P., and Jackson, R. R. 2000. Cues by which *Portia fimbriata*, an araneophagic jumping spider, distinguishes jumping-spider prey from other prey. J. Exp. Biol. 203:3485–3494.

Harland, D. P., and Jackson, R. R. 2002. Influence of cues from the anterior medial eyes of virtual prey on *Portia fimbriata*, an araneophagic jumping spider. J. Exp. Biol. 205:1861–1868

Hart, N. S. 2001. The visual ecology of avian photoreceptors. Prog. Ret. Eye Res. 20:675–703.

Herring, P. J. 1977. Bioluminescence of marine organisms. Nature 267:788–793.

Horridge, G. A. 1978. The separation of visual axes in apposition compound eyes. Phil. Trans. R. Soc. Lond. B 285:1–59.

Jablonski, P. G. 2001. Sensory exploitation of prey: manipulation of the initial direction of prey escapes by a conspicuous "rare enemy." Proc. R. Soc. Lond. B 268:1017–1022.

Jackson, R. R., Pollard, S. D., and Cerveira, A. M. 2002. Opportunistic use of cognitive smokescreens by araneophagic jumping spiders. Anim. Cogn. 5:147–157.

Jacobs, G. H. 1993. The distribution and nature of colour vision among the mammals. Biol. Rev. 68:413–471.

Jacobs, G. H., Deegan, J. F., Crognale, M. A., and Fenwick, J. A. 1993. Photopigments of dogs and foxes and their implications for canid vision. Vis. Neurosci. 10:173–180.

Jerlov, N. G. 1976. Marine Optics. Amsterdam: Elsevier.

Johnsen, S. 2001. Hidden in plain sight: the ecology and physiology of organismal transparency. Biol. Bull. 201:301–318.

Johnsen, S. 2002. Cryptic and conspicuous coloration in the pelagic environment. Proc. R. Soc. Lond. B 269:243–256.

Johnsen, S., Balser, E. J., and Widder, E. A. 1999. Light-emitting suckers in an octopus. Nature 398:113–114.

Johnsen, S., and Sosik, H. M. 2003. Cryptic coloration and mirrored sides as camouflage strategies in near-surface pelagic habitats: implications for foraging and predator avoidance. Limnol. Oceanogr. 48:1277–1288.

Johnsen, S., and Widder, E. A. 2001. Ultraviolet absorption in transparent zooplankton and its implications for depth distribution and visual predation. Mar. Biol. 138:717–730.

Land, M. F. 1965. Image formation by a concave reflector in the eye of the scallop, Pecten maximus. J. Physiol. 179:138–153.

Land, M. F. 1981. The eyes of Phronima, and other deep-sea amphipods. J. Comp. Physiol. 145:209–226.

Land, M. F. 1984a. Crustacea. In: Photoreception and Vision in Invertebrates (Ali, M. A., ed.). New York: Plenum Press; 401–438.

Land, M. F. 1984b. Molluscs. In: Photoreception and Vision in Invertebrates (Ali, M. A., ed.). New York: Plenum Press; 699–725.

Land, M. F. 1985a. Fields of view of the eyes of primitive jumping spiders. J. Exp. Biol. 119:381–384.

Land, M. F. 1985b. The morphology and optics of spider eyes. In: Neurobiology of Arachnids (Barth, F. G., ed.). Berlin: Springer-Verlag; 53–78.

Land, M. F. 1989. The eyes of hyperiid amphipods: relations of optical structure to depth. J. Comp. Physiol. A 164:751–762.

Land, M. F. 1990. Optics of the eyes of marine animals. In: Light and Life in the Sea (Herring, P. J., Campbell, A. K., Whitfield, M., and Maddock, L., eds.). Cambridge: Cambridge University Press; 149–166.

Land, M. F. 1999. The roles of head movements in the search and capture strategy of a tern (Aves, Laridae). J. Comp. Physiol. A 184:265–272.

Land, M. F. 2000. On the functions of double eyes in midwater animals. Phil. Trans. R. Soc. Lond. B 355:1147–1150.

Land, M. F. 2003. The spatial resolution of the pinhole eyes of giant clams (Tridacna maxima). Proc. R. Soc. Lond. B 270:185–188.

Land, M. F., and Barth, F. G. 1992. The quality of vision in the ctenid spider, Cupiennius salei. J. Exp. Biol. 164:227–242.

Land, M. F., and Eckert, H. 1985. Maps of the acute zones of fly eyes. J. Comp. Physiol. A 156:525–538.

Land, M. F., and Layne, J. E. 1995a. The visual control of behaviour in fiddler crabs. I. Resolution, thresholds, and the role of the horizon. J. Comp. Physiol. A 177:81–90.

Land, M. F., and Layne, J. E. 1995b. The visual control of behaviour in fiddler crabs. II.
Tracking control systems in courtship and defence. J. Comp. Physiol. A 177:91–103.

Land, M. F., Marshall, N. J., Brownless, D., and Cronin, T. W. 1990. The eye-movements of
the mantis shrimp Odontodactylus scyllarus (Crustacea: stomatopods). J. Comp. Physiol.
A 167:155–166.

Land, M. F., and Nilsson, D.-E. 2002. Animal Eyes. Oxford: Oxford University Press.

Latz, M. I. 1995. Physiological mechanisms in the control of bioluminescent countershading
in a midwater shrimp. Mar. Fresh. Behav. Physiol. 26:207–218.

Layne, J. E. 1998. Retinal location is the key to identifying predators in fiddler crabs (Uca
pugilator). J. Exp. Biol. 201:2253–2261.

Layne, J. E., Land, M. F., and Zeil, J. 1997. Fiddler crabs use the visual horizon to distinguish predators from conspecifics: a review of the evidence. J. Mar. Biol. 77:43–54.

Lindstrom, L., Alatalo, R. V., Lyytinen, A., and Mappes, J. 2001. Predator experience on
cryptic prey affects the survival of conspicuous aposematic prey. Proc. R. Soc. Lond. B
268:357–361.

Litwiler, T. L., and Cronin T. W. 2001. No evidence of accommodation in the eye of the
bottlenose dolphin, Tursiops truncatus. Mar. Mam. Sci. 17:508–525.

Lloyd, J. E. 1965. Aggressive mimicry in Photuris: firefly femmes fatales. Science 149:653–
654.

Lloyd, J. E. 1975. Aggressive mimicry in Photuris fireflies: signal repertoires by femmes
fatales. Science 187:452–453.

Loew, E. R., McAlary, F. A., and McFarland, W. N. 1996. Ultraviolet sensitivity in the lar-
vae of two species of marine atherinid fishes. In: Zooplankton Sensory Ecology and
Physiology (Lenz, P. H., Hartline, D. K., Purcell, J. E., and Macmillan, D. L., eds.).
Sydney: Gordon & Breach; 195–209.

Losey, G. S., Cronin, T. W., Goldsmith, T. H., Hyde, D., Marshall, N. J., and McFarland,
W. N. 1999. The UV visual world of fishes: a review. J. Fish Biol. 54:921–943.

Lythgoe, J. N. 1968. Visual pigments and visual range underwater. Vis. Res. 8:997–1012.

Lythgoe, J. N. 1979. The Ecology of Vision. Oxford: Clarendon Press..

Lythgoe, J. N., and Hemmings, C. C. 1967. Polarized light and underwater vision. Nature
213:893–894.

Marshall, N. J. 2000a. Communication and camouflage with the same "bright" colour in reef
fishes. Phil. Trans. R. Soc. Lond. B 355:1243–1248.

Marshall, N. J. 2000b. The visual ecology of reef fish colours. In: Animal Signals: Signalling
and Signal Design in Animal Communication (Espmark, Y., Amundsen, T., and
Rosenqvist, G., eds.). Trondheim, Norway: Tapir Academic Press; 83–120.

Marshall, N. J, Jones, J. P., and Cronin, T. W. 1996. Behavioural evidence for color vision in
stomatopod crustaceans. J. Comp. Physiol. A 179:473–481.

Marshall, N. J., and Land, M. F. 1993. Some optical features of the eyes of stomatopods. I.
Eye shape, optical axis and resolution. J. Comp. Physiol. A 173:565–582.

Marshall, N. J., Land, M. F., King, C. A., and Cronin, T. W. 1991a. The compound eyes of
mantis shrimps (Crustacea, Hoplocarida, Stomatopoda). I. Compound eye structure: the
detection of polarized light. Phil. Trans. R. Soc. Lond. B 334:33–56.

Marshall, N. J., Land, M. F., King, C. A., and Cronin, T. W. 1991b. The compound eyes of
mantis shrimps (Crustacea, Hoplocarida, Stomatopoda). II. Colour pigments in the eyes
of Stomatopod crustaceans: polychromatic vision by serial and lateral filtering. Phil.
Trans. R. Soc. Lond. B 334:57–84.

Marshall, N. J., and Messenger, J. B. 1996. Colour-blind camouflage. Nature 382:408–409.

Marshall, N. J., and Oberwinkler, J. 1999. The colourful world of the mantis shrimp. Nature 401:873–874.

Marshall, N. J., and Vorobyev, M. 2003. Color signals and color vision in fishes. In: Sensory Processing in Aquatic Environments (Collin, S. P., and Marshall, N. J., eds.). Berlin: Springer-Verlag; 194–222.

Martin, G. R. 1984. The visual fields of the tawny owl, *Strix aluco* L. Vis. Res. 24:1739–1751.

Martin, G. R. 1986. Total panoramic vision in the mallard duck, *Anas platyrhynchos*. Vis. Res. 26:1303–1306.

Martin, G. R. 1994. Visual fields in woodcocks *Scolopax rusticola* (Scolopacidae; Charadriiformes). J. Comp. Physiol. A 174:787–793.

Martin, G. R., and Katzir, G. 1994. Visual fields and eye movements in herons (Ardeidae). Brain Behav. Evol. 44:74–85.

Martin, G. R., and Katzir, G. 1999. Visual fields in short-toed eagles, *Circaetus gallicus* (Accipitridae), and the function of binocularity in birds. Brain Behav. Evol. 53:55–66.

Martin, G. R., and Prince, P. A. 2001. Visual fields and forging in procellariiform seabirds: sensory aspects of dietary segregation. Brain Behav. Evol. 57:33–38.

Mass, A. M., and Supin, A. Y. 1995. Ganglion cell topography of the retina in the bottlenosed dolphin, *Tursiops truncates*. Sensor. Sist. 11:256–293.

Mass, A. M., Supin, A. Y., and Severtsov, A. N. 1986. Topographic distribution of sizes and density of ganglion cells in the retina of a Porpoise, *Phocoena phocoena*. Aquat. Mam. 12:95–102.

McFarland, W. N., and Loew, E. R. 1994. Ultraviolet visual pigments in marine fishes of the family Pomacentridae. Vis. Res. 34:1393–1396.

McFarland, W. N., and Munz, W. R. 1975. The evolution of photopic visual pigments in fishes. Vis. Res. 15:1071–1080.

Mensinger, A. F., and Case, J. F. 1992. Dinoflagellate luminescence increases susceptibility of zooplankton to teleost predation. Mar. Biol. 112:207–210.

Merilaita, S., Lyytinen, A., and Mappes, J. 2001. Selection for cryptic coloration in a visually heterogeneous habitat. Proc. R. Soc. Lond. B 268:1925–1929.

Meyer, D. B. 1977. The avian eye and its adaptations. In: The Visual System in Vertebrates. Handbook of Sensory Physiology 7/5 (Crescitelli, F., ed.). Berlin: Springer-Verlag; 549–611.

Mizutani, A., Chahl, J. S., and Srinivasan, M. V. 2003. Motion camouflage in dragonflies. Nature 423:604.

Morin, J. G., Harrington, A., Nealson, K., Kreiger, N., Baldwin, T. O., and Hastings, J. W. 1975. Light for all reasons: versatility in the behavioral repertoire of the flashlight fish. Science 190:74–76.

Neitz, J., Geist, T., and Jacobs, G. H. 1989. Color vision in the dog. Vis. Neurosci. 3:119–25.

Nilsson, D.-E. 1994. Eyes as optical alarm systems in fan worms and ark clams. Phil. Trans. R. Soc. Lond. B 346:195–212.

Nilsson, D.-E. 1997. Eye design, vision and invisibility in planktonic invertebrates. In: Zooplankton: Sensory Ecology and Physiology (Lenz, P. H., Hartline, D. K., Purcell, J. E., and Macmillan, D. L., eds.). Sydney: Gordon & Breach; 149–162.

Nilsson, D.-E., and Modlin, R. F. 1994. A mysid shrimp carrying a pair of binoculars. J. Exp. Biol. 189:213–236.

Nilsson, D.-E., and Odselius, R. 1982. A pronounced fovea in the eye of a water flea, revealed by stereographic mapping of ommatidial axes. J. Exp. Biol. 99:473–476.

Nilsson, D.-E., and Pelger, S. 1994. A pessimistic estimate of the time required for an eye to evolve. Proc. R. Soc. Lond. B 256:54–58.

O'Day. W. T., and Fernandez, H. R. 1974. *Aristostomias scintillans* (Malacosteidae): a deep-sea fish with visual pigments apparently adapted to its own bioluminescence. Vis. Res. 14:545–550.

Olshausen, B. A., and Field, D. J. 1996. Emergence of simple-cell receptive field properties by learning a sparse code for natural images. Nature 381:607–609.

Osorio, D., Ruderman, D. L., and Cronin, T. W. 1998. Estimation of the effects of natural chromatic stimuli on luminance signals encoded by primate retina. J. Opt. Soc. Am. A 15:16–22.

Osorio, D., and Srinivasan, M. V. 1991. Camouflage by edge enhancement in animal coloration patterns and its implications for visual mechanisms. Proc. R. Soc. Lond. B 244:81–85.

Ott, M., and Schaeffel, F. 1985. A negatively powered lens in the chameleon. Nature 373:692–694.

Parker, A. R. 1998. Colour in Burgess Shale animals and the effect of light on evolution in the Cambrian. Proc. R. Soc. Lond. B 265:967–972.

Parker, A. R. 2000. 515 million years of structural colour. J. Opt. A Pure Appl. Opt. 2:R15–R28.

Peitsch, D., Fietz, A., Hertel, H., de Souza, J., Ventura, D. F., and Menzel, R. 1992. The spectral input systems of hymenopteran insects and their receptor-based colour vision. J. Comp. Physiol. A 170:23–40.

Pettigrew, J. D., Collin, S. P., and Ott, M. 1999. Convergence of specialized behaviour, eye movements and visual optics in the sandlance (Teleostei) and the chameleon (Reptilia). Curr. Biol. 9:421–424.

Prete, F. R. 1999. Prey recognition. In: The Praying Mantids (Prete, F. R., Wells, H., Wells, P. H., and Hurd, L. E., eds.). Baltimore, Md.: Johns Hopkins University Press; 141–179.

Prete, F. R., Hurd, L. E., Branstrator, D., and Johnson, A. 2002. Responses to computer-generated visual stimuli by the male praying mantis, *Sphodromantis lineola* (Burmeister). Anim. Behav. 63:503–510.

Rossel, S. 1980. Foveal fixation and tracking in the praying mantis. J. Comp. Physiol. 139:307–331.

Rossel, S. 1983. Binocular stereopsis in an insect. Nature 302:821–822.

Rossel, S., Corlija, J., and Schuster, S. 2002. Predicting three-dimensional target motion: how archer fish determine where to catch their dislodged prey. J. Exp. Biol. 205:3321–3326.

Ruderman, D. L. 1994. The statistics of natural images. Netw. Comp. Neur. Sys. 5:517–548.

Ruderman, D. L., and Bialek, W. 1994. Statistics of natural images: scaling in the woods. Phys. Rev. Lett. 73:814–817.

Ruderman, D. L., Cronin, T. W., and Chiao, C. C. 1998. Statistics of cone responses to natural images: implications for visual coding. J. Opt. Soc. Am. A 15:2036–2045.

Schaeffel, F., Murphy, C. J., and Howland, H. C. 1999. Accommodation in the cuttlefish (*Sepia officinalis*). J. Exp. Biol. 202:3127–3134.

Schwind, R. 1980. Geometrical optics of the *Notonecta* eye: adaptations to optical environment and way of life. J. Comp. Physiol. 140:59–68.

Shashar, N., Hagan, R., Boal, J. G., and Hanlon, R. T. 2000. Cuttlefish use polarization sensitivity in predation on silvery fish. Vis. Res. 40:71–75.

Shashar, N., Hanlon, R. T., and Petz, A. M. 1998. Polarization vision helps detect transparent prey. Nature 393:222–223.

Sherk, T. E. 1977. Development of the compound eyes of dragonflies (Odonata). I. Larval compound eyes. J. Exp. Zool. 201:391–416.

Sherk, T. E. 1978. Development of the compound eyes of dragonflies (Odonata). III. Adult compound eyes. J. Exp. Zool. 203:61–80.

Sillen-Tullberg, B. 1985. The significance of coloration per se, independent of background, for predator avoidance of aposematic prey. Anim. Behav. 33:1382–1384.

Sivak, J. G., and Warburg, M. R. 1983. Changes in the optical properties of the eye during metamorphosis of an anuran, *Pleobates syriacus*. J. Comp. Physiol. 150:329–332.

Summers, K., Bermingham, E., Weigt, L., McCafferty, S., and Dahlstrom, L. 1997. Phenotypic and genetic divergence in three species of dart-poison frogs with contrasting parental behaviour. J. Heredity 88:8–13.

Summers, K., Cronin, T. W., and Kennedy, T. 2003. Variation in spectral reflectance among populations of *Dendrobates pumilio*, the strawberry poison frog, in the Bocas del Toro Archipelago, Panama. J. Biogeogr. 30:35–53.

Summers, K., Symula, R., Clough, M., and Cronin, T. W. 1999. Visual mate choice in poison frogs. Proc. R. Soc. Lond. B 266:2141–2145.

Sumner, P., and Mollon, J. D. 2003. Colors of primate pelage and skin: objective assessment of conspicuousness. Am. J. Primatol. 59:67–91.

Théry, M., and Casas, J. 2002. Predator and prey views of spider camouflage. Nature 415:133.

Toh, Y., and Okamura, J.-Y. 2001. Behavioural responses of the tiger beetle larva to moving objects: role of binocular and monocular vision. J. Exp. Biol. 204:615–625.

Tolhurst, D. J., Tadmor, Y., and Chao, T. 1992. Amplitude spectra of natural images. Opthal. Physiol. Opt. 12:229–232.

Tucker, V. A. 2000. The deep fovea, sideways vision and spiral flight paths in raptors. J. Exp. Biol. 203:3745–3754.

van der Schaaf, A., and van Hateren, J. H. 1996. Modeling the power spectra of natural images: statistics and information. Vis. Res. 36:2759–2770.

van Hateren, J. H. 1993. Spatial, temporal, and spectral preprocessing for colour vision. Proc. R. Soc. Lond. B 251:61–68.

van Hateren, J. H., Hardie, R. C., Rudolph, A., Laughlin, S. B., and Stavenga, D. G. 1989. The bright zone, a specialized dorsal eye region in the male blowfly *Chrysomyia megacephala*. J. Comp. Physiol. A 164:297–308.

van Hateren, J. H., and van der Schaaf, A. 1996. Temporal properties of natural scenes. In: Proceedings of SPIE, vol. 2657: Human Vision and Electronic Imaging (Rogowitz, B. E., and Allebach, J. P., eds.). Bellingham, Wash.: SPIE Press; 139–143.

Via, S. E. 1977. Visually mediated snapping in the bulldog ant: a perceptual ambiguity between size and distance. J. Comp. Physiol. 121:33–51.

Viitala, J., Korpimäki, E., Palokangas, P., and Koivula, M. 1995. Attraction of kestrels to vole scent marks visible in ultraviolet light. Nature 373:425–527.

Walls, G. L. 1942. The Vertebrate Eye and Its Adaptive Radiation. Bloomington Hills, Mich.: Cranbrook Institute.

Warner, J. A., Latz, M. I., and Case, J. F. 1979. Cryptic bioluminescence in a midwater shrimp. Science 203:1109–1110.

Warrant, E. 2000. The eyes of deep-sea fishes and the changing nature of visual scenes with depth. Phil. Trans. R. Soc. Lond. B 355:1155–1159.

Warrant, E. J., Collin, S. P., and Locket, N. A. 2003. Eye design and vision in deep-sea fishes. In: Sensory Processing in Aquatic Environments (Collin, S. P., and Marshall, N. J., eds.). Berlin: Springer-Verlag; 303–322.

Webster, M. A., and Mollon, J. D. 1997. Adaptation and the color statistics of natural images. Vis. Res. 37:3283–3298.

Webster, M. A., Wade, A., and Mollon, J. D. 1996. Color in natural images and its implications for visual adaptation. In: Proceedings of SPIE, vol. 2657: Human Vision and Electronic Imaging (Rogowitz, B. E., and Allebach, J. P., eds.). Bellingham, Wash.: SPIE Press; 144–152.

Werner, C., and Himstedt, W. 1984. Eye accommodation during prey capture in salamanders (*Salamandra salamandra* L.). Behav. Brain Res. 12:69–73.

Widder, E. A., Latz, M. I., Herring, P. J., and Case, J. F. 1984. Far red bioluminescence from two deep-sea fishes. Science 225:512–514.

Williams, D. R., Sekiguchi, N., Haake, W., Brainard, D., and Packer, O. 1991. The cost of trichromacy for spatial vision. In: From Pigments to Perception (Valberg, A., and Lee, B. B., eds.). New York: Plenum Press; 11–22.

Young, R. E., and Mencher, F. M. 1980. Bioluminescence in mesopelagic squid: diel color change during counterillumination. Science 208:1286–1288.

Young, S. 1988. Chasing with a model eye. J. Exp. Biol. 137:399–409.

Zeil, J. 1998. Homing in fiddler crabs (*Uca lacteal annulipes* and *Uca vomeris*, Ocypodidae). J. Comp. Physiol. A 183:367–377.

Zeil, J., and Al-Mutairi, M. M. 1996. The variation of resolution and of ommatidial dimensions in the compound eyes of the fiddler crab *Uca lacteal annulipes* (Ocypodidae, Brachyura, Decapoda). J. Exp. Biol. 199:1569–1577.

Zeil, J., Nalbach, G., and Nalbach, H.-O. 1986. Eyes, eye stalks, and the visual world of semi-terrestrial crabs. J. Comp. Physiol. A 159:801–811.

Zeil, J., Nalbach, G., and Nalbach, H.-O. 1989. Spatial vision in a flat world: optical and neural adaptations in arthropods. In: Neurobiology of Sensory Systems (Singh, R. N., and Strausfeld, N., eds.). New York: Plenum Press; 123–137.

Zeil, J., and Zanker, J. M. 1997. A glimpse into crabworld. Vis. Res. 37:3417–3426.

7

The Production and Appropriation of Chemical Signals among Plants, Herbivores, and Predators

MATTHEW H. GREENSTONE
JOSEPH C. DICKENS

Interactions among plants, prey, and predators are shaped by a complex blend of interspecific chemical signals (allomones, synomones, and kairomones) and intraspecific chemical signals (pheromones). Prey-induced chemical cues are prominent among chemicals released by plants that mediate predator-prey interactions among arthropods. These chemical cues, like the alarm, marking, sex, and aggregation pheromones produced by prey and predators, may be appropriated for communication by other players in the system. Hence, there are trade-offs for the signalers in producing them.

The behavioral outcomes, mediated by predator-prey chemical communication, are determined largely by the capacities and constraints of sensory receptors. In addition, the role of chemical communication in predator-prey interactions is determined by the higher order processing of information in the brain. Further, traits involved in semiochemical-mediated interactions may be heritable and responsive to selection. However, conclusive evidence for coevolutionary interchanges in these interactions is lacking. Research addressing this issue has focused heavily on intensely managed agricultural and forested habitats.

The primary modality for communication among arthropods is olfactory (Greenfield 2002). Arthropod predators and their prey operate within a rich chemical environment comprising signals generated by them, host plants, and other organisms in all trophic levels (Vet and Dicke 1992). Plants, in addition to having constitutive defensive compounds (Nishida and Fukami 1990), also release volatile chemical signals upon attack by pathogens or herbivores that induce neighboring plants to trigger mechanisms that resist attack (Shulaev et al. 1997, Karban et al. 2000) or volatiles that attract and arrest arthropod natural enemies to aid in plant defense (Dicke and Sabelis 1988a, Turlings et al. 1990, Drukker et al. 1995, Guerreri et al. 1997). Not surprisingly, herbivores may also use these induced volatiles to locate host plants (Loughrin et al. 1995, Bolter et al. 1997) or to avoid them, possibly because natural enemies may have been recruited to protect the plant (Bernasconi Ockroy et al. 1998, Kessler and Baldwin 2001). Further, herbivores also may use constitutive odors of plants as host-finding cues (Kielty et al. 1996, Rapusas et al. 1996). Predators, in addition, may use other types of chemical messengers to their advantage. For example, they may use sex pheromones to locate or attract their prey (Stowe et al. 1987, Mendel et al. 1995). Though this suite of interactions has been described as tritrophic, the trophic position of any organism depends upon the role played by other species, and that role may vary in time and space (Sabelis et al. 1999).

Our purpose in this chapter is not to provide a comprehensive review of the semiochemical literature, which is vast; rather, it is to give the reader an appreciation of the scope of the field, discover unexplored subjects, and pose some questions for future research.

Terminology

Communication among organisms and even between the cells within organisms involves chemical signals that possess specificity and durability. At the interorganismal level, these chemical messengers are referred to as *semiochemicals,* and they convey specific information about the sender, information which increases the likelihood of a change in behavior or physiology in the receiver (Law and Regnier 1971). Another term, *infochemical* (Dicke and Sabelis 1988b), was coined essentially as a replacement for *semiochemical,* but according to key-word searches has not gained broad acceptance. (A search of Zoological Record 1990–2004 for the term *semiochemical* produces 112 hits, whereas *infochemical* produces 17 hits; a search of Cambridge Scientific Internet Database 1990–2004 for *semiochemical* produces 134 hits, whereas *infochemical* produces 15 hits.)

If a change in behavior or physiology in a receiver species is realized over the short term, the semiochemical is referred to as a *releaser,* but if the effect is realized over the long term it is referred to as a *primer* (Wilson and Bossert 1963). These signals have been further subdivided in accordance with whether the message is intraspecific (*pheromones*; Karlson and Butenandt 1959) or interspecific (*allelochemics*; Whittaker and Feeny 1971). Allelochemics are further classified in accordance with whether an adaptive advantage lies with the receiver (*kairomones*), the sender (*allomones*; Brown et al. 1970), or both (*synomones*; Nordlund and Lewis 1976). In the context of predator-prey relationships, pheromones released by prey to attract conspecifics for mating also may act as kairomones for predators and be used to locate a potential meal (Figure 7.1). Volatiles released by plants damaged by insect feeding attract predators and parasitoids that feed or oviposit on the herbivore (Dicke et al. 1990, Turlings et al. 1990, Dickens 1999). Because the plant is the sender, these volatiles are synomones, as both the plant and the natural enemy of the herbivore benefit. This terminology is useful for the classification of chemical signals, facilitates literature searches, and informs our search for evolutionary patterns. However, it is important to carefully note the effects of the signal and to specify the species involved in the communication. A recent attempt has been made to formalize additional classes of kairomones (Ruther et al. 2002), but this further elaboration seems unnecessary.

The enormous literature on the nature, timing, and function of pheromones has been thoroughly covered elsewhere and is beyond the scope of this chapter. Nevertheless, pheromones are the raw material for much of the ensuing discussion of the evolutionary interplay between signalers and receivers, and a general familiarity with them is essential to understanding what follows (Alves 1988, Cardé and Minks 1997, Greenfield 2002).

Overview of Multitrophic Semiochemical Interactions

Allomones and pheromones released by plants, predators, and herbivores may be exploited by other species in multitrophic systems. Hence, for each species there are trade-offs in producing these compounds.

Sources of Semiochemicals

The signaler may, de novo, sythesize semiochemicals, or sequester them, either unchanged or modified, from plants or other organisms in the diet (Nishida 2002). Taxonomy is not a reliable guide to whether an organism sequesters or manufactures semiochemicals. For example, leaf beetles (Chrysomelidae) are well known for sequestering host compounds, but *Linaeidea aenea*, an alder herbivore, produces a general defensive compound de novo (Kopf et al. 1997), and some, but not all, coccinellids synthesize their own defensive compounds (Creygnier and Hodek 1996).

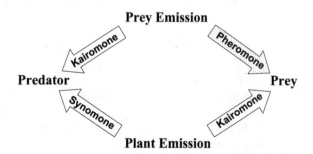

Figure 7.1. Diagram to show relationship among chemical signals, and terminology used to describe their effects, based on the receiver. For example, a blend of volatiles released by an insect and perceived as a *pheromone* by conspecifics is a *kairomone* when detected by a predator. Plant emissions received by a predator are *synomones* because both organisms benefit. These plant emissions may also enhance the pheromone of the prey and, thus, be received as a *kairomone* in this interaction. Open arrows = interspecific signal; shaded arrow = intraspecific signal.

The phenomenon of sequestration of plant compounds was recognized early in the study of insect chemical communication (Rothschild 1961) and is probably the rule rather than the exception for representatives of most major insect orders (Bowers 1990, Nishida 2002). The best-known cases of sequestration for defensive purposes by both larvae and adults involve acquisition of compounds by consumption of larval food plants. However, adult males of some lepidopteran species sequester chemicals from nonhost plants for use as sex pheromones (Eisner and Meinwald 1995). Such sequestration of chemicals from nonhost plants, known as *pharmacophagy*, has been recorded also in Diptera and Coleoptera (Nishida and Fukami 1990). One of the most dramatic cases involves diabroticite beetles consuming cucurbitacins. These chemicals are compounds that, though bitter and highly toxic to most invertebrate and vertebrate herbivores, are phagostimulants for diabroticite beetles (Tallamy and Halaweish 1993).

Chemical Signals Produced by Plants

Plants protect themselves from herbivores by producing defensive compounds (allomones) that are repellent, are lethally toxic, or reduce consumer fecundity (Schmutterer and Ascher 1987, Bowers 1990, Tallamy and Halaweish 1993, Miller and Borden 2000). Although these compounds are generally thought of as being constitutive, their concentration may be responsive to herbivore attack (Kielkiewicz and de Vrie 1990, Agrawal et al. 1999). Plant defensive compounds may have other important effects on herbivores. For example, they may enhance the activity of sex pheromones of moths and other herbivorous insects (Dickens et al. 1990, Dickens 1991, Dickens et al. 1993). Some constitutively produced plant chemicals attract generalist predators, as well (Flint et al. 1979, Reid and Lampman 1989). Others can act to increase or decrease the attractiveness of herbivore aggregation pheromones to the herbivore or its specialized predators (Dickens et al. 1991, Dickens et al. 1992, Erbilgin and Raffa 2001).

Even a plant that is extremely well protected chemically may suffer the depredations of an herbivore if feeding on alternative plants exposes the herbivore to intense natural enemy attack. For example, the sphingid moth *Manduca sexta* is mainly restricted to plants of the family Solanaceae, and is commonly found on *Datura wrightii* in the southwestern United States. *Datura wrightii* is suitable for oviposition and development of the moth, but eggs and larvae suffer heavy predation by birds and arthropods when associated with the plant. This fact may explain why *M. sexta* has made a recent shift to the myrtinaceous herb *Proboscidea parviflora*, on which it suffers very little predation, but on which both egg hatch rates and larval survivorship are sharply reduced because of the chemical properties of the plant (Mira and Bernays 2002). The role of chemical messengers affecting either herbivores or predators in these ecological interactions is probably significant but is unknown.

Plants being consumed release herbivore-induced volatile chemical bouquets, most often composed of mixtures of terpenes, short chain (C_5–C_8) alcohols, aldehydes, or benzenoid compounds (Takabayashi and Dicke 1996, Degenhardt et al.

2003, Schmelz et al. 2003). These volatiles may attract and arrest natural enemies that may help defend the plant. Because natural enemies of herbivores are drawn to these volatiles, they have been interpreted as synomones (Dicke and Sabelis 1988a). Alternatively, these compounds might function in direct chemical defense against herbivores and, thus, could be interpreted as plant allomones that are being exploited by natural enemies "eavesdropping" on plant signals (Stowe et al. 1995). Although some of the most thoroughly studied examples of these types of responses involve hymenopteran parasitoids (Turlings et al. 1990, Guerreri et al. 1997, De Moraes et al. 1998, Du et al. 1998, Turlings et al.1998), there is an expanding literature on similar behavior among arachnid and insect predators (Dicke and Sabelis 1988b, Dicke et al. 1990, Drukker et al. 1995, Mayland et al. 2000, Bernasconi Ockroy et al. 2001, Kessler and Baldwin 2001, Han and Chen 2002, James 2003).

Both *Podisus maculiventris* and *P. bioculatus* detect many of the same volatiles that are detected by their prey, *Leptinotarsa decemlineata*, but antennal receptors of the predators are more sensitive to systemic volatiles produced by the plant in response to feeding by their prey (Dickens 1999). *L. decemlineata* is more sensitive to constitutive green-leaf volatiles (GLVs) released at the site of feeding. The GLVs are six carbon alcohols, aldehydes, and their derivatives (e.g., acetates) that are produced by plants as a product of oxidation of surface lipids (Visser et al. 1979). A five-component blend of volatiles comprising (*E*)-2-hexen-1-ol, (*Z*)-3-hexen-1-ol, nonanal, (±)-linalool, and methyl salicylate attracted the generalist predator *P. maculiventris* in behavioral assays in the lab. This combination of odorants attracts *L. decemlineata* only when the amount of GLVs in the blend is reduced. Avoidance of high concentrations of GLVs, such as those likely to occur at feeding sites, would facilitate spacing of beetles on the plant for more efficient resource utilization and perhaps avoidance of predators. In contrast, antennal receptors of the specialist predator *P. bioculatus* were sensitive to low concentrations of sesquiterpenes emitted by damaged potato plants (Weissbecker et al. 2000).

Like parasitoids (Olson et al. 2003), predators are capable of associative learning and can differentiate plant species by the subtly different chemical blends they produce when under attack by the same herbivore species. This ability enables the predators to remain in a patch that has been especially productive for them (Takabayashi et al. 1994).

Herbivore-induced volatiles may attract other plant-feeding species (Harari et al. 1994), which could be viewed either as having misappropriated a synomone produced by a plant to recruit natural enemies, or as having exploited another organism's allomones for use as host-finding kairomones. In either case, this additional herbivory would exert countervailing selection against production of induced volatiles. Plants run the risk that induced volatiles may be used as kairomones by hyperpredators (i.e., top predators) and parasitoids of predators, which could reduce the effectiveness of herbivore natural enemies attracted by the volatiles (Sabelis et al. 1999). Plants also produce herbivore alarm pheromones that, by causing herbivores to scatter (Gibson and Pickett 1983), reduce damage directly. In the process, of course, they become allomones for the plant. Finally, plants might "eavesdrop" on plants of other species in their immediate environment, detecting damage-induced volatiles and ramping up production of protective chemicals of their own (Karban and Maron 2002).

Chemical Signals Produced by Herbivores and Predators

Sex Pheromones

Because sex pheromones attract mates over long distances, they put the signaler at risk of attack by predators that exploit them as prey-finding kairomones. Hendrichs et al. (1994) showed that yellowjackets, *Vespula germanica*, were able to locate lekking medflies, *Ceratitis capitata*, among dense foliage in the absence of visual cues. Subsequent research showed that females attracted to lekking males were also at increased risk of wasp predation (Hendrichs and Hendrichs 1998). Similarly, the lacewings *Chrysopa cognata* and *Chrysoperla carnea*, the lady beetle *Coleomegilla maculata*, and the click beetle *Elater ferruigineus* all respond to components of the sex pheromones of their prey (Boo et al. 1998, Zhu et al. 1999, Svensson et al. 2004).

Most studied cases of pheromones being used as predator kairomones involve movement by the predators along odor plumes to signaling individuals or groups of prey. Sedentary bolas spiders of the genus *Mastophora*, by generating the prey's own sex pheromones, cause receptive prey to move to them (Stowe et al. 1987). Adaptation can be exquisite: not only major pheromone components, but also the proper blends of pheromone components may be released (Haynes and Yeargan 1999).

Aggregation Pheromones

Aggregation pheromones are found in many insects but are especially well studied in beetles (Bartelt 1999, Peng et al. 1999, Cossé and Bartelt 2000). Bark beetles overcome defenses of potential host trees by attacking them en masse, using aggregation pheromones that are enhanced by monoterpenes released by the trees during their resinous defense. The simultaneous arrival of predaceous clerid beetles on these trees is mediated by aggregation-pheromone components of their prey (Wood et al. 1968, Vité and Williamson 1970, Bakke and Kvamme 1981). For example, *Thanasimus dubius* has receptor neurons housed within sensilla on the antennae that detect mainly (–)-frontalin, the principal component of the pheromone of its prey *Dendroctonus frontalis* (Payne et al. 1984). The ability of predators to respond differentially not only to pheromone enantiomers (compounds whose structures are mirror images), but also to enantiomers of host-tree monoterpenes acting as pheromonal synergists (Erbilgin and Raffa 2001), allows the predator to locate its cryptic prey in a variable forest environment.

While *T. dubius* responds preferentially to the *D. frontalis* pheromone, it also detects and responds behaviorally to pheromone components of cohabiting species (Mizelle and Nebeker, 1982, Payne et al. 1984). Its congener *T. formicarius* has individual receptor neurons tuned to host-tree monoterpenes and pheromone components of *Ips* bark beetles, as well as to (+)-lineatin (Tømmerås 1985), the principal pheromone component of the striped ambrosia beetle, *Trypodendron lineatum* (MacConnel et al. 1977). Both (+)-lineatin and a mixture of *Ips* pheromone components are attractive to the predator (Tømmerås 1985).

Alarm Pheromones

In the presence of a predator, many arthropods produce alarm pheromones that cause them to scatter or become less conspicuous. However, these compounds may be exploited by enemies as kairomones. The carabid beetle *Pterostichus melanarius* and the lacewing *Chrysoperla carnea* respond to components of the alarm pheromones of aphid prey (Kielty et al. 1996, Zhu et al. 1999). Larvae of the western flower thrips produce an alarm pheromone that alters foraging behavior of two arthropod predators, the anthocorid bug *Orius tristicolor* and the phytoseiid mite *Amblyseius cucumeris*, causing them to either narrow their search or avoid leaving the area, thereby increasing the likelihood that they will encounter prey (Teerling 1995). The spider *Habronestes bradleyi* uses the alarm pheromone of its ant prey, *Iridomyrmex purpureus*, to locate intraspecific battles where injured or preoccupied ants may be especially vulnerable to attack (Allan et al. 1996).

Predators may even produce alarm pheromones of their prey and thereby manipulate their behavior. For example, myrmecophilous staphylinid beetles produce ant alarm pheromones that enable them to stay in an ant column without being attacked (Kistner and Blum 1971). Because they steal prey from their ant hosts, this strategy can be considered a form of aggressive mimicry (Haynes and Yeargan 1999).

Although alarm pheromones may increase the probability that the signaler's kin will escape predation, if predators appropriate prey alarm pheromones as kairomones, there will be selection pressure on prey not to emit the alarm pheromone unless an individual is under direct attack. The tendency of many aphid species not to emit cornicle droplets until they are actually attacked is consistent with this hypothesis. Nevertheless, a specific test of the hypothesis (Mondor and Roitberg 2000) did not support it, suggesting another, unspecified reason for the delay in producing the droplets.

Alarm pheromones may be kairomones for predators, but the elicited behavior may appear paradoxical. Hence, although some staphylinids aggregate to aphids (Monsrud and Toft 1999, and references therein), linyphiid spiders may avoid them (Greenstone, M. H., unpublished; Figure 7.2), possibly because of the apparent nonoptimality of aphids for spider development and reproduction (Toft 1995, Toft 1996).

Marking Pheromones

Arthropods produce chemicals to mark the location of prey, to recruit kin for food retrieval, to mark oviposition sites, and to mark enemies for attack. Trail pheromones used by ants to recruit workers for food retrieval are exploited by a variety of predators to locate and ambush ants (Hölldobler and Wilson 1990). The aphid *Ceratovacuna lanigera* produces an alarm pheromone that doubles as a marking pheromone for antipredator defense. As an alarm pheromone, it causes advanced instars to depart but first instars to walk excitedly around the area. If then first instars encounter a

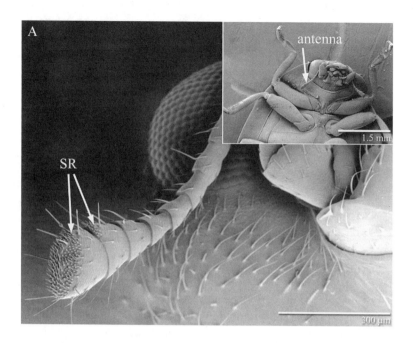

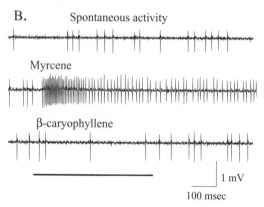

Figure 7.2. (A) Scanning electron micrographs of antenna of the predatory ladybeetle, *Harmonia axyridis*, showing olfactory sensilla arranged in sensory regions on the terminal and penultimate segments; (B) electrical response of neurons within a multiporous sensillum in *H. axyridis* to enantiomers of the prey-host monoterpene, myrcene and β-caryophyllene (Dickens, J. C., unpublished).

syrphid larva (a predator), they will apply additional pheromone to its surface, thereby marking it for attack (Arakaki 1989).

A spectacular example of misappropriation of a marking pheromone involves the hornet *Vespa mandarinia japonica*. Marking of a honey bee hive by a single hornet recruits nestmates in a mass attack that can kill all the members of a susceptible hive in a matter of hours. However, Japanese honey bees, *Apis cerana japonica*, respond to the predator's marking pheromone by massing in the entrance to the nest. There they ambush intruding hornets, enveloping them in a defensive ball of hundreds of bees in which the temperature is lethal to the hornet but not to the bees. A key part of the bees' defense appears to be the use of their own alarm pheromone to recruit workers to the nest entrance (Ono et al. 1995).

Defensive Compounds

Whether acting as contact or stomach poisons, or as volatile chemical signals, defensive compounds are allomones that send powerful messages to a signaler's enemies. However, they can be misappropriated by other species in the system. Pentatomid and coreid bugs eject defensive sprays when they are attacked (Eisner et al. 1991). When one of these bugs is trapped in the web of the orb-weaver *Nephila clavipes*, the spider will retreat momentarily to clean its mouthparts when sprayed, but is not otherwise deterred from killing and consuming the bugs. The sprays are also attractive to kleptoparasitic milichiid flies, which are able to feed, unmolested by the web's rightful inhabitant, on the bug carcasses. Depending on the size and nutritional status of the spider, size of prey, and number of flies recruited, the cost to the spider in lost prey biomass can be substantial.

Leaf beetles produce volatile defensive compounds that deter generalist invertebrate and vertebrate predators. Kopf et al. (1997) showed that larvae of a syrphid fly, *Parasyrphus nigritarsus*, use these secretions as kairomones to locate two leaf beetle species. Herbivores also sequester or egest plant defensive compounds for use in predator deterrence (Bowers 1990). In some cases, predators use these compounds as prey-finding signals, thereby turning recycled allomones into kairomones. For example, some cassidine chrysomelid beetles incorporate feces derived from feeding on a chemically defended plant, *Crysanthemum vulgare*, into a defensive shield that is attractive to a generalist predator, the ant *Myrmica rubra* (Müller and Hilker 1999).

A general constraint on the use of defensive compounds is the necessity of the signaler to protect itself by manufacturing detoxifying enzymes. For example, workers of two termite species that employ contact poisons for colony defense have substrate-specific alkene reductases to detoxify the poisons. Specificity is such that each species is immune to the poison of its own species but is killed by that of the other (Spanton and Prestwich 1981).

Inadvertent Kairomones

Animals and plants produce kairomones that advertise their presence to consumers, whether or not the kairomones appear to be allomones for the signaler. For example, chemicals in snail feces attract sciomyzid fly predators (Coupland 1996), and many plants produce volatile compounds used for host-finding by herbivores (reviewed in Visser 1986). Aquatic predators release chemicals into the water that advertise their presence to potential prey (Chivers and Smith 1998; see also Lima and Steury, ch. 8 in this volume, and Sih, ch. 11 in this volume). For example, freshwater gammarid amphipods reduce their risk of predation by becoming less active in the presence of chemical cues from dragonfly nymphs, which are ambush predators that respond to prey movement (Wudkevich et al. 1997). The wolf spider *Pardosa milvina* similarly reduces its activity level in the presence of excreta or silk from a larger co-occurring wolf spider species, *Hogna helluo* (Persons and Rypstra 2001).

How Are Volatile Semiochemicals Detected?

Mechanism

Predaceous insects detect kairomones released by their prey and synomones released by the prey-host complex (Tømmerås 1985, Dickens, J. C., unpublished). Insects, including predators, use multiporous cuticular structures called *sensilla*, principally located on the antennae and palps, to detect volatile chemical signals (Hansson 1999; Figure 7.3A). Within each sensillum are dendrites of sensory neurons tuned to individual chemicals of importance in feeding and reproductive behavior (Tømmerås 1985). Detection of an odor results in action potentials being elicited in the affected receptor neurons (Figure 7.3B). This message is then sent along axons to specific regions of the antennal lobe called *glomeruli*, before the message is modulated by inputs from sensory organs responsive to other stimulus modalities (Hildebrand 1995).

Although the exact mechanism for odor reception is not yet clear, several hypotheses exist and are important for understanding predator-prey interactions and the evolution of behavior mediating these interactions (Vogt et al. 1999, Vogt, in press). Odors are first adsorbed in the outer, waxy layer of the antenna, where they diffuse through pores in the walls of olfactory sensilla to the perireceptor lymph surrounding the dendrites of first-order receptor neurons housed within (Steinbrecht 1997). This lymph is rich in 16–20 kDa proteins, called *odorant-binding proteins* (OBPs), which represent the first level of biochemical specificity through which potential chemical messengers must pass (Vogt, in press; Figure 7.4). Several types of OBPs may be expressed within different sensilla, and their presence may be correlated with the specificity of neurons within sensillar types. The OBPs transport odor molecules across the perireceptor lymph to the neurons (Vogt et al. 1985, Vogt et al. 1999). At this point, the odor molecule is dropped off, where interaction with a seven-trans membrane domain, G-protein-coupled receptor protein leads to a biochemical cascade that results in a receptor current and subsequent action potentials. Release of the odor molecule by the OBP may be facilitated by pH changes occurring near the dendritic membrane (Wojtasek and Leal 1999). Alternatively, the OBP-odor complex may interact directly with the membrane receptor or other membrane protein (Rogers et al. 1997), which facilitates off-loading near the receptor (Vogt et al. 1999).

Insect Olfactory Sensillum

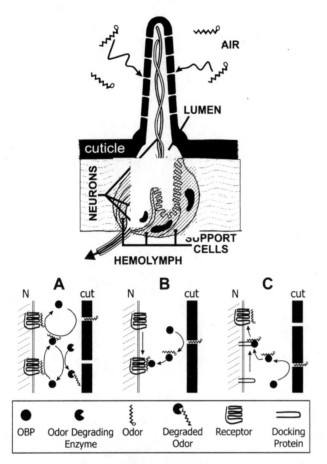

Figure 7.3. Cartoon representation (top) of an insect olfactory sensillum to show pores in the sensillum through which odors enter the sensillum, dendrites of neurons projected inside the sensillum lymph, and support cells; and models (bottom) for interaction of odors with receptors: (A) Odorant-binding protein (OBP) transports odor to 7-transmembrane-domain, G-protein-coupled receptor, where odor is dropped off, (B) complex of OBP-odor interacts directly as a pair with receptor, and (C) OBP-odor complex interacts with a docking protein before dropping off odor at receptor. [After Vogt, R. G. In press. Molecular basis of pheromone detection in insects. In: Comprehensive Insect Physiology, Biochemistry, Pharmacology and Molecular Biology (Gilbert, L. I., Iatro, K., and Gill, S., eds.). New York: Elsevier. With permission of Elsevier. See also Vogt et al. (1999).]

Sensitivity and Selectivity of Predator-to-Prey and Prey-Host Odors

Predators that rely on chemical signals to detect their prey depend on volatiles associated with the habitat of their prey (e.g., chemicals released by the host plant, including allomones or synomones) or on pheromones (which represent kairomones

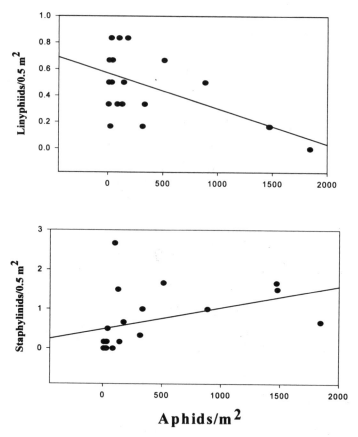

Figure 7.4. Relationships between cereal aphid predator densities in winter wheat, March 12–June 2, 1998, Oklahoma State University North Central Research Station in Lahoma, Oklahoma. Insects were sampled quantitatively bimonthly, aphids by counts on tillers and predators by D-vac suction sampler followed by hand counting (Greenstone 2001). Linyphiid spiders were negatively correlated ($r = -.580$, $p = .0074$), and staphylinid beetles were positively though less strongly correlated ($r = .412$, $p = .071$), with aphid densities (Greenstone, M. H., unpublished data).

for predators) released by the prey itself. Several studies have investigated reception by predaceous insects of pheromones of prey and host-plant compounds released in response to prey feeding (Table 7.1).

In general, specialized predators have receptor neurons that are tuned to phero-mone components of their prey. Outstanding examples of this phenomenon are preda-tory clerid beetles that feed on bark beetles. Both *T. dubius* and *T. formicarius* have receptors for bark beetle pheromone components (Hansen 1983, Payne et al. 1984, Tømmerås 1985). *Thanasimus dubius* detects and responds only to (1*S*, 5*R*)-(–)-frontalin, the aggregation pheromone released by females of its principal prey, the southern pine beetle *Dendroctonus frontalis*, in the southern United States (Payne et al. 1984). The *D. frontalis* pheromone plays a major role in the biology of *T. dubius*,

Table 7.1. Studies of Olfactory Receptor Systems of Insect Predators

Predator[a]	Prey	Type of Study[b]	Most Active
Thanasimus formicarius, Col.: Cleridae	Specialist; Col: Scolytidae, esp. *Ips typographus*	EAG, SCR	Bark beetle aggregation pheromones, esp. (–)-ipsenol, (+)-lineatin; Prey host volatiles: campher/pino-camphon (Hansen 1983, Tømmerås 1985)
Thanasimus dubius, Col.: Cleridae	Specialist; Col.: Scolytidae, esp. *Dendroctonus frontalis*	EAG	*Dendroctonus frontalis* aggregation pheromone: (–)-frontalin (Payne et al. 1984)
Teretiosoma nigrescens, Col.: Histeridae	Specialist; *Prostephanus truncatus*, Col.: Bostrichidae	EAG	*P. truncatus* aggregation pheromone: 1-methylethyl (2*E*)-2-methyl-2-pentenoate, 1-methylethyl (2*Z*)-2-methyl-2-pentenoate (Scholz et al. 1998)
Coleomegilla maculata, Col.: Coccinellidae	Generalist; *Aphids, lepidopterous larvae*	EAG	Prey host volatiles: (*E*)-β-farnesene, a-terpineol, 2-phenylethanol, β-caryophyllene (Zhu et al. 1999)
Chrysoperla carnea, Neuroptera: Chrysopidae	Generalist; *Aphids, lepidopterous larvae*	EAG	Prey host volatiles: 2-phenylethanol, (*E*)-β-farnesene; Aphid sex pheromone: (4a*S*,7*S*,7a*R*)-nepetalactone, (1*R*,4a*S*, 7*S*,7a*R*)-nepetalactonol (Zhu et al. 1999); L-tryptophan (Bakthavatsalam et al. 2000)
Perillus bioculatus, Heteroptera: Pentatomidae	Specialist; *Leptinotarsa decemlineata*, Col.: Chrysomelidae	GC/EAD, EAG	Prey host volatiles: Methyl salicylate, nonanal, (±)-linalool, indole (Dickens 1999); (*Z*)-3-hexenol, 2-phenylethanol > linalool, 4,8-dimethyl-1,3-(*E*),7-nonatriene, nonanal, decanal, (*R*)-(+)-limonene (Weissbecker et al. 1999); Sesquirerpenes-β-caryophyllene, (–)-germacrene D (Weissbecker et al. 2000)
Podisus maculiventris, Heteroptera: Pentatomidae	Generalist	GC/EAD, EAG	Prey host volatiles: Methyl salicylate, nonanal, (±)-linalool, indole (Dickens 1999); (*E*)-2-hexenol, (*E*)-2-hexanal, heptanol, nonanal, hexanal; Conspecific pheromone components: (±)-a-terpineol, benzyl alcohol (Sant'Ana and Dickens 1998)
Podisus nigrispinus, Heteroptera: Pentatomidae	Generalist	EAG	Prey host volatiles: (*E*)-2-hexenol, (*E*)-2-hexanal, heptanol, nonanal, hexanal; Conspecific pheromone components: (±)-a-terpineol, benzyl alcohol (Sant'Ana and Dickens 1998)

[a]*Podisus nigrispinus* are more sensitive to (+)-a-terpineol, (+)-linalool, benzyl alcohol, and nonanal than *P. maculiventris* (Sant'Ana and Dickens 1998); 2-phenylethanol is attractive to *C. maculata* and *C. carnea* in the field (Zhu et al. 1999).

[b]EAG = electroantennogram; SCR = single-cell recording; CG/EAD = coupled gas chromatography/electroantennogram detection.

because, upon landing on a prey-infested tree, *T. dubius* females deposit eggs in bark crevices and the developing larvae prey on *D. frontalis* eggs and larvae. Adults feed on *D. frontalis* adults. *Thanasimus dubius* adults have additional antennal receptors that respond to other bark beetle pheromones, including pheromones of *Ips* bark beetles and host-tree terpenes.

Single-cell recordings from neurons associated with antennal sensilla of *T. formicarius* and its prey *Ips typographus* demonstrated that receptor neurons of both predator and prey have similar specificity and sensitivity (Tømmerås 1985). This study also revealed in the predator receptor neurons responsive to additional bark beetle pheromones, as well as neurons responsive only to host-tree volatiles. Interestingly, receptor neurons for (–)-ipsenol and (+)-lineatin, the pheromones of two genera of bark beetles, were usually present in the same electrophysiological recordings, perhaps indicating co-compartmentalization of attractant receptors within the same sensilla.

Another specialist predator, *Teretiosoma nigrescens*, has antennal receptors sensitive to both pheromone components of its prey, *Prostephanus truncatus*, and responds behaviorally to both (Scholtz et al. 1998). There were no sexual differences in electroantennograms (EAGs) of either predator or prey (Bjostad and Roelofs 1980). The EAGs are thought to represent the summed receptor potentials of neurons responding on the antenna of an insect. However, the presence of a response to a given chemical stimulus shows only that the insect is able to detect the chemical and does not necessarily specify a role for the chemical in the insect's behavior.

Generalist predators have olfactory receptors for chemical signals released by prey and compounds released by host plants in response to prey feeding. For example, two generalist heteropteran predators of the family Pentatomidae, the spined soldier bug (SSB), *P. maculiventris*, and the Brazilian SSB, *P. nigrispinus,* detect a wide range of volatiles released by plants fed upon by prey (Sant'Ana and Dickens 1998, Dickens 1999; Table 7.1). Although both species have receptors for many of the same compounds, EAGs to some individual compounds differed: a prey host compound, nonanal, and pheromone components of the predator, (±)-terpineol, benzyl alcohol, and (±)-linalool. Not only did both sexes respond electrophysiologically to components of the male-produced aggregation pheromone, but individual components elicited EAGs from antennal receptors of immature forms (Sant'Ana and Dickens 1999), which responded behaviorally, as well (Sant'Ana et al. 1997). Interestingly, the largest EAGs recorded from both nymphs and adults were elicited by (*E*)-2-hexen-1-ol, a constitutive plant compound released locally in response to prey feeding. Thus, SSB could use the locally emitted constitutive volatile, together with visual and mechanical signals (Pfannenstiel et al. 1995), to locate prey at close range.

Olfactory receptors in both the predaceous lady beetle *C. maculata* and the lacewings *C. carnea* and *C. cognata* respond to pheromones of their aphid prey (Boo et al. 1998, Zhu et al. 1999). Although all three predators respond electrophysiologically to the sex pheromone components, in essence, (4a*S*, 7*S*, 7a*R*)-nepetalactone and (1*R*, 4a*S*, 7*S*, 7a*R*)-nepetalactol, only *C. maculata* responds electrophysiologically to the alarm pheromone, (E)-β-farnesene (Zhu et al. 1999).

Generalist and specialist predators may respond similarly to many of the same compounds released by the host plant of their prey. Coupled gas chromatography and EAGs from *P. maculiventris* and a specialist predator in the same family,

P. bioculatus, revealed that both predators respond selectively to the same compounds known to be released by a host plant of their prey, the Colorado potato beetle, *L. decemlineata* (Dickens 1999). *Perillus bioculatus* has receptors for several sesquiterpenes released by potato plants during feeding by *L. decemlineata* and may use these compounds to locate the beetle (Weissbecker et al. 2000). Both *P. maculiventris* and *P. bioculatus* are significantly more sensitive to the herbivore-induced volatile methyl salicylate than *L. decemlineata* (Dickens 1999).

Plants Turning the Tables on Insects: The Curious Incident of the Bog in the Nighttime

In our overview of the natural history and evolutionary behavioral ecology of multitrophic semiochemistry, we have described a variety of phenomena in the production and appropriation of chemical signals among plants, herbivores, and predators. However, the absence of expected phenomena can sometimes be as revealing as their presence (Doyle 1892).

In mostly damp, well-lighted, but nutrient-poor environments, some plants "climb the trophic ladder," luring and consuming arthropod prey, and interacting, as well, with commensals and pollinators (Juniper et al. 1989). The insectivorous habit in plants has evolved several times independently. It ought to provide a rich environment for the evolution of multitrophic chemical communication, involving, as it does, flowering plants that must attract pollinators, as well as prey. Furthermore, many of these plants also shelter small communities that include a few specific arthropod species (Fish and Hall 1978, Ratsirarson and Silander 1996), at least some of which may increase the plant's fitness by dismembering and macerating prey for digestion by commensal bacteria and absorption by the plant (Bradshaw 1983).

As insectivores that also harbor arthropod commensals, these plants could be expected to benefit from producing chemical lures. Some do produce odors that are generally attractive to insects (Joel 1988), and, within a given species, fragrant plants capture more arthropod prey than nonfragrant ones (Moran 1996). However, a detailed gas chromatographic study of volatiles produced by insectivorous plants (Jaffé et al. 1995) revealed only one known insect attractant, phenylacetaldehyde, which is attractive to many insect species (Cantelo and Jacobson 1979). Why have we not found more specific allomones, such as prey and inquiline sex pheromones, produced by insectivorous plants?

An obvious answer is that we have not looked hard enough. Another is that insects that visit the habitats of insectivorous plants do not occur with sufficient frequency and reliability to warrant the dedication of resources to the production of specific attractants. The fact that so many species exhibit an enormous taxonomic range of prey (they have been characterized as having diets that are "catholic in the extreme"; Juniper et al. 1989) supports this hypothesis. It does not, however, explain the absence of specific attractant pheromones for the arthropod inquilines, which, at least in the case of the pitcher plant *Sarracenia purpurea*, include a restricted and invariant, and hence frequent and reliable, set of fly species throughout its range (Fish and Hall 1978, Bradshaw 1983).

Another plausible hypothesis is that other signaling modalities are sufficient to lure prey to insectivorous plants. Many species are visually striking and occur in extensive and showy populations (Joel 1988). This signaling alone may be sufficient to attract large numbers of prey to the plants. Once there, they can be lured to traps by generally attractive odors (Jaffé et al. 1995) and visual cues (e.g., UV-absorbent or UV-reflective patterns; Joel et al. 1985), associated with nectar rewards (Kato 1993, Deppe et al. 2000).

Whether visual cues are attractive to inquilines appears to have been tested only once, for the pitcher plant mosquito *Wyeomyia smithii*, with negative results. However, a water-soluble extract of *Sarracenia* leaf is more attractive to ovipositing adult mosquitoes than a deionized water control, suggesting that it may produce an attractant (Istock et al. 1983). Because other insects were not tested, we do not know the specificity of the chemical cue. It would be worth revisiting this system with modern chemical techniques coupled with biological detectors (e.g., coupled gas chromatography and electroantennography).

Evolution of Multitrophic Semiochemical Interactions

The production of semiochemicals presents evolutionary opportunities for the signaler's enemies and, thus, evolutionary challenges for the signaler. This situation could lead to coevolution: "reciprocal genetic change in interacting species owing to natural selection imposed by each on the other" (Futuyma 1998, pp. 539–540). For example, for *T. dubius* and *D. frontalis,* Payne et al. (1984) hypothesized a semiochemical coevolutionary "arms race" in which the "first shot was fired" by the predator, which evolved antennal sensilla that detect mainly (–)-frontalin, the principal component of the pheromone of its prey (Payne et al. 1984). Because *D. frontalis* releases a mixture of 85% (–)- and 15% (+)-frontalin, production of the (+)-enantiomer might reflect the prey's reciprocal response, the evolution of an alternate chemical signal in an effort to avoid its predator.

After the publication of Ehrlich and Raven's (1964) seminal study on coevolution, the ubiquity of this process in ecological interactions was taken as a given, despite an almost total lack of supporting evidence (Janzen 1980, Jermy 1984, Berenbaum and Zangerl 1992). Hypothesis generation and testing in the ensuing decades indicated that coevolution could occur over a broad range of spatial and temporal scales, from a few interacting populations in a particular community, at a particular time, through mosaics of interacting populations dispersed in space and time (Futuyma and Keese 1992, Thompson 1999). A necessary precondition for evolution, and hence the coevolution of traits involved in semiochemical interactions, is natural heritable variation. However, this condition alone is not sufficient to demonstrate the existence of coevolution among interacting populations.

Affirmative answers to two questions will support the existence of coevolution within a local community in ecological time (Thompson 1999): (a) Is there reciprocal heritable variation for traits that affect the interactions? And (b) is there evidence for reciprocal natural selection on the interacting species? Not all variation in traits related to semiochemical-based trophic interactions is under genetic control. For example, the host plant on which an herbivore has been raised may affect a predator's prey preference, and the predator's nutritional state during exposure to plant volatiles may determine their attractancy or repellency (Sabelis et al. 1999, and references therein). Nevertheless, a number of relevant traits have heritable variation and are responsive to selection.

Plants exhibit heritable variation in the production of herbivore-induced volatiles (Takabayashi and Dicke 1996, Gouinguené et al. 2001), and heritable intraspecific variation in defensive plant chemistry and herbivore susceptibility to defensive plant chemicals have been demonstrated (Wallin and Raffa 2004). Molecular traits involved in arthropod olfaction are heritable, as well. The involvement of several proteins in the initial stages of the olfactory process decreases the noise and enhances the clarity of the semiochemical message. At the same time, changes in a single gene for an OBP may result in different patterns of behavior, as demonstrated for the fruit fly *Drosophila melanogaster* (Kim and Smith 2001) and the fire ant *Solenopsis invicta* (Krieger and Ross 2002).

Some relevant traits have also been shown to be responsive to selection. In the absence of herbivores, heavily defended plants experience lower fitness than less well defended ones (Berenbaum and Zangerl 1988). Further, the strength of response by the phytoseiid predator *Phytoseiulus persimilis* to the spider-mite-induced volatile linalool is responsive to selection (Margolies et al. 1997). Nevertheless, little evidence has been adduced for heritable variation on both sides of any two-species (let alone multitrophic) semiochemical interaction, or for reciprocal selection among the species (Berenbaum and Zangerl 1992, Futuyma and Keese 1992, Thompson 1999).

The mobility and dispersion of prey and predator, in space and time, could give rise to and shape coevolved semiochemical communication, as proposed in the geographic mosaic theory of coevolution (Thompson 1994). Adult males of the bark beetle *Ips pini* produce an aggregation pheromone, ipsdienol, which attracts adults to a tree that has been effectively colonized. Both *I. pini* and the predator *T. dubius* show local, as well as regional, variation in their responses to enantiomeric blends of the ipsdienol (Raffa and Kleipzig 1989, Herms et al. 1991). If this variation were heritable, one would expect bark beetles under the heaviest predation selection pressure to show the largest shifts in pheromone enantiomeric mixture from what is optimal for intraspecific signaling. In a reciprocal experiment that provides a test of this hypothesis, Raffa and Dahlsten (1995) found that when California and Wisconsin beetles were tested the most abundant predators responded most strongly to the enantiomeric mix of *Ips* populations from the other state, whereas *Ips* beetles always responded most strongly to the local enantiomeric blend. This finding suggests that there is a locally adapted optimal blend from which the beetles depart only in the face of very strong local selection.

Close concordance of the phylogenies of chemically protected plants and herbivore taxa specialized to feed upon them would be consistent with the operation of coevolution over evolutionary time. A well-documented example is the cladograms for some plant families characterized by unusual defenses (i.e., furanocoumarins in the Umbelliferae and Compositae, and latex canals combined with cardenolides or glycosides in the Asclepiadaceae and Apocynacea, and their associated adapted insect faunas; Farrell and Mitter 1993). Nevertheless, concordance of phylogenies can arise by means other than reciprocal natural selection, and alternative explanations must be ruled out before concordances can be ascribed to coevolution (Futuyma 1998).

Future Research

Plants and animals benefit by using chemical signals to fend off attack, find mates, send alarms about the presence of natural enemies, and signal the presence of one's own enemies to the enemies' enemies. But, as in other interactions (see Holland et al., ch. 2 in this volume), cost-benefit considerations will influence the likelihood of selection's leading to the evolution of new traits and interactions (Dicke and Sabelis 1989, Bergelson and Purrington 1996, Hoffmeister and Roitberg 1998). Specifically, the misappropriation of signals by various "eavesdroppers" may reduce the fitness of the signaler.

Current concepts concerning multitrophic semiochemical interactions are skewed by the literature's bias concerning crops and forest trees and their pests, and concerning the pests' predators. Agricultural and forestry systems, often considered to be relatively simple, are far more complex than the literature may lead us to believe. It is seldom true, for example, that a given pest experiences significant mortality from only a single predator (Kessler and Baldwin 2001). Instead, a quite diverse contingent of mostly generalist predators ranges widely over the available herbivores, moving among different crops in the landscape to find them (Greenstone and Sunderland 1999, Symondson et al. 2002). Nevertheless, less managed or more pristine ecosystems will have greater plant taxonomic diversity and, because herbivores tend to be relatively host-specific, greater herbivore diversity.

Studies of the semiochemistry of multitrophic predator-prey interactions in pristine or less managed habitats are relatively rare. A larger data set from such systems would enable tests of hypotheses that predict contrasts in the intensity or diversity of chemical signaling between intensely managed systems with few species and those that are more diverse and less managed. Here we offer two simple hypotheses that can be tested once we have a large data set from less managed systems. We would suggest that *investment in defensive and volatile plant chemistries will be more frequent in intensely managed systems*. Managed plant systems represent concentrated food resources for their herbivores (Root 1973). Because of their greater apparency (likelihood of being found; sensu Feeny 1976), plants in these systems will suffer earlier and more intense herbivory and will therefore be more likely to invest in the biochemical apparatus required to manufacture defensive compounds or herbivore-induced volatiles than those in less managed systems.

Second, we suggest that *diversity of plant chemistries and consumer chemical detectors will be greater in less managed systems*. That is, because of higher plant and insect diversity in such systems (Andow 1991), plants in less managed systems will exhibit a greater diversity of both constitutive defenses and herbivore-induced volatile chemical signals, or enantiomeric mixes of volatiles, and herbivores and predators will exhibit concomitantly greater olfactory versatility than those in more managed systems.

We invite critique, refinement, and tests of these hypotheses, as well as development of additional hypotheses. There are thousands of relevant studies in the chemical ecology literature, but many questions concerning the evolution and ecology of multitrophic semiochemical interactions have yet to be answered or, for that matter, even asked.

Acknowledgments

We are grateful to the editors for inviting us to prepare this review. We also thank S. Wilzer of the U.S. Department of Agriculture National Agricultural Library for assistance with literature searches; D. Vogt (University of South Carolina) for assistance with Figure 7.3; and C. Mitter (University of Maryland), M. Sabelis (Institute for Biodiversity and Ecosystem Dynamics, Amsterdam), E. Schmelz (U.S. Department of Agriculture, Agricultural Research Service), and J. Tumlinson (Pennsylvania State University) for helpful discussions and for alerting us to important references. Thanks also to J. Aldrich (U.S. Department of Agriculture, Agricultural Research Service), P. Barbosa (University of Maryland), S. Salom (Virginia Polytechnic Institute and State University), K. Sunderland (Horticulture Research International), and two anonymous reviewers for thoughtful comments on the chapter. This chapter is free from copyright and, as written by employees of the U.S. Department of Agriculture, Agricultural Research Service (a federal government agency), is in the public domain.

Literature Cited

Agrawal, A. A., Gorski, P. M., and Tallamy, D. W. 1999. Polymorphism in plant defense against herbivory: constitutive and induced resistance in *Cucumis sativus*. J. Chem. Ecol. 25:2285–2304.

Allan, R. A., Elgar, M. A., and Capon, R. J. 1996. Exploitation of an ant chemical alarm signal by the zodariid spider *Habronestes bradleyi* Walckenaer. Proc. R. Soc. Lond. Ser. B 263:69–73.

Alves, L. F. 1988. Chemical ecology and the social behavior of animals. Prog. Chem. Org. Nat. Prod. 53:1–85.

Andow, D. A. 1991. Vegetational diversity and arthropod population response. Annu. Rev. Entomol. 36:561–586.

Arakaki, N. 1989. Alarm pheromone eliciting attack and escape responses in the sugar cane woolly aphid, *Ceratovacuna lanigera* (Homoptera, Pemphigidae). J. Ethol. 7:83–90.

Bakke, A., and Kvamme, T. 1981. Kairomone response in *Thanasimus* predators to pheromone components of *Ips typographus*. J. Chem. Ecol. 7:305–312.

Bakthavatsalam, N., Singh, S. P., Tandon, P. L., Chaudhary, M., and Preethi, S. 2000. Electrophysiological responses of *Chrysoperla carnea* (Stephens) (Neuroptera: Chrysopidae) to some potential kairomonal substances. J. Entomol. Res. 24:109–114.

Bartelt, R. J. 1999. Sap beetles. In: Pheromones of Non-lepidopteran Insects Associated with Agricultural Plants (Hardie, J., and Minks, A. K., eds.). New York: CABI; 66–90.

Berenbaum, M. R., and Zangerl, A. R. 1988. Stalemates in the coevolutionary arms race: synthesis, synergisms, and other sundry sins. In: Chemical Mediation of Coevolution (Spencer, K. C., ed.). San Diego: Academic Press; 113–132.

Berenbaum, M. R., and Zangerl, A. R. 1992. Genetics of secondary metabolism and herbivore resistance in plants. In: Herbivores: Their Interactions with Secondary Plant Metabolites (Rosenthal, G. A., and Berenbaum, M. R., eds.). San Diego: Academic Press; 415–438.

Bergelson, J., and Purrington, C. B. 1996. Surveying patterns in the cost of resistance in plants. Am. Nat. 148:536–538.

Bernasconi Ockroy, M. L., Turlings, T. C. J., Ambrosetti, L., Bassetti, P., and Dorn, S. 1998. Herbivore-induced emissions of maize volatiles repel the corn leaf aphid, *Rhopalosiphum maidis*. Entomol. Exp. Appl. 87:133–142.

Bernasconi Ockroy, M. L., Turlings, T. C. J., Edwards, P. J., Fritzsche-Hoballah, M. E., Ambrosetti, L., Bassetti, P., and Dorn, S. 2001. Response of natural populations of predators and parasitoids to artificially induced volatile emissions in maize plants (*Zea mays* L.). Agric. For. Entomol. 3:201–209.

Bjostad, L. B., and Roelofs, W. L. 1980. An inexpensive electronic device for measuring electroantennogram responses to sex pheromone components with a voltmeter. Physiol. Entomol. 13:139–145.

Bolter, C. J., Dicke, M., Van Loon, J. J. A., Visser, J. H., and Posthumus, M. A. 1997. Attraction of Colorado potato beetle to herbivore-damaged plants during herbivory and after its termination. J. Chem. Ecol. 23:1003–1023.

Boo, K. S., Chung, I. B., Han, K. S., Pickett, J. A., and Wadhams, L. J. 1998. Response of the lacewing *Chrysopa cognata* to pheromones of its aphid prey. J. Chem. Ecol. 24:631–643.

Bowers, M. D. 1990. Recycling plant natural products for insect defense. In: Insect Defenses: Adaptive Mechanisms and Strategies of Prey and Predators (Evans, D. L., and Schmidt, J. O., eds.). Albany: State University of New York Press; 353–386.

Bradshaw, W. W. 1983. Interaction between the mosquito *Wyeomyia smithii,* the midge *Metriocnemus knabi,* and their carnivorous host *Sarracenia purpurea.* In: Phytotelmata: Terrestrial Plants as Hosts for Aquatic Insect Communities (Frank, J. H., and Lounibos, L. P., eds.). Medford, N.J.: Plexus; 161–189.

Brown, W. L., Jr., Eisner, T., and Whittaker, R. H. 1970. Allomones and kairomones: transpecific chemical messengers. BioScience 20:21–22.

Cantelo, W. M., and Jacobson, M. 1979. Corn silk volatiles attract many pest species of moths. Environ. Sci. Health A 14:695–707.

Cardé, R. T., and Minks, A. K. 1997. Insect Pheromone Research: New Directions. New York: Chapman & Hall.

Chivers, D. P., and Smith, R. J. F. 1998. Chemical alarm signaling in aquatic predator-prey systems: a review and prospectus. Ecosci. 5:338–352.

Cossé, A. A., and Bartelt, R. J. 2000. Male-produced aggregation pheromone of *Colopterus truncates:* structure, electrophysiological, and behavioral activity. J. Chem. Ecol. 26:1735–1748.

Coupland, J. B. 1996. Influence of snail feces and mucus on oviposition and larval behavior of *Pherbellia cinerella* (Diptera: Sciomyzidae). J. Chem. Ecol. 22:183–189.

Creygnier, P., and Hodek, I. 1996. Enemies of Coccinellidae. In: Ecology of Coccinellidae (Hodek, I., and Honek, A., eds.). Dordrecht, the Netherlands: Kluwer Academic; 319–350.

Degenhardt, J., Gershenzon J., Baldwin, I. T., and Kessler, A. 2003. Attracting friends to feast on foes: engineering terpene emission to make crop plants more attractive to herbivore enemies. Curr. Opin. Biotechnol. 14:169–176.

De Moraes, C. M., Lewis, W. J., Paré, P. W., Alborn, H. T., and Tumlinson, J. H. 1998. Herbivore-infested plants selectively attract parasitoids. Nature 393:570–573.

Deppe, J. L., Dress, W. J., Nastase, A. J., Newell, S. J., and Luciano, C. S. 2000. Diel variation of sugar amount in nectar from pitchers of *Sarracenia purpurea* L. with and without insect visitors. Am. Midl. Nat. 144:123–132.

Dicke, M., and Sabelis, M. W. 1988a. How plants obtain predatory mites as bodyguards. Neth. J. Zool. 38:148–165.

Dicke, M., and Sabelis, M. W. 1988b. Infochemical terminology: based on cost-benefit analysis rather than origin of compounds? Funct. Ecol. 2:131–139.

Dicke, M., and Sabelis, M. W. 1989. Does it pay plants to advertise bodyguards? Towards a cost-benefit analysis of induced synomone production. In: Cause and Consequences of Variation in Growth Rate and Productivity of Higher Plants (Lambers, H., Cambridge, M. L., Konings, H., and Pons, T. L., eds.). The Hague, the Netherlands: SPB Academic; 341–358.

Dicke, M., von Beek, T. A., Posthumus, M. A., Dom, N. B., von Bokhoven, H., and de Groot, A. 1990. Isolation and identification of volatile kairomone that affects acarine predator-prey interactions: involvement of host plant in its production. J. Chem. Ecol. 16:381–396.

Dickens, J. C. 1999. Predator-prey interactions: olfactory adaptations of generalist and specialist predators. Agric. For. Entomol. 1:47–54.

Dickens, J. C., Billings, R. F., and Payne, T. L. 1991. Green leaf volatiles: a ubiquitous chemical signal modifies insect pheromone responses. In: Insect Chemical Ecology (Hrdy, I., ed.). The Hague, the Netherlands: Academia Prague and SPB Academic; 277–280.

Dickens, J. C., Billings, R. F., and Payne, T. L. 1992. Green leaf volatiles interrupt aggregation pheromone response in bark beetles infesting southern pines. Experientia 48:523–524.

Dickens, J. C., Jang, E. B., Light, D. M., and Alford, A. R. 1990. Enhancement of insect pheromone responses by green leaf volatiles. Naturwissenschaften 77:29–31.

Dickens, J. C., Smith, J. W., and Light, D. M. 1993. Green leaf volatiles enhance sex attractant pheromone of the tobacco budworm, *Heliothis virescens* (Lep.: Noctuidae). Chemoecolology 4:175–177.

Doyle, A. C. 1892. The adventure of silver blaze. Strand 4:645–660.

Drukker, B., Scutareanu, P., and Sabelis, M. W. 1995. Do anthocorid predators respond to synomones from *Psylla*-infested pear trees in field conditions? Entomol. Exp. Appl. 77:193–203.

Du, Y., Poppy, G. M., Powell, W., Pickett, J. A., Wadhams, L. J., and Woodcock, C. M. 1998. Identification of semiochemicals released during aphid feeding that attract the parasitoid *Aphidius ervi*. J. Chem. Ecol. 24:1355–1368.

Ehrlich, P. R., and Raven, P. H. 1964. Butterflies and plants: a study in coevolution. Evolution 18:586–608.

Eisner, T., Eisner, M., and Deyrup, M. 1991. Chemical attraction of kleptoparasitic flies to heteropteran insects caught by orb-weaving spiders. Proc. Natl. Acad. Sci. U.S.A. 88:8194–8197.

Eisner, T., and Meinwald, J. 1995. The chemistry of sexual selection. Proc. Natl. Acad. Sci. U.S.A. 92:50–55.

Erbilgin, N., and Raffa, K. F. 2001. Modulation of predator attraction to pheromones of two prey species by stereochemistry of plant volatiles. Oecologia 127:444–453.

Farrell, B. D., and Mitter, C. 1993. Phylogenetic determinants of insect/plant community diversity. In: Species Diversity in Ecological Communities (Ricklefs, R. E., and Schluter, D., eds.). Chicago: University of Chicago Press; 253–256.

Feeny, P. 1976. Plant apparency and chemical defence. In: Biochemical Interactions between Plants and Insects (Wallace, J. W., and Mansell, R. L., eds.). Rec. Adv. Phytochem. 10:1–40.

Fish, D., and Hall, D. W. 1978. Succession and stratification of aquatic insects inhabiting the leaves of the insectivorous pitcher plant, *Sarracenia purpurea*. Am. Midl. Nat. 99:172–183.

Flint, H. M., Salter, S. S., and Walters, S. 1979. Caryophyllene: an attractant for the green lacewing. Environ. Entomol. 8:1123–1125.

Futuyma, D. J. 1998. Evolutionary Biology, 3rd ed. Sunderland, Mass.: Sinauer.

Futuyma, D. J., and Keese, M. C. 1992. Evolution of plants and phytophagous arthropods. In: Herbivores: Their Interactions with Secondary Metabolites (Rosenthal, G. A., and Berenbaum, M. A., eds.). New York: Academic Press; 439–475.

Gibson, R. W., and Pickett, J. A. 1983. Wild potato repels aphids by release of aphid alarm pheromone. Nature 302:608–609.

Gouinguené, S., Degen, T., and Turlings, T. C. J. 2001. Variability in herbivore-induced odour emissions among maize cultivars and their wild ancestors (teosinte). Chemoecology 11:9–16.

Greenfield, M. D. 2002. Signalers and Receivers: Mechanisms and Evolution of Arthropod Communication. Oxford: Oxford University Press.

Greenstone, M. H. 2001. Spiders in wheat: first quantitative data for North America. BioControl 46:439–454.

Greenstone, M. H., and Sunderland, K. D. 1999. Why a symposium on spiders in agroecosystems now? J. Arachnol. 27:267–269.

Guerreri, E., Pennacchio, F., and Tremblay, E. 1997. Effect of adult exposure on in-flight orientation to plant and plant-host complex volatiles in Aphidius ervi Halliday (Hymenoptera:Braconidae). Biol. Contr. 10:159–165.

Han, B. Y., and Chen, Z. M. 2002. Composition of the volatiles from intact and mechanically pierced tea aphid-tea shoot complexes and their attraction to natural enemies of the tea aphid. J. Agric. Food Chem. 50:2571–2575.

Hansen, K. 1983. Reception of bark beetle pheromone in the predaceous clerid beetle, Thanasimus formicarius. J. Comp. Physiol. A 150:371–378.

Hansson, B. S. 1999. Insect Olfaction. Berlin: Springer-Verlag.

Harari, A. R., Ben-Yakir, D., and Rosen, D. 1994. Mechanism of aggregation behavior in Maladera matrida Argaman (Coleoptera: Scarabeidae). J. Chem. Ecol. 20:361–371.

Haynes, K. F., and Yeargan, K. V. 1999. Exploitation of intraspecific communication systems: illicit signalers and receivers. Ann. Entomol. Soc. Am. 92:960–970.

Hendrichs, J., Katsoyannos, B. I., Wornoayporn, V., and Hendrichs, M. A. 1994. Odour-mediated foraging by yellowjacket wasps (Hymentopera: Vespidae): predation on leks of pheromone-calling Mediterranean fruit fly males (Diptera: Tephritidae). Oecologia 99:88–94.

Hendrichs, M. A., and Hendrichs, J. 1998. Perfumed to be killed: interception of Mediterranean fruit fly (Diptera: Tephritidae) sexual signaling by predatory foraging wasps (Hymenoptera: Vespidae). Ann. Entomol. Soc. Am. 91:228–234.

Herms, D. A., Haack, R. A., and Ayres, B. D. 1991. Variation in semiochemical-mediated predator-prey interaction: Ips pini (Scolytidae) and Thanasimus dubius (Cleridae). J. Chem. Ecol. 17:1705–1714.

Hildebrand, J. G. 1995. Analysis of chemical signals by nervous systems. Proc. Natl. Acad. Sci. U.S.A. 92:67–74.

Hoffmeister, T. S., and Roitberg, B. D. 1998. Evolution of signal persistence under predator exploitation. Ecosci. 5:312–320.

Hölldobler, B., and Wilson, E. O. 1990. The Ants. Cambridge, Mass.: Belknap Press.

Istock, C. A., Tanner, K., and Zimmer, H. 1983. Habitat selection by the pitcher-plant mosquito, Wyeomyia smithii: behavioral and genetic aspects. In: Phytotelmata: Terrestrial Plants as Hosts for Aquatic Insect Communities (Frank, J. H., and Lounibos, L. P., eds.). Medford, N.J.: Plexus; 191–204.

Jaffé, K., Blum, M. S., Fales, H. M., Mason, R. T., and Cabrera, A. 1995. On insect attractants from pitcher plants of the genus Heliamphora (Sarraceniaceae). J. Chem. Ecol. 21:379–384.

James, D. 2003. Field evaluation of herbivore-induced plant volatiles as attractants for beneficial insects: methyl salicylate and the green lacewing, *Chrysopa nigricornis*. J. Chem. Ecol. 29:1601–1609.

Janzen, D. H. 1980. When is it coevolution? Evolution 34:611–612.

Jermy, T. 1984. Evolution of insect/host plant relationships. Am. Nat. 124:609–630.

Joel, D. M. 1988. Mimicry and mutualism in carnivorous pitcher plants (Sarraceniaceae, Nepanthaceae, Cephalotaceae, Bromeliaceae). Biol. J. Linn. Soc. 35:185–197.

Joel, D. M., Juniper, D. E., and Dafni, A. 1985. Ultraviolet patterns in the traps of carnivorous plants. New Phytol. 101:585–593.

Juniper, B. E., Robins, R. J., and Joel, D. M. 1989. The Carnivorous Plants. London: Academic Press.

Karban, R., Baldwin, I. T., Baxter, K. J., Laue, G., and Felton, G. W. 2000. Communication between plants: induced resistance in wild tobacco plants following clipping of neighboring sagebrush. Oecologia 125:66–71.

Karban, R., and Maron, J. 2002. The fitness consequences of interspecific eavesdropping between plants. Ecology 83:1209–1213.

Karlson, P., and Butenandt, A. 1959. Pheromones (ectohormones) in insects. Annu. Rev. Entomol. 4:39–58.

Kato, M. 1993. Floral biology of *Nepenthes gracilis* (Nepenthaceae) in Sumatra. Am. J. Bot. 80:924–927.

Kessler, A., and Baldwin, I. T. 2001. Defensive function of herbivore-induced volatile emissions in nature. Science 291:2141–2144.

Kielkiewicz, M., and de Vrie, M. 1990. Within-leaf differences in nutritive value and defence mechanism in chrysanthemum to the two-spotted spider mite (*Tetranychus urticae*). Exp. Appl. Acarol. 10:33–43.

Kielty, J. P., Allen Williams, L. J., Underwood, N., and Eastwood, E. A. 1996. Behavioral responses of three species of ground beetle (Coleoptera: Carabidae) to olfactory cues associated with prey and habitat. J. Insect Behav. 9:237–250.

Kim, M. S., and Smith, D. P. 2001. The invertebrate odorant-binding protein LUSH is required for normal olfactory behavior in *Drosophila*. Chem. Senses 26:195–199.

Kistner, D. H., and Blum, M. S. 1971. Alarm pheromone of *Lasius* (*Dendrolasius*) *spathepus* (Hymenoptera: Formicidae) and its possible mimicry by two species of *Pella* (Coleoptera: Staphylinidae). Ann. Entomol. Soc. Am. 64:589–594.

Kopf, A., Rank, N. E., Roininen, H., and Tahvanainen, J. 1997. Defensive larval secretions of leaf beetles attract a specialist predator *Parasyrphus nigritarsis*. Ecol. Entomol. 22:176–183.

Krieger, M. J. B., and Ross, K. G. 2002. Identification of a major gene regulating complex social behavior. Science 295:328–332.

Law, J. H., and Regnier, F. E. 1971. Pheromones. Annu. Rev. Biochem. 40:533–548.

Loughrin, J. H., Potter, D. A., and Hamilton-Kemp, T. R. 1995. Volatile compounds induced by herbivory act as aggregation kairomones for the Japanese beetle (*Popillia japonica* Newman). J. Chem. Ecol. 21:1457–1467.

MacConnel, J. G., Borden, J. H., Silverstein, R. M., and Stokkink, E. 1977. Isolation and tentative identification of lineatin, a pheromone from the frass of *Trypodendron lineatum* (Coleoptera: Scolytidae). J. Chem. Ecol. 5:549–561.

Margolies, D. C., Sabelis, M. W., and Boyer, J. E., Jr. 1997. Response of a phytoseiid predator to herbivore-induced plant volatiles: selection on attraction and effect on prey exploitation. J. Insect Behav. 10:695–709.

Mayland, H., Margolies, D. C., and Charlton, R. E. 2000. Local and distant prey-related cues influence when an acarine predator leaves a prey patch. Entomol. Exp. Appl. 96:245–252.

Mendel, Z., Zegelman, L., Hassner, A., Assael, F., Harel, M., Tam, S., and Dunkelbaum, E. 1995. Outdoor attractancy of males of *Matsucoccus josephi* (Homoptera:Matsucoccidae) and *Elatophilus hebraicus* (Hemiptera:Anthocoridae) to synthetic female sex pheromone of *Matsucoccus josephi*. J. Chem. Ecol. 21:331–341.

Miller, D. R., and Borden, J. H. 2000. Dose-dependent and species-specific responses of pine bark beetles (Coleoptera: Scolytidae) to monoterpenes in association with pheromones. Can. Entomol. 132:183–195.

Mira, A., and Bernays, E. A. 2002. Trade-offs in host use by *Manduca sexta*: a plant character vs. natural enemies. Oikos 97:387–397.

Mizzell, R. F., and Nebeker, T. E. 1982. Preference and oviposition rates of adult *Thanasimus dubius* (F.) on three prey species. Environ. Entomol. 11:139–143.

Mondor, E. B., and Roitberg, B. D. 2000. Has the attraction of predatory coccinellids to cornicle droplets constrained aphid alarm signaling behavior? J. Insect Behav. 13:321–329.

Monsrud, C., and Toft, S. 1999. The aggregative numerical response of polyphagous predators to aphids in cereal fields: attraction to what? Ann. Appl. Ecol. 134:265–270.

Moran, J. A. 1996. Pitcher dimorphism, prey composition and the mechanisms of prey attraction in the pitcher plant *Nepenthes rafflesiana* in Borneo. J. Ecol. 84:515–525.

Müller, C., and Hilker, M. 1999. Unexpected reactions of a generalist predator towards defensive devices of cassidine larvae (Coleoptera, Chrysomelidae). Oecologia 118:166–172.

Nishida, R. 2002. Sequestration of defensive substances from plants by Lepidoptera. Annu. Rev. Entomol. 47:57–92.

Nishida, R., and Fukami, H. 1990. Sequestration of distasteful compounds by some pharmacophagous insects. J. Chem. Ecol. 16:151–164.

Nordlund, D. A., and Lewis, W. J. 1976. Terminology of chemical releasing stimuli in intraspecific and interspecific interactions. J. Chem. Ecol. 2:211–220.

Olson, D. M., Hodges, T. A., and Lewis, W. J. 2003. Foraging efficacy of a larval parasitoid in a cotton patch: influence of chemical cues and learning. J. Insect Behav. 16:613–624.

Ono, M., Igarashi, T., Ohno, E., and Sasaki, M. 1995. Unusual thermal defense by a honeybee against mass attack by hornets. Nature 377:334–336.

Payne, T. L., Dickens, J. C., and Richerson, J. V. 1984. Insect predator-prey coevolution via enantiomeric specificity in a kairomone-pheromone system. J. Chem. Ecol. 10:487–492.

Peng, C., Bartelt, R. J., and Weiss, M. J. 1999. Male crucifer flea beatles produce an aggregation pheromone. Physiol. Entomol. 24:98–99.

Persons, M. H., and Rypstra, A. L. 2001. Wolf spiders show graded antipredator behavior in the presence of chemical cues from different sized predators. J. Chem. Ecol. 27:2493–2504.

Pfannenstiel, R. S., Hunt, R. E., and Yeargan, K. V. 1995. Orientation of a hemipteran predator to vibrations produced by feeding caterpillars. J. Insect Behav. 8:1–9.

Raffa, K. F., and Dahlsten, D. L. 1995. Differential responses among natural enemies and prey to bark beetle pheromones. Oecologia 101:17–23.

Raffa, K. F., and Kleipzig, K. D. 1989. Chiral escape of bark beetles from predators responding to a bark beetle pheromone. Oecologia 102:17–23.

Rapusas, H. R., Bottrell, D. G., and Coll, M. 1996. Intraspecific attraction of rice to insect predators. Biol. Contr. 6:394–400.

Ratsirarson, J., and Silander, J. A., Jr. 1996. Structure and dynamics in *Nepenthes madagascariensis* pitcher plant microcommunities. Biotropica 28:218–227.

Reid, C. D., and Lampman, R. L. 1989. Olfactory responses of *Orius insidiosus* (Hemiptera: Anthocoridae) to volatiles of corn silk. J. Chem. Ecol. 15:1109–1115.

Rogers, M. E., Sun, M., Lerner, M. R., and Vogt, R. G. 1997. SNMP-1, a novel membrane protein of olfactory neurons of the silk moth *Antheraea polyphemus* with homology to the CD36 family of membrane proteins. J. Biol. Chem. 272:14792–14799.

Root, R. B. 1973. Organization of a plant-arthropod association in simple and diverse habitats: the fauna of collards (*Brassica oleraceae*). Ecol. Monogr. 43:95–124.

Rothschild, M. 1961. Defensive odours and Mullerian mimicry among insects. Trans. R. Entomol. Soc. Lond. 113:5–121.

Ruther, J., Meiners, T., and Steidle, J. L. M. 2002. Rich in phenomena—lacking in terms. A classification of kairomones. Chemoecol. 12:161–167.

Sabelis, M. W., Janssen, A., Pallini, A., Venzon, M., Bruin, J., Drukker, B., and Scutareanu, P. 1999. Behavioral responses of predatory and herbivorous arthropods to induced plant volatiles: from evolutionary ecology to agricultural applications. In: Induced Plant Defenses against Pathogens and Herbivores (Agrawal, A., Tuzun, S., and Bent, E., eds.). St. Paul, Minn.: APS Press; 269–298.

Sant'Ana, J., Bruni, R., Abdul-Baki, A. A., and Aldrich, J. R. 1997. Pheromone-induced movement of nymphs of the predator, *Podisus maculiventris* (Heteroptera: Pentatomidae). Biol. Contr. 10:123–128.

Sant'Ana, J., and Dickens, J. C. 1998. Comparative electrophysiological studies of olfaction in predaceous bugs, *Podisus maculiventris* and *P. nigrispinus*. J. Chem. Ecol. 24:965–984.

Sant'Ana, J., and Dickens, J. C. 1999. Olfactory reception of conspecific aggregation pheromone and plant odors by nymphs of the predator, *Podisus maculiventris*. J. Chem. Ecol. 24:1813–1826.

Schmelz, E. A., Alborn, H. T., and Tumlinson, J. H. 2003. Synergistic interactions between volicitin, jasmonic acid and ethylene mediate insect-induced volatile emission in *Zea mays*. Physiol. Plantarum 117:403–412.

Schmutterer, H., and Ascher, K. R. S., eds. 1987. Natural pesticides from the neem tree (*Azadirachta indica* A. Juss) and other tropical plants. Proc. Third Intern. Neem Conf., Nairobi, Kenya. GTX, Eschborn, 703 pp.

Scholz, D., Borgemeister, C., and Poehling, H.-M. 1998. EAG and behavioural responses of the larger grain borer, *Prostephanus truncatus*, and its predator, *Teretriosoma nigrescens*, to the borer-produced aggregation pheromone. Physiol. Entomol. 23:265–273.

Shulaev, V., Silverman, P., and Raskin, T. 1997. Airborne signaling by methyl salicylate in plant pathogen resistance. Nature 385:718–721.

Spanton, S. G., and Prestwich, G. 1981. Chemical self-defense by termite workers: prevention of autotoxication in two rhinotermitids. Science 214:1363–1365.

Steinbrecht, R. A. 1997. Pore structures in insect olfactory sensilla: a review of data and concepts. Intern. J. Insect Morphol. Embryol. 26:229–245.

Stowe, M. K., Tumlinson, J. H., and Heath, R. R. 1987. Chemical mimicry: bolas spiders emit components of moth prey specific sex pheromones. Science 236:964–967.

Stowe, M. K., Turling, T. C. J., Loughrin, J. H., Lewis, W. J., and Tumlinson, J. H. 1995. The chemistry of eavesdropping, alarm, and deceit. Proc. Natl. Acad. Sci. U.S.A. 92:23–28.

Svensson, G. P., Larsson, M. C., and Hedin, J. 2004. Attraction of the larval predator *Elater ferrugineus* to the sex pheromone of its prey, *Osmoderma ermita*, and its implications for conservation biology. J. Chem. Ecol. 30:353–363.

Symondson, W. O. C., Sunderland, K. D., and Greenstone, M. H. 2002. Can generalist predators be effective biocontrol agents? Annu. Rev. Entomol. 47:561–594.

Takabayashi, J., and Dicke, M. 1996. Plant-carnivore mutualism through herbivore-induced carnivore attractants. Trends Plant Sci. 1:109–113.

Takabayashi, J., Dicke, M., and Posthumus, M. A. 1994. Volatile herbivore-induced terpenoids in plant-mite interactions: variation caused by biotic and abiotic factors. J. Chem. Ecol. 20:1329–1354.

Tallamy, D. W., and Halaweish, F. T. 1993. Effects of age, reproductive activity, sex, and prior exposure on sensitivity to cucurbitacins in southern corn rootworm (Coleoptera: Chrysomelidae). Environ. Entomol. 22:925–932.

Teerling, C. R. 1995. Chemical ecology of western flower thrips. In: Thrips Biology and Management (Parker, B. L., Skinner, M., and Lewis, T., eds.). Nato ASI Series, Series A: Life Sciences 276. New York: Plenum Press; 439–447.

Thompson, J. N. 1994. The Coevolutionary Process. Chicago: University of Chicago Press.

Thompson, J. N. 1999. What we know and do not know about coevolution: insect herbivores and plants as a test case. In: Herbivores: Between Plants and Predators (Ollf, H., Brown, V. K., and Drent, R. H., eds.). Oxford: Blackwell Scientific Publications; 7–30.

Toft, S. 1995. Value of the aphid *Rhopalosiphum padi* as food for cereal spiders. J. Appl. Ecol. 32:552–560.

Toft, S. 1996. Indicators of prey quality for arthropod predators. In: Arthropod Natural Enemies in Arable Land, vol. 2 (Booij, K., and den Nijs, L., eds.). Oxford: Aarhus University Press; 107–116.

Tømmerås, B. Å. 1985. Specialization of the olfactory receptor cells in the bark beetle *Ips typographus* and its predator *Thanasimus formicarius* to bark beetle pheromones and host tree volatiles. J. Comp. Physiol. A 157:335–341.

Turlings, T. C. J., Lengwiler, U. B., Bernasconi Ockroy, M. L., and Wechsler, D. 1998. Timing of induced volatile emissions in maize seedlings. Planta 207:146–152.

Turlings, T. C. J., Tumlinson, J. H., and Lewis, W. J. 1990. Exploitation of herbivore-induced plant odors by host-seeking parasitic wasps. Science 250:1251–1253.

Vet, L. E. M., and Dicke, M. 1992. Ecology of infochemical use by natural enemies in a tritrophic context. Annu. Rev. Entomol. 37:141–172.

Visser. J. H. 1986. Host odor perception in phytophagous insects. Annu. Rev. Entomol. 31:121–144.

Visser, J. H., Van Straten, S., and Maarse, H. 1979. Isolation and identification of volatiles in the foliage of potato, *Solanum tuberosum*, a host plant of the Colorado potato beetle, *Leptinotarsa decemlineata*. J. Chem. Ecol. 5:11–23.

Vité, J. P., and Williamson, D. L. 1970. *Thanasimus dubius*: Prey perception. J. Insect Physiol. 16:233–239.

Vogt, R. G. In press. Molecular basis of pheromone detection in insects. In: Comprehensive Insect Physiology, Biochemistry, Pharmacology and Molecular Biology (Gilbert, L. I., Iatro, K., and Gill, S., eds.). New York: Elsevier.

Vogt, R. G., Callahan, F. E., Rogers, M. E., and Dickens, J. C. 1999. Odorant binding protein diversity and distribution among insect orders, as indicated by LAP, an OBP-related protein of the true bug, *Lygus lineolaris* (Hemiptera: Heteroptera). Chem. Senses 24:481–495.

Vogt, R. G., Riddiford, L. M., and Prestwich, G. D. 1985. Kinetic properties of a sex pheromone-degrading enzyme: the sensillar esterase of *Antheraea polyphemus*. Proc. Natl. Acad. Sci. U.S.A. 82:8827–8831.

Wallin, K. F., and Raffa, K. F. 2004. Feedback between individual host selection behavior and population dynamics in an eruptive herbivore. Ecol. Monogr. 74:101–116.

Weissbecker, B., Van Loon, J. J. A., and Dicke, M. 1999. Electroantennogram responses of a predator, *Perillus bioculatus*, and its prey, *Leptinotarsa decemlineata*, to plant volatiles. J. Chem. Ecol. 25:2313–2325.

Weissbecker, B., Van Loon, J. J. A., Posthumus, M. A., Bouwmeester, H. J., and Dicke, M. 2000. Identification of volatile sesquiterpenoids and their olfactory detection by the two-spotted stinkbug, *Perillus bioculatus*. J. Chem. Ecol. 26:1433–1455.

animals might use to update information about predation risk. Finally, recent neurological work suggests that the amygdala (in the vertebrate midbrain) is a key component in the assessment of predation risk and that differences in risk perception may reflect differences in the neural architecture of the amygdala. Collaborations between behavioral ecologists and neuroethologists could provide insights into the many aspects of risk perception.

Models of predator-prey population dynamics have traditionally focused on predators while treating prey as unresponsive entities (Murdoch and Oaten 1975, Taylor 1984, Turchin 2003). However, a great deal of behavioral research suggests that prey are anything but unresponsive. Many aspects of animal decision making (broadly defined) have been shown to respond swiftly and adaptively to changes in the risk of predation (Sih 1987, Lima and Dill 1990, Lima 1998b). Such behavioral responses often come at the cost of diminished growth, energetic state, and/or reproductive potential (Lima 1998a, Lima 1998b). Predators may influence prey populations and communities by means of these kinds of "nonlethal" effects as much as they do by means of their more obvious, lethal effects (Abrams 1995, Lima 1998b, Peacor and Werner 2001, Schmitz et al. 2003, Schmitz, ch. 12 in this volume).

The key to this flexible decision making by prey is the variable perception of risk, by which animals perceive greater risk when predation risk is indeed greater. This ability to perceive changes in one or more of the components of risk provides the foundation for the nonlethal ecological effects of predators. Hundreds of studies show that the variable perception of risk is taxonomically widespread, occurring in most (if not all) invertebrates and vertebrates (Lima 1998a). Animals of all kinds have little trouble perceiving large changes in the risk of predation, especially changes reflecting sudden or recent encounters with predators. Aquatic animals, in particular, appear able to discern a great deal about changes in the predatory environment through chemical signals (Chivers and Smith 1998, Kats and Dill 1998). Many animals also can make fine distinctions among predator types (Curio 1993) and retain the ability to do so even after a long (multigenerational) period of no contact with predators (see Blumstein et al. 2000).

But what do we actually understand about the perception of risk per se? That is, what do we know about the information available to prey regarding predation risk and the way that information is used to form an assessment of risk? The quick answer is, "not much" beyond the fact that animals act as though they know risk has changed in the local environment. To date, students of antipredator decision making have, implicitly or explicitly, usually assumed that animals have perfect knowledge about predation risk in the local environment. This assumption is perhaps most apparent in theoretical treatments of antipredator decision making, but it pervades, usually implicitly, empirical work, as well—perhaps only insofar as the issue is rarely discussed. This assumption of perfect information reflects the tenets underlying the adaptationist-optimality approach to the study of behavior, an approach dominating the study of antipredator decision making. We doubt that anyone would defend such notions of perfection as anything more than a good conceptual starting point, but behaviorally oriented ecologists have not moved much beyond this assumption to probe the nature of risk perception. To this end, we provide a wide-ranging treatment of the issues surrounding the perception of predation risk. We review the areas of risk perception that are understood relatively well and highlight the many important issues that are largely unexplored.

Basic Issues and Current Limitations

We begin with a useful abstraction for the formation of a perception of risk and its influence on behavior. Following Blumstein and Bouskila (1996), it is perhaps easiest to envision this process as a series of steps:

1. Information about risk is gained via one or more sensory modes.
2. Information so gained is processed into an assessment of the level of risk (the *informational state*, Blumstein and Bouskila 1996; or the perceived level of risk).
3. The animal combines this assessment, or perception, with information about the feeding environment, the animal's energetic or reproductive state, and the like, to produce observable behavior or decision making.

To accomplish steps 1 and 2, an animal must be able to sense the relevant information and then in some way assess the basic components and subcomponents that define the risk of being killed by a predator. The basic components of risk, as envisioned by Lima and Dill (1990), include time exposed to predation, the encounter rate with predators, and the conditional probability of being killed during an encounter. This conditional probability, in turn, is composed of the product of several probabilities of detection and escape, given the way in which predator-prey encounters unfold. An animal would, in principle, need to assess each of these components and subcomponents in forming its perception of risk.

Several limitations to studying risk perception are apparent under this 3-step scenario. The first is the fact that we usually can observe only the final outcome of step 3. Our primary window into the variable perception of risk is the fact that observed behavior changes with apparent changes in the risk of predation. An animal's actual perception of risk has never been measured per se (see also Blumstein and Bouskila 1996). Perhaps further advances in real-time neuroimaging will allow us to at least correlate specific brain activity with changes in various aspects of the predatory environment (Frings et al. 2002), but techniques that could be applied to unrestrained animals seem well into the future. In any case, we currently have no direct way to assess the accuracy of risk perception in a given animal. Accordingly, one must be careful when concluding that the lack of an overt behavioral response indicates a lack of a perceived change in risk (Brown, Adrian, Patton et al. 2001, Mirza and Chivers 2003). Our current inability to independently measure the perceived risk of predation is an obvious limitation.

The second major limitation is that behavioral ecologists have never measured all of the various components of risk for any study animal. Such would be a daunting task, one that would require a considerable understanding of predator behavior, as well as that of the prey (Lima 2002). One might more reasonably be able to measure the sum total risk of being killed (presumably, the outcome of the entire behavioral predator-prey interaction) and then relate that measurement to behavioral predictions, but few studies have made such an attempt (Gilliam and Fraser 1987, Godin and Smith 1988, Skalski and Gilliam 2002). We thus are not in a position to assess the accuracy of risk perception in any animal, even if we could measure such perceptions directly.

Finally, we currently understand little about the way information might be mapped into antipredator decision making. Although steps 1–3 must be accomplished in one way or another, they need not exist as discrete, sequential steps. In behaviorally simple animals, some "hardwired" neural structure likely maps a few key risk-related stimuli directly into behavior itself, without forming a readily observable perception or informational state. In other animals, decision making probably depends on both hardwired (genetically preprogrammed) neural processing and the formation of discrete informational states shaped by learning and experience with predators (see Brown and Chivers, ch. 3 in this volume).

Despite these limitations, much insight into behavioral predator-prey interactions, as well as their nonlethal effects, can still be gained from a thorough consideration of the perception of risk. We begin with a discussion of the kind of information that is available to prey animals and whether predators attempt to manage such information. We then consider accuracy in risk assessments, risk assessment under informational uncertainty, and the neurological bases for risk perception and assessment. Finally, we address the topics of individual differences in risk perception and the implications of imperfect risk assessment for population ecology.

Information about Risk

The information upon which animals base their perception of risk is an important piece of the perceptual puzzle (Blumstein and Bouskila 1996), one that is in principle more accessible than most other issues surrounding the perception of risk. An understanding of the information used by prey to assess risk is essential to the proper modeling of antipredator decision making, even if one must maintain, for practical reasons, the assumption of perfect risk perception.

What Do Animals Know about Predation Risk?

As mentioned, our understanding of what animals "know" about predation risk reflects the stimuli that overtly influence their antipredator behaviors. The often implicit assumption is that, if an animal does not respond to a potential indicator of risk, then that piece of information is unavailable to (i.e., is not perceived by) the animal in question. This assumption is eminently reasonable, for it would make little sense not to respond to, or in some way acknowledge, strong indicators of the risk of predation. However, a clear caveat applies to this assumption. The level of perceived risk associated with a given stimulus may be too low to warrant a response, especially if the animal in question is in a relatively poor energetic state (Lima 1998b).

Thus, one might mistake the lack of a behavioral response for the lack of perceived information. Such problematic cases are actually known to occur. Brown, Adrian, Patton et al. (2001) and Mirza and Chivers (2003) describe a situation in which fish do not respond overtly to low concentrations of a chemical alarm substance, even though more sophisticated testing showed that they did indeed perceive the information. There are also cases in which prey show few overt behavioral responses to threatening stimuli, even though the monitoring of physiological responses, such as changes in heart rate, clearly indicates the perception of the information (Jacobsen 1979). However, these kinds of problems should be minimal if one considers a reasonably wide range of stimulus strength and the more subtle ways in which an animal might respond to a given stimulus.

The information that animals may acquire about risk is much too varied to catalog here, but it is clear that animals respond to variable risk in the context of many different components and subcomponents of risk (Lima and Dill 1990). Much less clear, however, is the precise nature of the information on which behavioral decisions are based. For instance, models of flight-initiation distances suggest that the distance separating predator and prey should be the key determinant of prey behavior, which in fact appears to be the case in empirical studies (Ydenberg and Dill 1986). However, prey behavior might instead be informed by the angular dimension and looming rate of the predator (Dill 1974), rather than the distance of separation per se. Similarly, models suggest that antipredator vigilance should decline with increasing group size in socially feeding animals, a result that has been observed in a variety of species. However, group members might base their vigilance on information about nearest-neighbor distances or local density of group members rather than group size per se (Elgar 1989). Thus, even though it is clear that animals acquire considerable risk-related information, our view of that information might reflect more the structure of our conceptual thinking than the precise information to which prey animals respond.

The body of work on the chemical ecology of antipredator decision making deserves special mention regarding what animals know about risk. Not only do many aquatic animals sense the general presence of predators from chemical cues, but they may be able to determine the specific types of predators, the size of predators, and the types of prey in the predator's diet (Chivers and Smith 1998, Kats and Dill 1998, Chivers and Mirza 2001). Such chemicals may provide information across short (Turner and Montgomery 2003) or long distances (Lawrence and Smith 1989), or even an entire body of water (Hopper 2001). The chemical structure of cues also has been determined in some systems (e.g., Brown, Adrian, and Shih 2001). Most chemically oriented research has focused on fish and other aquatic animals (Chivers and Smith 1998, Kats and Dill 1998), but chemical cues are used to detect the recent presence of predators by, at least, some terrestrial reptiles. In addition, some mammals have been shown to detect specific information about predators from the predators' feces and urine (Kats and Dill 1998). In contrast, birds appear to lack a well-developed ability to sense and use chemical information about predation risk. There is also little work to suggest, one way or the other, that terrestrial insects use chemical information about risk (Kats and Dill 1998), but their aquatic counterparts very clearly do so (Chivers and Smith 1998, Kats and Dill 1998).

Information Pre- and Post-Encounter with Predators

Most studies that bear directly on information availability focus on the situation faced by animals during an encounter with a predator. This situation is relatively easy to study, because the information most relevant to a prey animal is more apparent and accessible to biologists, information such as the type of predator (Curio 1993) and its motivation (Magurran and Higham 1988), and distances separating predator, prey, and possible refuges (Ydenberg and Dill 1986). Such information, or surrogates thereof, is presumably readily available to prey animals, as well.

The information available to prey regarding the pre-encounter, or encounter, rate component of risk (Lima and Dill 1990) is much less clear. This situation reflects the lack of studies that relate antipredator decision making to quantitative measures of attack or encounter rates (Lima 1998b; but see Cresswell 1994 for a notable exception). Such a dearth of studies on this topic might seem surprising, but in fact information on attack or encounter rates is very difficult to obtain without detailed knowledge of the behavior of predators (Lima 2002). In any case, an animal's assessment of this component of risk is probably not based solely on encounters with predators; indications of recent kills and the like might provide some information, as well. Chemical cues also could provide a good deal of information about attack and encounter rates. There is, however, little work relating the concentration of chemical cues to predator abundance, attack rates, or prey mortality levels (Kats and Dill 1998). In fact, remarkably few studies have examined even whether animals respond to changes in the concentration of chemical cues (Kats and Dill 1998, Lima 1998b, Van Buskirk and Arioli 2002). Overall, we suspect that many classes of prey animals, just like the biologists who study them, have a difficult time assessing encounter and attack rates.

Sensory Compensation

An important and rarely acknowledged issue in the perception of risk is the number and nature of sensory modes available to the animal in question. In the study of antipredator decision making, there is a strong tendency to consider only one sensory mode at a time (often implicitly). However, two or more sensory modes are unlikely to be completely redundant, so they might provide complementary or compensatory sources of information. For example, Hartman and Abrahams (2000) suggest that a form of *sensory compensation* exists between chemical and visual sensory modes in risk perception by minnows. Minnows in turbid water rely more on their chemical senses than on their impaired visual sense. In clear water, chemical signals may be less important if they are not associated with visual cues of predator presence (Chivers et al. 2001). This situation might help explain why fish seem to show only a threshold response to increasing concentration of alarm substances (Brown, Adrian, Patton et al. 2001, Mirza and Chivers 2003), even though increases beyond the threshold probably indicate an increasing risk of predation. Mirza and Chivers (2003) suggest that an alarm substance may act mainly to alert the fish to the presence of danger once a threshold is exceeded, which then puts the fish in the mode of using visual cues over chemical cues. Without this insight into sensory compensation, one might conclude that the fish in question possess a rather crude perception of risk, one based on their response to chemical cues alone.

Sensory compensation is probably common in animals, for just about every species has at least two sensory modes that might be directed toward predator detection (but see Mathis and Vincent 2000). We expect that only species involved in very specialized predator-prey relationships would rely mainly on a single sensory mode. However, Freund and Olmstead (2000) suggest that some insects may devote a single sensory mode to predator detection while devoting another mode to feeding. The idea is that having two specialized sensory modes would avoid sensory interference that could detract from feeding and predator detection. In any case, multiple sensory modes and sensory compensation clearly deserve much more attention in the study of antipredator decision making.

Public Information in the Assessment of Predation Risk

Here we consider the extent to which information about a given animal's experience with predators might be available to other animals, essentially as "public information" about predation risk (sensu Valone 1989). Public information could conceivably be of great value from the antipredator perspective, because it is essentially cost-free compared to information gained through personal experience alone. Such information might often apply mainly to members of a given social group, but it might also extend across a much larger spatial scale to other animals in the local area.

The concept of public information has traditionally been associated with foraging problems faced by social animals (Valone 1989, Templeton and Giraldeau 1995), but public information, under many guises, has long been the subject of study in the antipredator behavior of social animals. For instance, public information about predator detection may be available in the form of alarm calls that signal the detection of a predator (Blumstein 1999). Similarly, in social minnows, predator inspectors may impart information about a potential threat to others in the group (Magurran and Higham 1988). The large literature on antipredator vigilance in social birds and mammals also focuses on the implications of public information about predation risk (Elgar 1989, Bednekoff and Lima 1998). One might envision many other situations in which the general degree of sociality, cohesion within of the group, and the like convey to group members public information about risk (Lima and Dill 1990).

Of potentially far-reaching ecological importance is public information that extends across much larger spatial scales than within a social group. The possibility of public information at a larger spatial scale is perhaps not as obvious as that within social groups. However, physically wide-ranging cues, such as chemicals in aquatic systems, can potentially diffuse over large volumes (Lawrence and Smith 1989, Mirza and Chivers 2003) and even entire bodies of water (Hopper 2001). Such chemical information might be particularly valuable when prey can also assess both the source of the cue (prey or predator) and the size of the local prey population (perhaps also from chemical cues; Peacor 2003), which could provide a better per capita assessment of risk. The long-distance effects of chemical cues need not be so great, however, and may not extend beyond a few meters (Turner and Montgomery 2003). The great majority of studies on the chemical ecology of antipredator behavior concern very short distance effects, so the degree of large-scale public (chemical) information is currently difficult to assess.

Ultimately, the degree to which there is significant large-scale, public information about risk will have many implications for the accuracy of risk estimates in prey and the speed with which changes in risk are perceived within a prey population. These issues of accuracy and speed ultimately have implications for population dynamics, as will be discussed. The notion of public information also has many implications for the way predators might "manage" the perception of risk in their prey.

Do Predators Manage the Information Available to Prey?

A predator would benefit from keeping prey "in the dark" about its presence in a given area. Prey would be more catchable if predators could keep the local perception of risk as low as possible; that is, if they could manage the information available to their prey. Such information management would, presumably, be most beneficial to predators with stable home ranges or complete (territorial) control over a prey population (Arcese et al. 1996). Transient predators might have relatively little to gain from information management, if short-term costs are involved.

Predators could take a number of steps to manage the information available to prey. One of the most important might be to restrict attacks in a given area. Repeated attacks could easily lead to the buildup of public information left by the predator. This information might be in the form of chemical cues resulting from successful acts of predation (Chivers and Smith 1998, Kats and Dill 1998). Public information might also spread quickly when several prey, perhaps in a social group, witness a given attack (Curio 1993, Pitcher and Parrish 1993), or when prey perceive the remains of recent kills as evidence of a predator's presence in the area (Curio 1993, Dukas 2001). Other problems for the predator occur when it is simply sighted by prey as it moves about in a given area. The proportion of prey involved in these nonlethal encounters should increase over time, making a patch relatively unprofitable as general antipredator behavior increases in the prey population (Charnov et al. 1976, Brown et al. 1999, Kotler et al. 2002, Lima 2002).

We thus expect predators to minimize the buildup of risk-related public information available to prey. Public information in chemical cues could be minimized with strategic defecation or urination patterns in space and time (Brown et al. 1995; see also Banks et al. 2000, for a study of prey managing chemical cues available to predators), or across a territory if such material is used for territory marking. The buildup of visual informational cues might be minimized by attacking straggler prey on the margins of groups or by consuming captured animals out of sight of potential future prey. Predators may manage informational cues simply by being spatially and temporally unpredictable, such that prey gain little information about future risk from a given sighting (Mitchell and Lima 2002, Hugie 2003). In an ongoing study of hawk-prey interactions, we have observed that bird-eating *Accipiter* hawks are essentially "random" (or maximally unpredictable) in their return times to a given hunting spot.

These hawks feed on highly social prey, so their "randomness" might reflect the management of public information about risk.

Distinguishing some of these possible information management behaviors from the generally stealthy existences of many predators will not be easy. Furthermore, behaviors such as eating prey in hidden spots could also reflect avoidance of aggression or kleptoparasitism. Similarly, attacking isolated stragglers might simply be the easiest short-term hunting strategy for predators. Still, such prey management is a real possibility, as has been seen in predation and brood parasitism by female cowbirds (Arcese et al. 1996). Ultimately, insights into information management by predators will require an explicit acknowledgment of the fact that both predator and prey are active participants in any behavioral predator-prey interaction, an acknowledgment we often fail to make in the study of antipredator decision making (Lima 2002).

Accuracy in Predation Risk Assessment

In principle, we could assess the accuracy of perceived risk in at least two ways. The first involves determining an animal's actual assessed risk of predation associated with a given behavioral situation (Bouskila and Blumstein 1992, Blumstein and Bouskila 1996). As mentioned, however, we are unable to do this at present. Perhaps a more promising approach would be to compare observed antipredator decision making to that predicted on theoretical grounds: any deviation could yield insight into the accuracy of risk perception. The problem with this method is our inability to make precise behavioral predictions, which reflects the difficulty in assessing the many components of risk (e.g., attack rates; Lima 1998b), as well as uncertainty about the fitness function under which a given animal operates (Brown 1992). Nevertheless, the issue of accuracy in the perception of risk is an important one to further consider.

Perception of Lethal Risks by Humans

Entire industries are devoted to the estimation of various lethal risks in humans, and humans can express perceptions directly in a way that other animals cannot. With humans, one may circumvent many of the problems inherent in studying risk perception in nonhumans. Thus, the results from studies on humans might yield some insight into risk perception in general. There are clear caveats here in that (a) many modern lethal risks are evolutionarily new to humans, and (b) the act of surveying people about risk perception might actually influence their expressed perceptions. Nevertheless, a consideration of human risk perception is both relevant and informative, and covers the best available information on risk perception.

Risk perception in humans has been an active area of research for decades, although there has been little cross-fertilization with behavioral biology. In an influential early essay on this topic, Liechtenstein et al. (1978) surveyed select groups for their estimates of a variety of lethal risks. Their results suggested that humans typically overestimate small risks and underestimate large risks. This result has been replicated in a variety of contexts (Slovic 1987) and has become a tenet of the field.

In another influential work, Slovic (1987) argued that risk assessment in humans is a largely rational affair that takes into consideration the degree to which a threat is both knowable and potentially serious. Slovic suggested that difficult-to-know risks are probably overestimated, especially when the potential consequences are perceived to be severe or the source of information is deemed unreliable. Slovic did not agree with the contention of some researchers that such overestimation is irrational, suggesting instead that such behavior can be interpreted as rational, given the degree of uncertainty involved. On the other hand, several studies suggest relative optimism in assessing risks (see van der Pligt 1998). Subjects sometimes see their personal risk as lower than that of others in the population. Whether these optimistic estimates are actual underestimates of risk, however, is unclear (van der Pligt 1998).

Much of this research on humans is based on the premise of rationality: the idea that a human's decisions are ultimately based on logic and mathematical probabilities, as well as a coherent and logically consistent view of the world (Sharif and LeBoeuf 2002). This premise is similar to the adaptationist program in the study of behavioral and evolutionary ecology, in which a given trait is assumed to be adaptive until one exhausts all reasonable hypotheses to that effect (Mitchell and Valone 1990). Just as the adaptationist approach has garnered its critics (e.g., Gould and Lewontin 1979), so too has the rationalist approach to the study of risk perception. Sharif and LeBoeuf (2002) provide a succinct summary of the arguments against the rationality paradigm, pointing out, among other things, that most humans, even many with a substantial background in statistics, harbor so many misconceptions about basic probability theory as to cast doubt on their ability to make accurate statements about the risks of various activities. For example, Wright et al. (2002) surveyed insurance claims adjusters (experts) and nonexperts for their estimates of various lethal risks. The results suggested that the experts were only slightly better at estimating risks than the nonexperts and that both overestimated risk. Other researchers defend the rationality approach as a logical way to proceed (Sharif and LeBouef 2002), much as in the defense of the adaptationist approach (Stephens and Krebs 1986, Mitchell and Valone 1990).

In perhaps the most relevant studies, Benjamin and Dougan (1997) and Benjamin et al. (2001) point out that in the studies inspired by Liechtenstein et al. (1978), including Wright et al. (2002), subjects were asked to estimate marginal probabilities of various events (risk without respect to class of risk taker). When subjects were asked to assess various lethal risks most relevant to their personal lives, the tendency to under- or overestimate risk largely disappeared. In the aggregate, estimates of risk were fairly accurate, suggesting that subjects with intimate knowledge of a given area could make relatively accurate estimates. It is worth noting, however, that, though the aggregate estimates were fairly accurate, individual subjects were often dramatically "off the mark" in their estimates. It is also clear the subjects in all of these studies must have based their estimates on a combination of personal and public information, but this distinction was not considered by researchers.

So far, the results of this controversial field paint a mixed picture for risk perception in nonhumans. We as humans are definitely deficient in our intuitive understanding of the basics of probability theory, and people may sometimes express wildly inaccurate estimates of certain lethal risks. On the other hand, humans tend to be able to at least rank risks with reasonable accuracy (Wright et al. 2002) and may be able to assess risk relatively accurately when they confront familiar issues. It is therefore tempting to conclude that animals could not be expected to estimate risks any better than behaviorally sophisticated humans, but this conclusion is by no means assured. Nonetheless, that humans tend to overestimate many risks, and tend to err on the side of overestimation when they face uncertainty, is also directly relevant to uncertainty and risk perception in nonhumans.

Risk Assessment in an Uncertain Environment

Prey animals will inevitably face much uncertainty about the level of predation risk in the local environment. This uncertainty may apply to any component or subcomponent of risk, but perhaps especially to estimates of encounter rates with predators. Bouskila and Blumstein (1992) pointed out that there are obvious dangers in attempting to gain more information about local risk, and thus suggested that animals are likely to use some sort of "rule of thumb" in estimating the risk of predation. In accordance with the notion that the fitness costs of overestimating risk should be lower than the costs of underestimating risk, they suggest that a good rule of thumb is to overestimate risk to some extent.

A rule of thumb based on overestimating risk is an intuitively appealing idea, one that fits well with the human adage of "erring on the side of caution." Nevertheless, Abrams (1994) showed that underestimating risk might lead to a lower fitness cost under some circumstances, although he concedes that risk overestimation generally may be a better strategy. Koops and Abrahams (1998) also questioned the general validity of the overestimation rule of thumb. These authors extended the simple stochastic dynamic model of Bouskila and Blumstein (1992) to include a variety of possible terminal fitness functions, which relate energetic state to fitness at the end of the time period in question. Their analysis suggests that the relative cost of overestimating versus underestimating risk depends on sometimes subtle differences in the shape of the terminal fitness functions. In our analysis of this issue, we have found that a critical point is the degree of tolerance that one specifies as an acceptable fitness cost (Figure 8.1; Steury, T. D., Mitchell, W. A., and Lima, S. L., unpublished data). Many different fitness functions may have specific levels of tolerance that favor underestimation, but the shape of most reasonable fitness functions generally favors overestimation as a good rule of thumb.

We suspect that overestimating risk is generally a good idea when one confronts significant uncertainty about risk (i.e., we expect that animals will appear to be overly cautious in many situations). However, given our present ignorance about the perception of risk in animals, nobody can be certain about such a conclusion. Some support for this idea comes from the work suggesting that humans often overestimate lethal risks, especially when they assess largely unknown risks. Further support comes from the fact that animals respond to many types of novel human disturbances as though such disturbances represent a risk of predation (Frid and Dill 2002), even though in many cases there really is no such threat. Finally, it might conceivably be better to overestimate high risks and underestimate low risks (or vice versa), but these more sophisticated rules of thumb have not been investigated. A more sophisticated rule of thumb might also recognize that there will likely be greater uncertainty about low levels of risk than about high levels.

Gaining Information about Risk

This issue of uncertainly about risk raises the question of just what risk an animal should accept to gain better information about the local level of risk. Information about risk is a valuable commodity, one that might be worth some risk to obtain. However, few studies have examined risk taking for the purpose of gaining information about risk (but see Koops 2004). This situation undoubtedly reflects our tendency to assume perfect information about risk.

Studies of predator inspection provide a notable exception to this tendency to ignore information-gathering in antipredator decision making (Dugatkin and Godin 1992). Predator inspection involves carefully approaching a predator, sometimes at considerable risk (Dugatkin 1992), in order to gather information about the identity of the potential predator (or about whether it is a predator), its motivation to hunt, and so forth (Magurran and Higham 1988). Predator inspection is particularly common in small fish (Lima 1998b) but has also been described in some mammals (FitzGibbon 1994). Vision is often limited in murky water; hence, fish, in particular, have a need to approach potential predators for further information (not all large objects or fish are predators). Accordingly, hungry fish that must feed within view of a large fish are more likely to inspect the potential predator than those less motivated to feed (Godin and Crossman 1994). Complicating this informational view are a few studies suggesting that predator inspection acts also to deter attacks, because inspection clearly signals that prey are aware of the predator's presence (FitzGibbon 1994, Godin and Davis 1995). Prey animals can nevertheless acquire much information about potential predators through inspection.

How Is Information Used to Reduce Uncertainty?

Again, there are many sources of information about predation risk in the environment. The question here is, how might such information be incorporated into some sort of assessment-perception of risk? Once again, the pervasive assumption of perfect information and perception of risk has worked against answering this question (or even asking it in the first place). In marked contrast, the literature on foraging behavior includes many studies that concern imperfect information and how animals might profitably use the information that they gather while foraging (Stephens and Krebs 1986).

A major source of information about risk is an animal's personal experience with predators. Such experience might come in the form of recent encounters or sightings of predators, and the like. There are many sorts of rules of information use that one might apply to such information (Beauchamp 2000), but to our knowledge only the Bayesian approach has been applied to the assessment of predation risk (Ydenberg 1998). All these studies address the encounter component of risk. Sih (1992) used the Bayesian approach to analyze an animal's decision to leave a refuge

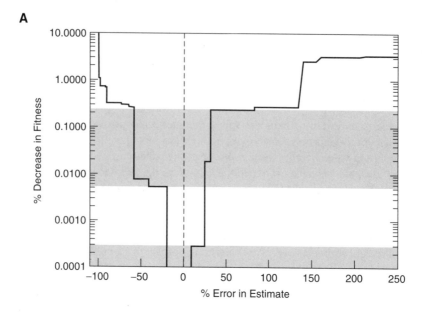

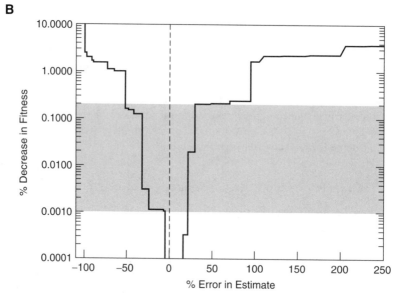

Figure 8.1. Representative theoretical relationship between the percent error in the estimate of predation risk (x-axis) and the % loss of fitness relative to that achieved with 0% error (y-axis). Animals are attempting to maximize fitness by choosing the habitat that maximizes fitness over a given period of time (as per Bouskila and Blumstein 1992, Koops and Abrahams 1998). Relationships were generated using stochastic dynamic programming. The terminal fitness functions used here were the (A) linear with saturation and (B) asymptotically increasing [see Koops and Abrahams (1998) for exact methods and parameter values used]. Positive % error indicates risk overestimation, and negative values indicate underestimation. Underestimation is limited to values up to 100%, but larger degrees of overestimation are possible.

(continued)

when it is unsure about the nearby presence of a predator. In this model, the refuging animal continually updates its prior estimates of predator presence, in accordance with the information that it receives over time (i.e., the detection of a nearby predator or lack thereof). Houtman (1995; see description in Ydenberg 1998) also used a Bayesian approach to model a similar situation regarding a fish's decision to feed (take a risk by moving) or remain motionless. Bayesian approaches have also been applied to informational uncertainty about predator detection in socially foraging animals (Lima 1994).

Animals almost certainly use some sort of updating rule, or rules, in assessing predator presence or the likelihood of attack during a particular time interval. However, whether or not Bayes's theorem provides an appropriate description of that process is an open question. Following the lead of the foraging literature (Beauchamp 2000), many other types of rules exist by which information can be incorporated into decision making. Conceivable rules might differ in how past information is discounted. The persistence of a predatory cue presents interesting problems for discounting, because chemical cues are likely to provide much information about risk, but also persist after the information is no longer valid. These updating rules could go beyond information gained from personal experience to consider public information about predation risk (as per Valone 1989) and the relative value of information from different sensory modes (Valone 1989). Incorporating both visual and chemical cues into such rules also would present an interesting challenge.

Importantly, these sorts of informational rules may be complicated by predators' active management of the risk-related information available to prey. As delineated earlier, predators may attempt to remain unpredictable both spatially and temporally so that a recent appearance of a predator may yield relatively little information about future attacks. Thus, any updating rule used by prey must reflect such predatory behavior and more generally the nature of the particular predator-prey behavioral game within which they evolved (Lima 2002). In any case, further work on rules that might prove useful for estimating the risk of predation (along the lines of Beauchamp 2000) would be valuable.

Whether prey should overestimate or underestimate predation risk depends on the fitness cost that is deemed tolerable (taking into account the cost of acquiring more information). For example, in (B), a tolerance level set at a fitness cost of 0.01% allows an overestimate of up to 29.7%, but a larger underestimate at 31.7% is possible. This situation favors the underestimation of risk. On the other hand, with a tolerable fitness cost set at 0.5 %, an overestimate of risk up to 95.5% is possible, but an underestimate of up to only 52.1% is allowable; here, overestimation is favored. Accordingly, areas shaded in gray indicate a range of fitness costs over which risk underestimation is favored; ranges favoring overestimation are indicated by the absence of shading. The overall situations depicted in (A) and (B) are representative of the terminal fitness functions considered in Koops and Abrahams (1998). That is, at very low tolerable fitness costs, there may be alternating ranges of fitness costs in which overestimation or underestimation is favored. Above a fitness cost of between 0.1–1.0%, however, overestimation is generally favored.

The Neurology of Predation Risk Assessment

The neurological circuitry and processing underlying fairly complex antipredator decision making has not been worked out for any animal. Neurobiologists have necessarily focused on specialized (reflex-like) behaviors to make progress. Many of these specialized behaviors have antipredator functions, such as escape behavior (Edwards et al. 1999, Eaton et al. 2001) and the acoustical startle response (Fendt and Fanselow 1999), but little neurological information exists on complex assessments of risk that probably occur routinely in many animals. Nevertheless, existing studies provide some insight into the neurological basis of risk perception.

Classic work on one of these specialized behavioral systems, the bat detection and evasion system in noctuid moths, illustrates some points about risk assessment (Roeder 1965; see also Triblehorn and Yager, ch. 5 in this volume). Here, "ears" along a moth's abdomen are tuned specifically to detect ultrasonic bats' feeding pulses. Sensory receptors respond to both the frequency and intensity with which bats emit feeding pulses. The processing of information from both the left and right ears (which takes place in thoracic ganglia) determines the location of the bat relative to the moth, as well as the most appropriate motor output, given the level of risk inferred (orienting away from the bat or immediate evasive maneuvering, etc.). Here, the three steps outlined earlier are taking place. The important sensory modes and information upon which the moths act is apparent, and the neural processing of information in the thoracic ganglia clearly takes care of steps 2 and 3 (although it is not clear that steps 2 and 3 are discrete entities in this process).

Risk perception in higher vertebrates presumably includes some simple assessment systems (like that in the noctuid moths) coupled to more complex cognitive processes. The nature of such neural systems is far from clear, but recent work on humans and rats may provide some insight. The human amygdala, a structure in the midbrain, is thought to be a key center of emotions (Davis 1992, Zald 2003) and probably acts very similarly in most other vertebrates (Martínez-García et al. 2002). Fear is clearly one of those emotions, so the amygdala is a prime candidate for the sorts of risk assessment and perception in which we are interested. Of great relevance is recent work which suggests that the amygdala contains in essence a "fear assessment module" designed to quickly detect threats and initiate appropriate responses (Lang et al. 2000, Öhman and Mineka 2001). In some cases, as few as three or four synapses may separate sensory receptors and a "decision" by the amygdala on an initial course of action. The quick and sometimes automatic responses induced by the human amygdala are nicely demonstrated in "masking" studies, in which subjects are exposed to a very brief image of a threatening stimulus, which is instantly followed by a persistent (masking) image of a nonthreatening stimulus (Lang et al. 2000, Öhman and Mineka 2001). The subjects in such masking studies were not consciously aware of the threatening image, but their amygdalas detected the image and immediately (within 100 milliseconds) initiated a heightened state of anxiety and vigilance. The subjects were aware only of their heightened state of anxiety, not the cause. Thus, the initiation of strong fear responses and the initial responses to it are not necessarily under conscious control in humans. This quick, subconscious activation of a fear assessment module is undoubtedly analogous to the specialized antipredator behaviors that have been studied in nonhumans.

There is obviously more to threat perception and decision making in vertebrates than the processes that take place in the amygdala. In particular, much of the more complex, trade-off-driven antipredator decision making undoubtedly involves higher brain centers in the cortex. Nevertheless, the amygdala is probably involved in this more complex decision making, as well. The amygdala projects neurologically to many areas of the vertebrate cortex (Martínez-García et al. 2002) and may also be the site at which fear-related memories are stored and processed (Fanselow and LeDoux 1999; but see Cahill et al. 1999). Of particular relevance to complex decision making is the observation that a single exposure to a severe threat can lead to relatively long-term changes in the amygdala and thus the decisions it influences (Wiedenmayer 2004).

The amygdala and associated neural structures are probably coupled also to the innate components of risk assessment, which undoubtedly influence many aspects of decision making. Animals may possess innate knowledge about the risk of predation, but it is probably more accurate to say that animals possess genetically programmed ways of processing information into specific assessments of risk. Neurologically, this innate information processing is probably expressed, in part, in the neural architecture and projections of the amygdala. Such neural architectures are undoubtedly the basis for species-specific abilities to recognize specific predators and the like (Curio 1978, Curio 1993), but the neural attributes underlying such abilities are largely unknown. Innate abilities to distinguish dangerous from safe habitats and to determine population-level differences in antipredator behavior also might be traced to the amygdala. Further neurological work in these and related areas promises many insights into risk assessment and antipredator decision making.

Individual Differences in the Perception of Predation Risk

Do some individuals consistently act as risk takers while others act as risk avoiders? Wilson et al. (1994) suggest that such differences do indeed exist and that they are inherent characteristics of the animals in question: some are relatively "bold," while others are "shy." But do bold and shy animals perceive low and high risk, respectively, in a given situation? Observed antipredator decision making reflects not only an animal's perception of risk, but also the value that it places on feeding or other activities. Hence, some animals might be consistently bold not because they perceive low risk but because they consistently place a higher value on a given activity. Boldness might also be associated with superior antipredator capabilities, which would make a given activity less risky for apparently bold individuals.

Recent work suggests that differences in boldness are indeed driven largely by differing perceptions of risk (or, more generally, differences in fearfulness). The most telling results come from artificial-selection experiments, in which experimenters select for relatively fearful or fearless behavioral phenotypes (Jones et al. 1994, Malmkvist and Hansen 2002, Sih et al. in press; see also Riechert, ch. 4 in this volume). After only a few generations, marked differences in fearfulness are apparent between the selected lines. These differences in fearfulness apply not merely to one specific behavioral situation (e.g., feeding, reproduction), but generalize across many situations (but see Wilson et al. 1993, Hedrick 2000). Parsimony thus suggests that the object of selection is fearfulness itself (or the perception of risk), rather than the tendency to place consistently high or low values on disparate entities such as food, social interactions, and the like.

In a wide-ranging review, Sih et al. (in press) refer to behavioral phenotypes driven by differing degrees of fearfulness as *behavioral syndromes*. These behavioral syndromes are undoubtedly related to the different personality types that have been studied for some time in animal and human psychology (Gosling 2001). Sih et al. suggest that behavioral syndromes reflect genetic correlations among behavioral traits such that strong selection for risk-taking in a given context will tend to produce (potentially maladaptive) risk-taking in other contexts. These genetic correlations may be mediated through the amygdala and its many projections to various areas of the brain (at least in vertebrates). This conjecture remains to be substantiated. However, amygdala-based effects are thought to underlie basic personality traits in humans (Öhman and Mineka 2001).

Future work in this emerging field will probably demonstrate the widespread existence of fear-driven behavioral syndromes and show that much of this variation has a genetic basis (Fraser et al. 2001, Sih et al. 2003, Riechert, ch. 4). Such an outcome naturally raises the question of just what forces maintain such variation in risk perception (fearfulness) in a population. Both Wilson (1998) and Sih et al. (in press) suggest that this variation could be maintained by some kind of frequency-dependent selection in which a mixture of behavioral types is ultimately favored over a single type (see also Abrams 2003). At an ecological level, this situation may be intimately related to the behavioral partitioning of the feeding niche in which a variety of distinct feeding types are maintained in a population (Swanson et al. 2003).

Perception of Risk and Population Dynamics

Current approaches to modeling the population-level effects of antipredator decision-making almost always assume that prey have perfect information about, or a perfect assessment of, risk. This simple assumption of perfect information or risk assessment is clearly a good starting point, but animals are not likely to be perfect. Do such realizations matter to population dynamics?

Imperfect information or assessment about the risk of encounters with predators may indeed have important implications for predator-prey dynamics. Both Brown et al. (1999) and Luttbeg and Schmitz (2000) found that imperfect assessment of the risk of encounters may act to stabilize predator-prey interactions, especially in spatially structured interactions. Clearly, if prey know that predators are likely to be in the immediate vicinity, then they will be more effective at predator avoidance than prey lacking such information. Thus, perfect knowledge has the effect of making prey more unavailable to predators, and the prey, in turn, do not pay high nonlethal costs (e.g., in terms of lowered reproduction) when predators are not actually present. Both of these factors have a destabilizing effect on predator-prey dynamics (Brown et al.

1999). Furthermore, the encounter component of risk may be difficult to assess for many animals, as discussed; hence, such imperfect assessments of risk might act to stabilize many types of predator-prey systems. Imperfect information about other components and subcomponents of risk could also have similar effects, but most attempts to include antipredator behavior into population models consider only the encounter component of risk (but see Abrams, ch. 13 in this volume).

Do certain predator-prey systems have available more risk-related information than others? Are these information-rich systems relatively prone to the destabilizing effects of well-informed antipredator decision making? We suggest that predator-prey systems with the greatest degree of public information about risk would be good candidates for such effects. Aquatic systems would seem to be the most likely candidates. Chemical cues in aquatic systems may be widely dispersed, contain much information, and invoke relatively quick changes in the perception of risk when risk changes. However, this suggestion is largely speculative, because few studies have investigated whether prey in aquatic systems even respond to variation in the concentrations of chemical cues of predation (Kats and Dill 1998).

Conclusions

The variable perception of risk forms the foundation on which the nonlethal effects of predators are transmitted to prey populations and communities. This variable perception of risk occurs across the broad spectrum of animal life, as evidenced by the fact that animals of all kinds respond to changes in predation risk. Beyond these facts, however, our present understanding of the perception of risk in prey animals is rudimentary at best. It is certainly no platitude to say that much more work is needed on just about every aspect of the study of risk perception in prey.

The outlook for progress in some areas in the study of risk perception is good. One of the most fruitful avenues of future research will be to characterize the risk-related information available to prey animals. Behavioral ecologists have done relatively little to characterize such information, relying mainly on an (often implicit) assumption of perfect information and risk assessment. Yet assessing information availability is relatively easy to do, and we have little doubt that doing so will provide significant new insights into behavioral predator-prey interactions. Another area that seems particularly fruitful is further work on behavioral syndromes based on differences in the perception of risk. In addition, collaborations between behavioral ecologists and neuroethologists should provide more insights into the neurological contributions to risk perception. Further theoretical work on information management by predators, as well as the population-level consequences of inaccurate risk assessments, should also prove insightful.

Progress in other areas is likely to remain stymied for some time, especially with regard to the accuracy of risk perception and, to a lesser extent, risk assessment under uncertainty. The main obstacle is the fact that risk perceptions per se cannot easily be measured (and may ultimately prove impossible to measure). Despite this obstacle, it is certainly worth pursuing additional theoretical work in these problem areas, especially with respect to the sorts of informational rules that animals might use to update estimates of risk, and the consequences of evolutionary games between predator and prey for such rules. From the empirical perspective, we certainly welcome any insights into how such perceptions might ultimately be measured.

Acknowledgments

This work was supported in part by National Science Foundation Grant IBN-0130758 to S.L.L. and by the Department of Life Sciences at Indiana State University. We thank P. Barbosa and two anonymous reviewers for many helpful comments on the original version of this work.

Literature Cited

Abrams, P. A. 1994. Should prey overestimate the risk of predation? Am. Nat. 144:317–328.

Abrams, P. A. 1995. Implications of dynamically variable traits for identifying, classifying, and measuring direct and indirect effects in ecological communities. Am. Nat. 146:112–134.

Abrams, P. A. 2003. Can adaptive behavior lead to diversification of traits determining a trade-off between foraging gain and predation risk? Evol. Ecol. Res. 5:653–670.

Arcese, P., Smith, J. N. M., and Hatch, M. I. 1996. Nest predation by cowbirds and its consequences for passerine demography. Proc. Natl. Acad. Sci. 93:4608–4611.

Banks, P. B., Norrdahl, K., and Korpimäki, E. 2000. Non-linearity in the predation risk of prey mobility. Proc. R. Soc. Lond. B Biol. Sci. 267:1621–1625.

Beauchamp, G. 2000. Learning rules for social foragers: implications for the producer-scrounger game and ideal free distribution theory. J. Theor. Biol. 207:21–35.

Bednekoff, P. A., and Lima, S. L. 1998. Randomness, chaos, and confusion in the study of anti-predator vigilance. Trends Ecol. Evol. 13:284–287.

Benjamin, D. K., and Dougan, W. R. 1997. Individuals' estimates of the risks of death: part I—a reassessment of the previous evidence. J. Risk Uncert. 15:115–133.

Benjamin, D. K., Dougan, W. R., and Buschena, D. 2001. Individuals' estimates of the risks of death: part II—new evidence. J. Risk Uncert. 22:35–57.

Blumstein, D. T. 1999. Alarm calling in three species of marmots. Behaviour 136:731–757.

Blumstein, D. T., and Bouskila, A. 1996. Assessment and decision-making in animals: a mechanistic model underlying behavioral flexibility can prevent ambiguity. Oikos 77:569–577.

Blumstein, D. T., Daniel, J. C., Griffin, A. S., and Evans, C. S. 2000. Insular tammar wallabies (*Macropus eugenii*) respond to visual but not acoustic cues from predators. Behav. Ecol.11:528–535.

Bouskila, A., and Blumstein, D. T. 1992. Rules of thumb for predation hazard assessment: predictions from a dynamic model. Am. Nat. 139:161–176.

Brown, G. E., Adrian, A. C., Jr., Patton, T., and Chivers, D. P. 2001. Fathead minnows learn to recognize predator odor when exposed to artificial alarm pheromone below their behavioural response threshold. Can. J. Zool. 79:2239–2245.

Brown, G. E., Adrian, J. C., Jr., and Shih, M. 2001. Behavioural responses of fathead minnows (*Pimephales promelas*) to hypoxanthine-3–N-oxide at varying concentration. J. Fish Biol. 58:1465–1470.

Brown, G. E., Chivers, D. P., and Smith, R. J. F. 1995. Localized defection by pike: a response to labeling by cyprinid alarm pheromone? Behav. Ecol. Sociobiol. 36:105–110.

Brown, J. S. 1992. Patch use under predation risk: I. models and predictions. Ann. Zool. Fennici 29:301–309.

Brown, J. S., Laundre, J. W., and Gurung, M. 1999. The ecology of fear: optimal foraging, game theory, and trophic interactions. J. Mammal. 80:385–399.

Cahill, L., Weinberger, N. M., Roozendaal, B., and McGaugh, J. L. 1999. Is the amygdala a focus of "conditioned fear"? Some questions and caveats. Neuron 23:227–228.

Charnov, E. L., Orians, G. H., and Hyatt, K. 1976. The ecological implications of resource depression. Am. Nat. 110:247–259.

Chivers, D. P., and Mirza, R. S. 2001. Predator diet cues and the assessment of predation risk by aquatic invertebrates: a review and prospectus. In: Chemical Signals in Vertebrates, vol. 9 (Marchlewska-Koj, A., Lepri, J. J., and Müller-Scharze, D., eds.). New York: Plenum Press; 27–284.

Chivers, D. P., Mirza, R. S., Bryer, P. J., and Kiesecker, J. M. 2001. Threat-sensitive predator avoidance by slimy sculpins: understanding the importance of visual vs. chemical information. Can. J. Zool. 79:876–873.

Chivers, D. P., and Smith, R. J. F. 1998. Chemical alarm signaling in aquatic predator-prey systems: a review and prospectus. Ecoscience 5:338–352.

Cresswell, W. 1994. Flocking is an effective anti-predation strategy in redshanks, *Tringa totanus*. Anim. Behav. 47:433–442.

Curio, E. 1978. The Ethology of Predation. New York: Springer-Verlag.

Curio, E. 1993. Proximate and developmental aspects of antipredator behavior. Adv. Study Behav. 22:135–238.

Davis, M. 1992. The role of the amygdala in fear and anxiety. Annu. Rev. Neurosci. 15:353–375.

Dill, L. M. 1974. Escape response of zebra danio. 1. Stimulus for escape. Anim. Behav. 22:711–722.

Dugatkin, L. A. 1992. Tendency to inspect predators predicts mortality risk in the guppy (*Poecilia reticulata*). Behav. Ecol. 3:124–127.

Dugatkin, L. A., and Godin, J.-G. J. 1992. Prey approaching predators: a cost-benefit perspective. Ann. Zool. Fennici 29:233–252.

Dukas, R. 2001. Effects of perceived danger on flower choice by bees. Ecol. Lett. 4:327–333.

Eaton, R. C., Lee, R. K., and Foreman, M. B. 2001. The Mauthner cell and other identified neurons of the brainstem escape network of fish. Prog. Neuorbiol. 63:467–485.

Edwards, D. H., Heitler, W. J., and Krasne, F. B. 1999. Fifty years of a command neuron: the neurobiology of escape behavior in the crayfish. Trends Neurosci. 22:153–161.

Elgar, M. A. 1989. Predator vigilance and group size in mammals and birds: a critical review of the empirical evidence. Biol. Rev. 64:13–33.

Fanselow, M. S., and LeDoux, J. E. 1999. Why we think plasticity underlying Pavlovian fear conditioning occurs in the basolateral amygdala. Neuron 23:229–232.

Fendt, M., and Fanselow, M. S. 1999. The neuroanatomical and neurochemical basis of conditioned fear. Neuro. Biobehav. Rev. 23:743–760.

FitzGibbon, C. D. 1994. The costs and benefits of predator inspection behaviour in Thomson's gazelles. Behav. Ecol. Sociobiol. 34:139–148.

Fraser, D. F., Gilliam, J. F., Daley, M. J., Le, A. N., and Skalski, G. T. 2001. Explaining leptokurtic movement distributions: intrapopulation variation in boldness and exploration. Am. Nat. 158:124–135.

Freund, R. L., and Olmstead, K. L. 2000. The roles of vision and antennal olfaction in enemy avoidance by three predatory heteropterans. Environ. Entomol. 29:733–742.

Frid, A., and Dill, L. M. 2002. Human-caused disturbance stimuli as a form of predation risk. Conserv. Ecol. 6(11): http://www.consecol.org/vol6/iss1/art11.

Frings, M., Maschke, M., Erichsen, M., Jentzen, W., Muller, S. P., Kolb, F. P., Diener, H. C., and Timmann, D. 2002. Involvement of the human cerebellum in fear-conditioned potentiation of the acoustic startle response: a PET study. Neuroreport 13:1275–1278.

Gilliam, J. F., and Fraser, D. F. 1987. Habitat selection when foraging under predation hazard: a model and a test with stream-dwelling minnows. Ecology 68:1856–1862.

Godin, J.-G. J., and Crossman, S. L. 1994. Hunger-dependent predator inspection and foraging behaviours in the threespine stickleback (*Gasterosteus aculeatus*) under predation risk. Behav. Ecol. Sociobiol. 34:359–366.

Godin, J.-G. J., and Davis, S. A. 1995. Who dares, benefits: predator approach behaviour in the guppy (*Poecilia reticulata*) deters predator pursuit. Proc. R. Soc. Lond. B 259:193–200.

Godin, J.-G. J., and Smith, S. A. 1988. A fitness cost of foraging in the guppy. Nature 333:69–71.

Gosling, S. 2001. From mice to men: what can we learn about personality from animal research? Psych. Bull. 127:45–86.

Gould, S. J., and Lewontin, R. C. 1979. Spandrels of San Marco and the panglossian paradigm—a critique of the adaptationist program. Proc. R. Soc. Lond. B 205:581–598.

Hartman, E. J., and Abrahams, M. V. 2000. Sensory compensation and the detection of predators: the interaction between chemical and visual information. Proc. R. Soc. Lond. B 267:571–575.

Hedrick, A. V. 2000. Crickets with extravagant mating songs compensate for predation risk with extra caution. Proc. R. Soc. Lond. B 267:671–675

Hopper, K. R. 2001. Flexible antipredator behavior in a dragonfly species that coexists with different predator types. Oikos 93:470–476.

Houtman, R. 1995. The influence of predation risk on within-patch foraging decisions of cryptic animals (Ph.D. dissertation, Simon Fraser University, Burnaby, British Columbia).

Hugie, D. M. 2003. The waiting game: a "battle of waits" between predator and prey. Behav. Ecol. 14:807–817.

Jacobsen, N. K. 1979. Alarm bradycardia in white-tailed deer (*Odocoileus virginianus*). J. Mammal. 60:343–349.

Jones, R. B., Mills, A. D., Faure, J. M., and Williams, J. B. 1994. Restraint, fear, and distress in Japanese quail genetically selected for long and short tonic immobility reactions. Physiol. Behav. 56:529–534.

Kats, L. B., and Dill, L. M. 1998. The scent of death: chemosensory assessment of predation risk by prey animals. Écoscience 5:361–394.

Koops, M. A. 2004. Reliability and the value of information. Anim. Behav. 67:103–111.

Koops, M. A., and Abrahams, M. V. 1998. Life history and the fitness consequences of imperfect information. Evol. Ecol. 12:601–613.

Kotler, B. P., Brown J. S., Dall, S. R. X., Gresser, S., Ganey, D., and Bouskila, A. 2002. Foraging games between gerbils and their predators: temporal dynamics of resource depletion and apprehension in gerbils. Evol. Ecol. Res. 4:495–518.

Lang, P. J., Davis, M., and Öhman, A. 2000. Fear and anxiety: animal models and human cognitive psychophysiology. J. Affect. Disord. 61:137–159.

Lawrence, B. J., and Smith, R. J. F. 1989. Behavioral response of solitary fathead minnows, *Pimephales promelas*, to alarm substance. J. Chem. Ecol. 15:209–219.

Lichtenstein, S., Slovic, P., Fischhoff, B., Layman, M., and Combs, B. 1978. Judged frequencies of lethal events. J. Exp. Psychol. 4:551–578.

Lima, S. L. 1994. Collective detection of predatory attack by birds in the absence of alarm signals. J. Avian Biol. 25:319–326.

Lima, S. L. 1998a. Nonlethal effects in the ecology of predator-prey interactions. BioScience 48:25–34.

Lima, S. L. 1998b. Stress and decision making under the risk of predation: recent developments from behavioral, reproductive, and ecological perspectives. Adv. Study Behav. 27:215–290.

Lima, S. L. 2002. Putting predators back into behavioral predator-prey interactions. Trends Ecol. Evol. 17:70–75.

Lima, S. L., and Dill, L. M. 1990. Behavioural decisions made under the risk of predation: a review and prospectus. Can. J. Zool. 68:619–640.

Luttbeg, B., and Schmitz, O. J. 2000. Predator and prey models with flexible individual behavior and imperfect information. Am. Nat. 155:669–683.

Magurran, A. F., and Higham, A. 1988. Information transfer across fish shoals under predation threat. Ethology 78:153–158.

Malmkvist, J., and Hansen, S. W. 2002. Generalization of fear in farm mink, *Mustela vison*, genetically selected for behaviour towards humans. Anim. Behav. 64:487–501.

Martínez-García, F., Martínez–Marcos, A., and Lanuza, E. 2002. The pallial amygdala of amniote vertebrates: evolution of the concept, evolution of the structure. Brain Res. Bull. 57:463–469.

Mathis, A., and Vincent, F. 2000. Differential use of visual and chemical cues in predator recognition and threat-sensitive predator avoidance responses by larval newts (*Notophthalmus viridescens*). Can. J. Zool. 78:1646–1652.

Mirza, R. S., and Chivers, D. P. 2003. Response of juvenile rainbow trout to varying concentrations of chemical alarm cue: response thresholds and survival during encounter with predators. Can J. Zool. 81:88–95.

Mitchell, W. A., and Lima, S. L. 2002. Predator-prey shell games: large-scale movement and its implications for decision-making by prey. Oikos 99:249–259.

Mitchell, W. A., and Valone, T. J. 1990. The optimization research program: studying adaptations by their function. Q. Rev. Biol. 65:43–65.

Murdoch, W. W., and Oaten, A. 1975. Predation and population stability. Adv. Ecol. Res. 9:2–132.

Öhman, A., and Mineka, S. 2001. Fears, phobias, and preparedness: towards an evolved module of fear and fear learning. Psychol. Rev. 108:483–522.

Peacor, S. D. 2003. Phenotypic modifications to conspecific density arising from predation risk assessment. Oikos 100:409–415.

Peacor, S. D., and Werner, E. E. 2001. The contribution of trait-mediated indirect effects to the net effects of a predator. Proc. Natl. Acad. Sci. 98:3904–3908.

Pitcher, T. J., and Parrish, J. K. 1993. Functions of shoaling behaviour in teleosts. In: Behaviour of Teleost Fishes (Pitcher, T. J., ed.). London: Chapman & Hall; 363–439.

Roeder, K. D. 1965. Moths and ultrasound. Sci. Amer. 212:94–102.

Schmitz, O. J., Adler, F. R., and Agrawal, A. A. 2003. Linking individual-scale trait plasticity to community dynamics. Ecology 84:1081–1082.

Sharif, E., and LeBoeuf, R. A. 2002. Rationality. Annu. Rev. Psychol. 53:491–517.

Sih, A. 1987. Predators and prey lifestyles: an evolutionary and ecological overview. In: Predation: Direct and Indirect Impacts on Aquatic Communities (Kerfoot, W. C., and Sih, A., eds.). Hanover, N.H.: University Press of New England; 203–224.

Sih, A. 1992. Prey uncertainty and the balancing of antipredator and feeding needs. Am. Nat. 139:1052–1069.

Sih, A., Bell, A. M., Johnson, J. C., and Ziemba, R. F. 2004. Behavioral syndromes: an integrative overview. Q. Rev. Biol. 79:241–277.

Sih, A., Kats, L. B., and Maurer, E. F. 2003. Behavioral correlations across situations and the evolution of anti-predator behavior in a sunfish-salamander system. Anim. Behav. 65:29–44.

Skalski, G. T., and Gilliam, J. F. 2002. Feeding under predation hazard: testing models of adaptive behavior with stream fish. Am. Nat. 160:158–172.

Slovic, P. 1987. Perception of risk. Science 236:280–285.

Stephens, D. W., and Krebs, J. R. 1986. Foraging Theory. Princeton, N.J.: Princeton University Press.

Swanson, B. O., Gibb, A. C., Marks, J. C., and Hendrickson, D. A. 2003. Trophic polymorphism and behavioral differences decrease intraspecific competition in a cichlid, *Herichthys minckleyi*. Ecology 84:1441–1446.

Taylor, R. J. 1984. Predation. New York: Chapman & Hall.

Templeton, J. J., and Giraldeau, L.-A. 1995. Patch assessment in foraging flocks of European starlings—evidence for the use of public information. Behav. Ecol. 6:65–72.

Turchin, P. 2003. Complex Population Dynamics. Princeton, N.J.: Princeton University Press.

Turner, A. M., and Montgomery, S. L. 2003. Spatial and temporal scales of predator avoidance: experiments with fish and snails. Ecology 84:616–622.

Valone, T. J. 1989. Group foraging, public information, and patch estimation. Oikos 56:357–363.

Van Buskirk, J., and Arioli, M. 2002. Dosage response of an induced defense: how sensitive are tadpoles to predation risk? Ecology 83:1580–1585.

van der Pligt, J. 1998. Perceived risk and vulnerability as predictors of precautionary behaviour. Brit. J. Health Psychol. 3:1–14.

Wiedenmayer, C. P. 2004. Adaptations or pathologies: long-term changes in brain and behavior after a single exposure to severe threat. Neuro. Biobehav. Rev. 28:1–12.

Wilson, D. S. 1998. Adaptive individual differences within single populations. Phil. Trans. R. Soc. Lond. B 353:199–205.

Wilson, D. S., Clark, A. B., Coleman, K., and Dearstyne, T. 1994. Shyness and boldness in humans and other animals. Trends Ecol. Evol. 9:442–446.

Wilson, D. S., Coleman, K., Clark, A. B., and Biederman, L. 1993. Shy-bold continuum in pumpkinseed sunfish (*Lepomis gibbosus*): an ecological study of a psychological trait. J. Comp. Psych. 107:250–260.

Wright, G., Bolger, F., and Rowe, G. 2002. An empirical test of the relative validity of expert and lay judgments of risk. Risk Anal. 22:1107–1122.

Ydenberg, R. C. 1998. Behavioral decisions about foraging and predator avoidance. In: Cognitive Ecology: The Evolutionary Ecology of Information Processing and Decision Making (Dukas, R., ed.). Chicago: University of Chicago Press; 343–378.

Ydenberg, R. C., and Dill, L. M. 1986. The economics of fleeing from predators. Adv. Study Behav. 16:229–249.

Zald, D. H. 2003. The human amygdala and the emotional evaluation of sensory stimuli. Brain Res. Rev. 41:88–123.

9

Constraints on Inducible Defenses

Phylogeny, Ontogeny, and Phenotypic Trade-Offs

RICK A. RELYEA

Antipredator defenses are widespread in the animal kingdom and include an overwhelming number of traits. Hundreds of studies have examined the inducible defenses of many prey taxa and documented a wide range of antipredator defenses. Most studies have focused on a single species of prey, observed prey responses at a single ontogenetic stage, and conducted the experiments under a single set of environmental conditions. Although this approach is useful in establishing which prey taxa respond to predators, the approach is limited, because it prevents us from identifying interacting environmental effects and makes it difficult to identify potential constraints. By taking a more comprehensive approach that includes multiple species, multiple ontogenetic stages, and multiple environmental conditions, one can identify a series of potential constraints that may prevent prey from exhibiting predator-induced plasticity. Such constraints are multifaceted and include historical constraints of phylogeny, restrictive developmental windows throughout ontogeny, and the presence of phenotypic trade-offs that occur when prey face conflicting pressures (e.g., from multiple predators, from predators and parasites, or from predators and competitors). In this chapter, I review what we know about these three potential constraints on inducible defenses and identify how these insights will likely change how we conduct experiments, interpret empirical results, and design future research.

Observing how predators and prey interact has always fascinated humans, and it is this fascination that has led countless biologists to study the ecological and evolutionary "arms race" between these two interacters. Although the study of predation has a long history (see Elton 1927, Leopold 1933), during the past decade we have witnessed an explosion of interest in predator-induced traits of prey (e.g., the ability of a single genotype to produce alternative phenotypes; Dill 1987, Kerfoot and Sih 1987, Lima and Dill 1990, Kats and Dill 1998, Tollrian and Harvell 1999, Relyea 2001b). This interest has documented a staggering diversity of antipredator strategies, including changes in prey behavior, morphology, and life history.

Although most of our attention has focused on which prey species respond to predators and why, there has been considerably less attention directed toward understanding why some prey species do not respond to predators when it appears that they should (at least at first glance). Thus, we can ask, what prevents prey from evolving or exhibiting predator-induced plasticity? DeWitt et al. (1998) have reviewed a number of costs and limitations associated with plastic responses, limitations that could hinder or prevent the evolution of plastic responses. These limitations include the following: (a) low reliability of environmental information (e.g., due to poor sensory ability or environmental cues that are of low quality), (b) long lag times between sensing of environmental cues and induction of appropriate phenotypes (e.g., slow induction of morphological traits), (c) the inability to develop plastic phenotypes that are as extreme as nonplastic phenotypes (i.e., "the jack-of-all-trades can be the master of none"), (d) the costs of plastic individuals' producing a phenotype that exceed the costs experienced by a nonplastic individual (i.e., the costs of plasticity per se), and (e) the high costs of acquiring the environmental information. These problems of sensing environmental change and paying the costs of responding to the environmental change can impact the magnitude of the plasticity expressed. However, there are three other major factors that can constrain predator-induced plasticity: (a) phenotypic trade-offs, (b) ontogenetic constraints, and (c) phylogenetic constraints. In this chapter, I review each of these factors within the realm of inducible defenses in animals and identify how they can constrain the evolution of antipredator responses or simply prevent responses from being expressed. At the end, I highlight the research path that lies before us in light of these constraints. Throughout the chapter, I draw upon a wealth of aquatic studies, because much of what we know about inducible defenses has come from these systems and because of my own familiarity with these systems.

Phenotypic Trade-Offs

To understand how inducible defenses evolve, we can consider a large body of theory on the evolution of phenotypic plasticity in general. Although there is a diversity of mathematical models with varying underlying assumptions, the models generally agree that phenotypic plasticity evolves because of the existence of phenotypic trade-offs in which a given phenotype may experience high fitness in one environment but low fitness in another environment (Via and Lande 1985, Via and Lande 1987, Gomulkiewicz and Kirkpatrick 1992, Moran 1992, Van Tienderen 1997, Tufto 2000, Sultan and Spencer 2002). If this were not true, a single phenotype would have a selective advantage and eventually dominate in a population (i.e., there would be no need for plasticity). When applied to predator-induced defenses, theory predicts that prey should experience phenotypic trade-offs such that a defensive phenotype produced in one environment may be well suited to that environment but not well suited to other environments (assuming similar frequencies of the different environments). In all of these models, the alternative environments are treated as mutually exclusive environments that prey experience in isolation. In reality, different environmental factors often occur in combination with each other. When these factors occur together, we can ask whether the presence of one environmental factor can constrain prey from responding to other environmental factors.

The Challenge of Multiple Predators

Reviews of the literature on inducible defenses indicate that, though scores of studies have been conducted, most have examined only one predator at a time (Sih 1987, Lima and Dill 1990, Kats and Dill 1998, Relyea 2004b). This approach has been highly successful in documenting the diversity of taxa that respond to predators and the wide array of traits that are plastic. In a substantial proportion of these studies, investigators have exposed prey to separate species of predators and demonstrated that most prey can discriminate between predator species. However, in nature, most prey experience combinations of predators (Schoener 1989, Polis 1991, Polis and Strong 1996, Relyea 2003a). Thus, we are left with the question, does living with multiple species of predators cause phenotypic trade-offs that constrain the realm of possible prey defenses?

When we ask how prey should respond to combinations of predators, we can start by examining the most straightforward situation, in which different predators induce defensive traits in the same direction but to different magnitudes. For example, the larvae of many amphibians alter their behavior and morphology (i.e., their relative shape) in response to a wide variety of predators (Relyea and Werner 2000, Van Buskirk 2000, Van Buskirk and Schmidt 2000, Relyea 2001a). In wood frogs (*Rana sylvatica*), predators induce tadpoles to exhibit decreased foraging activity and develop relatively deep tails and short bodies (which makes tadpoles harder to detect and harder to capture, respectively; Skelly 1994, Van Buskirk and Relyea 1998). However, when we expose wood frog tadpoles to different species of predators, we find that low-risk predators induce small changes in these traits, moderate-risk predators induce moderate changes in these traits, and high-risk predators induce large changes in these traits (i.e., the prey use a "scaled response strategy"; Relyea 2001b, Relyea 2001c). We exposed wood frog tadpoles to six different pairwise combinations of four predators (each predator ranging in risk from low to high) and observed how the tadpoles altered 12 behavioral, morphological, and life history traits. We found that wood frogs followed a general rule of thumb: respond to a pair of predators the same way that you respond to the more dangerous of the two predators alone (Relyea 2003a). That is, if prey use a scaled response strategy, when they respond to a highly dangerous predator, they are already defended against less dangerous predators.

In this case, the presence of multiple predators in the system can have a major impact on whether and how prey respond to a given predator. If high-risk predators are in the system and we add a low-risk predator, we will not observe any change in the prey's phenotype. However, if low-risk predators are in the system and we add a high-risk predator, we will observe a change in the prey's phenotype. Thus, high-risk predators can constrain prey from exhibiting the optimal (weaker) response to the low-risk predator. This result is of particular interest when we manipulate predators in experimental venues that contain predators in the surrounding area (e.g., cages in a field, cages in a pond).

The constraints that multiple predators place on prey responses is perhaps even more evident when predators induce prey traits in opposite directions. For example, pumpkinseed sunfish (*Lepomis macrochirus*) induce *Physa* snails to hide under rocks, but crayfish (*Orconectes rusticus*) induce the snails to emerge from the rocks. These behavioral responses appear to be adaptive, because the sunfish feed in the open water while the crayfish feed under the rocks (Turner et al. 1999, Turner et al. 2000). The trade-off for the prey defensive decision arises, of course, when both predators are present. *Physa* snails resolve this trade-off by exhibiting an intermediate behavioral phenotype, spending half their time under rocks and half the time in the open water. In these snails (and several other prey species; Martin et al. 1989, Peckarsky and McIntosh 1998, McIntosh and Peckarsky 1999), the presence of one predator places a constraint on the responses of prey to the second predator. If the two predators pose similar risks, then responding to only one predator increases the probability of death from the other predator. When the predators pose different risks, the prey can be even more constrained. If one predator poses a high risk, the addition of a less risky predator that favors the opposite phenotype will have no effect on the prey phenotype (Kotler et al. 1992). In summary, the existence of phenotypic trade-offs among different predators can pose substantial constraints on the ability (or willingness) of prey to respond to particular predators in the system.

The Challenge of Predation versus Parasitism

The impacts of combined predators parallel the impacts of combining predators and parasites. The optimal responses to each threat can be either in the same direction, or in opposite directions, and opposite directions can cause phenotypic trade-offs. For example, Thiemann and Wassersug (2000) found that fish predators (*Fundulus diaphanus*) induce green frog tadpoles (*Rana clamitans*) to become less active and spend more time resting on the bottom of aquaria. Because the parasitic cercariae preferentially occur on the bottom, the tadpoles' response to the predators causes an increase in parasitism. Similar effects occur in vertically migrating *Daphnia* that face fish predation when they migrate up in the water column to feed, and face parasitism when they migrate down to the benthos (Decaestecker et al. 2002). In freshwater snails (*Physa integra*), uninfected individuals spend more of their time hiding in covered habitats when fish (*Semotilus atromaculatus*) are present, but infected individuals spend less of their time in covered habitats (Bernot 2003). However, in this example, we must interpret the outcome cautiously, because it is not clear whether the infected snails' behavior is adaptive to the snail, or whether it is being manipulated by the parasite to promote transmission to a terminal host (e.g., aquatic waterfowl). In either case, the presence of the parasite appears to prevent the snails from exhibiting an adaptive response to the predatory fish. Thus, when prey coexist with both predators and parasites, their optimal response to one danger may constrain an optimal response to the other.

The Challenge of Eating versus Being Eaten

Most prey that respond to predators also face a trade-off between increased survival in the presence of predators and slower growth and development. For example, many species of rotifers and cladocerans develop spines in response to fish and invertebrate predators. However, these predator-induced individuals have lower intrinsic rates of growth or reproduction than the noninduced individuals (Havel 1987, Tollrian and Dodson 1999). Many protists defend themselves by changing their cell shape and developing "wings" to avoid being consumed by gape-limited predators, but they pay the price of slower development and longer generation times (Kuhlmann and Heckmann 1994, Kusch and Kuhlmann 1994, Kusch 1995). Likewise, tadpoles reduce their activity to become less detectable by predators, but they do so at the cost of lost foraging opportunities (Skelly 1994, Anholt and Werner 1995, Relyea and Werner 1999). In short, there appears to be a common trade-off for animal prey between the ability to resist predation and the ability to grow and develop.

If the ability to resist predation versus the ability to garner resources is a fundamental trade-off for prey, then it suggests that a prey's response to predators should be constrained by the amount of resources available. A number of theoretical models have examined this trade-off and have made similar general predictions for behavioral and life history traits. That is, when foraging ability and predator resistance are traded off, prey should exhibit stronger antipredator responses (take less risk) when there are plentiful resources (McNamara and Houston 1987, Abrams 1991, Werner and Anholt 1993). Although there are exceptions (Cerri and Fraser 1983, Fraser and Huntingford 1986, Horat and Semlitsch 1994), several empirical investigations of prey behavior have supported this prediction (Petranka 1989, Anholt et al. 1996, Van Buskirk and Yurewicz 1998, Relyea, 2004a). For example, Anholt et al. (1996) found that toad (*Bufo americanus*) tadpoles exhibited larger predator-induced reductions in swimming activity when food resources were plentiful than when food resources were scarce. Because behavioral decisions involve foraging, one would expect the food-dependent behavioral responses to predators to translate into food-dependent life history responses, as well. For instance, when, under high-food conditions (e.g., low competition), prey exhibit larger behavioral reductions in response to predators, they simultaneously experience larger reductions in growth (Werner and Anholt 1996, Van Buskirk and Yurewicz 1998, Relyea 2002a, Relyea 2002b). There are relatively few studies on the interactive effects of predators and resources (i.e., competition) on morphological defenses of prey, perhaps because morphological plasticity is relatively recently discovered in many systems. However, the available studies suggest that morphological traits follow the same patterns as behavior and life history traits (Relyea 2002a, Relyea 2002b, Relyea and Hoverman 2003, Relyea 2004a; but see Kuhlmann and Heckmann 1994).

Because most prey face a trade-off between predation resistance and foraging ability, low resource abundance (i.e., high competition) can serve as an effective constraint on the defensive options that are available to prey. In short, a wide variety of phenotypic trade-offs can constrain the expression of inducible defenses.

Ontogenetic Constraints

The importance of predator-induced traits across the ontogenetic stages of prey is one of the least explored aspects of predator-induced defenses. Interest in ontogeny has a long history that began with the recapitulation of phylogeny (von Baer 1828, Haeckel 1866) and investigations of how species have evolved accelerations and decelerations of different ontogenetic stages (e.g., paedomorphosis, neoteny, hererochrony; Haldane 1932, Gould 1977). Recently, there has been a growing interest in adding an ontogenetic perspective to the study of phenotypic plasticity, in general, and predator-induced plasticity, in particular (Schlichting and Pigliucci 1998, Pigliucci 2001, Relyea 2003b). To understand how ontogeny might constrain predator-induced defenses, we first need to understand how prey traits and prey strategies can change dramatically over ontogeny.

Given the hundreds of inducible defense studies, it is perhaps surprising that we know very little about whether and how prey defense strategies change over ontogeny. During the past two decades, the primary focus of ecologists has been to document the wide variety of organisms that change their traits in the presence of predators. As a result, our primary approach has been to expose prey to predators (or predator cues) early in ontogeny and then observe the induced phenotypes that are produced after an appropriate period of time (e.g., Pettersson and Brönmark 1994, Relyea 2000, Trussell and Nicklin 2002). This approach has certainly been very productive, providing us with a wealth of data about which prey taxa are plastic and what kinds of traits are capable of being induced. However, this approach has prevented us from understanding the dynamic nature of antipredator responses.

For example, gray tree frog larvae (*Hyla versicolor*) are known to respond to larval dragonfly predators by reducing their foraging activity and developing relatively deep tails and small bodies (Van Buskirk and McCollum 2000, Relyea 2001b, Relyea and Hoverman 2003). However, when we took an ontogenetic approach, we discovered that the defensive traits of tree frogs actually change over time (Relyea 2003b). Early in ontogeny, tadpoles rely on hiding and large reductions in activity to avoid predation. Midway through ontogeny, tadpoles no longer use hiding: they use smaller reductions in activity, and they develop large morphological defenses. Late in ontogeny, tadpoles no longer use behavioral defenses but rely entirely on morphological defenses (Figure 9.1).

Other studies have confirmed that prey rely more heavily on behavioral defenses early in ontogeny and more heavily on morphological defenses later in ontogeny (Pettersson et al. 2000). This change likely occurs because behavioral defenses are rapidly inducible and more costly, whereas morphological defenses are induced more slowly and at a lower cost (Relyea 2001b, Relyea 2003b). Had we not taken an ontogenetic approach, our interpretation of the prey's strategy would have depended on when we terminated the experiment.

Given that antipredator strategies can change over a lifetime, prey may possess developmental windows in which predator cues must be present during some critical stage in the prey's ontogeny. Wider developmental windows might be expected in prey that are vulnerable for a longer period of development (e.g., prey that do not quickly grow into a size refuge) and for prey that may experience predators anytime during the

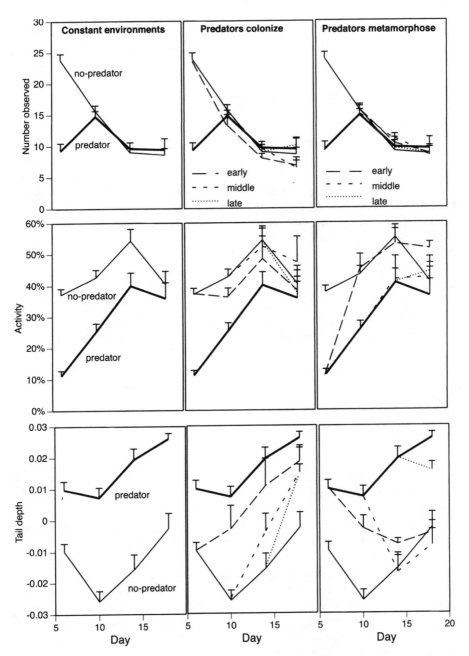

Figure 9.1. The number of gray treefrog tadpoles observed (top panels), the proportion of observed tadpoles that were active (middle panels), and the relative tail depth of tadpoles (bottom panels) reared in one of eight treatments. The treatments simulated either the constant presence or absence of a predator (left panels), colonization by a predator into a pond (center panels), or metamorphosis of a predator out of a pond (right panels). Data are means ±1 standard error. The experiment demonstrates that traits change over ontogeny and that predator-induced behavior and morphology are both reversible (reprinted from Relyea 2003b).

prey's vulnerable stage (Gabriel 1999). Further, developmental windows may be more common for morphological and life history traits than for behavioral traits (West-Eberhard 1989). Many behavioral defenses involve simply exhibiting a given behavior more or less often in the presence of a predator (e.g., foraging less or using a refuge more frequently), whereas morphological and life history traits may be possible only if preceded by a number of other morphological and life history decisions (e.g., prey cannot delay reproductive maturity if they have already reproduced).

The occurrence of developmental windows has been observed in several prey taxa. A classic example is the induction of defensive spines in zooplankton. If predator cues are present while *Daphnia pulex* are embryonic and still within their mother's brood pouch, they will develop defensive neckteeth that reduce their probability of being eaten by *Chaoborus* predators. In fact, the earlier in the embryonic stage that exposure begins, the greater the probability of being predator induced. In contrast, if the progeny are not exposed to the predator cues until after they are released from the brood pouch, they are no longer capable of being induced (Krueger and Dodson 1981). The physiological reasons for this developmental window are not well understood, but it appears that, once the exoskeleton is committed to a no-predator trajectory, the phenotype cannot be redirected to a predator trajectory.

A similar-life history pattern exists in marine bryozoans. Defensive spines are inducible when the individuals are exposed to predator cues early in ontogeny but not later in ontogeny (Harvell 1991). The cannibalistic morphs of tiger salamanders (*Ambystoma tigrinum*) follow a similar pattern: younger larvae are more likely to be induced than older larvae (Hoffman and Pfennig 1999). Although such narrow developmental constraints on morphological traits exist, this pattern is not universal. For example, gray tree frog tadpoles have wide developmental windows of inducible behavior and morphology. Tadpoles can be induced by predatory *Anax* dragonfly larvae throughout most of their larval period (Figure 9.1; Relyea 2003b).

Although we still have too few data on developmental windows to draw any strong generalities, these examples make it clear that prey can possess a range of developmental windows that can constrain the production of predator-induced defenses. In more practical terms, knowledge of developmental windows suggests that we need to take great care in deciding the stages at which to expose prey to predator cues and how we interpret the prey's responses. As we gather more data across systems, we will begin to establish patterns in the width of developmental windows in different types of traits and different taxonomic groups. Further, we should begin to gain insights into the selective forces that may have acted on developmental windows and test the following hypotheses: (a) prey with shorter time periods of vulnerability to predators should have narrower developmental windows than prey with longer periods of vulnerability to predators, and (b) prey that experience predators at multiple points in ontogeny should evolve wider developmental windows than prey that experience predators only early in ontogeny.

Taking an ontogenetic approach can also expose the constraints of phenotypic reversibility. Although predator-induced defenses are typically adaptive when the predator is present, they are often costly when the predator is absent (Tollrian and Harvell 1999). For example, many aquatic insect predators leave the aquatic realm when they metamorphose, converting a predator-filled habitat into a predator-free habitat (assuming a fairly synchronous metamorphosis of the predators). If there are substantial costs of carrying the predator-induced phenotype, selection should favor the reversibility of the phenotype (Gabriel 1999). However, there may be developmental constraints that do not permit an undoing of the defense. The conventional wisdom is that behavioral traits are rapidly reversible, whereas morphological traits are irreversible (West-Eberhard 1989).

Consistent with conventional wisdom, predator-induced behavioral plasticity is rapidly reversible, often on a time scale of minutes or hours (Forward and Hettler 1992, DeMeester et al. 1994, Wibe et al. 2001, Van Buskirk et al. 2002). For example, brine shrimp naupliar larvae (*Artemia*) exhibit diel vertical migration when predator cues are present, but within one day of removal of the predator cues, the shrimps no longer vertically migrate (Forward and Hettler 1992). The ability to rapidly reverse predator-induced behavior has led to a new theoretical model asking how prey should respond to predator environments that vary on a rapid time scale (Lima and Bednekoff 1999). The model predicts that, compared to prey in constant predator or no-predator environments, prey individuals facing rapidly fluctuating predator and no-predator environments (within a generation) should optimize their fitness by exhibiting stronger antipredator responses when the environment becomes high risk and stronger feeding behaviors when the environment becomes low risk. So far, there has been equivocal support for this predicted outcome (Hamilton and Heithaus 2001, Sih and McCarthy 2002, Van Buskirk et al. 2002, Pecor and Hazlett 2003). However, all of these studies provide excellent evidence that exhibiting predator-induced behavior rarely constrains individuals from reverting back to predator-free behavior. In fact, the rapid reversibility makes behavior such a special type of trait that many biologists do not even consider behavior to be a valid type of plasticity (Pigliucci 2001).

Although many researchers distinguish between behavioral and morphological plasticity in accordance with phenotypic reversibility, this basis turns out to be a false distinction: predator-induced morphology, too, can be reversible. For example, many protists can change their cell shape to avoid gape-limited predators and then revert back to their original shape when the predator cues are removed (Kuhlmann and Heckmann 1994). Similarly, the predator-induced crests and neckteeth of cladocerans can be reduced within 2–3 molts after the predator cues are absent (Tollrian and Dodson 1999). However, this phenomenon is not simply one of small organisms. Pettersson and Brönmark (1994) found that the predator-induced humpback shape of crucian carp (*Carassius carassius*) is partially reversible in the absence of predators. In a dramatic case of morphological reversibility, we have found that the dragonfly-induced shape changes of gray tree frog tadpoles is completely reversible, and the reversal can occur in less than 4 days (Figure 9.1; Relyea 2003b). Thus, contrary to conventional wisdom, morphological defenses can be reversed (albeit at a slower rate than behavioral responses and to different degrees), suggesting that predator-induced morphology does not necessarily pose a constraint on the future morphological phenotypes that an individual can exhibit. There are certain to be many morphological changes that are not reversible (e.g., animals that "decide" to grow to a large size may not be able to revert to a small size), but we currently have too few examples to draw many solid generalities.

It is important when we examine inducible defenses that we consider the role of ontogeny. By taking an ontogenetic perspective, we are sure to observe all of the trait changes that occur at all points in development. By switching prey between predator and no-predator environments throughout ontogeny, we can identify the presence of developmental windows that constrain prey from being induced, and identify constraints on phenotypic reversibility. While much of the current work has examined the impact of predators over ontogeny within a life stage (e.g., larval ontogeny or adult ontogeny), these issues should also be investigated across life stages to reveal the impact of early predator environments on the traits and performance of individuals after an ontogenetic niche shift (Relyea 2001a, Relyea and Hoverman 2003). That is, we need to continue to bring more ecology into the study of predator-induced responses of prey.

Phylogenetic Constraints

Biologists have pondered the relative importance of phylogeny and ecology on the traits of species for decades. That is, do species possess a particular phenotype because of their evolutionary history or because of the ecological conditions under which they currently live (or both)? This dual perspective has focused on morphological traits because they are readily measurable and typically are more accessible for a large number of taxa (e.g., from museum specimens) than behavior or life history traits are (Ricklefs and Travis 1980, Price 1991, Klingenberg and Ekau 1996). In addition, investigations of phylogenetic and ecological impacts on phenotypes have been conducted on nonplastic traits (or at least traits assumed to be nonplastic), with any plasticity in these traits being viewed as uninteresting developmental "noise" (Schlichting and Pigliucci 1998). However, we have now accumulated a vast collection of predator-induced-plasticity studies and have examined numerous species in various taxa (e.g., protists, rotifers, cladocerans, snails, and tadpoles), which can be used in phylogenetic comparisons. At the same time, phylogenetic methodologies have been developed that make the challenge of conducting the necessary independent contrasts an achievable goal (Felsenstein 1985, Harvey and Pagel 1991, McPeek 1995b). Thus, the time is becoming ripe for biologists to examine phylogenetic and ecological patterns in predator-induced plasticity and to determine whether the defenses of some prey species appear constrained by either their phylogenetic relationships, or the habitats in which they live.

Some ecologists have conducted plasticity experiments on small sets of related species (frequently belonging to a single genus, to help control for phylogenetic differences) and then have examined patterns of plasticity that appear to be related to each species' ecology (Harvell 1991, Kusch 1993a, Kusch 1993b, Kuhlmann and Heckmann 1994, Smith and Van Buskirk 1995, Lardner 2000, Relyea and Werner 2000, Relyea 2001b). However, some researchers have collected enough data from more diverse species assemblages to examine patterns of prey defense by using independent contrasts (e.g., McPeek 1995a, McPeek et al. 1996, McPeek 1998). One of the first attempts to explicitly examine predator-induced traits in a phylogenetic context was by Colbourne et al. (1997), who examined the occurrence of inducible head shapes in the cladoceran genus *Daphnia*. Invertebrate predators can induce "neckteeth" in *Daphnia* embryos and elongated "helmets" (e.g., head blades and head spines) in *Daphnia* adults. These morphological changes make the cladocerans less susceptible to predation (Havel 1987). Colbourne et al. (1997) mapped the occurrence of inducible head shapes (no change, helmets, or neckteeth) and the typical habitats in which the species live (ephemeral ponds, permanent fishless ponds, or permanent fish ponds) onto a phylogeny of 35 North American species, previously developed from molecular data (Figure 9.2; Colbourne and Hebert 1996). The formation of helmets was generally associated with species inhabiting permanent fish lakes, whereas the formation of neckteeth was associated with all pond types (the predator that induces neckteeth, *Chaoborus*, can live in all types of ponds).

According to the phylogeny, these inducible defenses appear to have been gained and lost multiple times over evolutionary time. For example, the ancestor of the subgenus *Ctenodaphnia* likely inhabited ephemeral ponds and had no helmet, yet the one species (out of seven) that evolved to inhabit permanent fish ponds also evolved an inducible helmet. In contrast, the ancestor of the subgenera *Hyalodaphnia* and *Daphnia* likely possessed inducible helmets and inhabited permanent fish lakes, yet the inducible helmets were lost in several species that evolved to inhabit fishless habitats. Thus, in these cases, selection for these morphological defenses because of a change in ecology (i.e., habitat) overcame the potential constraint of phylogeny. However, there are groups whose ancestor did not have inducible neckteeth, and the extant species appear to be phylogenetically constrained from evolving neckteeth (or ecologically constrained by some other factor that we have yet to identify), despite their co-occurrence with *Chaoborus*, which favor the evolution of neckteeth.

Phylogenetic approaches have been used also to look at the evolution of predator-induced behavior. For example, Richardson (2001a, 2001b) took a phylogenetic approach to predator-induced behavior in tadpoles. However, the prey had already spent much of their larval period living in ponds that likely varied in predator composition, and few (3 of 13) species exhibited predator-induced plasticity. Van Buskirk (2002) provides an excellent example of combining predator-induced behavior and morphology in a phylogenetic context. He pooled 9 years of experiments on tadpoles reared in outdoor pond mesocosms (i.e., cattle tanks) in the presence and absence of caged dragonfly predators. This data set included 15 species across four genera (*Rana, Hyla, Pseudacris,* and *Bufo*) from eastern North America and Europe (plus *Bombina,* a single species outgroup), allowing for the examination of predator-induced behavior and morphology with a combined phylogenetic and ecological perspective. He used previously published phylogenies to assess the phylogenetic relationships of the species, and a survey of 35 herpetologists to assess the range of pond types used by each species (from ephemeral wetlands to lakes containing fish). All 16 species exhibited lower activity with predators present, but this behavioral change differed in species from different genera. Activity reduction was strongest in *Rana* tadpoles, moderate in *Hyla* tadpoles, and weakest in *Bufo* tadpoles.

All 16 species also exhibited an increase in relative tail depth with predators present (which improves predator resistance; Van Buskirk et al. 1997, Van Buskirk and Relyea 1998). The magnitude of the increase exhibited phylogenetic patterns. The predator-induced tail increase was larger in *Hyla* tadpoles than in *Pseudacris* and *Bufo* tad-

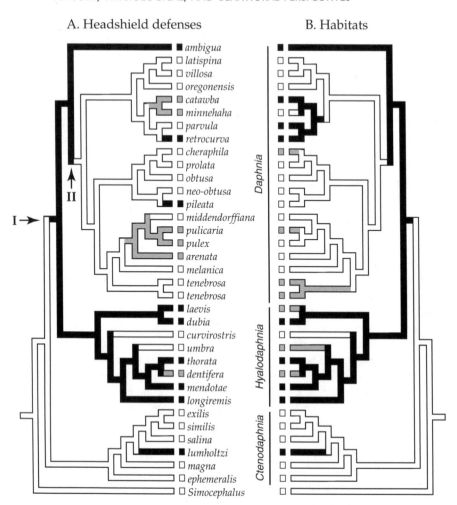

Figure 9.2. The phylogentic relationships among 35 species of caldocerans in the genus *Daphnia* as constructed by Colbourne and Hebert (1996) with an assumption of accelerated transformation. (A) Onto this phylogeny, the authors mapped the occurrence of predator-induced "helmets" (e.g., head blades and head spines; black shading) and predator-induced "neckteeth" (gray shading). (B) Onto this phylogeny, the authors also mapped the habitats use of each species: ephemeral ponds (white), permanent ponds or small lakes (gray), and larger lakes (black). Figure copied from Colbourne et al. (1997).

poles, whereas the induced tail changes in *Rana* tadpoles were wide-ranging. This result suggests that a species' phylogenetic history may constrain the magnitude of a species' predator-induced behavior and morphology. Furthermore, when the *Bufo* species were excluded from the analysis (because of their unique defense of being unpalatable), tadpole species with greater morphological plasticity possessed less behavioral plasticity, lending support to a pattern previously observed in just the *Rana* tadpoles (Relyea and Werner 2000, Relyea 2001b).

When comparing habitat use within a phylogenetic framework, Van Buskirk (2002) found that species from intermediate hydroperiod ponds showed both behavioral and morphological plasticity (i.e., the induction of reduced activity and relatively deeper tails), whereas species from more permanent ponds tended to rely on only behavioral plasticity (i.e., reduced activity). This differential pattern appears to have evolved because the costs of reduced foraging are too high for intermediate-pond species, which must metamorphose before the pond dries (Relyea 2001b, Relyea 2003b). For species occupying more permanent ponds, pond drying imposes no constraint, which allows these species to rely completely on behavioral defenses. Thus, the ecology of these species places particular constraints on the defensive options that are available.

When comparing the range of pond types used by each species, Van Buskirk (2002) found that species using a wider range of pond types (a reasonable index of predator variation that a species experiences) had greater morphological plasticity (as predicted by theory). However, there was no relationship between the range of pond types used and behavioral plasticity. This contrasting result is likely due to the broad scale of the predator variation examined (i.e., predator variation across ponds rather than within-pond predator variation). In short, taking a comparative, multispecies approach to predator-induced defenses helped develop insights that were largely unobtainable from single-species studies.

These studies highlight the value of examining antipredator responses of prey in a comparative, phylogenetic context. The phylogenetic approach allows us to examine how many times a predator-induced defense has evolved and the ecological forces that favor its evolution. In addition, it allows us to generate hypotheses as to why some species have not evolved inducible defenses even though their ecological conditions appear to favor them. Much of the leading work has been conducted on *Daphnia* and larval amphibians, probably because we have a large amount of data on numerous species in these systems, these systems have well-established phylogenies, and the adaptive value of the defenses is now well established. It should be clear from the examples offered here that we have only begun to explore this line of research (using few systems and few predator-induced traits). Much more needs to be done. We need to examine many more traits for which there exist ample data, including habitat use and life history decisions (e.g., Havel 1987, Spitze 1992, Weetman and Atkinson 2002). Numerous other predator-prey systems with well-defined phylogenies should also be examined to extend our breadth of insight and allow us to begin making strong generalities about underlying mechanisms. This emphasis will undoubtedly be one of the most productive research directions for an improved understanding of predator-induced plasticity.

Conclusions

As can be appreciated from this review, this is truly an exciting time for the study of predator-induced plasticity. We have spent decades studying how a diversity of prey species respond to their predators. Although we have come a long way in understanding how and why prey alter their defensive traits, there remains a great deal of work to do. Comparative approaches that control for phylogeny are just beginning to be explored, and they are providing improved insights into the potential constraining effects of phylogeny and ecology. We need to examine many more systems with a phylogenetic perspective, either by bringing together existing data from other systems and overlaying the data on known phylogenies (i.e., conducting meta-analyses), or by conducting numerous new experiments on a given taxonomic group. In this way, we will identify phylogenetic and ecological constraints on the ability of prey to exhibit antipredator responses.

We also need to stop ignoring the ontogenetic pathways that prey take in forming their defenses. It takes some foresight, but not a great deal of extra effort, to quantify prey phenotypes throughout an experiment rather than to observe phenotypes only at the end of the experiment. This extra effort will be rewarded with the realization that many prey do not have a set defense against predators throughout ontogeny, but instead have a continually changing strategy that provides insights into both the speed of induction and the potential costs of different types of antipredator responses. This additional effort will also help elucidate how traits change across development, which will aid interpretation of allometric versus nonallometric changes in morphological defenses (e.g., Relyea and Werner 2000). Coupled to any ontogenetic approach should be an examination of developmental windows that constrain the times of inducibility and phenotypic reversibility.

Finally, approaching inducible defenses from a more realistic perspective of multiple, simultaneous environments (different combinations of predators, parasites, and competitors) will shed light upon the constraints posed by the phenotypic trade-offs faced by prey. By exploring all of these constraints, we will develop a much better understanding of the ecology and evolution of predator-induced plasticity.

Acknowledgments

This chapter benefited immensely from numerous discussions with and revisions from J. Auld, J. Hoverman, A. Randle, C. Relyea, and N. Schoeppner. My research on phenotypic plasticity has been generously supported over the years by the University of Pittsburgh, the Pymatuning Laboratory of Ecology, and the National Science Foundation.

Literature Cited

Abrams, P. A. 1991. Life history and relationship between food availability and foraging effort. Ecology 72:1242–1252.

Anholt, B. R., Skelly, D. K., and Werner, E. E. 1996. Factors modifying antipredator behavior in larval toads. Herpetologica 52:301–313.

Anholt, B. R., and Werner, E. E. 1995. Interaction between food availability and predation mortality mediated by adaptive behavior. Ecology 76:2230–2234.

Bernot, R. J. 2003. Trematode infection alters the antipredator behavior of a pulmonate snail. J. No. Am. Benthol. Soc. 22:241–248.

Cerri, R. D., and Fraser, D. F. 1983. Predation and risk in foraging minnows: balancing conflicting demands. Am. Nat. 121:552–561.

Colbourne, J. K., and Hebert, P. D. N. 1996. The systematics of North American *Daphnia* (Crustacea: Anomopoda): a molecular phylogenetic approach. Phil. Trans. R. Soc. Lond. B 351:349–360.

Colbourne, J. K., Hebert, P. D. N., and Taylor, D. J. 1997. Evolutionary origins of pheno-typic diversity in *Daphnia*. In: Molecular Evolution and Adaptive Radiation (Givnish, T. J., and Sytsma, K. J., eds.). Cambridge: Cambridge University Press; 163–188.

Decaestecker, E., De Meester, L., and Ebert, D. 2002. In deep trouble: habitat selection con-strained by multiple enemies in zooplankton. Proc. Natl. Acad. Sci. 99:5481–5485.

DeMeester, L., Vandenberghe, J., Desender, K., and Dumont, H. J. 1994. Genotype-depen-dent daytime vertical distribution of *Daphnia magna* in a shallow pond. Belg. J. Zool. 124:3–9.

DeWitt, T. J., Sih, A., and Wilson, D. S. 1998. Costs and limits of plasticity. Trends Ecol. Evol. 13:77–81.

Dill, L. M. 1987. Animal decision making and its ecological consequences: future of aquatic ecology and behaviour. Can. J. Zool. 65:803–811.

Elton, C. 1927. Animal Ecology. New York: Macmillan.

Felsenstein, J. 1985. Phylogenies and the comparative method. Am. Nat. 125:1–15.

Forward, R. B., and Hettler, W. F. 1992. Effects of feeding and predator exposure on photoresponses during diel vertical migration of brine shrimp larvae. Limnol. Oceanogr. 37:1261–1270.

Fraser, D. F., and Huntingford, F. A. 1986. Feeding and avoiding predation hazard: behav-ioral response of prey. Ethology 73:56–68.

Gabriel, W. 1999. Evolution of reversible plastic responses: inducible defenses and environ-mental tolerance. In: The Ecology and Evolution of Inducible Defenses (Tollrian, R., and Harvell, D., eds.). Princeton, N.J.: Princeton University Press; 286–305.

Gomulkiewicz, R., and Kirkpatrick, M. 1992. Quantitative genetics and the evolution of re-action norms. Evolution 46:390–411.

Gould, S. J. 1977. Ontogeny and Phylogeny. Cambridge, Mass.: Harvard University Press.

Haeckel, E. 1866. Generelle Morphologie der Organismen: allgemeine Grundzuge der organischen Formen-Wissenschaft, mechanisch begrundet durch die von Charles Dar-win reformirte Descendenz-Theorie. Berlin: Reimer.

Haldane, J. B. S. 1932. The time action of genes, and its bearing on some evolutionary prob-lems. Am. Nat. 66:5–24.

Hamilton, I. M., and Heithaus, M. R. 2001. The effects of temporal variation in predation risk on antipredator behaviour: an empirical test using marine snails. Proc. R. Soc. Lond. B 268:2585–2588.

Harvell, C. D. 1991. Coloniality and inducible polymorphism. Am. Nat. 138:1–14.

Harvey, P. H., and M. D., Pagel. 1991. The Comparative Method in Evolutionary Biology. Oxford: Oxford University Press.

Havel, J. E. 1987. Predator-induced defenses: a review. In: Predation: Direct and Indirect Impacts on Aquatic Communities (Kerfoot, W. C., and Sih, A., eds.). Hanover, N.H.: University Press of New England; 263–278.

Hoffman, E. A., and Pfennig, D. W. 1999. Proximate causes of cannibalistic polyphenism in larval tiger salamanders. Ecology 80:1076–1080.

Horat, P., and Semlitsch, R. D. 1994. Effects of predation risk and hunger on the behaviour of two species of tadpoles. Behav. Ecol. Sociobiol. 34:393–401.

Kats, L. B., and Dill, L. M. 1998. The scent of death: chemosensory assessment of predation risk by prey animals. Ecoscience 5:361–394.

Kerfoot, W. C., and Sih, A. 1987. Predation: Direct and Indirect Impacts on Aquatic Com-munities. Hanover, N.H.: University Press of New England.

Klingenberg, C. P., and Ekau, W. 1996. A combined morphometric and phylogenetic analy-sis of an ecomorphological trend: pelagization in Antarctic fishes (Perciformes: Nototheniidae). Biol. J. Linn. Soc. 59:143–177.

Kotler, B. P., Blaustein, L., and Brown, J. S. 1992. Predator facilitation: the combined effect of snakes and owls on the foraging behavior of gerbils. Ann. Zool. Fennici 29:199–206.

Krueger, D. A., and Dodson, S. I. 1981. Embryological induction and predation ecology in *Daphnia pulex*. Limnol. Oceanogr. 26:219–223.

Kuhlmann, H.-W., and Heckmann, K., 1994. Predation risk of typical ovoid and "winged" morphs of *Euplotes*. Hydrobiologica 284:219–227.

Kusch, J. 1993a. Behavioral and morphological changes in ciliates induced by predator *Amoeba proteus*. Oecologia 96:354–359.

Kusch, J. 1993b. Induction of defensive morphological changes in ciliates. Oecologia 94:571–575.

Kusch, J. 1995. Adaptation of inducible defenses in *Euplotes diadaleos* (Ciliophora) to predation risks by various predators. Micro. Ecol. 30:79–88.

Kusch, J., and Kuhlmann, H. W. 1994. Cost of *Stenostomum*-induced morphological defense in ciliate *Euplotes octocarinatus*. Arch. Hydrobiol. 130:257–267.

Lardner, B. 2000. Morphological and life history responses to predators in larvae of seven anurans. Oikos 88:169–180.

Leopold, A. 1933. Game Management. New York: Scribner.

Lima, S. L., and Bednekoff, P. A. 1999. Temporal variation in danger drives antipredator behavior: the predation risk allocation hypothesis. Am. Nat. 153:649–659.

Lima, S. L., and Dill, L. M. 1990. Behavioral decisions made under the risk of predation: a review and prospectus. J. Zool. 68:619–640.

Martin, T. H., Wright, R. A., and Crowder, L. B. 1989. Non-additive impact of blue crabs and spot on their prey assemblages. Ecology 70:1935–1942.

McIntosh, A. R., and Peckarsky, B. L. 1999. Criteria determining behavioural responses to multiple predators by a stream mayfly. Oikos 85:554–564.

McNamara, J. M., and Houston, A. I. 1987. Starvation and predation as factors limiting population size. Ecology 68:1515–1519.

McPeek, M. A. 1995a. Morphological evolution mediated by behavior in damselflies of two communities. Evolution 49:749–769.

McPeek, M. A. 1995b. Testing hypotheses about evolutionary change on single branches of a phylogeny using evolutionary contrasts. Am. Nat. 145:686–703.

McPeek, M. A. 1998. The consequences of changing top predator in a food web: a comparative experimental approach. Ecol. Monogr. 68:1–23.

McPeek, M. A., Schrot, A. K., and Brown, J. M. 1996. Adaptation to predators in a new community: swimming performance and predator avoidance in damselflies. Ecology 77:617–629.

Moran, N. A. 1992. The evolutionary maintenance of alternative phenotypes. Am. Nat. 139:971–989.

Peckarsky, B. L., and McIntosh, A. R. 1998. Fitness and community consequences of avoiding multiple predators. Oecologia 113:565–576.

Pecor, K. W., and Hazlett, B. A. 2003. Frequency of encounter with risk and the tradeoff between pursuit and antipredator behaviors in crayfish: a test of the risk allocation hypothesis. Ethology 109:97–106.

Petranka, J. W. 1989. Response of toad tadpoles to conflicting chemical stimuli: predator avoidance versus "optimal" foraging. Herpetologica 45:283–292.

Pettersson, L. B., and Brönmark, C. 1994. Chemical cues from piscivores induce a change in morphology in crucian carp. Oikos 70:396–402.

Pettersson, L. B., Nilsson, P. A., and Brönmark, C. 2000. Predator recognition and defence strategies in crucian carp, *Carassius carassius*. Oikos 88:200–212.

Pigliucci, M. 2001. Phenotypic Plasticity: Beyond Nature and Nurture. Baltimore, Md.: Johns Hopkins University Press.

Polis, G. A. 1991. Complex trophic interactions in deserts: an empirical critique of food-web theory. Am. Nat. 138:123–155.

Polis, G. A., and Strong, D. R. 1996. Food web complexity and community dynamics. Am. Nat. 14:813–846.

Price, T. 1991. Morphology and ecology of breeding warblers along an altitudinal gradient in Kashmir, India. J. Anim. Ecol. 60:643–664.

Relyea, R. A. 2000. Trait-mediated effects in larval anurans: reversing competition with the threat of predation. Ecology 81:2278–2289.

Relyea, R. A. 2001a. The lasting effects of adaptive plasticity: predator-induced tadpoles become long-legged frogs. Ecology 82:1947–1955.

Relyea, R. A. 2001b. Morphological and behavioral plasticity of larval anurans in response to different predators. Ecology 82:523–540.

Relyea, R. A. 2001c. The relationship between predation risk and antipredator responses in larval anurans. Ecology 82:541–554.

Relyea, R. A. 2002a. Competitor-induced plasticity in tadpoles: consequences, cues, and connections to predator-induced plasticity. Ecol. Monogr. 72:523–540.

Relyea, R. A. 2002b. The many faces of predation: how selection, induction, and thinning combine to alter prey phenotypes. Ecology 83:1953–1964.

Relyea, R. A. 2003a. How prey respond to combined predators: a review and an empirical test. Ecology 84:1827–1839.

Relyea, R. A. 2003b. Predators come and predators go: the reversibility of predator-induced traits. Ecology 84:1840–1848.

Relyea, R. A. 2004a. Fine-tuned phenotypes: tadpole plasticity under 16 combinations of predators and competitors. Ecology 85:172–179.

Relyea, R. A. 2004b. Integrating phenotypic plasticity when death is on the line: insights from predator-prey systems. In: The Evolutionary Biology of Complex Phenotypes (Pigliucci, M., and Preston, K., eds.). Oxford: Oxford University Press; 176–194.

Relyea, R. A., and Hoverman, J. T. 2003. The impact of larval predators and competitors on the morphology and fitness of juvenile tree frogs. Oecologia 134:596–604.

Relyea, R. A., and Werner, E. E. 1999. Quantifying the relation between predator-induced behavioral responses and growth performance in larval anurans. Ecology 80:2117–2124.

Relyea, R. A., and Werner, E. E. 2000. Morphological plasticity of four larval anurans distributed along an environmental gradient. Copeia 2000:178–190.

Richardson, J. M. L. 2001a. A comparative study of activity levels in larval anurans and response to the presence of different predators. Behav. Ecol. 12:51–58.

Richardson, J. M. L. 2001b. Relative roles of adaptation and phylogeny in determination of larval traits in diversifying anuran lineages. Am. Nat. 157:282–299.

Ricklefs, R. E., and Travis, J. 1980. A morphological approach to the study of avian community organization. Auk 97:321–338.

Schlichting, C. D., and Pigliucci, M. 1998. Phenotypic Evolution: A Reaction Norm Perspective. Sunderland, Mass.: Sinauer.

Schoener, T. W. 1989. Food webs from the small to the large. Ecology 70:1559–1589.

Sih, A. 1987. Predators and prey lifestyles: an evolutionary and ecological overview. In: Predation: Direct and Indirect Impacts on Aquatic Communities (Kerfoot, W. C., and Sih, A., eds.). Hanover, N.H.: University Press of New England; 203–224.

Sih, A., and McCarthy, T. M. 2002. Prey responses to pulses of risk and safety: testing the risk allocation hypothesis. Anim. Behav. 63:437–443

Skelly, D. K. 1994. Activity level and susceptibility of anuran larvae to predation. Anim. Behav. 47:465–468.

Smith, D. C., and Van Buskirk, J. 1995. Phenotypic design, plasticity, and ecological performance in two tadpole species. Am. Nat. 145:211–233.

Spitze, K. 1992. Predator-mediated plasticity of prey life-history and morphology: *Chaoborus americanus* predation on *Daphnia pulex*. Am. Nat. 139:229–247.

Sultan, S. E., and Spencer, H. G. 2002. Metapopulation structure favors plasticity over local adaptation. Am. Nat. 160:271–283.

Thiemann, G. W., and Wassersug, R. J. 2000. Patterns and consequences of behavioural responses to predators and parasites in *Rana* tadpoles. Biol. J. Linn. Soc. 71:513–528.

Tollrian, R., and Dodson, S. I. 1999. Inducible defenses in cladocera: constraints, costs, and multipredator environments. In: The Ecology and Evolution of Inducible Defenses (Tollrian, R., and Harvell, D., eds.). Princeton, N.J.: Princeton University Press; 177–202.

Tollrian, R., and Harvell, D. 1999. The Ecology and Evolution of Inducible Defenses. Princeton, N.J.: Princeton University Press.

Trussell, G. C., and Nicklin, M. O. 2002. Cue sensitivity, inducible defense, and trade-offs in a marine snail. Ecology 83:1635–1647.

Tufto, J. 2000. The evolution of plasticity and nonplastic spatial and temporal adaptations in the presence of imperfect environmental cues. Am. Nat. 156:121–130.

Turner, A. M., Bernot, R. J., and Boes, C. M. 2000. Chemical cues modify species interactions: the ecological consequences of predator avoidance by freshwater snails. Oikos 88:148–158.

Turner, A. M., Fetterolf, S. A., and Bernot, R. J. 1999. Predator identity and consumer behavior: differential effects of fish and crayfish on habitat use of a freshwater snail. Oecologia 118:242–247.

Van Buskirk, J. 2000. The costs of an inducible defense in anuran larvae. Ecology 81:2813–2821.

Van Buskirk, J. 2002. A comparative test of the adaptive plasticity hypothesis: relationships between habitat and phenotype in anuran larvae. Am. Nat. 160:87–102.

Van Buskirk, J., and McCollum, S. A. 2000. Functional mechanisms of an inducible defence in tadpoles: morphology and behaviour influence mortality risk from predation. J. Evol. Biol. 13:336–347.

Van Buskirk, J., McCollum, S. A., and Werner, E. E. 1997. Natural selection for environmentally induced phenotypes in tadpoles. Evolution 51:1983–1992.

Van Buskirk, J., Muller, C., Portmann, A., and Surbeck, M. 2002. A test of the risk allocation hypothesis: tadpole responses to temporal change in predation risk. Behav. Ecol. 13:526–530.

Van Buskirk, J., and Relyea, R. A. 1998. Selection for phenotypic plasticity in *Rana sylvatica* tadpoles. Biol. J. Linn. Soc. 65:301–328.

Van Buskirk, J., and Schmidt, B. R. 2000. Predator-induced phenotypic plasticity in larval newts: trade-offs, selection, and variation in nature. Ecology 81:3009–3028.

Van Buskirk, J., and Yurewicz, K. L. 1998. Effects of predators on prey growth rate: relative contributions of thinning and reduced activity. Oikos 82:20–28.

Van Tienderen, P. H. 1997. Generalists, specialists, and the evolution of phenotypic plasticity in sympatric populations of distinct species. Evolution 51:1372–1380.

Via, S., and Lande, R. 1985. Genotype-environment interaction and the evolution of phenotypic plasticity. Evolution 39:502–522.

Via, S., and Lande, R. 1987. Evolution of genetic variability in a spatially heterogeneous environment: effects of genotype-environment interaction. Genet. Res. 49:147–156.

von Baer, K. E. 1828. Entwicklunsgeschichte der Thiere: Beobachtung und Reflexion. Konegsberg, F.R.G.: Borntrager.

Weetman, D., and Atkinson, D. 2002. Antipredator reaction norms for life history traits in *Daphnia pulex*: dependence on temperature and food. Oikos 98:299–307.

Werner, E. E., and Anholt, B. R. 1993. Ecological consequences of the trade-off between growth and mortality rates mediated by foraging activity. Am. Nat. 142:242–272.

Werner, E. E., and Anholt, B. R. 1996. Predator-induced behavioral indirect effects in anuran larvae. Ecology 77:157–169.

West-Eberhard, M. J. 1989. Phenotypic plasticity and the origins of diversity. Annu. Rev. Ecol. Syst. 20:249–278.

Wibe, A. E. T., Nordtug, T., and Jenssen, B. M. 2001. Effects of bis (tributyltin) oxide on antipredator behavior in threespine stickleback *Gasterosteus aculeatus* L. Chemosphere 44:475–481.

PART III

POPULATION- AND COMMUNITY-LEVEL INTERACTIONS

The discussions in parts I and II have clearly demonstrated that assessment of the biotic environment, whether it be an assessment of the presence of prey items or of predation risk, is a critical component of the life of traditional predators, omnivorous predators, and prey animals. Further, they have illustrated how perception and assessment are not only constrained by sensory capacities (as reflected in animal physiology and behavior), but by ontogeny, phylogeny, and ecology. In part III, researchers provide specific, dramatic examples of the consequences of the capacities and constraints in predators' perception of potential prey and in prey assessments of risk on population- and community-level predator-prey interactions. Even though predator-prey interactions often are ultimately reduced to the interactions between two individuals, a great deal of important research on predator-prey interactions focuses on the interactions between predator and prey populations. The characterization of these populations may be predicted in accordance with our understanding of the interactions of individuals, or populations may be characterized according to traits unique to them.

In chapter 10, Robert Denno, Deborah Finke, and Gail Langellotto review the direct and indirect effects of vegetation and habitat structural complexity on predator-prey interactions, intraguild predation, and cannibalism, and explore the consequences of these effects for food web dynamics. They provide convincing evidence for the importance of habitat complexity and its consequences to the abundance and diversity of invertebrate predators in a habitat. The authors particularly stress the importance of complexity in reducing intraguild predation. The role of higher spatial scale (i.e., landscape effects) is highlighted, and they provide examples of its importance in altering numerical responses, dispersal patterns, availability of refuges for

prey, intraguild predation, and other aspects of the ecology of predator-prey interactions. In general, they effectively argue that vegetation and habitat complexity
mediate the consequences of predator-prey interactions at local, habitat, and landscape spatial scales.

Andrew Sih continues in chapter 11 with the theme of the importance of the habitat
by considering the patterns of spatial overlap between predators and prey and how
these patterns affect encounter rates, predation rates, and, ultimately, predator-prey
population and community dynamics. That is, he discusses whether habitat complexity or other factors determine the presence and abundance of predator and prey
species in any given habitat, and how the interplay determines the spatial ecology
of both species. In particular, Sih proposes that it is critically important to understand how the interplay between predator and prey, and their behavioral responses
to each other, determine patterns of predator-prey spatial overlap. Sih urges this
approach because, though in nature both predators and prey are mobile and have
the potential to respond to each other, most studies have investigated the spatial
ecology of predator-prey interactions by keeping one interacting partner immobile
and then assessing the responses of the other. To illustrate, Sih focuses on one key
factor, resource availability, and shows that its influence is significant and contrary
to the prediction of current models because of the spatial dynamics at play.

In chapter 12, Oswald Schmitz explores how knowledge of predator-prey interactions informs our understanding of population and community structure and function. He argues that the trade-offs between foraging and avoiding predators are a
central force in shaping prey population demography and predator-prey dynamics.
He further argues that the key determinants of how these trade-offs shape populations and communities are the portion of the habitat used by prey and the prey's
antipredator behavior in relation to predator hunting mode. Also, Schmitz introduces
a concept, *habitat domain,* which may influence our understanding of the patterns
of spatial overlap between predators and prey, as discussed by Sih. Schmitz describes
habitat domain as the proportion of the entire available habitat used by a species.
Thus, where a species is distributed may be described as a narrow or broad habitat
domain. Schmitz ends by providing interesting speculation on the behavioral interactions that are likely to be usefully incorporated into population and community
models.

Peter Abrams carries this speculation one step further in chapter 13 by actually
exploring the modeling of the adaptive changes in the interactions between species
in different trophic levels. Indeed, Abrams suggests that discrepancies between some
current models and empirical tests of the predictions of the models are, in part, associated with failure to consider adaptive responses of predators (consumers) to prey
(resources), or vice versa. Thus, Abrams develops and assesses a model that includes
a simple representation of adaptive processes, in order to see how all trophic levels
respond to a perturbation affecting one level.

10

Direct and Indirect Effects of Vegetation Structure and Habitat Complexity on Predator-Prey and Predator-Predator Interactions

ROBERT F. DENNO
DEBORAH L. FINKE
GAIL A. LANGELLOTTO

Our major objective here is to review the direct and indirect effects of vegetation and habitat structural complexity on predator-prey and predator-predator (intraguild predation and cannibalism) interactions and to explore the consequences of such habitat-mediated interactions on prey and food web dynamics. For this assessment, we emphasize interactions involving invertebrate predators (insects, arachnids, and crustaceans) in both terrestrial and aquatic systems. Structural complexity effects on predators were examined at the spatial scales of microhabitat (e.g., leaf surface), habitat (e.g., plant species diversity, plant density, litter diversity, and substrate heterogeneity), and landscape (e.g., patch size and background matrix composition).

We found widespread evidence that many species of invertebrate predators occur at elevated densities in complex-structured vegetation or habitats. The primary mechanisms underlying their accumulation in such habitats are refuge from intraguild predation, including cannibalism; access to alternative food resources (prey and plant food such as pollen); and refuge from physical disturbance (stream discharge and drought). A relatively favorable microclimate and enhanced searching efficiency do not often explain why predators occur more abundantly in complex-structured habitats.

There is also evidence suggesting that the accumulation of predators in complex-structured habitats results in enhanced prey suppression, an effect that occasionally cascades to basal resources, where primary productivity increases. However, if predators are excluded from complex habitats or their foraging efficiency is reduced, prey gain refuge from predation and often occur at higher densities than they do in structurally simple habitats. Refuges from predation and intraguild predation in complex-structured habitats often result because predators

211

are too large to effectively maneuver in such habitats, their vision and search are impaired, their foothold (ability to grip substrate) is adversely affected, or potential prey more effectively detect predators and escape.

Landscape-level complexity influences predator-prey interactions by means of strong habitat edge effects, altered numerical responses of predators, patch-size effects, and the nature of the background matrix in which focal habitats are nested. In general, habitat characteristics that decouple predator-prey interactions (e.g., selectively impermeable habitat edges, a matrix structure that selectively hinders predator dispersal and thus patch connectivity, and refuges for prey and intraguild prey) may lead to unstable dynamics and prey outbreaks. Also at the landscape scale, predator impacts on herbivores can vary directly or inversely with habitat patch size or show no relationship, depending on factors that (a) promote or impede (generalist) predator incursions from the surrounding matrix or (b) alter the aggregative response of predators to changes in prey density.

Currently, data suggest that vegetation and habitat complexity mediate the consequences of predator-prey interactions at local, habitat, and landscape spatial scales. Nevertheless, experiments that isolate the effects of habitat edge, patch size, patch isolation, and background matrix on predator dispersal are sorely needed. Such experiments will be essential to gain a more in-depth understanding of how we can manipulate natural landscapes to preserve or restore trophic structure or alter agricultural systems to maximize predator impacts.

The notion that vegetation structure and habitat complexity mediate species interactions at higher trophic levels is not new. Decades ago, several authors suggested that plants set the stage on which natural enemies interact (Price et al. 1980, Denno and McClure 1983). At the time, though, the complex ways variation in basal resources and habitat structure influence multitrophic interactions were not well understood. Subsequently, variability in plants (e.g., in architecture, chemistry, and phenology) and in the structural complexity of habitats where they grow have been shown to have dramatic consequences for predator-prey, host-parasitoid, and predator-predator interactions (Root 1973, Kareiva 1987, Hunter and Price 1992, Bottrell et al. 1998, Denno et al. 2002, Finke and Denno 2002, Cronin 2003, Langellotto and Denno 2004). In the context of ecological theory, vegetation and habitat structure can affect the stability of predator-prey dynamics (Kareiva 1987, Cantrell et al. 2001, Vandermeer and Carvajal 2001, Cronin et al. 2004), the dynamics of metapopulations (Holyoak and Lawler 1996, Lei and Hanski 1997, Fagan et al. 1999, Holyoak 2000, Cronin 2003), the strength of top-down impacts on herbivores (Denno et al. 2002, Lewis and Eby 2002), the dynamics of food webs via altered intraguild predation (Finke and Denno 2002), the probability for trophic cascades (Agrawal and Karban 1997, Halaj et al. 2000), and the consequences of spatial subsidies for multitrophic interactions (Polis and Hurd 1996, Fagan et al. 1999, Post et al. 2000). Moreover, habitat structure has been shown to influence the effectiveness of the natural-enemy complex, including predators in agricultural cropping systems and ornamental plant landscapes (Riechert and Bishop 1990,

Settle et al. 1996, Thies and Tscharntke 1999, Landis et al. 2000, Symondson et al. 2002). The extent that habitat structure influences spatial processes (e.g., numerical responses of predators, their interhabitat dispersal, edge effects, and thus the coupling of predator-prey interactions) in habitats compromised by human activity (e.g., fragmented landscapes) is also of immediate concern in conservation biology (Bolger et al. 2000, Cronin 2003) and restoration ecology (Bond and Lake 2003, Brown 2003). Thus, understanding the direct and indirect effects of habitat structural complexity on natural enemy interactions across a range of spatial scales is important in both ecological and applied contexts.

In this chapter, we review the direct and indirect effects of habitat structural complexity (vegetation and substrate complexity, as well as landscape heterogeneity) on predator-prey and predator-predator interactions, with consideration of extended consequences for herbivores and food web dynamics. We address the effects of variation in habitat structure on predators at several spatial scales: (a) microhabitat variation in architectural complexity (e.g., within-plant variation in trichomes, leaf-surface texture, leaf domatia, and branch density) and (b) within-habitat variation in structural complexity (e.g., plant species diversity, plant density and height, litter/debris volume and complexity, and substrate heterogeneity). Although there is a paucity of data, we also address (c) the consequences of patch or habitat size, edge-area ratio, and matrix heterogeneity (i.e., the structure of the background landscape in which focal habitats are nested) on predator-prey interactions at a landscape spatial scale. For our assessment, we emphasize experimental studies in which habitat structure was manipulated and predator effects were measured; however, we also include comparative efforts in which predator effects were determined across a range of habitats varying in some structural element. The studies we discuss document the effects of habitat structure on invertebrate predators (insects, arachnids, and crustaceans). We emphasize habitat-mediated predator effects in terrestrial systems, but not to the exclusion of representative aquatic examples.

Our specific objectives are first to examine the direct effects of vegetation and habitat structure on invertebrate predators (their abundance, dispersion, spatial dynamics, and diversity) and explore the mechanisms underlying their accumulation or aggregation in vegetation or habitats varying in structural complexity. Potential mechanisms include altered rates of cannibalism or intraguild predation, prey abundance and availability, modified capture efficiency of prey, and refuge from harsh climatic conditions and physical disturbance. Second, we examine how habitat complexity mediates predator-prey and predator-predator interactions and indirectly affects consumer (e.g., herbivore and detritivore) abundance and primary productivity. At the individual level, we focus on experimental studies that elucidate how predator and prey (including intraguild prey) behavior changes in habitats varying in structural complexity and how altered behavior influences risk of attack. We then scale these findings up to the population level to examine how habitat structure at the landscape spatial scale influences predator-prey population dynamics, levels of intraguild predation, and ultimately food web dynamics, including spatial subsidies (i.e., incursions of predators from neighboring habitats) and their consequences.

Habitat Complexity: The Context
for Assessing Predator Effects

The relative complexity of a habitat can be quantified largely by the number of different structural elements per unit habitat volume (McCoy and Bell 1991, Langellotto and Denno 2004). How habitat complexity is defined, as well as its mediating effects on natural enemies, depends, in large part, on the spatial scale of the assessment (McCoy and Bell 1991, Landis et al. 2000). From a predator's perspective, habitat structure can vary from simple to complex at several spatial scales, all of which can influence interactions with prey (Table 10.1). From small to large, these spatial scales are microhabitat, habitat, and landscape.

Microhabitat Scale Complexity

At the smallest spatial scale of microhabitat, there is variation in the structure of the surface or substrate that can influence predator abundance and prey capture. For example, when plants are the habitat of an organism, there can be variation in the density of trichomes per leaf, the presence or abundance of leaf domatia (hair tufts associated with the major leaf veins or elsewhere), or the density of foliage and branches, all of which are known to influence the abundance or foraging efficiency of invertebrate predators (e.g., Gunnarsson 1990, Agrawal and Karban 1997, Halaj et al. 2000, McNett and Rypstra 2000, Stavrinides and Skirvin 2003). In such cases, an increase in the density of structures per unit volume or the presence versus the absence of a structure has been interpreted as enhanced microhabitat complexity (see Langellotto and Denno 2004). In several cases, predator search and consequences for prey were compared on structurally simple and complex varieties of plants that differed in the density or presence of plant surface structures (e.g., Kareiva and Sahakian 1990, Bottrell et al. 1998, Legrand and Barbosa 2003) or on plants where surface structure was manipulated experimentally (e.g., Agrawal and Karban 1997, Roda et al. 2000, Norton et al. 2001).

Habitat Scale Complexity

Structural complexity can also vary at the larger spatial scale of habitat, where it also affects predator abundance and interactions with prey (Brust 1994, Lewis and Eby 2002, Langellotto and Denno 2004). For instance, within a habitat, structure can vary from simple to complex, as reflected by changes in plant species diversity (e.g., monoculture vs. polyculture or weedy vs. pure cropping systems; Russell 1989, Godfrey and Leigh 1994, Coll and Bottrell 1995, Balfour and Rypstra 1998, Siemann et al. 1998), vegetation architecture (e.g., low- vs. high-profile vegetation, low- vs. high-stem density; Leber 1985, Snyder and Ives 2001, Brose 2003), litter/debris complexity (e.g., absent vs. present, compact vs. loose, conventional vs. no-till agricultural fields; Uetz 1991, Brust 1994, Döbel and Denno 1994, Settle et al. 1996, Clark et al. 1997, Usio and Townsend 2002), or substrate heterogeneity in aquatic habitats (e.g., variation in rubble, debris, or macrophyte presence and diversity; Palmer et al. 1996, Stewart et al. 2003). Numerous studies have com-

Table 10.1. Elements of Habitat Complexity Affecting Predator-Prey Interactions at Three Spatial Scales

Element of Habitat Complexity	References
Microhabitat scale	
Leaf surface structure	
Trichomes/pubescence (±)	Roda et al. 2000, Yang 2000, Fordyce and Agrawal 2001, Stavrinides and Skirvin 2003
Domatia (±)	Grostal and O'Dowd 1994, Agrawal 1997, Agrawal and Karban 1997, Norton et al. 2001, English-Loeb et al. 2002
Leaf-surface waxes (±)	Eigenbrode et al. 1996, Rutledge et al. 2003
Plant architecture	
Branch/foliage density (low vs. high)	Gunnarsson 1990, Halaj et al. 2000
Whole-plant structure (simple vs. complex)	Kareiva and Sahakian 1990, Clark and Messina 1998a, Legrand and Barbosa 2003
Habitat scale	
Plant species diversity	
Mono- vs. polyculture	Russell 1989, Godfrey and Leigh 1994, Coll and Bottrell 1995
Weedy vs. pure	Horn 1981, Balfour and Rypstra 1998
Woodlot vs. urban	Hanks and Denno 1993, Leddy 1996, Tooker and Hanks 2000
Stand architecture	
Vegetation height (low vs. high)	Snyder and Ives 2001
Stem density (low vs. high)	Leber 1985, Lewis and Eby 2002, Brose 2003
Litter/debris	
Presence vs. absence	Everett and Ruiz 1993, Brust 1994, Döbel and Denno 1994, Palmer et al. 1996, Settle et al. 1996
Depth/compact vs. loose	Bultman and Uetz 1984, Uetz 1991
Conventional vs. no-till crops	Hammond and Stinner 1987, Carcamo 1995, Clark et al. 1997
Landscape scale	
Habitat edge effects (permeability)	Bohlen and Barrett 1990, Thomas 1992, Cantrell et al. 2001, Bommarco and Fagan 2002, Frampton et al. 1995, Wratten et al. 2003
Patch size (edge-to-area ratio) or fragmentation (interpatch distrance)	Kareiva 1987, Döbel and Denno 1994, Gibb and Hochuli 2002, With et al. 2002, Braschler et al. 2003, Cronin et al. 2004
Background matrix (±, cover, structure)	Kareiva 1987, Döbel and Denno 1994, Collinge and Palmer 2002, Cronin 2003

pared the effects of habitat manipulations (such as alterations in structural diversity) on a predator's searching behavior, on density, or on consequences for prey (Bultman and Uetz 1982, Riechert and Bishop 1990, Brust 1994, Döbel and Denno 1994), whereas other studies have measured predator effects across a range of habitats that varied naturally in their structural complexity (Hanks and Denno 1993, McNett and Rypstra 2000, Lewis and Eby 2002). We took advantage of both experimental studies and surveys to assess the consequences of habitat structure on predators.

Landscape Scale Complexity

At the largest spatial scale of landscape, variation in structural complexity across habitats also can influence predator-prey interactions and their extended consequences for lower trophic levels. Thus, understanding the effects of neighboring habitat structure on predator-prey interactions in the focal habitat becomes key at this spatial scale. Two perspectives are relevant here to potential predator effects. First, the discrepancy in structure between two neighboring habitats may differentially affect predator and prey movement across habitat edges (Kareiva 1987, Fagan et al. 1999, Frampton et al. 1995, Bommarco and Fagan 2002). If the habitat edge acts as a selective filter for prey, the consequence might be a differential subsidy (i.e., incursion) of predators into the focal habitat that, in turn, may alter predator-prey interactions and food web dynamics (Fagan et al. 1999, Cantrell et al. 2001). Second, the proportional cover of the focal habitat, relative to the neighboring habitat matrix, may be important and may change as a consequence of habitat fragmentation. Here issues of focal patch size, changes in edge-area ratio, and interpatch distance come into play, as they interact with discrepancies in the structure of the background matrix to affect predator-prey dynamics (Cronin 2003). However, for many studies involving arthropod predators, effects of patch size (i.e., edge-area ratio), habitat fragmentation (as reflected in interpatch distance), and the cover and structure of the background matrix are often confounded, making an explicit effect of each difficult to detect.

Vegetation and Habitat Complexity Enhance Predator Density and Diversity

There is widespread evidence (39 of 50 studies) that invertebrate predators selectively accumulate and reach higher densities in structurally complex habitats, a response that occurs among a wide range of predator taxa (spiders, mites, heteropterans, neuropterans, beetles, syrphid flies, and ants) and at several spatial scales (see meta-analytical synthesis by Langellotto and Denno 2004). Specifically, the higher densities of predators observed in complex-structured habitats result from increases in plant species diversity, the structural diversity of living vegetation and litter/debris, and substrate heterogeneity in aquatic habitats. Many of the studies that we include here are experimental and involve manipulations of habitat complexity with measured responses of predators.

Microhabitat Spatial Scale

At the microhabitat scale, a greater number of predaceous mites occur on plants with leaf domatia than on plants lacking them, on plants with domatia experimentally blocked, or on plants with smaller domatia (Grostal and O'Dowd 1994, Agrawal 1997, Agrawal and Karban 1997, Norton et al. 2001, English-Loeb et al. 2002). In addition, predatory mites reach higher densities on pubescent plant varieties than on glabrous ones (Roda et al. 2000). At a small spatial scale, spiders are more abundant on structurally complex conifers with a high needle density than on trees with reduced needle density (Gunnarsson 1990, Halaj et al. 2000). Thus, within a single individual plant, leaf-surface heterogeneity and architectural complexity can promote higher densities of invertebrate predators. Notably, though, complex-structured plant surfaces with trichomes or certain cuticular waxes can interfere with the foraging behavior of predators, such as some coccinellids, lacewings, and heteropterans, resulting in lower densities (Kareiva and Sahakian 1990, Eigenbrode et al. 1996, Bottrell et al. 1998, Fordyce and Agrawal 2001, Rutledge et al. 2003, Stavrinides and Skirvin 2003).

Habitat Spatial Scale

At the larger scale of habitat, invertebrate predators often occur more abundantly in taxonomically diverse vegetation (19 of 25 studies; Langellotto and Denno 2004). For instance, there are numerous examples of predators reaching higher densities in polycultures than in monocultures (Russell 1989, Letourneau 1990, Godfrey and Leigh 1994, Mensah 1999), in weedy crops than in weed-free monocultures (Horn 1981, Balfour and Rypstra 1998), and in floristically diverse woodlots than in simplified urban plantings (Hanks and Denno 1993, Leddy 1996, Tooker and Hanks 2000; but see for counterexamples see Russell 1989, Andow 1991). Predator taxa frequently exhibiting this positive response include spiders, heteropterans, and beetles (Brust 1994, Brose 2003, Langellotto and Denno 2004). Enhancing the structural complexity of either aboveground or aquatic vegetation (Halaj et al. 1998, Rypstra et al. 1999, McNett and Rypstra 2000, Sunderland and Samu 2000, Brose 2003); or substrate litter, debris, or vegetation cover (Uetz 1979, Bultman and Uetz 1982, Bultman and Uetz 1984, Riechert and Bishop 1990, Uetz 1991, Brust 1994, Döbel and Denno 1994, Settle et al. 1996, Finke and Denno 2002, Langellotto 2002, Stewart et al. 2003); also promotes higher densities of predators (hunting and web-building spiders, heteropterans, carabid beetles, dragonfly naiads, and predaceous caddisflies)—a response that occurs in both terrestrial (Langellotto and Denno 2004) and aquatic systems (Crowder and Cooper 1982, Habdija et al. 2002, Stewart et al. 2003).

In one case, structural complexity of aboveground vegetation explained more variation in carabid beetle density than plant species diversity (Brose 2003). In other studies, it was not possible to extract the relative contributions of plant species diversity and structural diversity to enhanced predator densities (Hanks and Denno 1993, Tooker and Hanks 2000). In addition to promoting a higher density of invertebrate predators, enhanced habitat complexity often encourages higher predator diversity and species richness (Uetz 1979, Balfour and Rypstra 1998, Siemann et al. 1998, Halaj et al. 2000, Tooker and Hanks 2000, Brose 2003). Although the positive relationship between habitat, or vegetation, complexity and both the density and diversity of invertebrate predators is robust (reviewed in Langellotto and Denno 2004), there are certainly examples of predators (e.g., some spiders and ants) that either do not respond to habitat complexity or show a negative association (Russell 1989, Uetz 1991, Perfecto and Sediles 1992, Costello and Daane 1998), particularly when they are physically excluded (Messina et al. 1997, Lewis and Eby 2002).

Landscape Spatial Scale

Although studies on the effects of landscape-level complexity on natural enemies are few, ecologists generally recognize the potential contribution of landscape complexity to the density and diversity of natural enemy complexes (Thies and Tscharntke 1999, Langellotto and Denno 2004). To assess landscape-level complexity and its effects on predators, it is important to consider how complexity varies across several different habitats (e.g., agricultural plantings, old fields, and woodlots; Landis et al. 2000).

The issue at hand is whether a focal habitat, surrounded by a diverse landscape of neighboring habitat types, carries a higher density of predators than the same type of focal habitat with less diverse adjoining habitats. Some evidence exists suggesting that landscape-level complexity does promote higher predator densities. For example, in the intertidal marshes of eastern North America, the density of a major (wolf spider) predator in low-marsh habitats (dominated by the cordgrass *Spartina alterniflora*) is much higher in coastal areas with extensive upland expanses of another cordgrass (*Spartina patens*), where this predator also resides. The spider is much less abundant in areas where *S. alterniflora* is not abutted on the upland edge by *S. patens* (Lewis, D., and Denno, R. F., unpublished data). Likewise, some parasitoids can be more abundant and parasitism rates of hosts can be significantly higher within a complex mosaic of agricultural fields embedded with hedgerows and woodlots, as compared to simple agricultural fields lacking diverse landscape features (Marino and Landis 1996; but see Menalled et al. 1999 for less convincing support). Notably, at the scale of landscape, enhancing habitat diversity around agricultural fields is being used as a means of retaining and enhancing invertebrate predator densities and thus improving conservation biological control (Thies and Tscharntke 1999, reviewed in Landis et al. 2000). For now, limited evidence suggests that increasing landscape-level complexity can promote higher predator densities in certain focal habitats.

Mechanisms Underlying Elevated Predator Density in Complex-Structured Habitats

Possible reasons that predators accumulate in complex-structured habitats or achieve higher densities there are diverse and include (a) refuge from intraguild predation and cannibalism, (b) increased availability of food resources (e.g., prey, pollen, and nectar) and other requisites (e.g., web attachment sites for spiders), (c) a more equitable microclimate, (d) refuge from physical disturbance (e.g., stream discharge), and (e) enhanced foraging efficiency and capture success (Table 10.2).

Refuge from Intraguild Predation and Cannibalism

There is widespread evidence that both intraguild predation and cannibalism are reduced when invertebrate predators occur in complex-structured habitats. Examples in which predators gain refuge from intraguild predation in complex habitats include hunting spiders (Langellotto 2002), web-building spiders (Gunnarsson 1990), mites (Roda et al. 2000, Norton et al. 2001), predaceous heteropterans (Agrawal and Karban

Table 10.2. Mechanisms Underlying Predator Accumulation in Structurally Complex Habitats

Mechanism	References
Refuge from intraguild predation or cannibalism	Gunnarsson 1990, Agrawal and Karban 1997, Roda et al. 2000, Norton et al. 2001, Finke and Denno 2002, Langellotto 2002
Increased resources Prey	Döbel and Denno 1994, Halaj et al. 1998; counterexample: Langellotto and Denno 2004
Nonprey food (e.g., nectar and pollen)	Patt et al. 1997, Barbosa and Wratten 1998, Bugg and Pickett 1998, Eubanks and Denno 1999, Eubanks and Denno 2000, Harmon et al. 2000, Landis et al. 2000, Roda et al. 2003
Other requisites (e.g., web attachment sites)	McNett and Rypstra 2000
Favorable microclimate	Uetz 1979, Riechert and Bishop 1990, Grostal and O'Dowd 1994, Agrawal 1997, Pratt et al. 2002; counterexample: Langellotto 2002
Refuge from physical disturbance	Palmer et al. 1992, Palmer 1996, Lancaster and Belyea 1997, Matthaei et al. 2000, Negishi et al. 2002, Gjerlov et al. 2003, Magoulick and Kobza 2003
Enhanced foraging efficiency or capture success	Denno et al. 2002; counterexamples: Belcher and Thurston 1982, Kaiser 1983, Leber 1985, Kareiva 1987, Agrawal and Karban 1997, Babbit and Tanner 1998, Clark and Messina 1998a, Clark and Messina 1998b, Roda et al. 2000, Langellotto 2002, Lewis and Eby 2002

1997, Finke and Denno 2002), thrips (Agrawal and Karban 1997), and dragonfly naiads (Crowder and Cooper 1982). Invertebrate predators such as hunting spiders also find refuge from cannibalism in complex-structured habitats (Langellotto 2002). Notably, intraguild predation is diminished by refuges provided at both microhabitat and habitat spatial scales. For example, mites, thrips, and small heteropteran predators are provided refuge from larger predators on plants with leaf domatia or trichomes (Agrawal and Karban 1997, Roda et al. 2000, Norton et al. 2001). At the scale of habitat, dense vegetation provides spiders refuge from bird predation (Gunnarsson 1990), and leaf litter offers wolf spiders and predaceous heteropterans refuge from spider predation (Finke and Denno 2002, Langellotto 2002). In virtually all of these cited studies, smaller predator species or smaller individuals of the same species gain refuge from larger predators that are either physically excluded from complex-structured habitats or whose foraging behavior is impaired.

Increased Availability of Prey and Other Resources

Several studies suggest that predators aggregate in complex habitats because prey, alternative food resources (e.g., pollen and nectar), or other requisites such as web-attachment sites are more abundant there (Sheehan 1986, Patt et al. 1997, Wise and

Chen 1999, McNett and Rypstra 2000, Langellotto and Denno 2004). A meta-analysis of insect herbivore responses to manipulated habitat structure found no overall effect suggesting that the strong pattern of predator accumulation in structurally diverse habitats was not driven by prey (herbivore) abundance (Langellotto and Denno 2004). There are cases, however, in which prey availability does appear to promote predator increase. For instance, prey availability in addition to structural complexity explains much of the variation in spider abundance in conifer trees (Halaj et al. 1998). Also, wolf spiders aggregate in habitats with abundant prey, but their numerical response is dramatically enhanced when leaf litter is present (Döbel and Denno 1994).

Overall, though, there is a paucity of data demonstrating that predators consistently accumulate in complex-structured habitats in response to elevated prey density there. There are more studies showing that, when enhanced habitat complexity results in more nonprey food or structural resources for predators, they indeed achieve higher densities there (Patt et al. 1997, Barbosa and Wratten 1998, Bugg and Pickett 1998, Landis et al. 2000). For instance, flowering dandelions in close proximity to alfalfa encourages higher densities of coccinellid beetles, because they aggregate in response to pollen resources (Harmon et al. 2000). Additionally, the presence of pods (alternative plant food) in lima bean fields encourages much higher densities of big-eyed bug predators (Eubanks and Denno 1999, Eubanks and Denno 2000; see Eubanks, ch. 1 in this volume). Web-building spiders are more abundant in diverse vegetation, not because more prey are captured there, but because it provides them with more structural sites for web attachment (McNett and Rypstra 2000).

Favorable Microclimate

It has been suggested that predators aggregate in complex-structured habitats because they encounter a more favorable microclimate (Uetz 1979, Riechert and Bishop 1990, Grostal and O'Dowd 1994, Agrawal 1997, Pratt et al. 2002). There are a few cases involving small mites and leaf domatia supporting this hypothesis. For example, under conditions of low humidity, access to leaf domatia increases the fitness of predaceous mites by enhancing reproduction (Grostal and O'Dowd 1994). However, more studies clearly suggest that leaf domatia provide small arthropod predators (mites, thrips, and heteropterans) refuge from intraguild predation (Agrawal and Karban 1997, Roda et al. 2000, Norton et al. 2001). Experimental studies with wolf spiders show that they do not select complex-structured habitats on the basis of microclimate, but indeed gain refuge from intraguild predation and cannibalism (Langellotto 2002). Overall, evidence is scarce, although studies on the subject are also limited, that invertebrate predators aggregate in complex-structured habitats because they gain refuge from harsh ambient climate.

Refuge from Physical Disturbance

In contrast, there is extensive data suggesting that invertebrate predators aggregate and persist in stream habitats where substrate heterogeneity provides refuge from physical disturbance such as discharge events and periods of drought (Palmer et al. 1996, Lancaster and Belyea 1997). In stream systems, spatial heterogeneity is often positively correlated with the number of spatial refuges (Brown 2003). Numerous studies show that the presence of woody debris, or substrate structure or stability, confers greater resistance of fauna (both predators and prey) to flooding events and promotes recovery after such perturbations (Palmer et al. 1996, Matthaei et al. 2000). Structural diversity in the habitat may provide spatial refuges from high-velocity flow in the water column of streams, as well as in the hyporheic zone (i.e., the substrate under a stream channel; Palmer et al. 1992). Specifically, high-refuge sites can be characterized as having a high proportion of low-shear areas during high-discharge events (Gjerlov et al. 2003). Also, backwaters and inundated habitats can provide flow refuges, because they accumulate macroinvertebrates, including predators, during a spate (Negishi et al. 2002). Variation in stream-bottom topography that includes deep pools provides refuges for predators during periods of drought (Magoulick and Kobza 2003).

Enhanced Foraging Efficiency and Capture Success

There is little data to support the notion that predators achieve high densities in structurally complex, as opposed to structurally simple, habitats because they search and capture prey more efficiently there. In one mesocosm study, in which dispersal was constrained, wolf spiders captured and killed herbivore prey more frequently in complex-structured arenas with leaf litter than in ones lacking litter, suggesting more efficient prey capture (Denno et al. 2002). By contrast, most studies show that predators exhibit reduced searching efficiency and prey capture rate in complex vegetation, both at the microhabitat (Belcher and Thurston 1982, Agrawal and Karban 1997, Clark and Messina 1998a, Clark and Messina 1998b) and habitat spatial scales (Kareiva 1987, Lewis and Eby 2002), and across a wide range of predator taxa, including insects (Babbit and Tanner 1998, Roda et al. 2000), arachnids (Kaiser 1983, Langellotto 2002), and crustaceans (Leber 1985, Lewis and Eby 2002).

Consensus

Of the potential factors underlying the accumulation of predators in complex-structured habitats, there is extensive support for three: refuge from intraguild predation, including cannibalism; acquisition of additional resources (e.g., pollen, nectar, alternate prey, and sites for web attachment); and refuge from physical perturbations in stream habitats. There are few data to support the view that buffered microclimate and enhanced searching efficiency explain the robust pattern of enhanced predator abundance in complex habitats.

Consequences of Habitat Structure for Herbivore Suppression and Food Web Dynamics

Given that many predators accumulate at high densities in complex-structured habitats and that intraguild predation is often reduced there, a logical extension is that enemy impacts should be greater on consumers such as herbivores and detritivores.

Moreover, if the consumers involved have the potential to significantly affect basal resources (primary productivity/yield or detritus), then habitat complexity might determine the extent that predator effects cascade by way of consumers to influence basal resources. However, the potential for this habitat-mediated trophic cascade will be affected by several factors, including the number of trophic levels in the system (three or four), predator density and voraciousness, the strength of predator-prey and predator-predator interactions, and the relative refuge that habitat complexity provides for both herbivorous prey and intraguild prey (Brust 1994, Agrawal and Karban 1997, Finke and Denno 2002).

In multitrophic systems, there is data to suggest that habitat or vegetation complexity can mediate food web dynamics and affect the probability for trophic cascades. For instance, compared to cotton plants lacking complex leaf surfaces with domatia (i.e., hair tufts), domatia-bearing cotton plants support higher densities of arthropod predators, have lower densities of herbivores (mites), and produce 30% more fruit (Agrawal and Karban 1997). The mechanism underlying this trophic cascade is likely reduced intraguild predation on plants with structural refuges (domatia) and enhanced top-down impacts on herbivores (Agrawal and Karban 1997, Norton et al. 2001, Roda et al. 2000). Likewise, in a potato system, increasing the structure of the litter layer promotes high densities of carabid predators, causes low densities of herbivorous beetles, and results in a 35% increase in tuber yield (Brust 1994). However, if herbivore effects on plants are weak as a result of high tolerance to herbivory, then enhanced predator densities in complex vegetation may not cascade to affect basal resources, even if herbivore densities are significantly reduced (Halaj et al. 2000).

In a natural salt marsh system, habitat complexity influences species interactions and trophic dynamics. In this system, predatory wolf spiders accumulate in litter-rich cordgrass habitats where herbivore (planthopper) densities are significantly suppressed compared to those in litter-poor habitats (Döbel and Denno 1994). The presence of leaf litter in this system enhances the numerical response of spiders to planthopper prey (Döbel and Denno 1994), provides mirid bugs (planthopper egg predators) with a refuge from intraguild predation by spiders (Finke and Denno 2002), boosts top-down effects on herbivores, and elevates plant productivity (Denno et al. 2002). The strong mediating effect of vegetation structure on planthopper suppression results because (a) both spiders and mirids accumulate in litter-rich habitats, (b) both predators are voracious and alone can devastate planthopper populations, (c) the intraguild predation of mirids by spiders is intense and asymmetric in simple-structured habitats, (d) intraguild predation is significantly dampened in the presence of litter, (e) mirids are much more at risk from spider attack than are planthoppers in litter-free habitats, and (f) planthoppers do not find significant refuge from predation in litter-rich habitats (Döbel and Denno 1994, Denno et al. 2002, Finke and Denno 2002, Langellotto 2002, Finke and Denno 2003). In three studies (cotton, potato, and cordgrass), interactions between herbivores and plants, and predators and prey (including intraguild prey), are potentially strong, and intraguild prey, more than herbivorous prey, are selectively afforded a spatial refuge from predation in complex-structured habitats.

There are additional studies in which elevated predator densities in complex-structured habitats result in herbivore suppression, although the details of trophic structure and species interactions are not outlined. For instance, armored scale insects, herbivorous mites, and herbivore-inflicted damage are less abundant in structured vegetation where generalist arthropod predators abound (Hanks and Denno 1993, Halaj et al. 2000, Tooker and Hanks 2000, Pratt et al. 2002). In contrast, there are other cases in which herbivore populations (mites, leafhoppers, and red grouse) are not suppressed in complex-structured vegetation or habitats, even though predators are more abundant there (Letourneau 1990, Agrawal 1997, English-Loeb et al. 2002). Nonetheless, there is substantial evidence suggesting that the elevated densities of predators in complex habitats, coupled with the ability to successfully forage there, often results in less abundant consumers.

By contrast, if complex-structured habitats or microhabitats selectively interfere with predator foraging, herbivores in such habitats often achieve higher densities due to spatial refuges and relaxed predation (Kareiva 1987, Messina et al. 1997, Lewis and Eby 2002). Predators suffering diminished foraging success in complex habitats include a wide range of invertebrate taxa, such as mites (Kaiser 1983), dragonfly naiads (Babbit and Tanner 1998), heteropterans (Eigenbrode et al. 1996), thrips (Hoddle 2003), carabid beetles (Snyder and Ives 2001), coccinellid beetles (Kareiva 1987, Messina et al. 1997, Clark and Messina 1998b, Rutledge et al. 2003), neuropterans (Clark and Messina 1998a, Clark and Messina 1998b), shrimp (Leber 1985), and crabs (Lewis and Eby 2002). Spatial refuges and the subsequent decoupling of predator-prey interactions occur in complex habitats when (a) predators but not prey are physically excluded by virtue of body size differences (Leber 1985, Lewis and Eby 2002), (b) predators experience reduced searching efficiency (Kareiva 1987, Grevstad and Klepetka 1992, Eigenbrode et al. 1996, Clark and Messina 1998b, Rutledge et al. 2003), or (c) prey actively switch to habitats that provide refuge when predators are present (Lima and Dill 1990, Eubanks and Miller 1992, Schmitz et al. 1997). When predator search is selectively hindered in complex-structured habitats, herbivores often reach high densities (Crowder and Cooper 1982, Leber 1985, Messina et al. 1997) and can even exhibit outbreak and unstable dynamics (Kareiva 1987, Döbel and Denno 1994).

Body size differences between predators and prey (including intraguild prey), visual impairment, and foothold (i.e., the ability to grip substrate) appear to play key roles in the reduced foraging efficiency of some predators in complex-structured vegetation or habitats, and thus in creating spatial refuges for prey.

Habitat Complexity Alters Behavior and Mediates Predator Search and Prey Refuge

Complex habitats can diminish predator foraging efficiency and provide spatial refuges for prey by inhibiting predator movement and search. Predator-mediated shifts to more complexly structured habitats have been documented for a variety of prey taxa, including spiders (Eubanks and Miller 1992), mayflies (Soluk and Collins 1988), grasshoppers (Schmitz and Suttle 2001), and shrimp (Everett and Ruiz 1993), suggesting that complex habitats reduce predation. It is important to note, however, that a habitat or microhabitat shift in response to the presence of one predator can actually increase the probability of falling prey to other predators (Soluk and Collins 1988,

Losey and Denno 1998). In this section, we investigate how predator and prey behaviors interface with habitat structure to create refuges from predation. Specifically, we explore the role of body size differences between predators and prey, visual impairment, and foothold in mediating predator foraging success in complex habitats.

Body Size Effects on Maneuverability

Habitat complexity can selectively impair larger predators by impeding their ability to maneuver through a habitat (Frampton et al. 1995) or by excluding them from foraging sites altogether (Clark and Messina 1998a, Lewis and Eby 2002). Invertebrate predators that exhibit reduced mobility in dense vegetation include dragonfly naiads (Babbit and Tanner 1998), carabids (Frampton et al. 1995), and crabs (Lewis and Eby 2002). However, the degree to which complexity hinders predator movement can vary as a function of predator size. As a result, the impact of habitat structure on predator foraging ability and capture success can change throughout ontogeny (Yang 2000).

Complex habitats also provide concealed feeding locations or hiding sites for prey, from which larger predators are excluded (Leber 1985, Clark and Messina 1998a). Aphids feed at leaf blade junctions where leaf rolls are too narrow to permit entry by coccinellid and neuropteran predators (Messina et al. 1997, Clark and Messina 1998a, Clark and Messina 1998b). Likewise, eggs and nymphs of many predators, including mites, thrips and heteropterans, are commonly found within leaf domatia, where intraguild predation is reduced (Agrawal and Karban 1997, Roda et al. 2000, English-Loeb et al. 2002). However, in some cases, the prey themselves can be too large to access hiding sites (Leber 1985). For instance, small amphipods find refuge from shrimp predation in architecturally complex vegetation, whereas larger decapod prey are physically excluded and fall prey to shrimp (Leber 1985).

Visual Impairment of Predator Search

In addition to being an obstacle to predator movement, habitat complexity can act as a visual barrier that interferes with the ability of predators to locate prey (Heck and Crowder 1991, Finke and Denno 2002). The presence of leaf litter has been shown to provide mirid bugs with a refuge from intraguild predation by spiders (Finke and Denno 2002). This refuge was attributed to the fact that litter clutters the visual field of the spider, rendering the mobile mirid bug less conspicuous to the visually searching spider (Finke and Denno 2002). The visual impairment of predators in complex habitats can be compounded by complementary prey behaviors. In the presence of predators, caridean shrimps escape to the opposite side of the leaf blade, with only their mobile eyestalks extending beyond the leaf margin. From this secluded position, shrimps observe predators yet remain undetected (Main 1985, Main 1987).

Foothold and the Mechanics of Predator Movement

Small-scale features of the habitat, such as leaf surface texture and plant architectural complexity, can impede the mechanics of predator movement and the ability of predators to gain a foothold or grip (Kareiva and Sahakian 1990, Eigenbrode et al.

1996, Fordyce and Agrawal 2001, Stavrinides and Skirvin 2003). For example, the presence of wax bloom on cabbage leaves interferes with foraging in generalist invertebrate predators such as neuropterans, heteropterans, and coccinellids (Eigenbrode et al. 1996), although the mechanism underlying this response varies with predator species. Neuropterans and heteropterans that forage on standard cabbage varieties with wax bloom accumulate wax crystals in their tarsi. As a result, these predators spend more time grooming and less time foraging on standard cabbage as compared to cabbage with glossy leaf wax (Eigenbrode et al. 1996). Coccinellids show a similar decline in foraging success on cabbage with wax bloom; however, wax debris does not collect in the tarsi of these beetles. Instead, coccinellids change their foraging mode from walking to a less efficient scrambling movement in the presence of wax bloom (Eigenbrode et al. 1996). Wax bloom on pea plants also reduces the efficiency of coccinellid search by diminishing the ability of coccinellid larvae to adhere to the plant (Rutledge et al. 2003).

In addition to leaf surface characteristics, plant architecture (leafiness or degree of branching) can also affect the ability of coccinellids to remain on the plant. Coccinellids can cling to stems and tendrils of leafless pea plants but have difficulty maintaining a grasp on the smooth, slippery leaves of the normal plants, which causes them to fall off twice as frequently on normal pea plants as on the leafless variety (Kareiva and Sahakian 1990). Coccinellids also can show reduced foraging efficiency on more structurally complex varieties of pea plants characterized by a reticulate branching architecture (Legrand and Barbosa 2003).

Although reduced foraging efficiency of predators is the most frequent result, complexity has been shown to enhance the foraging success of enemies, as well. Trichomes on cucumber plants slow the walking speed of parasitoids, which actually enhances foraging success by enabling them to locate more hosts (van Lenteren and de Ponti 1991, but see Hua et al. 1987). In streams, substrate rugosity positively influences the ability of invertebrate predators such as dytiscid diving beetles to retain their holds and forage, especially during high-velocity discharges (Lancaster and Mole 1999).

By influencing movement and access to spatial refuges, habitat complexity not only determines how predators and prey maneuver in and cope in their focal habitats, but also mediates landscape-level processes, such as dispersal across habitat boundaries, movement through the interhabitat matrix, and thus the immigration-emigration dynamic among habitat patches. As a result, it is the interplay between the behavior of predators and their prey and habitat structure that determines their relative interhabitat movements, the strength of predator-prey interactions, and ultimately food web dynamics in the landscape.

Landscape-Level Complexity, Predator-Prey Interactions, and Food Web Dynamics

There is growing awareness that food web dynamics can be strongly influenced by the movement of predators and resources across habitat boundaries (Polis et al. 1996, Post et al. 2000, Holt 2002, Murakami and Nakano 2002). Moreover, the incursion of predators from one habitat into another can destabilize food web dynamics and alter species interactions dramatically (Fagan et al. 1999, Jeffries 2000, Cantrell et al. 2001, Nakano and Murakami 2001). Thus, it becomes critical to understand how landscape structure and habitat boundaries influence the interhabitat movements of predators and what the consequences of such incursions are for predator-prey interactions and food web dynamics. In this section we consider the effects of habitat edges (often associated with abrupt spatial changes in structure), as well as the nature of the landscape matrix in which focal habitats are nested, on predator movement and thus the probability for predator subsidy. We also examine how landscape structure and background matrix influence a predator's numerical response and potential decoupling effects on predator-prey interactions. Last, we assess how predator-prey interactions and food web dynamics change across a gradient of habitat patch sizes and under conditions of habitat fragmentation.

Edge Effects

Habitat edges (boundaries between distinct habitat types) might influence predator-prey interactions in four ways: as barriers to dispersal; as influences on mortality; as involvement in spatial subsidies; and as generators of novel interactions, such as intraguild predation (see Fagan et al. 1999). There is evidence that habitat edges can act like cell membranes that are differentially permeable to predator movement (Stamps et al. 1987). Some species of invertebrate predators, such as spiders, carabid beetles, coccinellids, and syrphid flies, disperse freely across habitat boundaries (Bommarco and Fagan 2002, Langellotto 2002, Cronin et al. 2004), whereas others are impeded to varying degrees (Kareiva 1987, Thomas 1992, Frampton et al. 1995, Wratten et al. 2003). Thus, what constitutes an edge to one predator species may not to another species. Also, some arthropod predators, such as wolf spiders, cross habitat boundaries and travel far into neighboring habitats (>100 m; Lewis, D., and Denno, R. F., unpublished data), while other species venture out only a few meters (Halaj et al. 2000). If "impermeable edges" are encountered, individuals of both predator and prey often move parallel to the edge, where they accumulate (Bohlen and Barrett 1990, Frampton et al. 1995). This phenomenon may explain why encounter rates of some predators and prey and predation rates can be higher in edge habitats (Kaiser 1983).

In some cases, predator dispersal across habitat edges is asymmetric (immigration is greater than emigration), leading to "supersaturation" of habitats (Fagan et al. 1999). This phenomenon occurs in some lycosid spiders where they selectively colonize litter-rich vegetation, exhibit reduced emigration, and more effectively suppress prey populations (Döbel and Denno 1994, Langellotto 2002). For invertebrate predators, reduced emigration likely results from diminished risks of intraguild predation or cannibalism (Langellotto 2002), or a greater encounter rate with prey or other food resources, such as pollen (Wetzler and Risch 1984). In contrast, other cases of selective permeability involve aquatic invertebrates where complex habitat structure deters predators but not prey from crossing habitat edges, resulting in a refuge from predation (Lewis and Eby 2002). Intra- and interhabitat movement is also affected by the structure of the habitat, as well as by the landscape matrix in which habitats are nested (Frampton et al. 1995, Cronin 2003, Cronin et al. 2004). In general, dense vegetation or habitats with otherwise contrasting structure can hinder predator movement, and both the structural complexity and expanse of the interhabitat matrix can alter interpatch dispersal and connectivity (Kaiser 1983, Collinge and Palmer 2002, Frampton et al. 1995, Wratten et al. 2003). For example, dense and wide grassy banks can act as edge barriers to carabid beetle movement between barley fields (Frampton et al. 1995). In contrast, other studies show that carabid beetles move freely from one habitat type to another (Bommarco and Fagan 2002). Overall, however, there is extensive evidence that habitat edges can alter predator-prey interactions by means of selective effects on the dispersal of predators, prey, or both.

Habitat edges can also affect predator-prey interactions by inducing differences in mortality (Fagan et al. 1999). If predators and prey accumulate near habitat edges, predation rates can increase (Kaiser 1983). Similarly, in small habitat fragments with a high edge-to-area ratio, prey mortality can be intensified as a result of the predator incursion from the surrounding matrix (Bonte et al. 2002, Cronin et al. 2004). Also, at the interface between streams and riparian vegetation, allochthonous inputs of emerging aquatic insects promote high densities of insectivorous birds that, in turn, depress herbivorous insects more in riparian edges than in the upland forest, where birds are less common (Murakami and Nakano 2002).

Habitat edges are also involved in spatial subsidies when predators venture across habitat boundaries and influence predator-prey dynamics in neighboring habitats. One case of an extensive spatial incursion of predators involves the annual proliferation of a hunting spider from overwintering habitats in the upland to neighboring intertidal habitats where it selectively accumulates at high densities in structured vegetation and dramatically reduces prey populations (Döbel and Denno 1994, Lewis, D., and Denno, R. F., unpublished data). Novel interactions are generated because this spider is an intraguild predator and food web dynamics are altered (Döbel and Denno 1994, Denno et al. 2002). Because these spiders, as well as other predators, collectively colonize litter-rich habitats, and because intraguild predation is relaxed there, herbivore suppression is much greater than in simple-structured habitats (Finke and Denno 2002, Finke and Denno 2003). At a smaller spatial scale, invertebrate predators cross habitat edges into neighboring habitats, where they significantly reduce prey populations (Halaj et al. 2000) or eliminate them entirely (Cronin et al. 2004). Deploying refuge habitats in agricultural ecosystems assumes that significant predator incursions will occur across habitat boundaries and result in enhanced biological control (Landis et al. 2000, Frampton et al. 1995, Symondson et al. 2002).

Matrix Effects on the Numerical Response of Predators

By hindering dispersal and thus a predator's numerical response to increasing prey density, diverse landscapes can decouple predator-prey interactions, promote outbreaks of prey, and encourage unstable dynamics (Kareiva 1987, Döbel and Denno 1994, With and King 1999). Understanding how habitat edge effects and the structure of the landscape matrix interact to influence the relative dispersal of both predators and prey is key to elucidating the impacts of landscape diversity on predator-prey dynamics (Döbel and Denno 1994, Cronin et al. 2004). In fragmented salt marsh habitats characterized by archipelagoes of suitable habitat patches nested within a matrix of flooded mudflats, hunting spiders exhibit a weak numerical response to planthopper prey populations that ultimately erupt (Döbel and Denno 1994). However, in open expanses of continuous habitat, spiders rapidly colonize areas of increasing prey density and deter outbreaks. A similar scenario occurs on goldenrod (*Solidago*) in complex-structured landscapes where coccinellids exhibit a slow numerical response to increases in aphid prey density and populations erupt, a situation that does not arise in monocultures of goldenrod, where predators effectively track prey (Kareiva 1987). Similarly, the ability of coccinellids to track aphid populations on clover is more adversely affected in fragmented landscapes than in less fragmented ones (With et al. 2002). In all these cases, prey are highly mobile and predators are less so, a life history difference that combined with landscape complexity to further exacerbate prey tracking and destabilizes population dynamics (Döbel and Denno 1994). Alternatively, and at a smaller spatial scale, others have also argued that complex habitats promote the persistence of predators and prey by providing refuges from predation and disrupting predator movement (Huffaker 1958). Thus, much as diverse vegetation, by interfering with specialist herbivores' searching behavior, can dissuade them from concentrating on scattered resources (Root 1973, Kareiva 1983, Connor et al. 2000), diverse landscapes can hinder the ability of predators to aggregate effectively in response to prey.

Patch-Size and Habitat Fragmentation Effects

Examining predator-prey interactions across a gradient of patch sizes incorporates elements of habitat edge, landscape matrix, refuge from predation, the differential dispersal ability of predators and prey, and numerical response. Admittedly, however, patch-size effects are often confounded with patch-isolation (interpatch distance) effects in the context of habitat fragmentation impacts on predator-prey interactions (see With et al. 2002 for an explicit study of distance effects). From a predator's perspective, increasing patch size can result in higher densities of prey and other food resources (Root 1973, Connor et al. 2000), as well as can an increase in the structural complexity of the habitat (Irlandi 1997). If predators exhibit strong numerical responses to concentrated prey and aggregate in complex-structured habitats, then one might predict higher predator densities and larger predator effects in large patches (Kareiva 1987, Döbel and Denno 1994, Thies and Tscharntke 1999, Langellotto and Denno 2004). Alternatively, edge-related phenomena, such as predator incursion, might have greater impact in small patches because of an increased edge-area ratio (Fagan et al. 1999, Cantrell et al. 2001). The key to determining predator impacts along a patch-size gradient is knowing how food resources, habitat edges, and background matrix integrate to affect the immigration-emigration dynamic of both predators and prey (Cantrell et al. 2001, Bowman et al. 2002, Cronin 2003, Cronin et al. 2004). There are data to support both scenarios of increasing and decreasing predator impacts across a patch-size gradient.

Predator impacts on prey can be greater in large patches if predators aggregate there in response to high prey density or more complex structure (Döbel and Denno 1994, Langellotto and Denno 2004), if a predator's numerical response to prey is selectively hindered in small or isolated patches by the structure of the interpatch matrix habitat (Kareiva 1987, Döbel and Denno 1994, With et al. 2002), or if refuges from their own predators are relatively more available in large patches, so that intraguild predation is reduced (Döbel and Denno 1994). Thus, patch connectivity becomes critical in selectively affecting the dispersal dynamics of predators and prey (Döbel and Denno 1994, Cronin 2003). Matrix effects that deter predator aggregation seem particularly strong when small ground-foraging predators (such as coccinellid beetles and hunting spiders) track very mobile and fecund prey (e.g., aphids and planthoppers) in broken landscapes. Under these circumstances, effective tracking and prey suppression occur primarily within large, contiguous patches of the focal habitat (Kareiva 1987, Kreuss and Tscharntke 1994). Notably, there can be a threshold effect whereby fragmentation effects (e.g., interpatch distance) on predator dispersal, and thus prey tracking, are felt more when the abundance of the focal habitat falls below a certain fraction of the total habitat (With and King 1999, With et al. 2002). Thus, for arthropod-based systems, unstable predator-prey dynamics can result in small or isolated patches, because predators form weak numerical responses in complex-structured landscapes where matrix effects hinder dispersal (Kareiva 1987, Döbel and Denno 1994, With et al. 2002).

By contrast, predator impacts can be greater in small patches or fragmented landscapes. This result can occur when habitat edges are permeable and generalist predators freely colonize from the surrounding matrix, or when predators and prey are concentrated together. In such cases, predators can reach higher densities in small patches or habitat fragments and impose greater mortality on prey than they can in larger patches. For example, in prairie grasslands, spiders occur far more abundantly in small host-plant patches, where herbivorous prey are less abundant and frequently go extinct (Cronin et al. 2004). The rarity of prey on small host-plant patches results from a combination of spider-inflicted mortality and notably the nonlethal effect of enhanced emigration. Similarly, in a salt marsh grassland, the rarity of herbivores (planthoppers and leafhoppers) and specialist predators (heteropterans) on small host-plant patches is associated with the incursion of intraguild predators (lycosid spiders) from neighboring habitats (Denno 1980, Denno et al. 1981, Denno, R. F., unpublished data). Such incursions of generalist predators (such as spiders) can even induce critical patch-size effects for prey, alter source-sink dynamics, and lead to the extinction of local patches, with regional consequences for prey persistence (Cantrell et al. 2001, Cronin et al. 2004).

Based on the few existing studies, there is no general spatial pattern of predator density and consequences for prey along patch-size gradients. Density patterns emerge that result from system-dependent responses of predators and prey to the spatial configuration of focal habitat patches and the landscape matrix in which they are nested. Of critical importance are predator and prey life history traits (e.g., dispersal ability, body size, and foraging behavior) that interface with landscape and habitat structure to influence the probability for patch location, the permeability of habitat edges, and refuge from predation. In general, strong and immediate numerical responses by voracious predators are thought to promote effective temporal and spatial tracking of prey and enhance suppression (Murdoch 1990, Döbel and Denno 1994). Any factor, including any of those associated with habitat structure, that diminishes such responses will decouple the predator-prey interaction and potentially lead to relaxed predation and unstable prey dynamics (Kareiva 1987, Döbel and Denno 1994). Such factors include selectively impermeable habitat edges, the constraining nature of the matrix habitat as it affects patch connectivity, refuges for prey and intraguild prey, and inherent differences in mobility and body size between predators and prey (Döbel and Denno 1994, Tilman et al. 1997, Cronin 2003).

Habitat fragmentation can have relatively greater direct effects on predators than on lower trophic groups (Gibb and Hochuli 2002, Braschler et al. 2003), effects that often cascade to lower trophic levels (Crooks and Soulé 1999). For instance, there is evidence that habitat fragmentation can enhance predator subsidies that subsequently influence "critical patch size" and the long-term persistence of prey metapopulations (Hanski and Ovaskainen 2000, Cantrell et al. 2001). Other studies suggest that, when habitat fragmentation alters trophic structure, food web dynamics change dramatically. For example, the loss of top carnivores in fragmented landscapes can result in the release of mesopredators (intraguild prey), often with devastating consequences for shared prey (Litvaitis and Villafuerte 1996, Rogers and Caro 1998, Crooks and Soulé 1999). Thus, the number of trophic levels in the system (four in this last case) and which ones are selectively affected by landscape structure are critical to predicting food web dynamics in fragmented habitats.

Conclusions

Understanding how habitat structure influences the movement and interaction of predators and prey across heterogeneous landscapes is critical to agriculture, conservation biology, and restoration ecology (Settle et al. 1996, Thies and Tscharntke 1999, Bolger et al. 2000, Landis et al. 2000, Symondson et al. 2002, Tscharntke et al. 2002, Brown 2003, Cronin 2003). Although the subdiscipline of spatial ecology is recognized as central in these applied contexts (Fagan et al. 1999, Hanski 1999, Landis et al. 2000, Cantrell et al. 2001), we can make depressingly few specific recommendations that will enhance pest suppression or promote species' persistence and stable food web dynamics across complex-structured landscapes (Landis et al. 2000). The crux of the issue defaults to our inadequate knowledge of predator dispersal and foraging efficiency as they interface with habitat heterogeneity (as reflected in within-patch structure, patch size, and fragmentation at the scale of landscape) to affect predator-prey interactions, species persistence, and community dynamics (Fagan et al. 1999, Hanski and Ovaskainen 2000).

From our review of the literature, we have identified both within-habitat and between-habitat effects of structural complexity on predators and their interaction with prey (e.g., on predator search and dispersal, prey tracking and suppression, and food web dynamics; Table 10.3). We envision collation of these known effects as useful for guiding future research. Yet we are currently struggling to make specific recommendations regarding the placement and size of predator refuges in agricultural landscapes that might maximize dispersal to and pest impact in neighboring cropping systems (Settle et al. 1996, Thies and Tscharntke 1999, Landis et al. 2000). Moreover, because natural enemies can select for plant morphologies that increase their

Table 10.3. Within-Habitat and Between-Habitat Effects of Structural Complexity on Predator Search and Dispersal, Prey Tracking and Suppression, and Food Web Dynamics

Impact	References
Within-habitat effects	
Individual predator foraging behavior	
Physical exclusion/interference	
Predator size and maneuverability	Leber 1985, Frampton et al. 1995, Babbitt and Tanner 1998, Clark and Messina 1998a, Clark and Messina 1998b, Lewis and Eby 2002
Foothold on substrate	Kareiva and Sahakian 1990, Eigenbrode et al. 1996
Visual impairment of search	Main 1985, Main 1987, Heck and Crowder 1991, Lancaster and Mole 1999
Prey suppression and food web dynamics	
Enhanced prey suppression due to predator aggregation	Hanks and Denno 1993, Halaj et al. 2000, Tooker and Hanks 2000, Pratt et al. 2002; counterexamples: Letourneau 1990, Agrawal 1997, English-Loeb et al. 2002
Trophic cascade occurs	Brust 1994, Agrawal and Karban 1997, Denno et al. 2002
Diminished prey suppression due to spatial refuge	Kareiva 1987, Messina et al. 1997, Lewis and Eby 2002
Between-habitat effects: landscape-level impacts	
Edge effects	
Barrier to predator dispersal	Bohlen and Barrett 1990, Thomas 1992, Collinge and Palmer 2002, Frampton et al. 2002, Wratten et al. 2003; counterexamples: Bommarco and Fagan 2002, Langellotto 2002, Cronin et al. 2004
Increased prey mortality at edge	Kaiser 1983, Murakami and Nakano 2002
Spatial incursion of predators	Döbel and Denno 1994, Halaj et al. 2000, Cronin et al. in press
Matrix-diminished numerical response	Huffaker 1958, Kareiva 1987, Döbel and Denno 1994, With et al. 2002
Patch-size-enhanced impacts on prey	
Large patches	Kareiva 1987, Döbel and Denno 1994
Small patches	Cronin et al. 2004
Habitat fragmentation (inter-patch distance) diminishes predator impact on prey	With et al. 2002

accessibility to herbivores (Main 1985, Marquis and Whelan 1996), incorporating such traits into crop breeding and engineering programs offers new possibilities for plant-mediated predator manipulation and biological control (see Bottrell et al. 1998). Finally, maximizing substrate heterogeneity is thought to be critical in stream restoration programs to ensure the recovery of degraded drainage systems (Bond and Lake 2003, Brown 2003), but just how to specifically do so remains an open question.

Melding theoretical developments in spatial ecology (Hanski and Ovaskainen 2000, Cantrell et al. 2001, Cronin 2003) with a practical understanding of predator dynamics in natural and managed systems should allow us to better address and solve applied problems in agriculture, conservation, and restoration. Toward this end, the rapidly growing literature demonstrating that vegetation and habitat complexity mediate the consequences of predator-prey interactions at local, habitat, and regional spatial scales suggests that this finding should be a primary focus for future research and implementation. Critical to our understanding of habitat structure-mediated effects on specific predator-prey interactions at the landscape spatial scale will be replicated field experiments that isolate the effects of edge, patch size, interpatch distance, and background matrix. To date, no such inclusive experiments exist for arthropod-based systems.

Acknowledgments

P. Barbosa, J. Cronin, M. Eubanks, and two anonymous referees reviewed earlier drafts of this chapter, and we hope to have incorporated their many insightful suggestions. We are most grateful to these colleagues for their advice and support. This research was supported by National Science Foundation Grants DEB-9903601 and DEB-0313903 to R.F.D.

Literature Cited

Agrawal, A. A. 1997. Do leaf domatia mediate a plant–mite mutualism? An experimental test of the effects on predators and herbivores. Ecol. Entomol. 22:371–376.

Agrawal, A. A., and Karban, R. 1997. Domatia mediate plant-arthropod mutualism. Nature 387:562–563.

Andow, D. A. 1991. Vegetational diversity and arthropod population response. Annu. Rev. Entomol. 36:561–568.

Babbit, K. J., and Tanner, G. W. 1998. Effects of cover and predator size on survival and development of *Rana utricularia* tadpoles. Oecologia 114:258–262.

Balfour, R. A., and Rypstra, A. L. 1998. The influence of habitat structure on spider density in a no-till soybean agroecosystem. J. Arachnol. 26:221–226.

Barbosa, P., and Wratten, S. D. 1998. Influence of plants on invertebrate predators: implications to conservation biological control. In: Conservation Biological Control (Barbosa, P., ed.). San Diego: Academic Press; 83–100.

Belcher, D. W., and Thurston, R. 1982. Inhibition of movement of larvae of the convergent lady beetle by leaf trichomes of tobacco. Environ. Entomol. 11:91–94.

Bohlen, P. J., and Barrett, G. W. 1990. Dispersal of the Japanese beetle (Coleoptera Scarabaeidae) in strip-cropped soybean agroecosystems. Environ. Entomol. 19:955–960.

Bolger, D. T., Suarez, A. V., Crooks, K. R., Morrison, S. A., and Case, T. J. 2000. Arthropods in urban habitat fragments in southern California: area, age, and edge effects. Ecol. Appl. 10:1230–1248.

Bommarco, R., and Fagan, W. F. 2002. Influence of crop edges on movement of generalist predators. Agric. For. Entomol. 4:21–30.

Bond, N. R., and Lake, P. S. 2003. Characterizing fish-habitat associations in streams as the first step in ecological restoration. Austral. Ecol. 28:611–621.

Bonte, D., Baert, L., and Maelfait, J.-P. 2002. Spider assemblage structure and stability in a heterogeneous coastal dune system (Belgium). J. Arachnol. 30:331–343.

Bottrell, D. G., Barbosa, P., and Gould, F. 1998. Manipulating natural enemies by plant variety selection and modification: a realistic strategy? Annu. Rev. Entomol. 43:347–367.

Bowman, J., Cappuccino, N., and Fahrig, L. 2002. Patch size and population density: the effect of immigration behavior. Conserv. Ecol. 6:9.

Braschler, B., Lampel, G., and Baur, B. 2003. Experimental small-scale grassland fragmentation alters aphid population dynamics. Oikos 100:581–591.

Brose, U. 2003. Bottom-up control of carabid beetle communities in early successional wetlands: mediated by vegetation structure or plant diversity? Oecologia 135:407–413.

Brown, B. L. 2003. Spatial heterogeneity reduces temporal variability in stream insect communities. Ecol. Lett. 6:316–325.

Brust, G. E. 1994. Natural enemies in straw-mulch reduce Colorado potato beetle populations and damage in potato. Biol. Control 4:163–169.

Bugg, R. L., and Pickett, C. H. 1998. Habitat management to enhance biological control: a concept and its applications. In: Enhancing Biological Control: Habitat Management to Promote Natural Enemies of Agricultural Pests (Pickett, C. H., and Bugg, R. L., eds.). Berkeley: University of California Press; 1–23.

Bultman, T. L., and Uetz, G. W. 1982. Abundance and community structure of forest floor spiders following litter manipulation. Oecologia 55:34–41.

Bultman, T. L., and Uetz, G. W. 1984. Effect of structure and nutritional quality of litter on abundances of litter-dwelling arthropods. Am. Mid. Nat. 111:165–172.

Cantrell, R. S., Cosner, C., and Fagan, W. F. 2001. How predator incursions affect critical patch size: the role of the functional response. Am. Nat. 158:368–375.

Carcamo, H. A. 1995. Effects of tillage on ground beetles (Coleoptera: Carabidae): a farmscale study in central Alberta. Can. Entomol. 127:631–639.

Clark, M. S., Gage, S. H., and Spence, J. R. 1997. Habitats and management associated with common ground beetles (Coleoptera: Carabidae) in a Michigan agricultural landscape. Environ. Entomol. 26:519–527.

Clark, T. L., and Messina, F. J. 1998a. Foraging behavior of lacewing larvae (Neuroptera: Chrysopidae) on plants with divergent architectures. J. Insect Behav. 11:303–317.

Clark, T. L., and Messina, F. J. 1998b. Plant architecture and the foraging success of ladybird beetles attacking the Russian wheat aphid. Entomol. Exp. Appl. 86:153–161.

Coll, M., and Bottrell, D. G. 1995. Predator-prey association in mono- and bicultures: effect of maize and bean vegetation. Agric. Ecosyst. Environ. 54:115–125.

Collinge, S. J., and Palmer, T. M. 2002. The influences of patch shape and boundary contrast on insect response to fragmentation in California grasslands. Landsc. Ecol. 17:647–656.

Connor, E. F., Courtney, A. C., and Yoder, J. M. 2000. Individuals-area relationships: the relationship between animal population density and area. Ecology 81:734–748.

Costello, M. J., and Daane, K. M. 1998. Influence of ground cover on spider populations in a table grape vineyard. Ecol. Entomol. 23:33–40.

Cronin, J. T. 2003. Matrix heterogeneity and host-parasitoid interactions in space. Ecology 84:1506–1516.

Cronin, J. T., Haynes, K. J., and Dillemuth, F. 2004. Spider effects on planthopper mortality, dispersal and spatial population dynamics. Ecology 85:2134–2143.

Crooks, K. R., and Soulé, M. E. 1999. Mesopredator release and avifaunal extinctions in a fragmented system. Nature 400:563–566.

Crowder, L. B., and Cooper, W. E. 1982. Habitat structural complexity and the interaction between bluegills and their prey. Ecology 63:1802–1813.

Denno, R. F. 1980. Ecotope differentiation in a guild of sap-feeding insects on the salt marsh grass, Spartina patens. Ecology 61:702–714.

Denno, R. F., Gratton, C., Peterson, M. A., Langellotto, G. A., Finke, D. L., and Huberty, A. F. 2002. Bottom-up forces mediate natural-enemy impact in a phytophagous insect community. Ecology 83:1443–1458.

Denno, R. F., and McClure, M. S. 1983. Variable Host Plants and Herbivores in Natural and Managed Systems. New York: Academic Press.

Denno, R. F., Raupp, M. J., and Tallamy, D. W. 1981. Organization of a guild of sap-feeding insects: equilibrium vs. nonequilibrium coexistence. In: Insect Life History Patterns (Denno, R. F., and Dingle, H., eds.). New York: Springer-Verlag; 151–181.

Döbel, H. G., and Denno, R. F. 1994. Predator-planthopper interactions. In: Planthoppers: Their Ecology and Management (Denno, R. F., and Perfect, T. J., eds.). New York: Chapman & Hall; 325–399.

Eigenbrode, S. D., Castagnola, T., Roux, M. B., and Steljes, L. 1996. Mobility of three generalist predators is greater on cabbage with glossy leaf wax than on cabbage with a wax bloom. Entomol. Exp. Appl. 81:335–343.

English-Loeb, G., Norton, A. P., and Walker, M. A. 2002. Behavioral and population consequences of acrodomatia in grapes on phytoseiid mites (Mesostigmata) and implications for plant breeding. Entomol. Exp. Appl. 104:307–319.

Eubanks, M. D., and Denno, R. F. 1999. The ecological consequences of variation in plants and prey for an omnivorous insect. Ecology 80:1253–1266.

Eubanks, M. D., and Denno, R. F. 2000. Host plants mediate omnivore-herbivore interactions and influence prey suppression. Ecology 81:865–875.

Eubanks, M. D., and Miller, G. L. 1992. Life-cycle and habitat preference of the facultatively arboreal wolf spider Gladicosa pulchra (Araneae: Lycosidae). J. Arachnol. 20:157–164.

Everett, R. A., and Ruiz, G. M. 1993. Coarse woody debris as a refuge from predation in aquatic communities—An experimental test. Oecologia 93:475–486.

Fagan, W. F., Cantrell, R. S., and Cosner, C. 1999. How habitat edges change species interactions. Am. Nat. 153:165–182.

Finke, D. L., and Denno, R. F. 2002. Intraguild predation diminished in complex-structured vegetation: implications for prey suppression. Ecology 83:643–652.

Finke, D. L., and Denno, R. F. 2003. Intra-guild predation relaxes natural enemy impacts on herbivore populations. Ecol. Entomol. 28:67–73.

Fordyce, J., and Agrawal, A. A. 2001. The role of plant trichomes and caterpillar group size on growth and defence of the pipevine swallowtail, Battus philenor. J. Anim. Ecol. 70:997–1005.

Frampton, G. K., Cligli, T., Fry, G. L. A., and Wratten, S. D. 1995. Effects of grassy banks on the dispersal of some carabid beetles (Coleoptera: Carabidae) on farmland. Biol. Cons. 71:347–355.

Gibb, H., and Hochuli, D. F. 2002. Habitat fragmentation in an urban environment: large and small fragments support different arthropod assemblages. Biol. Cons. 106:91–100.

Gjerlov, C., Hildrew, A. G., and Jones, J. I. 2003. Mobility of stream invertebrates in relation to disturbance and refugia: a test of habitat templet theory. J. No. Am. Benthol. Soc. 22:207–223.

Godfrey, L. D., and Leigh, T. F. 1994. Alfalfa harvest strategy effect on lygus bug (Hemiptera: Miridae) and insect predator population density: implications for use as trap crop in cotton. Environ. Entomol. 23:1106–1118.

Grevstad, S. S., and Klepetka, B. W. 1992. The influence of plant architecture on the foraging efficiencies of a suite of ladybird beetles feeding on aphids. Oecologia 92:399–404.

Grostal, P., and O'Dowd, D. J. 1994. Plants, mites and mutualism: leaf domatia and the abundance and reproduction of mites on Viburnum tinus (Caprifoliaceae). Oecologia 97:308–315.

Gunnarsson, B. 1990. Vegetation structure and the abundance and size distribution of spruce-living spiders. J. Anim. Ecol. 59:743–752.

Habdija, I., Radanovic, I., Prime-Habdija, B., and Spoljar, M. 2002. Vegetation cover and substrate type as factors influencing the spatial distribution of trichopterans along a Karstic River. Intern. Rev. Hydrobiol. 87:423–437.

Halaj, J., Ross, D. W., and Moldenke, A. R. 1998. Habitat structure and prey availability as predictors of the abundance and community organization of spiders in western Oregon forest canopies. J. Arachnol. 26:203–220.

Halaj, J., Ross, D. W., and Moldenke, A. R. 2000. Importance of habitat structure to the arthropod food web in Douglas-fir canopies. Oikos 90:139–152.

Hammond, R. B., and Stinner, B. R. 1987. Soybean foliage insects in conservation tillage systems: effects of tillage, previous cropping history, and soil insecticide application. Environ. Entomol. 16:524–531.

Hanks, L. M., and Denno, R. F. 1993. Natural enemies and plant water relations influence the distribution of an armored scale insect. Ecology 74:1081–1091.

Hanski, I. 1999. Metapopulation ecology. Oxford: Oxford University Press.

Hanski, I., and Ovaskainen, O. 2000. The metapopulation capacity of a fragmented landscape. Nature 404:755–758.

Harmon, J. P., Ives, A. R., Losey, J. E., Olson, A. C., and Rauwald. 2000. *Coleomagilla maculata* (Coleoptera: Coccinellidae) predation on pea aphids promoted by proximity to dandelions. Oecologia 125:543–548.

Heck, K. L., and Crowder, L. B. 1991. Habitat structure and predator-prey interactions in vegetated aquatic systems. In: Habitat Structure: The Physical Arrangement of Objects in Space (Susan, B., McCoy, E., and Mushinsky, H., eds.). London: Chapman & Hall; 281–299.

Hoddle, M. 2003. The effect of prey species and environmental complexity on the functional response of *Franklinothrips orizabensis:* a test of the fractal foraging model. Ecol. Entomol. 28:309–318.

Holt, R. D. 2002. Food webs in space: on the interplay of dynamic instability and spatial processes. Ecol. Res. 17:261–273.

Holyoak, M. 2000. Habitat patch arrangement and metapopulation persistence of predators and prey. Am. Nat. 156:378–389.

Holyoak, M., and Lawler, S. P. 1996. The role of dispersal in predator-prey metapopulation dynamics. J. Anim. Ecol. 65:640–652.

Horn, D. J. 1981. Effect of weedy backgrounds on colonization of collards by green peach aphid, *Myzus persicae,* and its major predators. Environ. Entomol. 10:285–289.

Hua, L. Z., Lammes, F., van Lenteren, J. C., Huisman, W. P. T., van Vianen, A., and de Ponti, O. M. B. 1987. The parasite-host relationship between *Encarsia formosa* Gahan (Hymenoptera, Aphelinidae) and *Trialeurodes vaporariorum* (Westwood) (Homoptera, Aleyrodidae). XXV. Influence of leaf structure on the searching activity of *Encarsia formosa*. J. Appl. Entomol. 104:297–304.

Huffaker, C. B. 1958. Experimental studies on predation: dispersion factors and predator-prey oscillations. Hilgardia 27:343–383.

Hunter, M. D., and Price, P. W. 1992. Playing chutes and ladders: heterogeneity and the relative roles of bottom-up and top-down forces in natural communities. Ecology 73:724–732.

Irlandi, E. A. 1997. Seagrass patch size and survivorship of an infaunal bivalve. Oikos 78:511–518.

Jeffries, R. L. 2000. Allochthonous inputs: integrating population changes and food web dynamics. TREE 15:19–22.

Kaiser, H. 1983. Small spatial scale heterogeneity influences predation success in an unexpected way: model experiments on the functional response of predatory mites (Acarina). Oecologia 56:249–256.

Kareiva, P. 1983. Influence of vegetation texture on herbivore populations: resource concentration and herbivore movement. In: Variable Plants and Herbivores in Natural and Managed Systems (Denno, R. F., and McClure, M. S., eds.). New York: Academic Press; 259–289.

Kareiva, P. 1987. Habitat fragmentation and the stability of predator-prey interactions. Nature 326:388–390.

Kareiva, P., and Sahakian, R. 1990. Tritrophic effects of a single architectural mutation in pea plants. Nature 345:433–434.

Kreuss, A., and Tscharntke, T. 1994. Habitat fragmentation, species loss, and biological control. Science 264:1581–1584.

Lancaster, J., and Belyea, L. R. 1997. Nested hierarchies and scale-dependence of mechanisms of flow refugium use. J. No. Am. Benthol. Soc. 16:221–238.

Lancaster, J., and Mole, A. 1999. Interactive effects of near-bed flow and substratum texture on the microdistribution of lotic macroinvertebrates. Arch. Hydrobiol. 146:83–100.

Landis, D. A., Wratten, S. D., and Gurr, G. M. 2000. Habitat management to conserve natural enemies of arthropod pests in agriculture. Ann. Rev. Entomol. 45:175–201.

Langellotto, G. 2002. The aggregation of invertebrate predators in complex habitats: ecological mechanisms and practical applications (Ph.D. dissertation, University of Maryland, College Park).

Langellotto, G. A., and Denno, R. F. 2004. Responses of invertebrate natural enemies to complex-structured habitats: a meta-analytical synthesis. Oecologia 139:1–10.

Leber, K. M. 1985. The influence of predatory decapods, refuge, and microhabitat selection on seagrass communities. Ecology 66:1951–1964.

Leddy, P. M. 1996. Factors influencing the distribution and abundance of azalea lacebug, *Stephanitis pyrioides,* in simple and complex landscape habitats (Ph.D. dissertation, University of Maryland, College Park).

Legrand, A., and Barbosa, P. 2003. Plant morphological complexity impacts foraging efficiency of adult *Coccinella septempunctata* L. (Coleoptera: Coccinellidae). Environ. Entomol. 32:1219–1226.

Lei, G. C., and Hanski, I. 1997. Metapopulation structure of *Cotesia melitaearum,* a specialist parasitoid of the butterfly *Melitaea cinxia.* Oikos 78:91–100.

Letourneau, D. K. 1990. Abundance patterns of leafhopper enemies in pure and mixed stands. Environ. Entomol. 19:505–509.

Lewis, D. B., and Eby, L. A. 2002. Spatially heterogeneous refugia and predation risk in intertidal saltmarshes. Oikos 96:119–129.

Lima, S. L., and Dill, L. M. 1990. Behavioral decisions made under risk of predation—A review and perspectus. Can. J. Zool. 68:619–640.

Litvaitis, J. A., and Villafuerte, R. 1996. Intraguild predation, mesopredator release, and prey stability. Conserv. Biol. 10:676–677.

Losey, J. E., and Denno, R. F. 1998. Positive predator-predator interactions: enhanced predation rates and synergistic suppression of aphid populations. Ecology 79:2143–2152.

Magoulick, D. D., and Kobza, R. M. 2003. The role of refugia for fishes during drought: a review and synthesis. Fresh. Biol. 48:1186–1198.

Main, K. L. 1985. The influence of prey identity and size on selection of prey by two marine fishes. J. Exp. Mar. Biol. Ecol. 98:145–152.

Main, K. L. 1987. Predator avoidance in seagrass meadows: prey behavior microhabitat selection and cryptic coloration. Ecology 68:170–180.

Marino, P. C., and Landis, D. A. 1996. Effect of landscape structure on parasitoid diversity and parasitism in agroecosystems. Ecol. Appl. 61:276–284.

Marquis, R. J., and Whelan, C. J. 1996. Plant morphology and recruitment of the third trophic level: subtle and little-recognized defenses? Oikos 75:330–334.

Matthaei, C. D., Arbuckle, C. J., and Townsend, C. R. 2000. Stable surface stones as refugia for invertebrates during disturbance in a New Zealand stream. J. No. Am. Benthol. Soc. 19:82–93.

McCoy, E. D., and Bell, S. S. 1991. Habitat structure: the evolution and diversification of a complex topic. In: Habitat Structure: The Physical Arrangement of Objects in Space (Bell, S. S., McCoy, E. D., and Mushinsky, H. R., eds.). London: Chapman & Hall; 3–27.

McNett, B. J., and Rypstra, A. L. 2000. Habitat selection in a large orb-weaving spider: vegetational complexity determines site selection and distribution. Ecol. Entomol. 25:423–432.

Menalled, F. B., Marino, P. C., Gage, S. H., and Landis, D. A. 1999. Does agricultural landscape structure affect parasitism and parasitoid diversity? Ecol. Appl. 9:634–641.

Mensah, R. K. 1999. Habitat diversity: implications for the conservation and use of predatory insects of *Heliocoverpa* spp. in cotton systems in Australia. Intern. J. Pest Manage. 45:91–100.

Messina, F. J., Jones, T. A., and Nielson, D. C. 1997. Host-plant effects on the efficacy of two predators attacking Russian wheat aphids (Homoptera: Aphidae). Environ. Entomol. 26:1398–1404.

Murakami, M., and Nakano, S. 2002. Indirect effect of aquatic insect emergence on a terrestrial insect population through bird predation. Ecol. Lett. 5:333–337.

Murdoch, W. W. 1990. The relevance of pest-enemy models to biological control. In: Critical Issues in Biological Control (Mackaur, M., Ehler, L. E., and Roland, J., eds.). Intercept, U.K.: VCH Publishers.

Nakano, S., and Murakami, M. 2001. Reciprocal studies: dynamic interdependence between terrestrial and aquatic food webs. Proc. Natl. Acad. Sci. U.S.A. 98:166–170.

Negishi, J. N., Inoue, M., and Nunokawa, M. 2002. Effects of channelisation on stream habitat in relation to a spate and flow refugia for macroinvertebrates in northern Japan. Fresh. Biol. 47:1515–1529.

Norton, A. P., English-Loeb, G., and Belden, E. 2001. Host plant manipulation of natural enemies: leaf domatia protect beneficial mites from insect predators. Oecologia 126:535–542.

Palmer, M. A., Arensburger, P., Martin, A. P., and Denman, D. W. 1996. Disturbance and patch-specific responses: the interactive effects of woody debris and floods on lotic invertebrates. Oecologia 105:247–257.

Palmer, M. A., Bely, A. E., and Berg, K. E. 1992. Response of invertebrates to lotic disturbance—test of the hyporheic refuge hypothesis. Oecologia 89:182–194.

Patt, J. M., Hamilton, G. C., and Lashomb, J. H. 1997. Impact of strip-insectary intercropping with flowers on conservation biological control of the Colorado potato beetle. Adv. Hortic. Sci. 11:175–181.

Perfecto, I., and Sediles, A. 1992. Vegetational diversity, ants (Hymenoptera: Formicidae), and herbivorous pests in a neotropical agroecosystem. Environ. Entomol. 21:61–67.

Polis, G. A., Holt, R. D., Menge, B. A., and Winemiller, K. O. 1996. Time, space, and life history: influences on food webs. In: Food Webs: Integration of Patterns and Dynamics (Polis, G. A., and Winemiller, K. O., eds.). New York: Chapman & Hall; 435–460.

Polis, G. A., and Hurd, S. D. 1996. Allochthonous input across habitats, subsidized consumers, and apparent trophic cascades: examples from the ocean-land interface. In: Food Webs: Integration of Patterns and Dynamics (Polis, G. A., and Winemiller, K. O., eds.). New York: Chapman & Hall; 275–285.

Post, D. M., Conners, M. E., and Goldberg, D. S. 2000. Prey preference by a top predator and the stability of linked food chains. Ecology 81:8–14.

Pratt, D. P., Rosetta, R., and Croft, B. A. 2002. Plant-related factors influence the effectiveness of *Neoseiulus fallacies* (Acari: Phytoseiidae), a biological control agent of spider mites on landscape ornamental plants. J. Econ. Entomol. 95:1135–1141.

Price, P. W., Bouton, C. E., Gross, P., McPheron, B. A., Thompson, O. N., and Weis, A. E.

1980. Interactions among three trophic levels: influence of plants on interactions between insect herbivores and natural enemies. Ann. Rev. Ecol. Syst. 11:41–65.

Riechert, S. E., and Bishop, L. 1990. Prey control by an assemblage of generalist predators: spiders in garden test systems. Ecology 71:1441–1450.

Roda, A., Nyrop, J., Dicke, M., and English-Loeb, G. 2000. Trichomes and spider-mite webbing protect predatory mite eggs from intraguild predation. Oecologia 125:428–435.

Rogers, C. M., and Caro, M. J. 1998. Song sparrows, top carnivores, and nest predation: a test of the mesopredator release hypothesis. Oecologia 116:227–233.

Root, R. B. 1973. Organization of a plant-arthropod association in simple and diverse habitats: the fauna of collards. Ecol. Monogr. 43:95–124.

Russell, E. P. 1989. Enemies hypothesis: a review of the effect of vegetation diversity on predatory insects and parasitoids. Environ. Entomol. 18:590–599.

Rutledge, C. E., Robinson, A. P., and Eigenbrode, S. D. 2003. Effects of a simple plant morphological mutation on the arthropod community and the impacts of predators on a principal insect herbivore. Oecologia 135:39–50.

Rypstra, A. L., Carter, P. E., Balfour, R. A., and Marshall, S. D. 1999. Architectural features of agricultural habitats and their impact on the spider inhabitants. J. Arachnol. 27:371–377.

Schmitz, O. J., Beckerman, A. P., and O'Brien, K. M. 1997. Behaviorally mediated trophic cascades: effects of predation risk on food web interactions. Ecology 78:1388–1399.

Schmitz, O. J., and Suttle, K. B. 2001. Effects of top predator species on direct and indirect interactions in a food web. Ecology 82:2072–2081.

Settle, W. H., Ariawan, H., Astuti, E. T., Cahyana, W., Hakim, A. L., Hindayana, D., Lestari, A. S., and Sartanto, P. 1996. Managing tropical rice pests through conservation of generalist natural enemies and alternative prey. Ecology 77:1975–1988.

Sheehan, W. 1986. Response by specialist and generalist natural enemies to agroecosystem diversification: a selective review. Environ. Entomol. 15:456–461.

Siemann, E., Tilman, D., Haarstad, J., and Ritchie, M. 1998. Experimental tests of the dependence of arthropod diversity on plant diversity. Am. Nat. 152:738–750.

Snyder, W. E., and Ives, A. R. 2001. Generalist predators disrupt biological control by a specialist parasitoid. Ecology 82:705–716.

Soluk, D. A., and Collins, N. C. 1988. Synergistic interactions between fish and stoneflies: facilitation and interference among stream predators. Oikos 52:94–100.

Stamps, J. A., Buechner, M., and Krishnan, V. V. 1987. The effects of edge permeability and habitat geometry on emigration from patches of habitat. Am. Nat. 129:533–552.

Stavrinides, M., and Skirvin, D. 2003. The effect of chrysanthemum leaf trichome density and prey spatial distribution on predation of *Tetranychus urticae* (Acari: Tetranychidae) by *Phytoseiulus persimilis* (Acari: Phytoseiidae). Bull. Entomol. Res. 93:343–350.

Stewart, T. W., Shumaker, T. L., and Radio, T. A. 2003. Linear and nonlinear effects of habitat structure on composition and abundance in the macroinvertebrate community of a large river. Am. Mid. Nat. 149:293–305.

Sunderland, K. D., and Samu, F. 2000. Effects of agricultural diversification on the abundance, distribution, and pest control potential of spiders: a review. Entomol. Exp. Appl. 95:1–13.

Symondson, W. O. C., Sunderland, K. D., and Greenstone, M. H. 2002. Can generalist predators be effective biological control agents? Ann. Rev. Entomol. 47:561–594.

Thies, C., and Tscharntke, T. 1999. Landscape structure and biological control in agroecosystems. Science 285:893–895.

Thomas, C. F. G. 1992. Spatial dynamics of spiders in farmland (dissertation, University of Southampton, Southampton, U.K.).

Tilman, D., Lehman, C. L., and Thomson, K. T. 1997. Plant diversity and ecosystem productivity: theoretical considerations. Proc. Natl. Acad. Sci. U.S.A. 94:1857–1861.

Tooker, J. F., and Hanks, L. M. 2000. Influence of plant community structure on natural enemies of pine needle scale (Homoptera: Diaspididae) in urban landscapes. Environ. Entomol. 29:1305–1311.

Tscharntke, T., Steffan-Dewenter, I., Kruess, A., and Thies, C. 2002. Contribution of small habitat fragments to conservation of insect communities of grassland-cropland landscapes. Ecol. Appl. 12:354–363.

Uetz, G. W. 1979. The influence of variation in litter habitats on spider communities. Oecologia 40:29–42.

Uetz, G. W. 1991. Habitat structure and spider foraging. In: Habitat Structure and Diversity (McCoy, E. D., Bell, S. S., and Mushinsky, H. R., eds.). London: Chapman & Hall.

Usio, N., and Townsend, C. R. 2002. Functional significance of crayfish in stream food webs: roles of omnivory, substrate heterogeneity and sex. Oikos 98:512–522.

Vandermeer, J., and Carvajal, R. 2001. Metapopulation dynamics and the quality of the matrix. Am. Nat. 158:211–220.

van Lenteren, J. C., and de Ponti, O. M. B. 1991. Plant leaf morphology, host plant resistance and biological control. Symp. Biol. Hung. 1991:365–386.

Wetzler, R. E., and Risch, S. J. 1984. Experimental studies of beetle diffusion in simple and complex crop habitats. J. Anim. Ecol. 53:1–19.

Wise, D. H., and Chen, B. 1999. Impact of intraguild predators on survival of a forest-floor wolf spider. Oecologia 121:129–137.

With, K. A., and King, A. W. 1999. Dispersal success on fractal landscapes: a consequence of lacunarity thresholds. Landsc. Ecol. 14:73–82.

With, K. A., Pavuk, D. M., Worchuck, J. L., Oates, R. K., and Fisher, J. L. 2002. Threshold effects of landscape structure on biological control in agroecosystems. Ecol. Appl. 12:52–65.

Wratten, S. D., Bowie, M. H., Hickman, J. M., Evans, A. M., Sedcole, J. R., and Tylianakis, J. M. 2003. Field boundaries as barriers to movement of hover flies (Diptera: Syrphidae) in cultivated land. Oecologia 134:605–611.

Yang, L. 2000. Effects of body size and plant structure on the movement ability of a predaceous stinkbug, *Podisus maculiventris* (Heteroptera: Pentatomidae). Oecologia 125:85–90.

11

Predator-Prey Space Use as an Emergent Outcome of a Behavioral Response Race

ANDREW SIH

Key to the interaction of predator and prey is their use of space (patch use, habitat use). The pattern of spatial overlap between predators and prey affects their encounter rates, predation rates, and, ultimately, predator-prey population and community dynamics. Hundreds of studies have shown that prey tend to avoid areas with more predators. Prey and predators would, then, be negatively associated. Conversely, numerous studies taking a predator perspective have shown that predators tend to prefer areas with more prey—a positive spatial association. These responses clearly conflict.

Interestingly, a recent review found that surprisingly few theoretical or empirical studies have examined how the interplay between predator and prey behavioral responses to each other determines patterns of predator-prey spatial overlap (Lima 2002). Instead, theoretical and experimental studies on predator-prey behaviors typically hold one side fixed (e.g., using caged predators or immobile prey), in order to focus on the behavior of the other. In nature, in many systems, both predators and prey are mobile and have the potential to engage in a behavioral response race. If prey win the race, the outcome is a negative spatial association between the two, whereas, if predators win, the result is a positive spatial association. The goal of this chapter is to provide an overview of factors that might influence the outcome of this race.

Extant models predict that, when patches vary in resource availability, then both predators and prey should be more abundant in high-resource patches. That is, given a spatially variable resource base, predators and prey should exhibit a positive spatial association, and predators should win the race. No published behavioral study appears to address this prediction directly. Here, I present a new experimental study on the space race between predatory salamander larvae and tadpole prey. The key result was that predators and prey showed a significant

240

negative spatial association: in essence, contrary to the prediction of the models, prey won the race. This result stimulated a reconsideration of the logic underlying the basic prediction of the models. In brief, I suggest that, in existing models, predators win the race because prey are constrained by a "spatial anchor," which is essentially the fixed distribution of their resources, whereas predators have no corresponding spatial anchor. In contrast, in nature, the space use of both predators and prey might be influenced by various constraints, costs, and benefits. In the subsequent discussion, I outline ideas and predictions about how some of these factors might affect the outcome of the predator-prey space race. Finally, I suggest directions for future study.

Understanding animal space use (habitat use, patch use) is a fundamental issue in ecology. Among other things, space use influences interactions among members of a given species, competition between species, and exposure to abiotic stressors. Most important, in the current context, space-use decisions by predators and prey determine the pattern of spatial overlap between the two that, in turn, affects predator-prey encounter rates, predation rates, and, ultimately, predator-prey population and community dynamics (Murdoch and Stewart-Oaten 1989, Krivan 1997, van Baalen and Sabelis 1999; see also Schmitz, ch. 12 in this volume).

Given the importance of predator-prey space use, it seems reasonable to expect behavioral ecologists to know a great deal about the predator and prey behavioral decisions underlying their joint space use. In fact, though we know much about either predator or prey space-use decisions, we know surprisingly little about the behavioral ecology of their joint space use. This chapter summarizes our extant knowledge, and presents new data and ideas on this critical issue.

From the predator view, two large bodies of work address predator patch or habitat use. Optimal-patch-use studies address space use for individual predators (Stephens and Krebs 1986), and ideal-free-distribution (IFD) studies examine space use for groups of competing predators (Fretwell and Lucas 1970, Kacelnik et al. 1992, Kennedy and Gray 1993). In either case, theory predicts and empirical studies show that predators generally concentrate their efforts in areas with more prey. Predator-prey population ecologists refer to this pattern as the *aggregative response* (Hassell 1978). These bodies of work, however, assume that prey do not respond to predators. Indeed, many of the classic experimental studies examining predator patch decisions used immobile or barely mobile prey (e.g., flowers, eggs, pupae, or mealworms). In essence, what we know about predator behavioral decisions on space use comes largely from situations in which only predators (not prey) are free to choose among patches.

From the prey view, innumerable studies show that prey tend to avoid areas with higher predation risk (Sih 1987, Lima 1998). Theories on prey avoidance of high-risk sites, however, almost always assume fixed predation regimes (Houston and McNamara 1999, Clark and Mangel 2000) and typically feature one habitat with more food (for prey) and more predators, as opposed to a safer, but lower food, habitat. Similarly, experimental studies on prey avoidance of areas with high risk typically use constrained predators (Gilliam and Fraser 1987, Abrahams and Dill 1989). For example, predators are often caged to one side of an experimental arena (usually the side with more food for prey), while only prey are free to choose among patches.

In reality, in many natural situations, both predators and prey are free to exercise patch or habitat choice. Predators can respond to prey, and prey can respond to predators (see Schmitz, ch. 12). The pattern of spatial coincidence between the two is an emergent outcome of a behavioral response race between predators and prey (Sih 1984, Lima 2002). If predators win the race, the outcome is a positive association between the two (i.e., more predators are found in areas with more prey; Figure 11.1A). If, however, prey win the race, then the two are negatively associated (i.e., prey are more abundant in areas with fewer predators; Figure 11.1A). The two counteracting responses might have canceling effects (i.e., there might be no winner). The outcome would then be no significant spatial association, despite active behavioral responses by both sides.

The "winner-loser" terminology deserves some clarification. In isolation from other fitness factors, predators and prey have opposite interests. Predators do best by foraging where there are more prey, and prey do best by avoiding areas with more predators. Thus, in the absence of other considerations, it is reasonable to say that if predators and prey are found together, predators have "won" the race, whereas, if they are found apart, then prey have "won" the race. The "loser" has lower fitness than it would have had with another pattern of space use. However, if other important fitness factors are included, then either a positive or negative pattern of spatial association can arise even when both predators and prey exhibit optimal space use. For example, some sites might have an abiotic environment that is highly stressful for prey, but not for predators. The optimal prey behavior might be to stay in nonstressful sites even if doing so allows predators to aggregate with them. One view of this situation might be that, because both sides are exhibiting optimal space use, there is no winner or loser. I will take a different view. In my use of the terminology, I will say that, even if prey are exhibiting optimal space use, the external constraint

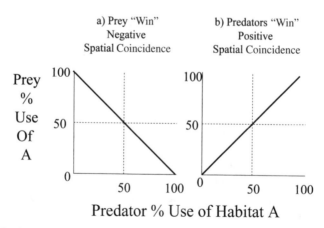

Figure 11.1. Spatial correlations between predators and prey as an emergent outcome of a predator-prey space race. (A) If prey are successful in avoiding predators (i.e., if prey win the space race between predators and prey), then the outcome is a negative association between the two. (B) If predators are successful in aggregating in areas with more prey (i.e., if predators win the race), then the result is a positive spatial association.

(here, the stressful abiotic conditions) causes prey to lose the race. That is, regardless of the factors that explain the outcome, if predator-prey patch-use decisions result in a positive association, I will say that predators have won the race, whereas, if the outcome is a negative association, prey have won the race. A major theme of this chapter will be to organize our thinking on types of factors, including external constraints, that influence the outcome of this race.

Lima (2002) recently reviewed the literature on predator-prey games, including the predator-prey space race. He found eight theoretical papers and, remarkably, only three experimental studies that focused on the behavioral ecology of predator-prey space use when both are free to choose among habitats (Sih 1984, Formanowicz and Bobka 1989, Bouskila 2001). Of course, the ecological literature is teeming with field surveys that document predator and prey densities in multiple samples. Analyses of correlations between predator and prey densities show that all patterns of spatial association (negative, positive, or random) between predators and prey occur in nature (Stiling 1987, Walde and Murdoch 1988, Rose and Leggett 1990, Mehlum et al. 1999, Fauchald et al. 2000, Benoit-Bird and Au 2003). These field surveys, however, generally include little if any information on behavioral mechanisms underlying the observed spatial patterns.

Below, I first briefly summarize the results of earlier models and experiments on the predator-prey space race. In particular, I focus on the key logic underlying the main predictions of extant models. I then describe the main results of a new experimental study on the joint space use of predatory salamander larvae and tadpole prey. The results of this study went directly against the main prediction of most existing models. This exciting outcome pointed me toward some plausible hypotheses to explain the discrepancy and suggested new directions for future study.

Brief Summary of Extant Models

Models of the predator-prey space game are built on the IFD. Standard IFD models focus on two trophic levels with consumers feeding on resources. Resource renewal rates are assumed to vary among patches: some patches have high resources (HR), while others have low resources (LR). A key assumption is that consumers compete; thus, a higher density of consumers in a given patch reduces the mean feeding rate in that patch. The simplest models predict that consumers should then match resources. The ratios of consumers in HR and LR habitats should be the same, respectively, as the productivity ratios for each of the two habitats. For example, if one habitat is four times more productive than another, then four times as many consumers should be in the more productive habitat. These models assume that resources cannot switch habitats and that consumers have no predators.

Models of the predator-prey space race add a third trophic level (van Baalen and Sabelis 1993, Hugie and Dill 1994, Sih 1998, Alonzo 2002). The basic scenario then involves predators that attack prey that consume resources. Predators compete for prey, and prey compete for resources. In most models, predators do not consume resources (but see Heithaus 2001), and resources still do not move (but see Schwinning and Rosenzweig 1990). The scenario might represent, for example, carnivores feeding on herbivores that feed on plants. The models seek a joint IFD in which both predators and prey are at their evolutionarily stable strategy (ESS): essentially, in which neither predators nor prey can improve their fitness by switching habitats. Predators follow the standard "equal feeding rate in all patches" criterion. Prey balance risk and feeding needs. Of course, the value of any given patch for predators depends on prey decisions, and vice versa. At first glance, it might seem that such a system would have no equilibrium. Predators should move to aggregate where there are more prey, forcing prey to leave, inducing predators to leave following prey, and so on. In fact, the models find a joint ESS.

Although the models differ substantially in details, some simple general results emerge. Notably, in a broad range of scenarios, given a patchy distribution of resources, predators are predicted to aggregate in more productive patches with more resources (van Baalen and Sabelis 1993, Hugie and Dill 1994, Sih 1998, van Baalen and Sabelis 1999, Alonzo 2002). Given that predators do not consume these resources, this is a fascinating result that Sih (1998) called a "leapfrog effect." In contrast, although prey are also predicted to be more abundant in more productive patches, they are typically expected to be more uniformly distributed than predators. The basic logic is that, without carnivores, herbivores should prefer areas with higher plant productivity. Carnivores should then also prefer those highly productive sites. This preference should tend to drive herbivores out of those patches; however, given no constraints on carnivore space use, herbivores cannot effectively escape carnivores in space. Because herbivores must still feed, under a broad range of conditions, they should ultimately at least somewhat prefer the more productive sites. In turn, carnivores should prefer those sites. Note that the result is a positive association between predators and prey. Both prefer patches with more resources (plant productivity). In the terminology of contests, predators are predicted to "win" the race.

Most IFD models solve for the evolutionarily stable outcome, but do not address the movements of animals in and out of patches that underlie the equilibrium outcome. Interestingly, models that track animal movements following simple, sensible (but not necessarily optimal) rules also predict that both carnivores and herbivores should be more abundant in patches with greater plant availability (Schwinning and Rosenzweig 1990, Nisbet et al. 1997). For example, Nisbet et al. (1997) derived this result with models in which all animals immigrate passively into patches, but carnivore emigration rates are higher when herbivores are scarce. Herbivore emigration rates are higher when either carnivores are more abundant, or plants are less abundant.

More complex models have examined numerous aspects of reality beyond the simplest scenario, including (a) more complex and variable degrees of competition among predators or among prey (Hugie and Dill 1994, Sih 1998, Bouskila 2001), (b) nonlinear functional responses (Hugie and Dill 1994, Sih 1998, Bouskila 2001), (c) metabolic costs (Hugie and Dill 1994), (d) state dependence (Alonzo 2002), (e) two types of predators (Bouskila 2001) and intraguild predation (Heithaus 2001), and (f) mobile resources that also avoid consumption (Schwinning and Rosenzweig 1990). Finally, van Baalen and Sabelis (1993, 1999) have explicitly addressed the population-dynamic consequences of the predator-prey space race.

The notion that predators should win the race is intriguing. Why should this outcome occur? My interpretation of the basic intuition is as follows. In the absence of external constraints or costs, conflict games (predator-prey games, male-female conflict games) often have no equilibrium (Parker 1979, Schwinning and Rosenzweig 1990). External constraints, however, stabilize the system. If only one side has a constraint (or has a stronger constraint), then the other side wins the race. In the predator-prey space race, a key type of external constraint is a spatial anchor, which is to say, essentially, any reason, outside the predator-prey race per se, for which either predators or prey should prefer some patches over others. In the models already described, prey have a spatial anchor: the spatial distribution of their resources. Predators have no spatial anchor. As a result, predators win the theoretical race.

In some existing models, predators also have a spatial anchor. Hugie and Dill (1994) considered the situation in which predators have higher inherent attack success in some patches than in others. The sites with low predator attack success then function as refuges for prey. The models predicted that, if patches differ in predator attack success (i.e., in prey safety) but not in resource value, then predators and prey should exhibit a negative spatial association: prey should win the race. Most prey should hide in the refuge sites. Because predators suffer poor attack success in refuge sites, they do not aggregate there, even though prey are more abundant in those sites. Thus, when predators have a spatial anchor, prey win the space race.

Overall, the most basic prediction emerging from most of the models is that, if patches vary only in resource value, then predators and prey should both aggregate in patches with high resource productivity. Predators and prey should exhibit a positive spatial association (i.e., predators should win the predator-prey space race). If, however, patches vary in inherent predator attack success (or, conversely, in prey safety) and not in resource productivity, then predators and prey should be negatively associated. Prey should be in refuge sites where predators suffer low attack success, whereas predators should be in sites where they enjoy higher attack success, even though most prey are found in the other patches.

Brief Summary of Experimental Studies

I suggest that, to address the predator-prey space race, a relevant experimental design should include the following treatments:

1. Prey only
2. Predators only
3. Free-ranging prey with caged predators
4. Free-ranging predators with caged prey, and
5. Free-ranging prey and predators

Because almost all of the models anchor the space race by having a patchy distribution of resources, the most basic experiment should offer different resource levels in different patches. Thus, experimental arenas should have at minimum two patches: one with HR, and the other with LR. Treatment 1 addresses whether prey prefer the HR patch in the absence of predators. Treatment 2 tests whether predators have spatial preferences in the absence of prey and whether they ignore resources (that they do not consume). Treatment 3 is the standard treatment, testing whether prey avoid predators. Treatment 4 is the standard setup for testing whether predators tend to aggregate in areas with more prey. In practice, because predators are often larger than prey, treatment 4 might be more difficult to implement than 3. Nonetheless, in principle, both treatments should be useful to fully elucidate predator-prey behavioral decisions. Treatment 5 is the obvious situation seen in nature but that is rarely studied by behavioral ecologists. According to theory, if patches vary in resource availability (and not in inherent safety), then we expect both predators and prey to be positively associated with areas with high resource productivity.

In fact, to my knowledge, no published study has included all of these treatments. Furthermore, none of the three experimental studies cited by Lima (2002) in his review of predator-prey games included measurements or manipulations of the resource base. Thus, extant experimental studies have not explicitly addressed the focal scenario or predictions of most of the described models.

Sih (1982, 1984) looked at how prey mobility and the presence of refuges for prey influenced the spatial association between predatory notonectids (aquatic insects known as *backswimmers*) and their prey. With immobile prey (fruit flies trapped on the water surface), notonectids aggregated in areas with more prey (Sih 1982). That is, not surprisingly, predatory notonectids won their race against immobile prey. In contrast, with mobile prey (mosquito larvae) and no spatial refuge, Sih (1984) detected no significant correlation between the spatial distributions of predators and prey. Apparently, the notonectids' preference for areas with more prey was offset by the tendency for mobile prey to avoid notonectids. Finally, as later predicted by Hugie and Dill (1994), with spatial refuge available, the distributions of notonectids and mosquito larvae were negatively correlated (Sih 1984).

Bouskila (2001) examined effects of moonlight on rodent and snake microhabitat use in the field. In the fall, when snakes were inactive, on moonlit nights, rodents avoided the open microhabitat, presumably because owls were particularly dangerous in open areas on well-lighted nights. In contrast, as predicted by his IFD model, during seasons when snakes were active, the presence of snakes dampened the reaction of rodents to moonlight. That is, the benefit of avoiding owl predation risk by moving out of the open into the bush habitat was reduced by the fact that snakes also shifted into the bush habitat, perhaps in response to rodents. As emphasized by Bouskila (2001), the predator-prey space race caused a nonintuitive result, a lack of rodent response to moonlight.

The one extant study that included most of the treatments suggested here is by Formanowicz and Bobka (1989) on spatial associations between predatory dytiscid beetles and either of two tadpole prey (*Rana, Hyla*). Formanowicz and Bobka included three of the five treatments described: prey alone, predators alone, and prey and predators together. Their experimental arena included four patches that varied in habitat complexity (no substrate, sand, artificial vegetation, and sand with artificial vegetation). However, the habitat complexity did not represent refuge for prey. When alone, both prey and predators preferred the most complex habitat. Thus, the null expectation (if predators and prey do not respond to one another) is a positive spatial association between the two. In fact, when they were put together, predators and prey exhibited spatial distributions that were not significantly different from random, and a pattern of spatial association that was not significantly different from random. Formanowicz and Bobka noted that this result concurred with Sih's (1984) observation of no significant spatial association between notonectids and mobile mosquito larvae. Although these studies provide a start, the field clearly needs additional explicit tests of theory on predator-prey spatial races.

A New Experimental Study

Together with colleagues, T. Garcia and P. Rehage, I examined the space race between predatory salamander larvae (*Ambystoma tigrinum*) and herbivorous Pacific treefrog tadpoles (*Hyla regilla* or *Pseudacris regilla*). *Hyla regilla* are common, abundant inhabitants of both streams and ponds throughout the Pacific coast of the United States and southern Canada. In ponds without fish, they can suffer heavy predation from tiger salamander larvae, *A. tigrinum*. Tiger salamander larvae are commonly found in ponds throughout much of the United States. I describe the basic methods and major results relevant to issues here, with details to be left to a forthcoming essay.

We quantified predator and prey space use in four treatments: #1—prey only; #2—predators only; #3—free-ranging prey with predators fenced to one side; and #4—free-ranging predators and prey. Because predators were considerably larger than prey (salamanders ranged from 4.5 to 6.0 g wet weight, whereas tadpoles were about 0.3 g wet weight), we were unable to devise a caging method that allowed us to run the treatment with free-ranging predators and caged prey. In both treatments 3 and 4, predators could encounter, attack, and consume prey. We ran 4 replicates of treatments 1, 2, and 3, and 12 replicates of the most interesting situation (i.e., treatment 4, where both predators and prey are freely roaming).

Experiments were conducted in 90 × 45 cm plastic arenas filled to a depth of 5 cm (above the substrate) with well water. The substrate in each arena consisted of a 2.5 cm-deep layer of washed sand. Each arena was divided into two equal 45 × 45 cm halves by a wire metal frame. In treatment 3, two layers of 1.2 cm mesh wire were hung from the frame to cage predators to the high-resource side of the tank. Tadpole movement was not impeded by this mesh. In all other treatments, animals were free to roam throughout the arena. A 25 cm diameter feeding patch (plastic dish) was buried in the center of each half of the arena so that the lip was barely visible but provided no barrier to predator or prey movement. Each patch contained 25 strands of nylon rope (artificial vegetation). Although this rope added some habitat complexity, it did not provide prey refuge from salamander attack.

To standardize resources, we used commercially available algal discs (1.4 cm diameter) that are a suitable dietary source to support rapid growth of *H. regilla* tadpoles from hatching to metamorphosis (Benard, M., personal communication). Tadpoles fed avidly on these discs. In each arena, we placed four discs in one feeding patch (randomly chosen) and one disc in the other. The discs gradually fell apart during the experiment but remained coherent enough to clearly stay within the feeding patch. Both predators and prey were fed ad libitum (tadpoles were fed algal discs; salamanders were fed bloodworms and *H. regilla* tadpoles) for at least 24 hours, and then held without food for 24 hours before the experiment.

Algal discs were added to arenas at 1,000 hours. Ten tadpoles were added at 1,015 hours, and two salamander larvae were added at 1,100 hours. Every 15 minutes for the next 6 hours, we recorded data on prey numbers and activity in five microhabitat categories within each of four main habitats (inside or outside the feeding patch, on the high- or low-resource side of the arena), and on main-habitat use data for predators. Here, I present data on our most basic result, the proportion of animals in the high-resource side.

Our predictions were as follows. In treatment 1 (prey alone), in accordance with standard IFD theory, we predicted that tadpoles should match resources. Given a resource ratio of 4:1, we predicted that 80% of the tadpoles should be on the high-resource side. Salamander larvae do not feed on algal discs. Thus in treatment 2 (predators alone), we expected salamanders to not differ from random (50% on the high-resource side). Alternatively, predators might use algal resources as a cue for locating tadpole prey (for examples of predators' or parasitoids' aggregating in areas with their prey's host plants, see Vinson 1985, Thomas 1989). In that case, even in the absence of tadpoles, salamanders might prefer the high-resource side. Following the usual expectation in predator-prey ecology, we expected prey to avoid predators; thus, we predicted that the presence of salamanders caged into the high-resource side should cause tadpoles to reduce their use of that side as compared to treatment 1. Finally, with both predators and prey present and freely roaming, in accordance with extant models, in treatment 4 we expected (a) both predators and prey to be more abundant on the high-resource side, (b) predators to aggregate in the high-resource side even more than prey do, and (c) a positive spatial association between predators and prey.

To quantify the spatial association between predators and prey, we plotted the proportion of prey on the high-resource side against the proportion of predators on the high-resource side. If predators win the race, then whenever prey are more abundant on the high-resource side, predators should also be more abundant on that side (and vice versa, if prey are more abundant on the low-resource side). The result should be a positive correlation between predator and prey use of the high-resource side (Figure 11.1A). This is the pattern predicted by the three trophic-level predator-prey space race models. If, on the other hand, prey avoid predators (prey win the race), then predator and prey distributions should be negatively associated (Figure 11.1B). Whenever predators aggregate on one side, prey should be on the other side. Finally, it is possible that counteracting predator-prey behavioral responses will result in no significant spatial association between predators and prey.

Results and Specific Discussion

Figure 11.2 shows the proportion of predators or prey on the high-resource side in the different treatments. A one-way ANOVA showed that tadpole space use was significantly influenced by treatments ($F = 22.47$; df = 2, 17; $p < .001$). In the absence of predators, tadpoles spent about 75% of their time on the high-resource side. This result differs significantly from random (50%; $t = 3.16$, df = 3, $p = .05$), but does not differ significantly from matching, as predicted by simple IFD theory (80%; $t = 0.59$,

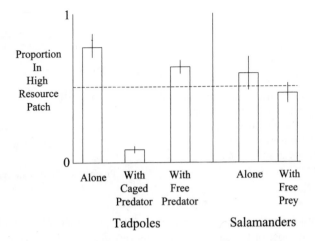

Figure 11.2. The proportion of predatory salamanders or tadpole prey in the high-resource side of an experimental arena when prey or predators are held in single-species groups, or with the other species. Predators were either caged or free-ranging. Shown are means and standard errors. The horizontal dashed line is the null expectation (0.50).

$p > .50$). As expected, the presence of salamanders fenced into the high-resource side caused a large, significant decrease in tadpole use of the high-resource side (Dunnet's test: $p < .05$). In the absence of prey, salamander use of the two halves of the tank did not differ significantly from random ($t = 0.83$, df $= 3$, $p > .40$).

As predicted, when both predators and prey were free-ranging, tadpoles showed a preference for the high-resource side ($t = 2.67$, df $= 11$, $p = .01$). However, contrary to predictions, salamander larvae did not aggregate on the side with high resources (and, on average, higher tadpole densities). Instead, salamander patch use did not significantly differ from random ($t = 0.74$, df $= 11$, $p > .40$).

Figure 11.3 shows the relationship between the proportion of prey in the high-resource side and the proportion of predators in the high-resource side. Each point represents the mean value for a given replicate when observations are pooled over time. The y-intercept, 0.825, is the proportion of prey on the high-resource side if all predators are on the low-resource side. As expected, this value is significantly greater than random ($t = 4.45$, $p < .01$). According to the regression line, if predators are uniformly distributed (50% on each side), then, as one might expect, prey show a slight preference for the high-resource side. Most interestingly, contrary to the prediction of extant models, predator and prey distributions were negatively associated ($r = -.69$, $N = 12$, $p < .001$). If predators spent more time on the high-resource side, then prey spent less time on that side. Of course, this result is not entirely unanticipated. It simply means that prey avoid predators (i.e., that prey win the predator-prey space race). The result is only unexpected in the sense that existing models predict that in this experimental scenario, predators should win the race.

Overall, prey behavior generally fitted adaptive expectations. In the absence of predators, prey space use matched prey resource base. When predators were present

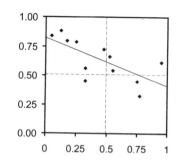

Figure 11.3. For the treatment in which both predators and prey are free-ranging, the correlation between predator and prey use of the high-resource side. Each point shows the mean space use for one replicate. Pearson's $r = -0.69$, $p < .001$.

and fenced into the high-resource side, prey avoided predators. When freely roaming predators spent most of their time on the low-resource side, prey heavily favored the side that had both more resources and greater safety. When freely roaming predators spent more time on the high-resource side, prey spent less time on that side (i.e., prey avoided freely roaming predators). In contrast, predator behavior did not match adaptive expectations. Predators did not tend to aggregate in areas with more prey.

Following the intuition outlined earlier, several types of anchors or constraints could possibly explain why prey won this race. First, predatory salamander larvae interfered with each other, thus increasing the cost of predator aggregation in areas with more prey. Second, because predators were much larger and more active than prey, prey might have had better information about predator space use than vice versa. Third, after predators had consumed a few prey, and prey had fed on algal disks, both were probably not very hungry. In that case, the prey's incentive to avoid predators was probably greater than the predator's incentive to aggregate with prey. Below, I discuss these ideas in more detail.

Discussion: A Broader Look at Factors Influencing the Space Race

In a broad view, I suggest that the outcome of the predator-prey space race should depend on (a) the relative abilities of predators and prey to respond to each other spatially, (b) the relative costs of responding, and (c) the relative benefits of responding (Table 11.1). As noted earlier, in the absence of external costs, constraints, or anchors, the race might have no clear, logical winner. Instead, patterns of spatial association might fluctuate with no stable equilibrium (Schwinning and Rosenzweig 1990). In nature, however, there might often be factors present that can, at least in principle, anchor the predator-prey space race (see Denno et al., ch. 10 in this volume, and Schmitz, ch. 12). I next provide a conceptual overview, together with novel hypotheses, on how various factors might affect the outcome of the predator-prey space race.

Table 11.1. Factors That Should Influence the Outcome of the Predator-Prey Race

Relative Abilities to Respond	Relative Costs of Responding	Benefits of Responding
Movement ability	Movement costs	"Life-dinner" principle
Information availability/ processing	Conflicting fitness needs– spatial anchors	Energy state or risk of starvation
		Prey or predator density

In extant models that include patchy resource distributions, prey have an anchor, the spatial distribution of their resources. This anchor produces a cost for prey responding to predators: prey that avoid predators by leaving the high-resource patches pay a cost in terms of reduced feeding (growth or reproductive) rates. In the simple models, predators have no comparable anchor. As a result, predators win the race. Thus, I suggest that most models predict a positive spatial association between predators and prey because of the assumed scenario and not because of first principles inherent in the predator-prey interaction.

If, in contrast, predators have a spatial anchor and prey do not, then prey should win the race. For example, in models that include safe sites where predators are inherently inefficient (Hugie and Dill 1994, Heithaus 2001), predators have a reason to favor patches that are not safe sites. That is, predators have a spatial anchor. If the difference in inherent attack efficiency in the two patches is large, then predators avoid sites that are inherently safe for prey, even if most prey are found in those patches. If patches do not vary in resource levels (i.e., prey have no resource anchor), then prey prefer safe sites, both because they are inherently safe (low risk per predator) and because predators avoid those patches. The overall expectation is a negative association between predators and prey, with prey concentrated in safe sites and predators concentrated in nonsafe sites.

More generally, I hypothesize that any environmental factor that influences prey or predator fitness and that varies across space can serve as a spatial anchor that influences the predator-prey space race. Obvious possibilities include other predators (that interfere with each other) or prey (e.g., Bouskila 2001, Heithaus 2001); other competitors; nesting, mating, or parental care considerations; abiotic factors (e.g., temperature, moisture, light levels), and so on. With regard to effects on the outcome of the predator-prey race, the key is whether the spatial anchors have greater effects on predators or on prey.

Relative movement costs should also influence the race. At one level, movement costs depend on relative mobility per se. In energetic terms, movement costs (energy cost over distance traveled) should often be lower for more mobile organisms. Thus, if one side is inherently more mobile than the other, the more mobile side should tend to win the response race. Above and beyond mobility and energy costs per se, in many systems, it is likely that the risk of movement is much higher for prey than for predators (i.e., active prey tend to be killed by predators; Lima 1998, Sih et al. 2000). Prey might then reduce their activity in response to predators without shifting in space (Sih and McCarthy 2002). In this case, predators should win the space race.

Differences between predators and prey in activity bring up another potentially important factor, information constraints (Luttbeg and Schmitz 2000; see Lima and Steury, ch. 8 in this volume). To respond appropriately, both predators and prey need information about the spatial distribution of the opponent (Sih 1992). Information requires cues. In many systems, cue generation appears to be proportional to activity. More active animals are more conspicuous in terms of visual, sound, or chemical cues (see, in this volume, Triblehorn and Yager, ch. 5, Cronin, ch. 6, and Greenstone and Dickens, ch. 7). If prey are inactive (e.g., hiding in refuge) and inconspicuous, it should be difficult for predators to gauge prey space use. In contrast, so long as prey can detect predators while hiding in refuge, actively searching predators might be relatively easy to monitor. This consideration should favor prey's winning the space race.

Finally, the behavioral analogue of the life/dinner principle (Dawkins and Krebs 1979) suggests that prey should win the space race. For predators, the difference between success and failure (in the race) is a dinner (or not), whereas for prey it is a matter of life and death. Selection should be much stronger on prey than on predators. The relevance of this logic depends, however, on several factors. First, the life/dinner principle seems most reasonable if both predators and prey are well fed. The logic changes, however, if either side is starving. For starving predators, losing the space race can be critical (dinner can matter a lot), whereas, for starving prey, winning the race against predators can be futile if prey then starve to death. Second, the relative selection pressures on predators and prey should depend on their relative densities. If, for example, predators are relatively rare (and thus pose little total risk), then prey can lose the space race and still not suffer high predation risk. On the other hand, if prey are relatively rare and thus predators are often starving, then predators are under strong selection to win the race. Finally, the outcome could depend on the spatio-temporal scale of interest. In the short term, predators can afford to lose some meals without jeopardizing their fitness. However, at larger scales, predators cannot afford to lose track of prey spatial aggregations. Even if missing a meal or two is not too costly, in the long term, or on larger spatial scales, predators cannot afford to be caught in large areas with few prey over long periods. Thus, all else the same, on a larger spatio-temporal scale, predators might be more likely to win the race.

Conclusions

There are many unanswered questions about the spatial ecology of predators and prey, and many of the ideas discussed in the previous section have not been modeled. Explicit models of these scenarios should generate useful predictions to guide further study. As noted earlier, very few experiments have tested predictions on outcomes of the predator-prey space race in either the field or the laboratory. In general, tests should attempt to include all five treatments: prey alone, predators alone, prey with fixed predators, predators with fixed prey, and free-ranging predators and prey. Tests of extant models should examine effects of variations in resource ratios or safety ratios on patterns of space use. Further tests could examine effects of other aspects of reality [i.e., multiple predators or multiple prey, total predator or prey densities (see Barbosa et al., ch. 16 in this volume), predator or prey state, movement costs, or other spatial anchors] on the outcome of the race.

Although hundreds of field surveys include information on the spatial coincidence of predators and prey, most of these do not distinguish between demographic (birth/death), trophic (predation per se), and behavioral mechanisms. We focus on the behavioral response race; however, ultimately, it should be useful to integrate the effects of these different kinds of mechanisms on overall spatial pattern. In addition, field surveys on predators and prey typically include little or no information on spatial patchiness of the underlying resource base, or on the inherent safety of different patches.

Literature Cited

Abrahams, M. V., and Dill, L. M. 1989. A determination of the energetic equivalence of the risk of predation. Ecology 70:999–1007.

Alonzo, S. H. 2002. State-dependent habitat selection games between predators and prey: the importance of behavioural interactions and expected lifetime reproductive success. Evol. Ecol. Res. 4:759–778.

Benoit-Bird, K. J., and Au, W. W. L. 2003. Prey dynamics affect foraging by a pelagic predator (*Stenella longirostris*) over a range of spatial and temporal scales. Behav. Ecol. Sociobiol. 53:364–373.

Bouskila, A. 2001. A habitat selection game of interactions between rodents and their predators. Ann. Zool. Fenn. 38:55–70.

Clark, C. W., and Mangel, M. 2000. Dynamic State Variable Models in Ecology: Methods and Applications. New York: Oxford University Press.

Dawkins, R., and Krebs, J. R. 1979. Arms races between and within species. Proc. R. Soc. Lond. 205:489–511.

Fauchald, P., Erikstad, K. E., and Skarsfjord, H. 2000. Scale dependent predator-prey interactions: the hierarchical spatial distribution of seabirds and prey. Ecology 81:773–783.

Formanowicz, D. R., Jr., and Bobka, M. S. 1989. Predation risk and microhabitat preference: an experimental study of the behavioural responses of prey and predator. Am. Midl. Nat. 121:379–386.

Fretwell, S. D., and Lucas, H. L. 1970. On territorial behaviour and other factors influencing habitat distribution in birds. I. Theoretical development. Acta Biotheoret. 19:16–36.

Gilliam, J. F., and Fraser, D. F. 1987. Habitat selection under predation hazard: a test of a model with foraging minnows. Ecology 68:1856–1862.

Hassell, M. P. 1978. The Dynamics of Arthropod Predator-Prey Systems. Princeton, N.J.: Princeton University Press.

Heithaus, M. R. 2001. Habitat selection by predators and prey in communities with asymmetrical intraguild predation. Oikos 92:542–554.

Houston, A., and McNamara, J. 1999. Models of Adaptive Behaviour. Cambridge: Cambridge University Press.

Hugie, D. M., and Dill, L. M. 1994. Fish and game: a game theoretic approach to habitat selection by predators and prey. J. Fish Biol. 45:151–169.

Kacelnik, A., Krebs, J. R., and Bernstein, C. 1992. The ideal free distribution and predator-prey populations. Trends Ecol. Evol. 7:50–55.

Kennedy, M., and Gray, R. D. 1993. Can ecological theory predict the distribution of foraging animals? A critical analysis of experiments on the ideal free distribution. Oikos 68:158–166.

Krivan, V. 1997. Dynamic ideal free distribution: effects of optimal patch choice on predator-prey dynamics. Am. Nat. 149:164–178.

Lima, S. L. 1998. Stress and decision making under the risk of predation: recent developments from behavioral, reproductive, and ecological perspectives. Adv. Study Behav. 27:215–290.

Lima, S. L. 2002. Putting predators back into behavioural predator-prey interactions. Trends Ecol. Evol. 17:70–75.

Luttbeg, B., and Schmitz, O. J. 2000. Predator and prey models with flexible individual behavior and imperfect information. Am. Nat. 155:669–683.

Mehlum, F., Hunt, G. L. J., Klusek, Z., and Decker, M. B. 1999. Scale-dependent correlations between the abundance of Brunnich's guillemots and their prey. J. Anim. Ecol. 68:60–72.

Murdoch, W. W., and Stewart-Oaten, A. 1989. Aggregation by parasitoids and predators: effects on equilibrium and stability. Am. Nat. 134:288–310.

Nisbet, R. M., Diehl, S., Wilson, W. G., Cooper, S. D., Donalson, D. D., and Kratz, K. 1997. Primary-productivity gradients and short-term population dynamics in open systems. Ecol. Monogr. 67:535–553.

Parker, G. A. 1979. Sexual selection and sexual conflict. In: Reproductive Competition and Sexual Selection (Blum, M. A., and Blum, N. A., eds.). New York: Academic Press; 123–166.

Rose, G. A., and Leggett, W. C. 1990. The importance of scale to predator-prey spatial correlations: an example of Atlantic fishes. Ecology 71:33–43.

Schwinning, S., and Rosenzweig, M. L. 1990. Periodic oscillations in an ideal-free predator-prey distribution. Oikos 59:85–91.

Sih, A. 1982. Optimal patch use: variations in selective pressure for efficient foraging. Am. Nat. 120:666–685.

Sih, A. 1984. The behavioral response race between predator and prey. Am. Nat. 123:143–150.

Sih, A. 1987. Predator and prey lifestyles: an evolutionary and ecological overview. In: Predation: Direct and Indirect Impacts on Aquatic Communities (Kerfoot, W. C., and Sih, A., eds.). Hanover, N.H.: University Press of New England; 203–224.

Sih, A. 1992. Forager uncertainty and the balancing of antipredator and feeding needs. Am. Nat. 139:1052–1069.

Sih, A. 1998. Game theory and predator-prey response races. In: Game Theory and Animal Behavior (Dugatkin, J. A., and Reeve, H. K., eds.). New York: Oxford University Press; 221–238.

Sih, A., Kats, L. B., and Maurer, E. F. 2000. Does phylogenetic inertia explain the evolution of ineffective antipredator behavior in a sunfish-salamander system? Behav. Ecol. Sociobiol. 49:48–56.

Sih, A., and McCarthy, T. M. 2002. Prey responses to pulses of risk versus pulses of safety: testing the risk allocation hypothesis. Anim. Behav. 63:437–443.

Stephens, D., and Krebs, J. R. 1986. Foraging Theory. Princeton, N.J.: Princeton University Press.

Stiling, P. D. 1987. The frequency of density-dependence in insect host-parasitoid systems. Ecology 68: 844–856.

Thomas, C. D. 1989. Predator-herbivore interactions and the escape of isolated plants from phytophagous insects. Oikos 55:291–298.

van Baalen, M., and Sabelis, M. W. 1993. Coevolution of patch selection strategies of predator and prey and the consequences for ecological stability. Am. Nat. 142:646–670.

van Baalen, M., and Sabelis, M. W. 1999. Nonequilibrium population dynamics of "ideal and free" prey and predators. Am. Nat. 154:69–88.

Vinson, S. B. 1985. The behavior of parasitoids. In: Comprehensive Insect Physiology, Biochemistry, and Pharmacology (Kerkut, G. A., and Gilberg, L. I., eds.). Oxford: Pergamon Press; 417–469.

Walde, S. J., and Murdoch, W. W. 1988. Spatial density dependence in parasitoids. Annu. Rev. Entomol. 33:441–466.

12

Behavior of Predators and Prey and Links with Population-Level Processes

OSWALD J. SCHMITZ

An important challenge in ecology is to discern the level of biological detail one needs to know about prey and predator interactions in order to make accurate predictions about population and community structure and function. One candidate detail is the predation risk response of prey to predators. Prey must often trade off foraging with avoiding predators. Such trade-offs change individual performance in ways that ultimately affect prey population demography and predator-prey dynamics. At the same time, the nature of the trade-off may vary, depending on predator species. Different predator species may display disparate hunting modes and locate in different habitats used by a prey species. As a consequence, the nature of the foraging/predator-avoidance trade-off played by individual prey may depend on the portion of the habitat used (habitat domain) by the prey in relation to predator species hunting mode (predator identity) and habitat domain. The contingencies between habitat domain and hunting mode may lead to different population-level phenomena in different predator-prey systems. In this chapter, I synthesize information on the nature of prey and predator habitat use and predator hunting mode to develop a general working hypothesis for understanding the nature of predator-prey interactions. This framework helps to reconcile when prey and predator behavior matters, requiring behavior to be included explicitly in considerations of population and community dynamics and, when prey and predator behavior does not matter, allowing this detail to be abstracted in considerations of dynamics.

Ecologists have amassed a large body of evidence demonstrating that predators can have tremendous influences on the structure and dynamics of ecological populations and communities. This evidence derives from two separate but complementary avenues of research, population ecology and evolutionary ecology. Population ecology has attempted to understand structure and dynamics by studying the direct lethal effects that predators inflict on their prey and how these lethal effects translate into changing numbers of prey over time (see Stireman et al., ch. 14 in this volume). Evolutionary ecology, on the other hand, has explored how the mere presence of predators in a community can force prey to make behavioral choices between vital activities, such as feeding, and avoiding contact with predators by reducing activity or seeking out refuges (see, in this volume, Lima and Steury, ch. 8, and Sih, ch. 11). Avoiding predators detracts from individual nutritional intake. Foregoing feeding to avoid predators, in turn, can have profound implications for life history traits, such as survival, growth, and reproduction, which have consequent ramifications for population demography of prey. The difference in perspective between population and evolutionary ecology is not the factor causing changes at the population and the individual levels, but assumptions about the mechanistic linkage between them are.

Classical population ecology, for instance, envisions predator-prey interactions in terms of two components (Oaten and Murdoch 1975, Real 1979). That is, the behavioral act of finding prey (which varies with prey abundance) and the behavioral act of capturing and consuming prey (which reduces prey abundance). This perspective which is encapsulated in functional response (Oaten and Murdoch 1975, Real 1979) unilaterally places full control over the outcome of the interaction with the predator. The victims are assumed to submit passively to their fate. Evolutionary ecology, on the other hand, argues that it makes little evolutionary sense for prey simply to allow themselves to be picked off by their predators. Rather, prey should adopt antipredator strategies to reduce or even eliminate predation risk (Mangel and Clark 1986, Lima and Dill 1990, Lima 1998a, Lima 1998b). Such a perspective, which is encapsulated in theory on adaptive behavior, is likewise unilateral, because predators are viewed simply as a stimulus that elicits antipredator responses of prey.

In reality, predator-prey interactions are dynamic (Lima 2002). Prey that respond adaptively to their predators reduce predator hunting success. Predators that, in turn, respond adaptively to prey evasion tactics increase their hunting success. Such changes in predator and prey behavior necessarily influence per capita resource intake and corresponding life history development of hunter and victim (see Abrams, ch. 13 in this volume). This, in turn, influences local population densities of each, which then subsequently feeds back to influence predator and prey behavior. Fundamentally then, the act of hunting by predators, as well as predator evasion by prey, results in a dynamical game (Sih 1984, Schwinning and Rosenzweig 1990, van Baalen and Sabelis 1993, Hugie and Dill 1994, Bouskila 2001, Alonzo 2002, Kotler et al. 2002). Thus, a complete picture of the link between predator-prey interactions and community structure and function must involve an understanding of the way behavior- and population-level interactions influence the dynamical interplay between predators and prey (Brown et al. 1999, Lima 2002).

In this chapter, I explore how prey antipredator behavior in relation to predator hunting mode is linked to population- and community-level processes. I focus here on identifying the nature of predator-prey behavioral interactions and show how the nature of those interactions determines whether or not predators cause behavioral effects of prey on community structure (e.g., level of plant damage by herbivore prey) to propagate or attenuate over time. The intent is to contribute toward developing a conceptualization of predator-prey interactions (Sih 1984, Brown et al. 1999, Lima 2002) in which both predator and prey are viewed as active participants in the interplay. My synthesis of the information, however, will largely involve a reductionist evaluation focused on individual predator species and their prey species. This seems somewhat antithetical to current progress in ecology, which has recognized the importance of emergent multiple predator effects on prey populations and communities (Sih et al. 1998). However, different predator species can have different indirect effects on plant species because of the degree to which they influence the abundance and behavior of their herbivore prey (Schmitz and Suttle 2001). Consequently, there are contingent ways that predator effects could be transmitted through a community. These alternatives influence subsequent interpretations about important drivers of community structure and function. For example, several coexisting predator species could theoretically have no net direct effect on prey density or behavior and hence no net indirect effects on plants. In this case, one concludes that the system is entirely bottom-up controlled. Alternatively, different predator species may exert strong but antagonistic direct effects on their prey, leading to similar weak or no indirect effects on plants. However, it would be incorrect to conclude that the system is bottom-up controlled in this case.

Basically, the potential for these two contingencies to arise in a system means that predator species effects must be examined in isolation from each other before we can draw reliable conclusions about their integrated effects on populations and communities. The philosophical point here is that, by taking a "building blocks" approach to understanding the mechanisms underlying species-specific predator-prey interactions, we can begin to predict the emergent, aggregate effects of multiple predator species. When this approach is taken, we may find that nature is surprisingly more linear than would be expected, given the disparate makeup of predator species effects on their prey and the prey's resources (e.g., Turner et al. 1999, Schmitz and Sokol-Hessner 2002). But now we can begin to offer a mechanistic reason for the outcome.

A Predator Is a Predator—or Is It?

The prevailing assumption in most theory in population and evolutionary ecology is that all predator species elicit qualitatively similar risk responses in their prey. Classical population ecology takes one extreme view and assumes that predators cause no risk responses. Evolutionary ecology takes the other extreme and assumes that predators uniformly cause vigilant prey to reduce activity that makes them conspicuous to predators, or to hide outright. Engaging in such behavior comes at the expense of maximizing resource intake rate or some other currency important to fitness. Nevertheless, both population and evolutionary ecology have each had successes in predicting population- and community-level processes when they have implemented their respective assumptions. This observation begs the question: When is it safe to abstract predator effects on prey trade-off behavior in conceptualizations of population- and community-level processes?

The key to answering this question lies in looking for patterns that emerge from mechanistic considerations of predator-prey interactions. These patterns cannot, however, be identified in most studies of predator-prey interactions. Namely, studies have focused on single predator and single prey species, because prey responses to predators may be contingent on specific environmental conditions. Also, prey must deal with multiple predators and other natural enemies. So, syntheses of studies involving single predators and prey could be confounded by different local environmental conditions of the studies themselves. To avoid confounding effects, one must look to studies that explicitly examine effects of several predator species on the same prey species in a single system but that break down the predator effects by predator species.

I illustrate this point with a set of recent studies that aim to explain the direct and indirect effects of multiple predator species on the functioning of a meadow ecosystem (Schmitz and Suttle 2001, Schmitz and Sokol-Hessner 2002, Sokol-Hessner and Schmitz 2002). I start with this case study from my own research because the outcome caused me to wonder whether the observed pattern occurred broadly among different study systems. This question precipitated a literature review, also presented in this chapter. This literature review leads to a broad conceptualization of the interplay between predators and prey. That conceptualization is that predator habitat domain determines the nature of the prey response to predators, and that predator hunting mode determines whether effects of antipredator behavior of prey persist or attenuate at the community level.

Predator Species Effects on a Single Prey Species in an Old-Field System

I focus on a series of studies completed in an old-field ecosystem comprising 18 species of grasses and herbs, a dominant generalist insect herbivore (the grasshopper *Melanoplus femurrubrum*, which was the focal prey species), and three numerically and biomass-dominant arthropod predators, with a demonstrated ability to capture and subdue the grasshopper species in the study field (Schmitz and Suttle 2001). In old fields, *Melanoplus* grasshoppers tend to spend the nighttime in the lower canopy and litter layer for thermal protection. They move toward the upper canopy during the day to seek out younger plant tissue that has nutritionally higher quality than leaves in the lower canopy (Pitt 1999).

During the course of their daily activity cycle, grasshoppers in the study ecosystem face a virtual gauntlet of predator species. The predator species are the nursery web spider *Pisaurina mira* (Pisauridae), the jumping spider *Phidippus rimator* (Salticidae), and the wolf spider *Rabidosa rabida* (Lycosidae). The predators occupy a different part of the canopy and display a different hunting mode.

Pisaurina is commonly found on stalks and under leaves of tall grasses and herbs (Figure 12.1). It is a sit-and-wait predator, remaining motionless at a fixed location and ambushing prey organisms when they approach within striking distance. Because individuals remain steadfastly perched at one location in the canopy, *Pisaurina* individuals appear to be a threat only during parts of the daytime when grasshoppers feed in the upper canopy of the old field. *Phidippus* actively hunts its prey, but it does so primarily within the emergent vegetation in a field (Figure 12.1). This spider routinely jumps among blades of grass and leaves of herbs and pounces on its prey. *Rabidosa* is a member of a family of spiders previously thought to be cursorial, primarily ground-dwelling, hunters (Kaston 1981). It appears, however, that they employ more of a sit-and-pursue strategy for hunting (Ford 1978, Wagner and Wise 1997) and are cursorial when prey abundances fall below acceptable minimum thresholds, forcing them to seek new feeding sites (Wagner and Wise 1997). *Rabidosa* presents a continuous predation threat as it feeds during daytime and nighttime in the litter layer and in the lower part of the canopy (Figure 12.1). In summary, these three species span the gamut of hunting modes, from completely stationary to highly mobile. They also span widely different habitat domains, where habitat domain is defined as the proportion of the entire available habitat used. For example, habitat in the old-field system can be defined as the entire vertical canopy created by patches of herb species growing adjacent to patches of grass species. These patches are juxtaposed on a 0.25 m. sq. scale, creating a mosaic of herb and grass patches across the field.

It is noteworthy that habitat domain and microhabitat use are not synonymous. For example, microhabitats in the old-field system could be patches of grass and patches of herbs. However, the spider species do not all move throughout the full grass or herb canopy, even though they may use both herb and grass patches (Figure 12.1). Thus, habitat domain is the extent to which individuals move throughout a habitat or microhabitat. Because *Phidippus* roams freely throughout the vertical canopy, it has a broad habitat domain. However, *Pisaurina* uses only the upper canopy leaves and flower heads of plants and thus has a narrow habitat domain. *Rabidosa* uses only the lower canopy leaves and the litter layer and thus also has a narrow habitat domain. Do predator species hunting mode and habitat domain have any bearing on prey behavior?

The relationship between predator hunting mode, habitat domain, and grasshopper behavior was assayed by measuring, in the presence and absence of each spider species, (a) grasshopper utilization of grass and herb resources (habitats), (b) the vertical location of grasshoppers in the canopy, and (c) the degree of grasshopper vigilance (inactivity). Earlier work (Beckerman et al. 1997, Schmitz et al. 1997) indicated that this kind of assay would be a meaningful probe for risk responses. Predator presence caused grasshoppers largely to forego feeding on a preferred resource, grass, and to seek refuge in herbs such as the goldenrod *Solidago rugosa*, whose complex leafy structure offers a refuge from predation risk.

Individual grasshoppers responded differently to each predator species, despite having a broad habitat domain. In comparison to their use of habitat in the absence of predators, grasshoppers exhibited significant habitat shifts from grasses to herbs in the presence of the comparatively sedentary predators *Pisaurina mira* and *Rabidosa rabida* (Schmitz and Suttle 2001). Moreover, the degree to which they used different grass and herb habitats was statistically similar in the presence of either predator species (Schmitz and Suttle 2001). However, grasshoppers did not alter their habitat selection, relative to a no-predator control, in the presence of the actively hunting *Phidippus rimator* (Schmitz and Suttle 2001). In addition to shifting from grasses to

Figure 12.1. Use of the vegetation canopy by different species of spider predators in the New England meadow ecosystem, represented by elipses. *Pisaurina mira* (top species) is commonly found on stalks and under leaves in the upper canopy of the field. It is a sit-and-wait predator, remaining motionless at a fixed location and ambushing prey organisms when they approach within striking distance. *Phidippus rimator* (middle species) actively hunts its prey throughout the middle canopy of the field. *Rabidosa rabida* (bottom species) is a sit-and-pursue predator that hunts on the litter layer and in the lower part of the canopy. The figure is modified from that presented in Schmitz and Suttle (2001).

herbs, grasshoppers were lower in the canopy in the presence of the upper canopy predator *Pisaurina*, higher in the canopy in the presence of the lower canopy predator *Rabidosa*, and displayed no significant bias in vertical location in the presence of the widely roaming *Phidippus*, when comparisons were made to conditions without predators (Figure 12.2). Also, in comparison with no-predator conditions, grasshoppers increased the proportion of time resting in the safe herbs (an index of vigilance) in the presence of *Pisaurina* and *Rabidosa* but exhibited no such change in vigilance in the presence of *Phidippus*.

Habitat shift to avoid predation risk is costly, however. For example, in the presence of *Pisaurina* predators, grasshopper foraging activity was reduced by 65 minutes, which represents an 18% reduction in daily feeding time (Rothley et al. 1997). The shift from consuming grass to consuming herbs, coupled with a 10% decrease in total daily dry mass intake due to lowered feeding time, translates into a 25% reduction in estimated daily energy intake (Rothley et al. 1997), thereby increasing the chance of starvation and lowered offspring production.

The nature of the predator avoidance behavior by the grasshoppers appears to be related to the breadth of the predator's habitat domain (see Sih, ch. 11). There were significant shifts in grasshopper behavior associated with predator species that resided exclusively within narrow domains of the entire field canopy. Here, grasshoppers could

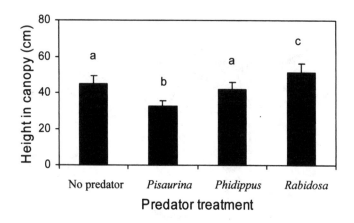

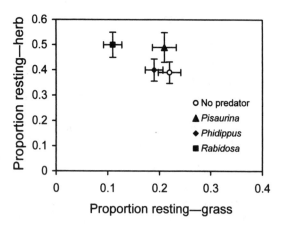

Figure 12.2. Habitat use by grasshoppers in the New England meadow ecosystem in the presence and absence of different spider species. Upper panel presents data on the height in the canopy in relation to predators that use different portions of the canopy, and the lower panel presents data on vigilance activity in grass and herb habitats. Compared to a no-predator control, grasshoppers were lower in the canopy in the presence of the upper-canopy predator *Pisaurina*, were higher in the canopy in the presence of the lower canopy predator *Rabidosa*, and displayed no significant bias in vertical location in the presence of the widely roaming *Phidippus* (upper panel). Lowercase letters distinguish treatments that differ at $p <$ .05. Compared to a control without predators, grasshoppers exhibited significant habitat shifts from grasses to herbs in the presence of the comparatively sedentary predators *Pisaurina* and *Rabidosa*, but they did not alter their habitat selection, when compared to a no-predator control, in the presence of the actively hunting *Phidippus* (lower panel). Moreover, compared to a no-predator control, grasshoppers increased the proportion of time spent resting in the safe herbs (an index of vigilance) in the presence of *Pisaurina* and *Rabidosa*, but exhibited no such change in vigilance in the presence of *Phidippus*. Values are mean and standard errors in two dimensions.

decrease risk by moving away to other locations in the canopy, especially into leafy *S. rugosa*. Such behavioral shifts were not observed when prey were faced with a predator species that posed a threat throughout the entire canopy. The differences in the nature of the predator-avoidance behavior of the grasshoppers may represent different degrees of risk aversion that may be related to the amount of information they have about predator presence (Bouskila and Blumstein 1992, Sih 1992, Luttberg and Schmitz 2000). *Pisaurina* and *Rabidosa* have a continuous presence within a specific habitat domain, so they may provide a persistent cue of high risk to grasshoppers in those locations in which they hunt (Schmitz and Suttle 2001). Alternatively, *Phidippus* may provide persistent, moderate cues throughout the canopy. In this case, grasshoppers must weigh a considerable energetic and associated survival cost of remaining continuously vigilant to this predator, especially when direct threat may not be imminent, against the likelihood of encountering and being captured by *Phidippus* in any one time period. Thus, they may become chronically much less risk averse when faced with this highly mobile predator that moves throughout the vegetation canopy, considering the fitness costs associated with continuous heightened predator avoidance (Lima and Bednekoff 1999, Bouskila 2001).

These data illustrate that not all predator species elicit the same risk responses in a prey species (see also Lima and Steury, ch. 8). Grasshoppers seem to differentiate among the predator species and associated risk. The differences in prey response can be categorized according to predator habitat domain. Moreover, the different risk responses elicited by the different predators correspond to the end points already described. In certain cases, there is no response as is assumed by classical population ecology. In other cases, there is a similar kind and degree of habitat shift, as is assumed by theory on adaptive behavior.

Prey Responses to Predator Hunting Mode and Habitat Domain: A Generalizable Pattern?

I surveyed the literature to determine whether there was an emerging pattern consistent with that observed in the old-field system. Although studies of multiple predator effects in communities are an emerging industry, I used only studies that met the following, specific criteria:

1. The studies had to be experimental and involve two or more predator species (or predation cues) and a single common prey species under identical experimental or study conditions; moreover, the experiment had to have a treatment that did not include predators (i.e., a predation control).
2. The studies had to provide some natural history insight into the hunting behavior (mode) and the habitat domain of each predator.
3. The experiments had to have treatments in which a single predator or predation cue was presented to a single prey species—in other words, the studies could not have confounding effects of interspecific interactions among predator species or among prey species.
4. The experimental systems had to offer prey the opportunity to choose among habitats that were normally part of their natural environment.

I found nine studies that met these criteria, including the old-field case study detailed already. The studies included in the sample (Table 12.1) covered both terrestrial and aquatic systems and included vertebrate and arthropod predators and prey. In some studies, the investigators used cues of predator presence (chemical signal of predator presence or chemical signal of feeding by a specific predator). In other studies, freely moving predators and prey were observed in experimental arenas or in their natural field state. Incidentally, many of these studies attempted to draw linkages between "predator identity" and prey responses. In most cases, the term *identity* was largely a taxonomic designation. I propose here, however, that it is not that the taxonomic status of the predator per se influences the outcome of the interaction, but rather that taxonomy is a surrogate for the nature of the predator's habitat domain and hunting mode.

Predator species were classified according to three hunting modes. First, a sit-and-wait predator is one that remains at a fixed location for prolonged periods, whether it is hunting or not (days to weeks). This predator ambushes prey whenever they approach within close striking distance. They basically wait for prey to come to them. Second, a sit-and-pursue strategy is one in which the predator remains at a fixed feeding location when actively engaged in hunting. The predator either waits for prey and ambushes them whenever they approach within close striking distance, or it rushes toward prey and pounces on them when they are in the predator's vicinity. Sit-and-pursue predators move to new feeding locations when prey abundances fall below acceptable minimum thresholds. In other words, these predators change locations frequently to enhance the likelihood that a moving prey will encounter them. Finally, actively hunting predators are continuously on the prowl, seeking prey.

The predator species used in the studies were generally sit-and-pursue or actively hunting species. They usually occupied a specific habitat domain. For example, in pond systems where the entire habitat comprises the water column, the pond bottom (benthic zone) and the shoreline (littoral zone), fish predators moved throughout the open water column, and insect predators such as dragonfly larvae moved along the pond bottom. In a grassland system, lizards were present in the ground layer, and birds hunted in the vegetation in the upper part of the grassland canopy. This restriction of predators to a narrow habitat domain created a situation in which prey had a choice between locations within the broader habitat: a safe region in which predators were absent, and a risky region that was the predator's domain. Where such options were available (i.e., prey had a broad habitat domain), prey routinely switched their location within the habitat to use the predator-free domain and often reduced their feeding activity, as compared to experimental controls that excluded predators. It appears that, whenever predators occupy a portion of the entire habitat used by prey, prey switch their locations within the habitat and decrease their activity in response to predator presence (see Sih, ch. 11).

The survey also demonstrates that whenever predator habitat use overlaps completely with prey habitat use (e.g., sidewinder snakes in deserts or jumping spiders in old fields), prey do not display visible behavioral responses to their predators (Table 12.1). The reason, again, may lie in the cost of predator avoidance compared to the likelihood of being captured. Actively hunting predators may present a persistent cue to prey. In such cases, heightened risk responses incur a high fitness penalty (Lima and Bednekoff 1999), leaving few, if any, options except to respond only to imminent threat whenever a predator is directly encountered (Bouskila 2001).

There are also cases in which prey size influences the strength of response to predator (Relyea, ch. 9 in this volume). For example, small grasshoppers display stronger or different responses to a predator species than larger grasshoppers (Beckerman et al. 1997, Schmitz et al. 1997, Ovadia and Schmitz 2002). Likewise, small frog tadpoles typically exhibit stronger responses to their predators than do larger tadpoles (Table 12.1). In some cases, this size-dependent response arises because larger prey individuals have reached a predation-size refuge in which predators are no longer perceived as a serious threat (see review by DeWitt and Langerhans 2003, Lima and Steury, ch. 8). In other cases, such as short-lived species in seasonal environments, individuals must develop, mature, and reproduce within a fixed time window. Individuals that are initially smaller may take longer to reach maturity than initially larger individuals. If the cost of avoiding predators results in further developmental delays, small individuals may altogether fail to become reproductively mature. Thus, smaller individuals may be expected to be more risk prone than larger individuals (Rowe and Ludwig 1991, Abrams and Rowe 1996, Houston and McNamara 1999, Clark and Mangel 2000, Ovadia and Schmitz 2002). In essence, the relative costs and benefits of responding to predators may change, thereby changing the outcome of the predator-prey game.

A Prey Is a Prey—or Is It?

From a broad perspective, prey species display a wide range of predator avoidance and escape tactics (Endler 1986). So, the immediate answer to the question whether or not all prey can be treated the same way in conceptualizations of predator-prey interactions would be a definite *no*. But in order to develop some generalizations about the interplay between specific predators and prey, we ask here a more focused question: whether or not individuals of different prey species respond in qualitatively similar ways to the same predator species. Answering this question again requires a gathering of information from a subset of studies on predator-prey interactions, namely those that have examined the effect of a single predator on several prey species in a single system. The selected studies had to meet the following, specific criteria: (a) the studies had to be experimental and involve two or more prey species and a single common predator species (or predation cue) in identical experimental or study conditions, (b) the studies had to provide some insight into prey use of the natural habitat, (c) the experiments had to have treatments in which the predator or predation cue was presented to each prey species individually (i.e., there could be no confounding effects of interspecific interactions among prey species), and (d) the experimental systems had to offer prey the opportunity to choose among habitats that were normally part of their natural environment.

I identified six studies in this survey (Table 12.2). The studies were from terrestrial systems and included vertebrate and arthropod predators and prey. In all cases, the investigators observed freely moving predators and prey in experimental arenas

Table 12.1. Studies That Explicitly Explore the Effects of Multiple Predator Species with Different Hunting Modes and Habitat Domains on the Antipredator Behavior of a Single Prey Species.

Predator	Hunting Mode	Habitat	Habitat Domain	Prey	Prey Response	Source
Cerastes (sand viper)	Sit-and-wait	Desert	Under shrub	Gerbillus (desert gerbil)	Habitat shift to open/ activity reduction	1
Bubo b. (eagle owl)	Sit and-pursue	Desert	Open	Gerbillus (desert gerbil)	Hiding under cover/ activity reduction	1
Lepomis (green sunfish)	Sit-and-pursue	Stream	Stream pools	Aquarius (water strider)	Hiding in stream riffles and on shoreline/activity reduction	2
Dolomedes (fishing spider)	Sit-and-pursue	Stream	Shoreline	Aquarius (water strider)	Hiding in center of stream pools	2
Salvelinus (brook trout)	Active	Stream	Water column	Baetis (mayfly nymph)	Hiding/less drifting	3
Megarcys (stone fly larvae)	Active	Stream	Bottom substrate	Baetis (mayfly nymph)	Escape by drifting away	3
Eumeces (prairie skink)	Sit-and-pursue	Grassland	Ground	Melanoplus (grasshopper)	No response	4
Pooecetes and Tyrannus (sparrows and kingbirds)	Active	Grassland	Above field and upper canopy	Melanoplus (grasshopper)	Hiding in lower canopy/ activity reduction	4
Orconectes (crayfish)	Sit-and-pursue	Pond	Bottom substrate	Physella (snails)	Hiding near water surface	5
Lepomis (pumpkinseed sunfish)	Active	Pond	Open water	Physella (snails)	Hiding in cover	5
Lepomis (bluegill sunfish)	Active	Pond	Open water	Rana (sm. tadpoles) Rana (lg. tadpoles)	Weak activity reduction No response	6
Anax (dragonfly larva)	Sit-and-pursue	Pond	Substrate	Rana (sm. tadpoles) Rana (lg. tadpoles)	Strong activity reduction No response	6

Predator	Foraging mode	Habitat	Microhabitat	Prey	Response	Source
Crotalus (sidewinder)	Sit-and-pursue	Desert	Shrub and open	*Dipodomys* (kangaroo rats)	No response	7
Bubo v. (great-horned owl)	Sit-and-pursue	Desert	Open	*Dipodomys* (kangaroo rats)	Use cover	7
Pisaurina (nursery web spider)	Sit-and-wait	Old field	Upper canopy	*Melanoplus* (grasshopper)	Hiding in lower canopy/activity reduction	8
Phidippus (jumping spider)	Active	Old field	Entire canopy	*Melanoplus* (grasshopper)	No response	8
Rabidosa (wolf spider)	Sit-and-pursue	Old field	Lower canopy	*Melanoplus* (grasshopper)	Hiding in upper canopy/activity reduction	8
Nonecta (backswimmer)	Active	Pond open	Water	*Rana* (sm. tadpoles) / *Rana* (lg. tadpoles)	Hiding/activity reduction / Some hiding/activity reduction	9
Aeschna (dragonfly larva)	Sit-and-pursue	Pond	Substrate	*Rana* (sm. tadpoles) / *Rana* (lg. tadpoles)	Activity reduction / Hiding/some activity reduction	9
Anax (dragonfly larva)	Sit-and-pursue	Pond	Substrate	*Rana* (sm. tadpoles) / *Rana* (lg. tadpoles)	Activity reduction / Hiding/some activity reduction	9
Triturus c/u I (newt)	Sit-and-pursue	Pond	Vegetation	*Rana* sm. tadpoles / *Rana* (lg. tadpoles)	Activity reduction / No response	9

Note: Source codes are as follows: (1) Kotler et al. 1992, (2) Krupa and Sih 1998, (3) McIntosh and Peckarsky 1999, (4) Pitt 1999, (5) Turner et al. 1999, (6) Eklov 2000, (7) Bouskila 2001, (8) Schmitz and Suttle 2001, and (9) Van Buskirk 2001.

Table 12.2. Studies That Explicitly Explored the Responses of Different Prey Species with Different Habitat Use to the Same Predator Species

Predator	Hunting Mode/Habitat	Prey	Prey Habitat Use	Prey Response	Source
Bubo v. (great horned owl)	Sit-and-pursue in open and shrub	*Dipodomys* (kangaroo rat)	Open	Reduced activity/evasive leap) under imminent risk	1
		Microdipodops (kangaroo mouse)	Open	Reduced activity/evasive leap under imminent risk	1
		Chaetodipus b. (pocket mouse)	Shrub	Reduced activity/evasive leap under imminent risk	1
		Chaetodipus f. (poket mouse)	Shrub	Reduced activity/evasive leap under imminent risk	1
Bubo b. (eagle owl)	Sit-and-pursue in open	*Gerbillus a.* (desert gerbil)	Shrub and open	Hiding in shrub cover	2
		Gerbillus p. (desert gerbil)	Shrub and open	Hiding in shrub cover	2
Pisaurina (nursery web spider)	Sit-and-wait in upper canopy and on grass	*Melanoplus* (grasshopper)	Grass and herb	Hiding in herb cover/reduced activity	3
		Chorthippus (grasshopper)	Grass	Reduced activity/remaining in place/hiding behind leaves and stems	3
Micropterus (largemouth bass)	Sit-and-pursue and active in stream pools (mainly downstream)	*Campostoma* (minnows)	Stream pools	Schooling/switching to upstream pools to avoid predators	4
Canis (coyote)	Active in open	*Oronectes v.* (crayfish)	Substrate	Reducing activity/hiding	4
		Odocoileus h. (black-tailed deer)	Open range	Change use of slope and aspect/grouping to stand ground	5
		Odocoileus v. (white-tailed deer)	Open range and shrub	Fleeing, and seeking cover in shrub	5
Pardosa (wolf spider)	Sit-and-pursue in lower canopy and on ground	*Prokelisia m.* (planthopper)	*Spartina* lower canopy	Reducing activity/remaining in place	6
		Prokelisia d. (planthopper)	*Spartina* lower canopy	Reducing activity/remaining in place/jumping of plant	
		Delpahacodes (planthopper)	*Spartina* lower canopy	Hiding behind leaves and stems	
		Sanctanus (leafhopper)	*Spartina* lower canopy	Jumping off plant	
		Trigonotylus (plantbug)	*Spartina* upper canopy	Reducing activity/remaining in place	6

Note: Source codes are as follows: (1) Longland and Price 1991, (2) Kotler et al. 1992, (3) Schmitz 1998, (4) Gelwick 2000, (5) Lingle 2002, and (6) Denno et al. 2003.

or in their natural field state. The cases included predators with each of the three hunting modes (sit-and-wait, sit-and-pursue, and active).

Whenever prey and predator occupy the same portion of the entire habitat (each have a narrow domain), prey routinely respond by reducing their activity (Table 12.3). Likewise, prey with a narrow habitat domain that face widely roaming predators (i.e., with broad habitat domain) have no recourse but to reduce conspicuous activity when that predator is in the prey's habitat (Table 12.3). Prey that use a variety of habitats (broad domain) undergo habitat shift when they are facing predators with a narrow habitat domain (Table 12.3). Finally, when prey and predators both have broad habitat domains, prey rarely exhibit habitat shifts or activity reductions (Table 12.3). Instead, prey respond behaviorally when they are faced with imminent threat and usually do so by fleeing (e.g., Denno et al. 2003).

The empirical data show that there are two general risk responses of prey to predators. The first is an activity reduction when habitat refuges are unavailable because of a broad habitat domain of the predator, or when prey and predators have the same, narrow habitat domain (e.g., sit-and-wait predators and prey that specialize on resources in the same habitat). The second is seeking refuge in alternative habitats when predators have a narrow habitat domain but prey have a broad habitat domain. Acute responses such as fleeing rapidly under imminent threat occurred only under conditions in which both predators and prey had broad habitat domains.

The synthesis is based on a limited sample size in the empirical data. As a consequence, it may be difficult to explain certain outcomes at this time. For example, prey with narrow habitat domains seem to respond behaviorally to predators with overlapping, narrow habitat domains. Yet prey with broad habitat domains do not respond behaviorally when they overlap with predators that have equally broad domains. Further research on predator hunting mode and prey and predator habitat domains will help to resolve whether or not this kind of finding can be explained away as idiosyncratic to the small data set, or indeed whether it is a general outcome requiring explanation.

This synthesis suggests that prey always undergo some form of predator avoidance behavior, even if it is a last-minute response. Nevertheless, the extent to which behavior is manifest at the population and community level depends on the nature of the predator hunting mode, and prey and predator habitat domain. Thus, for practical modeling purposes, functions describing predator and prey behavior are not always required in order to predict population- and community-level dynamics. For example, the last-minute escape tactics used by prey, when both predators and prey have a broad habitat domain, can be easily captured in a population-scale model by characterizing this behavioral detail as a parameter that specifies simply the probability of prey capture by predators. In other cases, prey exhibit some form of chronic or persistent antipredator behavior, such as activity reduction or shift in habitat location. If such behavior matters at the population and community scale, then functions describing this kind of behavior must be included in models specifying population and community dynamics (Abrams 1995, and Abrams, ch. 13).

If behavioral effects attenuate on the time scale of population and community dynamics, then behavior can be abstracted. However, it is impossible to tell from a synthesis of behavioral interactions whether behavioral responses matter at the

Table 12.3. Synthesis of Prey Antipredator Responses in Relation to Predators' Hunting Mode and the Habitat Domain of Prey and Predator Species

Predator Habitat Domain	Hunting Mode	Prey Habitat Domain	
		Narrow	Broad
Narrow	Sit-and-wait	Activity reduction	Habitat shift
	Sit-and-pursue	Activity reduction	Habitat shift
	Active	Activity reduction	Habitat shift
Broad	Sit-and-wait		
	Sit-and-pursue	Activity reduction	Activity reduction/ habitat shift
	Active	Activity reduction	No response except evasion under imminent threat

Note: Sit-and-wait predators do not have a board habitat domain by definition. The nature of the prey behavioral responses was determined by empirical data presented in Tables 12.1 and 12.2. This represents a working hypothesis of the kind of effects to be expected in different field systems.

population and community scale. One must seek evidence from experiments that explicitly test for the consequences of behavior on population and community processes.

Linking Behavior with Population- and Community-Level Processes

In order to explore the potential link between predator-prey behavior and population and community processes, I return to the case study of multiple predator species effects on the New England meadow ecosystem (Schmitz and Suttle 2001, Schmitz and Sokol-Hessner 2002, Sokol-Hessner and Schmitz 2002). These studies were designed to examine the effect of different spider predators on survival of the generalist grasshopper and on the nature of trophic interactions. Specifically, the studies explored how predator hunting mode was related to the likelihood that the habitat shift exhibited by the grasshopper in response to predation risk led to trait- or behavior-mediated, as opposed to density-mediated, trophic interactions. If behavioral effects (trait-mediated effects) dominate, then predators should have no significant effect on grasshopper survival (Schmitz 1998). But, if the habitat shift by grasshoppers mediates the indirect effects of predators on plants, then predators should have a positive indirect effect on the grasshoppers' preferred resource, grasses, and a negative indirect effect on the refuge resource, herbs (Schmitz 1998). If, on the other hand, behavioral effects attenuate (i.e., density-mediated effects dominate), then predators should cause a significance reduction in grasshopper density as compared to density in conditions without predators. That is, they should have an indirect positive effect on both grasses and herbs (Schmitz 1998).

The studies showed that the nature of the population- and community-level processes depended on the predator hunting mode. The sit-and-wait spider *Pisaurina* had no significant effect on *M. femurrubrum* grasshopper survival as compared to a control containing no predators (Schmitz and Suttle 2001, Sokol-Hessner and Schmitz 2002). However, the habitat shift by grasshoppers led to a positive indirect effect on grass and a negative indirect effect on herbs, a trait-mediated indirect effect (Schmitz and Suttle 2001, Schmitz and Sokol-Hessner 2002). The sit-and-pursue spider *Rabidosa* caused grasshoppers to undergo a habitat shift (Schmitz and Suttle 2001). At the same time, the predator caused a significant reduction in grasshopper density compared to a no-predator control (Schmitz and Suttle 2001, Sokol-Hessner and Schmitz 2002). This result, in turn, led to a positive indirect effect of the spider species on both grasses and herbs: a density-mediated indirect effect (Schmitz and Suttle 2001, Schmitz and Sokol-Hessner 2002). In this case, the density effects of the predator "swamped out" the behavioral effects. Finally, the active hunting *Phidippus* caused a significant reduction in grasshopper density, relative to a no-predator control (Schmitz and Suttle 2001, Sokol-Hessner and Schmitz 2002) that also led to a density-mediated positive indirect effect on both grasses and herbs (Schmitz and Suttle 2001, Schmitz and Sokol-Hessner 2002). This purely density-based effect is expected, given that grasshoppers displayed little, if any, behavioral responses to this predator (Table 12.1).

The series of studies in the old-field system identified three different scenarios related to predator hunting mode. The sit-and-wait predator caused purely trait-mediated effects at both the population and the community scale. Thus, behavioral details must be considered explicitly in models describing population- and community-level phenomena. The sit-and-pursue predator caused a habitat shift at the behavioral scale but caused strong density reductions at the population scale. As a consequence, the density effects "swamped out" the behavioral effects, resulting in a net density-mediated effect at the community level. In this case, behavior can be abstracted in considerations of population- and community-level phenomena. Finally, the actively hunting predator caused little or no behavioral responses of prey; all effects were the consequences of changes in prey density. Thus, consideration of behavior is not required in this case.

There are a few other studies that explicitly examined indirect effects of different predators in the context of food web interactions. These studies also revealed that the nature of the indirect effects on plants seems to be related to the specific predator hunting mode, and prey and predator habitat domain. In a stream system, brook trout (*Salvelinus fontenalis*) and stonefly (*Megarcys signata*) are both active hunters, but trout prowl the water column, whereas stoneflies crawl on the stream bottom. Mayfly (*Baetis bicaudatus*) prey avoided contact with trout by hiding under rocks and becoming inactive (activity reduction). Mayflies avoided stoneflies by drifting away to other locations (habitat shift; McIntosh and Peckarsky 1999). These differential responses of mayflies to predators altered the extent and spatial distribution of mayfly impacts on their algal food resources. In a pond system (Bernot and Turner 2001), pumpkinseed sunfish (*Lepomis gibbosus*) and crayfish (*Orconectes rusticus*) hunt *Physa* snails. Pumpkinseed are active hunters that continually prowl the water column, whereas crayfish wait for prey at a fixed location and ambush them whenever they approach within striking distance. When faced with predation risk from pumpkinseed, snails decreased their activity and hid in covered habitats. When faced with predation risk from crayfish, snails moved to the surface of the water (Bernot and Turner 2001). Consequently, sunfish caused a reduction in periphyton biomass in covered habitats and an increase in periphyton in the snail's normal open-water habitat. Crayfish presence caused reductions in near-surface periphyton, and increases in periphyton in the normal snail habitat. Different prey in the same system may respond differently to the same predator, leading to different community structure. Herbivorous minnows (*Campostoma anomalum*) and crayfish (*O. virilis*) differ in their avoidance of bass (*Micropterus salmoides*) predators, thereby changing the nature of the trait effect (Gelwick 2000). Bass congregate within specific pools, and minnows avoid those pools. Crayfish avoid bass by feeding at night when bass are inactive; they hide in burrows during daytime. Predators have a positive and negative indirect effect on algae by causing minnows to move out of pools to damage algae in refuges; predators have a positive indirect effect on algae by causing crayfish to reduce foraging.

In the examples from aquatic systems, the effects of behavior were identified but not contrasted with density effects. In the terrestrial examples, behavioral effects were isolated from density effects, but the predator hunting modes were not replicated. Consequently, it is not possible to draw general conclusions about the relationship between predator hunting mode and trait- versus density-based direct and indirect effects based on the empirical data presented in this chapter. There is some difficulty in completing such a synthesis, because most studies exploring trait effects were designed in such a way that density effects were precluded from operating (Werner and Peacor 2003). Thus, although many studies amply demonstrate the occurrence of trait effects, they are rarely designed to quantify their importance relative to density effects.

In the absence of such data, I propose a simple working hypothesis to advance empirical research on predator-prey interactions. That is, that predator hunting mode in relation to habitat domain determines whether direct and emergent indirect effects at the population and community level are trait- or density-determined (Table 12.4). Specifically, at the one extreme, sit-and-wait predators should cause only trait effects. At the other extreme, actively hunting predators should generally cause density effects to "swamp out" trait effects, except in circumstances where predators have a narrow habitat domain and prey have a broad habitat domain (i.e., prey can switch habitats to seek refuge from predation). Finally, sit-and-pursue predators should cause a mixture of trait and density effects. Whichever effect dominates depends on the ability of prey to detect and either avoid contact with these predators, or escape from them. Trait effects should certainly dominate whenever there is recourse for prey to seek refuge by switching their habitat use. Given that the majority of predator species surveyed in this study exhibit a sit-and-pursue strategy (Tables 12.1 and 12.2), there may be a rich amount of interaction between trait and density effects in many systems.

This hypothesis takes a prey-centric perspective on predator-prey interactions. However, predators are also known to adjust their hunting modes and habitat domains in response to changing environmental conditions, physiological state, and prey antipredator behavior (Lima 2002). Broad empirical evidence for switches in community-level effects resulting from altered predator hunting strategies in response to prey antipredator behavior remains nonexistent (Lima 2002). There is one case in

Table 12.4. Hypothesized Nature of Direct and Emergent Indirect Effects at the Population and Community Levels, Arising from Prey Behavioral Responses to Predators with Different Hunting Modes and Habitat Domains

		Prey Habitat Domain	
Predator Habitat Domain	Hunting Mode	Narrow	Broad
Narrow	Sit-and-wait	Trait effects	Trait effects
	Sit-and-pursue	Trait/density effects	Trait/density effects
	Active	Trait	Trait effects
Broad	Sit-and-wait		
	Sit-and-pursue	Trait/density effects	Trait/density effects
	Active	Trait effects	Density effects

Note: Sit-and-wait predators do not have a broad habitat domain by definition. *Trait effects* refers to indirect effects arising from antipredator behavior of prey. *Density effects* refers to indirect effects arising from numerical reductions in prey abundance due to direct predation. The nature of the emergent indirect effect was determined by the summary presented in Table 12.3. This represents a working hypothesis of the nature of the emergent indirect effects to be expected under different envionmental conditions.

which altered predator hunting behavior has been shown to alter both the nature and strength of community-level effects. On Isle Royale, Michigan, the effects of wolves (*Canis lupus*) on balsam fir (*Abies balsamea*) mediated by moose (*Alces alces*) are linked to winter snowfall levels. In high-snowfall winters, wolves hunt in larger packs than in low-snowfall winters. Wolves prey on moose that are forced to aggregate along lakeshores because the snow is less deep there than it is elsewhere on the landscape (Post et al. 1999). Wolves are extremely efficient at killing moose in these conditions, because moose have little recourse to escape once encountered. Thus, wolves reduce moose populations to levels where they cause limited damage to balsam fir. This is largely a density-based effect. In years when snowfall levels are low, moose scatter more widely across the landscape. In this case, it becomes inefficient for wolves to hunt in large packs, so they divide into smaller packs and become more confined into local territories (Post et al. 1999). Moose population density remains high in these years, because moose more freely escape predation by fleeing (Post et al. 1999) or by seeking refuge habitats (Edwards 1983). This density then leads to a stronger impact of moose on balsam fir than in high snowfall years (Post et al. 1999). The ability of moose to evade wolves, in this case, ultimately leads to lower predation mortality and in turn larger effects on balsam fir across the landscape. This effect then translates largely into a trait-based effect (see Gittleman and Gompper, ch. 17 in this volume, for other examples).

Clearly, there is a critical need for ecologists to be more systematic about studying the interplay between trait and density effects and for quantifying their relative importance in different systems (Werner and Peacor 2003). Such systematic studies should include three key elements. First, behavioral responses of prey to predators (e.g., activity budgets, habitat utilization) should be measured in experiments in which predators are prevented from killing prey. Second, the survivorship of prey should be measured in the presence and absence of predators (predation control). Ideally, such a study should include treatments that measure prey survival as a consequence of risk alone (i.e., predators are present but prevented from capturing and subduing prey) and predation alone to tease apart behavioral and density effects on population demography (Beckerman et al. 1997, Peacor and Werner 2001, Werner and Peacor 2003). Finally, one should measure the indirect effect of the predator on the prey's resource to elucidate the strength and nature of the indirect effect. This should again be done by including both risk-only and predation treatments to allow for comparisons.

Implications for Multiple Predator Effects

I now briefly return to the issue that prey in natural systems rarely live exclusively with one predator species. In reality, they routinely face multiple predators and must cope with these situations by trading off their responses to the different predator species (Sih et al. 1998, Lima and Steury, ch. 8). The insight about habitat domain and hunting mode can be extended to explain whether multiple predator effects on prey populations and on communities might be linear or nonlinear. Nonlinear effects usually arise when a prey species, in attempting to avoid one predator species by moving to other parts of its habitat, becomes more vulnerable to other predators species in other locations of its habitat (Chang 1996, Losey and Denno 1998; see also Denno et al., ch. 10 in this volume). In the old-field system, however, there was limited possibility for this nonlinear effect to happen. *Phidippus rimator* and *R. rabida*, which occupy the middle and lower parts of the canopy, respectively, caused the same level of mortality risk to grasshoppers (Sokol-Hessner and Schmitz 2002). Thus, switching habitat locations under this condition will have no net effect on mortality risk. There is the potential for nonlinearity when grasshoppers, in attempting to avoid *P. mira* in the upper canopy, move to the middle canopy only to encounter *P. rimator*. However, this will be a trivial effect (Fauth 1990, Sih et al. 1998). Even minor, positive mortality levels would be considered infinite nonlinearity, because *P. mira* has no significant direct effect on grasshopper mortality, relative to nonpredator control conditions.

Nonlinearities may also arise because intraguild predation between the spider species lowers the density of predators hunting prey. However, predator habitat domain prevents such intraguild effects from happening, because the predators occupy complementary habitat domains and do not overlap strongly with each other (Figure 12.1). Consequently, the net effects on prey populations and on plants of predator species in combination are simply the average of the individual predator species effects (Schmitz and Sokol-Hessner 2002, Sokol-Hessner and Schmitz 2002). A similar linear effect of multiple predators was observed in the pond system composed of pumpkinseed sunfish and crayfish preying on *Physa* snails (Turner et al. 1999), where, again, the predators have different hunting modes and occupy complementary habitat domains. The logical extension of this line of reasoning is that nonlinear effects are likely to emerge when predators overlap in their habitat domains or possess hunting modes that induce different kinds and levels of risk. This notion, however, remains a testable hypothesis (Sih et al. 1998).

Conclusions

The synthesis presented here suggests that predator habitat domain determines the nature of the prey response to predators, and predator hunting mode determines whether effects of antipredator behavior of prey persist or attenuate at the community level. Whenever predators have a narrow habitat domain and prey have a broad domain that allows habitat switching, behavioral effects are likely to persist from the individual level to the population and community scale (Table 12.4). Thus, antipredator behavior needs to be considered explicitly in models of population and community dynamics. With sit-and-pursue predators that have a broad habitat domain, there will be interplay between density and behavioral effects. Finally, when predators actively hunt and overlap completely with prey habitat use, behavioral effects will likely attenuate, causing density effects to dominate at the population and community scale. Thus, behavioral effects can be abstracted in models of population and community dynamics.

The fact that a single prey species responds differently to different predators (Table 12.1) suggests that prey are able to distinguish among predator species and discriminate the level of threat they impose. The differential responses of prey to predators may then be viewed as representing a continuum of prey responses that vary with the costs and benefits of responding to predators of a given size and propensity to use different habitat domains. In essence, different predators create different rules of engagement based on their hunting mode and habitat domain. Predator and prey thus play out their life histories on a continuous "landscape of fear" (Brown et al. 1999, Laundre et al. 2001). And so, a predator-prey dynamical game may present the unifying perspective needed for understanding the nature of predator-prey interactions at the individual, population, and community scales (Sih 1984, Brown et al. 1999, Lima 2002).

Literature Cited

Abrams, P. A. 1995. Implications of dynamically-variable traits for identifying, classifying and measuring direct and indirect effects in ecological communities. Am. Nat. 146:112–134.

Abrams, P. A., and Rowe, L. 1996. The effects of predation on the age and size of maturity of prey. Evolution 50:1052–1061.

Alonzo, S. H. 2002. State-dependent habitat selection games between predators and prey: the importance of behavioral interactions and expected lifetime reproductive success. Evol. Ecol. Res. 4:759–778.

Beckerman, A. P., Uriarte, M., and Schmitz, O. J. 1997. Experimental evidence for a behavior-mediated trophic cascade in a terrestrial food chain. Proc. Natl. Acad. Sci. U.S.A. 94:10735–10738.

Bernot, R. J., and Turner A. M. 2001. Predator identity and trait-mediated indirect effects in a littoral food web. Oecologia 129:139–146.

Bouskila, A. 2001. A habitat selection game of interactions between rodents and their predators. Ann. Zool. Fennici 38:55–70.

Bouskila, A., and Blumstein, D. T. 1992. Rules of thumb for predation hazard assessment: predictions from a dynamics model. Am. Nat. 139:161–176.

Brown, J. S., Laundre, J. W., and Gurung, M. 1999. The ecology of fear: optimal foraging, game theory and trophic interactions. J. Mammal. 80:385–399.

Chang, G. C. 1996. Comparison of single vs. multiple species of generalist predators for biological control. Environ. Entomol. 25:207–212.

Clark, C. W., and Mangel, M. 2000. Dynamic State Variable Models in Ecology. Oxford: Oxford University Press.

Denno, R. F., Gratton, C., Döebel H., and Finke, D. L. 2003. Predation risk affects the relative strength of top-down and bottom-up impacts on insect herbivores. Ecology 84:1032–1044.

DeWitt, T. J., and Langerhans, R. B. 2003. Multiple prey traits, multiple predators: keys to understanding complex community dynamics. J. Sea Res. 49:143–155.

Edwards, J. 1983. Diet shifts in moose due to predator avoidance. Oecologia 60:185–189.

Eklov, P. 2000. Chemical cues from multiple predator-prey interactions induce changes in behavior and growth of anuran larvae. Oecologia 123:192–199.

Endler, J. A. 1986. Defense against predators. In: Predator-Prey Relationships (Feder, M. E., and Lauder, G. V., eds.). Chicago: University of Chicago Press; 109–134.

Fauth, J. E. 1990. Interactive effects of predators and early larval dynamics of the tree frog *Hyla chrysoscelis*. Ecology 71:1609–1616.

Ford, M. J. 1978. Locomotory activity and the predation strategy of the wolf spider *Pardosa amentata* (Clerk) (Lycosidae). Anim. Behav. 26:31–35.

Gelwick, F. P. 2000. Grazer identity changes the spatial distribution of cascading trophic effects in stream pools. Oecologia 125:573–583.

Houston, A. I., and McNamara, J. M. 1999. Models of Adaptive Behavior: An Approach Based on State. Cambridge: Cambridge University Press.

Hugie, D. M., and Dill, L. M. 1994. Fish and game: a game theoretic approach to habitat selection by predators and prey. J. Fish Biol. 45:151–169.

Kaston, B. J. 1981. How to Know the Spiders. Dubuque, Iowa: Brown.

Kotler, B. P., Blaustein, L., and Brown, J. S. 1992. Predator facilitation—the combined effects of snakes and owls on the foraging behavior of gerbils. Ann. Zool. Fennici 29:199–206.

Kotler, B. P., Brown, J. S., Dall, S. R. X., Gresser, S., Ganey, D., and Bouskila, A. 2002. Foraging games between gerbils and their predators: temporal dynamics of resource depletion and apprehension in gerbils. Evol. Ecol. Res. 4:495–518.

Krupa, J. J., and Sih, A. 1998. Fishing spiders, green sunfish, and a stream-dwelling water strider: male-female conflict and prey responses to single versus multiple predator environments. Oecologia 117:258–265.

Laundre, J. W., Hernandez, L., and Altendorf, K. B. 2001. Wolves, elk, and bison: reestablishing the "landscape of fear" in Yellowstone National Park, USA. Can. J. Zool. 79:1401–1409.

Lima, S. L. 1998a. Nonlethal effects in the ecology of predator-prey interactions—what are the ecological effects of anit-predator decision-making? BioScience 48:25–34.

Lima, S. L. 1998b. Stress and decision-making under the risk of predation: recent developments from behavioral, reproductive, and ecological perspectives. Adv. Study Behav. 27:215–290.

Lima, S. L. 2002. Putting predators back into behavioral predator-prey interactions. Trends Ecol. Evol. 17:70–75.

Lima, S. L., and Bednekoff, P. A. 1999. Temporal variation in danger drives antipredator behavior: the predation risk allocation hypothesis. Am. Nat. 153:649–659.

Lima, S. L., and Dill, L. M. 1990. Behavioral decisions made under the risk of predation: a review and prospectus. Can. J. Zool. 68:619–640.

Lingle, S. 2002. Coyote predation and habitat segregation of white-tailed deer and mule deer. Ecology 83:2037–2048.

Longland, W. S., and Price, M. V. 1991. Direct observations of owls and heteromyid rodents—can predation risk explain microhabitat use? Ecology 72:2261–2273.

Losey, J. E., and Denno, R. F. 1998. Interspecific variation in the escape response of aphids: effect on risk of predation from foliar-foraging and ground-foraging predators. Oecologia 115:245–252.

Luttbeg, B., and Schmitz, O. J. 2000. Predator and prey models with flexible individual behavior and imperfect information. Am. Nat. 155:669–683.

Mangel, M., and Clark, C. C. 1986. Dynamic Modeling in Behavioral Ecology. Princeton, N.J.: Princeton University Press.

McIntosh, A. R., and Peckarsky, B. L. 1999. Criteria determining behavioral responses to multiple predators by a stream mayfly. Oikos 85:554–564.

Oaten, A., and Murdoch, W. W. 1975. Functional response and stability in predator-prey systems. Am. Nat. 109:289–298.

Ovadia, O., and Schmitz, O. J. 2002. Linking individuals with ecosystems: experimentally identifying the relevant organizational scale for predicting trophic abundances. Proc. Natl. Acad. Sci. U.S.A. 99:12927–12931.

Peacor, S. D., and Werner, E. E. 2001. The contribution of trait-mediated indirect effects to the net effects of a predator. Proc. Natl. Acad. Sci. U.S.A. 98:3904–3908

Pitt, W. C. 1999. Effects of multiple vertebrate predators on grasshopper habitat selection: trade-offs due to predation, foraging and thermoregulation. Evol. Ecol. 13:499–515.

Post, E., Peterson, R. O., Stenseth, N. C., and McLaren, B. E. 1999. Ecosystem consequences of wolf behavioural response to climate. Nature 401:905–907.

Real, L. A. 1979. Ecological determinants of functional response. Ecology 60:481–485.

Rothley, K. D., Schmitz, O. J., and Cohon, J. L. 1997. Foraging to balance conflicting demands: novel insights from grasshoppers under predation risk. Behav. Ecol. 8:551–559.

Rowe, L., and Ludwig, D. 1991. Size and timing of metamorphosis in complex life-cycles—time constraints and variation. Ecology 72:413–427.

Schmitz, O. J. 1998. Direct and indirect effects of predation and predation risk in old-field interaction webs. Am. Nat. 151:327–342.

Schmitz, O. J., Beckerman, A. P., and O'Brien, K. M. 1997. Behaviorally mediated trophic cascades: effects of predation risk on food web interactions. Ecology 78:1388–1399.

Schmitz, O. J., and Sokol-Hessner, L. 2002. Linearity in the aggregate effects of multiple predators on a food web. Ecol. Lett. 5:168–172.

Schmitz, O. J., and Suttle, K. B. 2001. Effects of top predator species on direct and indirect interactions in a food web. Ecology 82:2072–2081.

Schwinning, S., and Rosenzweig, M. L. 1990. Periodic oscillations in an ideal-free predator-prey distribution. Oikos 59:85–91.

Sih, A. 1984. The behavioral response race between predator and prey. Am. Nat. 123:143–150.

Sih, A. 1992. Prey uncertainty and the balancing of anti-predator and feeding needs. Am. Nat. 139:1052–1069.

Sih, A., Enlund, G., and Wooster, D. 1998. Emergent impacts of multiple predators on prey. Trends Ecol. Evol. 13:350–355.

Sokol-Hessner, L., and Schmitz, O. J. 2002. Aggregate effects of multiple predator species on a shared prey. Ecology 83:2367–2372.

Turner, A. M., Fetterolf, S. A., and Bernot, R. J. 1999. Predator identity and consumer behavior: differential effects of fish and crayfish on habitat use of a freshwater snail. Oecologia 118:242–247.

van Baalen, M., and Sabelis, M. 1993. Coevolution of patch selection—strategies of predator and prey and the consequences for ecological stability. Am. Nat. 142:646–670.

Van Buskirk, J. 2001. Specific induced responses to different predator species in anuran larvae. J. Evol. Biol.14:482–489.

Wagner, J. D., and Wise, D. H. 1997. Influence of prey availability and conspecifics on patch quality for a cannibalistic forager: laboratory experiments with the wolf spider *Schizocosa.* Oecologia 109:474–482.

Werner, E. E., and Peacor, S. D. 2003. A review of trait-mediated indirect interactions in ecological communities. Ecology 84:1083–1100.

13

The Consequences of Predator and Prey Adaptations for Top-Down and Bottom-Up Effects

PETER A. ABRAMS

Early theory regarding top-down and bottom-up effects in food chains and ecosystems was based on very simple models. The predictions of that theory have been successful in describing the responses of some systems to impacts on top or bottom levels, but there are at least as many cases that are inconsistent with the theory. In this chapter, this theory is extended to include the possibility of adaptive change in the interactions between levels. The models have two or three trophic levels and include direct density dependence on all levels. The analysis considers the change in abundance of each level in response to a standardized perturbation (i.e., an increase in the per capita mortality rate). The combination of adaptive change and direct density dependence greatly increases the range of potential responses of trophic-level abundances to altered mortalities. The equilibrium abundance of the top level in a consumer-resource system may increase in response to an increase in its own per capita mortality rate, because of either its foraging adaptations or the resource species' antipredator adaptations. In three-level systems, the interaction of direct density dependence with adaptive foraging makes it possible for most of the responses of abundance of one species or level to mortality, at its own or another level, to be positive or negative. Whether a given level increases or decreases often can be determined, with sufficient knowledge about the shapes of the relationships between densities and fitness. In particular, knowing the sign of the immediate response of foraging effort to a change in food density allows the signs of several effects to be determined. The ability of this analysis to account for observations that are inconsistent with earlier theory is reviewed.

In an extremely influential paper, Oksanen et al. (1981) extended ideas of Hairston et al. (1960) and Fretwell (1977) to predict the changes in abundance of different trophic levels in an ecosystem with increased inputs to the bottom (producer) level. Their model was based on the assumption that each level in the system acted as though it were a single population whose abundance was solely limited by food and predators, and whose individuals had characteristics that did not change over time. The model was later analyzed further to assess the impact of top predators on the abundances of lower levels. The predictions of this basic model received remarkable confirmation in the experiments of Carpenter and Kitchell (1993). The general predictions from the Oksanen et al. (1981) models were that inputs to the bottom level would increase the top level and even-numbered levels below the top, while leaving the abundances of odd-numbered levels below the top unchanged. Conversely, removal of individuals from the top level would decrease their abundance and the abundances of odd-numbered levels below the top, while increasing the abundances of even-numbered levels below the top. Most discussion in recent years has focused on which of these two sets of changes in abundance (bottom-up or top-down) is larger in magnitude.

The Oksanen et al. (1981) models have been elaborated in many ways over the past two decades (reviewed briefly in DeAngelis et al. 1996). The predictions of the models have been the subject of numerous experimental studies (reviewed in Brett and Goldman 1996, Leibold et al. 1997, Persson 1999, and Shurin et al. 2002). Determining the nature of top-down and bottom-up effects continues to be the focus of considerable research: recent experimental studies include Dyer and Letourneau (1999), Menge (2000), Halaj and Wise (2001), Moran and Scheidler (2002), and Benndorf et al. (2002). However, one topic that has received only limited attention is the role of adaptive processes in determining the responses of trophic-level abundances to environmental factors that have a direct impact on the top, middle, or bottom levels. Experimental studies are often inconsistent with the predictions of the Oksanen et al. (1981) models (Leibold et al. 1997). Could it be that adaptive responses of predators (consumers) to prey (resources), or vice versa, account for many of these discrepancies? At present it is difficult to answer this question, because of the relatively small body of theory suggesting how trophic-level abundances are likely to change, given different types of adaptive adjustment of predatory or antipredatory characteristics (for a review, see Bolker et al. 2003). Every naturalist knows that predator and prey generally respond to each other, but the logical consequences of this simple observation seem to be missing from most formal analyses of top-down and bottom-up effects. Work on models with two or more species per trophic level has shown that adaptive replacement of species within a trophic level can alter top-down and bottom-up effects (Leibold 1989, Abrams 1993, Leibold 1996).

In this chapter, I develop and analyze models that include a simple representation of adaptive processes to reassess the predicted responses of all trophic levels to a perturbation affecting one level. Simple models of food chains are used. These models differ from the Oksanen et al. (1981) models in that they allow adjustment of defensive characteristics in prey (resource) species, or of foraging characteristics in predator (consumer) species, or both. Some previous comparisons of top-down and bottom-up effects have been criticized (Chase et al. 2000) because of the qualitatively different perturbations that are usually used on different trophic levels (i.e., fertilization of the bottom level vs. removal or addition of the top level). Here, I examine the changes in population size of each trophic level to a standard perturbation applied to each one of the levels in turn. The perturbation is an increase in per capita mortality. This approach has the advantage that it is usually possible to vary this parameter experimentally by harvesting individuals. It has the additional advantage that it is consistent with many of the types of perturbations employed in the past. Fertilization of the bottom level is defined by its positive effect on the immediate per capita growth rate of that level (Abrams 1993), and a decrease in the per capita mortality of the bottom level increases its per capita growth rate. Removal of all the top predators has the same effect as a predator per capita mortality increase that is large enough to cause extinction. Abrams and Vos (2003) have a more detailed discussion of the advantages of using per capita mortality rates to quantify top-down and bottom-up effects.

The models discussed here, like those in Oksanen et al. (1981), assume that the system in question can be represented with a food chain model. In other words, there is only one "type" of organism on each level, or, alternatively, the different types can be characterized by their mean characteristics. Thus, the results should apply to food chains or larger systems with relatively homogeneous trophic levels. Because larger systems can adapt by species replacement at each level, the models are likely to be better representations, at least of multispecies systems, than comparable models without adaptive change. The underlying population dynamics are described by ordinary differential equations. However, to the extent possible, the component functions of these equations (e.g., the predator's numerical response) are not given specific forms. Most of the analysis is directed toward systems that are stable. This emphasis allows application of the type of general equilibrium analysis pioneered by Puccia and Levins (1986), and employed by Bender et al. (1984) and Yodzis (1988) in their analyses of indirect effects. Models with adaptive foraging or antipredator characters assume that the population can be adequately characterized by its mean trait value. This assumption is met when the range of traits present at any given time is narrow, but it is also often met when there is considerable between-individual variation, when the fitness does not change in a highly nonlinear manner with the trait (Abrams et al. 1993). The dynamics of adaptive traits are described below in equations (2). The analysis concentrates on systems with two or three trophic levels.

Will traditional trophic cascades or the changes in abundances due to enhancement of the bottom trophic level be changed by adaptive change in foraging or antipredator traits? Answering this question requires a review of the responses expected in the absence of such adaptive processes. This review also introduces the methods used to analyze models that include adaptation. My "standard" model without adaptations differs in some respects from those proposed by Oksanen et al. (1981), and these differences alter the Oksanen et al. (1981) predictions of zero effects of some perturbations on trophic-level abundances. These zero effects have long been known to be sensitive to the assumption that consumer species have per capita growth rates that are unaffected by their own density. If direct negative effects of density are present on all trophic levels, then, provided that all of the consumer species have linear functional responses, the predictions regarding mortality are as follows (Abrams

1993, Abrams and Vos 2003). An increase in the per capita death rate of a given species (a) decreases its own equilibrium density, (b) decreases the density of all species above it in the chain, (c) increases the density of the species immediately below it and odd numbers of levels below it, and (d) decreases the densities of species that are even numbers of levels below it. Thus, for example, in a three-level system, an increase in the top predator's per capita death rate decreases the resource density because it increases consumer density. An increase in the resource population's per capita death rate decreases the populations of all higher level species. Note that the responses of these systems to mortality applied at one level are in concordance with the intuitive notions that (a) the density of a species decreases when mortality is applied to it; and (b) mortality imposed on a predator increases its prey population, whereas mortality imposed on a prey decreases its predator.

Setting the Stage: Two-Level (Consumer-Resource) Models without Adaptive Change

In both two- and three-level models, the population size of the bottom level is referred to as the *resource*, and is denoted by R. The second level is the *consumer* and has population size N, while the top level is the *predator* with population size P. (Of course, the consumer is itself a predator.) The general two-level model is built up of the following component functions: f, the per capita growth rate of the resource (density dependence makes this a decreasing function of resource density, although f may increase at low densities because of an Allee effect); the satiation function of the consumer, g, measures the proportional reduction in an average instantaneous capture rate of prey due to handling time and satiation; $g = 1$ when $R = 0$, and g decreases as the encounter rate with resource, CR, increases. The parameter C is the slope of the consumer's functional response to resource density and is often referred to as the consumer's *attack rate* on resources. This parameter can be influenced by foraging traits of the consumer and antipredator traits of the resource.

The resource conversion function of the consumer, b, specifies the relationship between food intake and per capita growth rate, for a given consumer density. This relationship is often assumed to be simple proportionality, but is more likely to increase nonlinearly. The interference function of the consumer, d, specifies the increase in consumer per capita death rate as a function of its own density. This value may reflect competition for types of resource other than food, or aggressive behavior between individuals. The parameters d_1 and d_2 are the per capita death rates of resource and consumer, respectively. Putting these functions and parameters into a standard predator-prey model yields

$$\frac{dR}{dt} = R[f(R) - CNg(CR) - d_1]$$

$$\frac{dN}{dt} = N[b(CRg(CR)) - \delta(N) - d_2] \qquad (1a,b)$$

Well-known methods can be used to determine (a) whether an equilibrium point with positive densities of both species is locally stable, and (b) how the equilibrium densities of resource and consumer change as a function of the per capita death rates. Some details are given in the Appendix. One finding from this analysis is far from new (Rosenzweig and MacArthur 1963) but is still not widely appreciated. That is, the equilibrium consumer population can increase with an increase in its own death rate. If the consumer interference function, δ, is absent, then the condition for such an increase to occur is precisely the same condition needed for the equilibrium point to be an unstable focus. Satisfaction of this condition means that there will be cycles of the consumer and resource populations around this point. Some appreciation for why this increase occurs can be gained when one looks at the isocline diagram for this system. The isoclines for systems with and without interference (but with satiation) are shown in Figure 13.1. Increasing mortality moves the consumer isocline to the right, which initially increases the consumer density at the equilibrium point (i.e., the intersection of the two isoclines). However, having an intersection of a vertical consumer isocline to the left of the hump in the resource isocline implies cycles (Rosenzweig and MacArthur 1963). This may be why the possibility of an increase in the consumer is not more widely known. There are no empirical examples of such an increase that I am aware of. On the other hand, the average population sizes of cycling predator populations apparently have not been calculated, even in laboratory systems, where it is certainly feasible to do so for organisms with short generation times.

The analysis in the Appendix demonstrates that, if there is consumer interference, increasing the per capita death rate of the consumer may increase its density in a stable system. Conditions for stability and for N to increase with d_2 are no longer identical in this case, because the consumer isocline is no longer vertical (see the dashed line in Figure 13.1). However, the conditions required for an increase in density with higher mortality in stable systems with interference are usually rather restrictive. The problem is that the consumer isocline must be sloped significantly to the right in order to

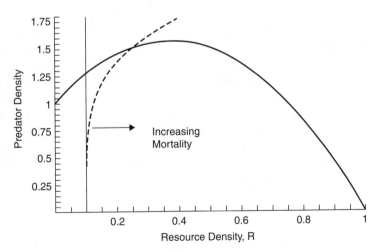

Figure 13.1. Isoclines for the consumer-resource system, showing the effects of greater consumer mortality on the position of the consumer isocline. The dashed line is the consumer isocline when there is consumer interference (direct density dependence).

have stability (see Figure 13.1). This means that there is generally only a small distance that it can be moved to the right (because of higher mortality) before the consumer isocline intersects the resource isocline to the right of the hump in the resource isocline, and the traditional result for stable systems (N decreases with d_2) is restored. The isocline diagrams show that the resource will always increase as consumer death rate is increased. Although not as tidy as the result for systems with linear functional responses, the basic finding is that, unless one is dealing with a cycling system, the standard food web results (based on models with linear functional responses) are adequate. This finding is no longer true when adaptive adjustment of foraging or defensive traits is possible.

In all cases I have discussed, increasing the resource mortality decreases the equilibrium consumer population, and it decreases the equilibrium resource population if the consumer experiences any interference. Thus, the change to a nonlinear functional response does not alter the predictions of simpler models, so long as the system is stable. In cycling systems, increasing the death rate of the resource is likely to alter the amplitude of the cycles, which can have a variety of effects on the mean consumer population (Abrams et al. 1997).

Consequences of Consumer Adaptations to Consuming Resources in Two-Level Systems

This section extends the approach used in the previous section to examine the consequences of altered mortality in a system in which the consumer can adaptively adjust a costly foraging trait, depending on resource density. Increasing the attack rate, C, generally entails reductions in some other component of fitness. To simplify the analysis, I begin by assuming that the consumer does not exhibit satiation. Satiation was responsible for the counterintuitive result of the consumer's increasing with its own death rate when there was no adaptation. Here, the goal is to see if adaptive processes by themselves can produce counterintuitive effects, so the logical starting point is a model in which such effects do not occur. This condition is true of the basic model with linear functional responses. The foraging adaptation involves a trade-off between a higher attack rate, C, and a decrease in some other component of fitness; here it is assumed to be an increase in the per capita death rate. This relationship can be represented by expressing the per capita mortality rate as the sum of mortality that is not influenced by the foraging trait and mortality that increases with increases in the foraging trait C. These components are denoted respectively by d_2 and $D_2(C)$. The trait, C, is assumed to change at a rate that is proportional to the fitness reward from a unit change in C (i.e., dW/dC). All this leads to the following dynamic system:

$$\frac{dR}{dt} = R[f(R) - CN - d_1]$$

$$\frac{dN}{dt} = N[b(CR) - \delta(N) - D_2(C) - d_2]$$

$$\frac{dC}{dt} = V\frac{dW}{dC} = V[Rb'(CR) - D_2'(C)] \qquad (2a\text{--}c)$$

where V is a factor that scales the rate of adaptive change in C; V may be a constant or a function of C. The primes in equations (2) denote derivatives of functions with respect to their arguments. Equation (2c) for the dynamics of C states that traits climb hills in the adaptive landscape at a rate that increases with the steepness of the gradient. This equation arises from several different types of adaptive change: (a) evolutionary change in asexual populations with limited mutational variation (Geritz et al. 1998), (b) evolutionary change in polygenic traits in sexual populations when the variance in the trait is sufficiently small (Iwasa et al. 1991, Abrams et al. 1993), and (c) some models of behavioral change (Taylor and Day 1997). Abrams and Vos (2003) show that equation (2c) also describes the change in trait values when C changes because of shifts in the relative abundances of two species or clones. The value of V is clearly much larger for behavioral than for evolutionary change, but the equilibrium requires that $dW/dC = 0$, and the value of C that satisfies this condition is not affected by the rate of change.

If equations (2) exhibit cycles that involve extreme values of C, then an additional term may be needed in the equation for trait dynamics to keep C from becoming negative or unreasonably large (see Abrams and Matsuda 1997, Ma et al. 2003). It is reasonable to assume that the equilibrium trait value maximizes fitness (i.e., that $R^2b'' - D_2'' < 0$ at the equilibrium value of C) (see Abrams and Chen 2002 or Abrams 2003 for more detailed discussions of the assumptions underlying the trait dynamics). This system can be subjected to the same type of equilibrium analysis employed above, and the results are (a) an increase in d_1 decreases both populations, and (b) an increase in d_2 always increases the resource population, but may increase or decrease the consumer population. (The Appendix gives formulas for the changes in densities with d_2.)

The equilibrium consumer population in equations (2) increases as its own death rate increases, provided that the system is stable, and the following condition is satisfied:

$$f'(R^2b'' - D'') + N(b' + CRb'') < 0 \qquad (3)$$

This expression is the sum of two terms. The first is usually positive; $f' < 0$ because of density dependence in resource growth, and $R^2b'' - D'' < 0$ if the trait equilibrium is to maximize fitness locally. An Allee effect in the resource population could cause f' to be positive at low resource densities. The second term in equation (3) has the sign of the quantity $b' + CRb''$, which is the derivative of CRb' with respect to CR. This is a measure of the curvature in the birthrate function b. If the consumer birthrate increases at a strongly decelerating rate as food intake increases, the quantity will be pessimistic; otherwise, it is positive. The quantity $b' + CRb''$ also determines whether the initial response of the foraging trait to an increase in resource density (R) is to increase ($b' + CRb'' > 0$) or decrease ($b' + CRb'' < 0$; Abrams 1991). A large magnitude, negative second derivative of b means that the numerical response saturates. Because of this, when food (resource) increases in abundance, it is better to reduce risky foraging and get nearly the same amount of food at a lower fitness cost. Empirical work, mainly on larval amphibians, has found that foraging effort declines with increasing food availability (Werner and Anholt 1993), strongly suggesting a negative value for $b' + CRb''$ in these organisms. The expression $b' + CRb''$ also determines the sign of the eventual response of the equilibrium C to a change in the equilibrium resource density in equations (2). Thus, if C decreases in response to R, and if the density dependence in the resource growth function (f) is weak or consumer population (N) is high, one expects that increasing mortality on the top level will cause it to increase, along with the bottom level.

The equilibrium of this system can be an unstable focus, meaning that cycles around the equilibrium are possible. Stability conditions are somewhat complicated, but necessary conditions for an unstable equilibrium include the requirement that $b' + CRb'' < 0$ and that the rate of adaptive change, V, be sufficiently large (Abrams 1992a). If V is small enough, the population dynamics are essentially independent of the dynamics in C, and those dynamics must be stable, given that the functional response is linear. If V is very large, so that change in the trait is effectively instantaneous, then the adaptive adjustment of the trait, C, results in a rapid reduction in foraging effort with increased food density. This reduction moderates the increase in food intake with increasing food availability, essentially producing a type-2 (saturating) functional response. Such responses are well known for their ability to generate population cycles.

In the system described, the immediate response of the consumer is always to decrease in abundance in response to an increase in its own per capita mortality. However, this short-term decline provokes both an increase in resource and a decrease in the rate of harvesting resource, resulting in an eventual increase in the consumer. Figure 13.2 provides an example in which the density-independent death rate of the consumer in the system is roughly doubled at time 0. The time course of both populations and the trait are shown. This example involves a rate of change of the trait that is intermediate between what one might expect for behavior and for evolution. Interestingly, an order of magnitude reduction in V has very little effect on the length of the initial period of decrease in the consumer population. The consumer begins to increase as soon as its increased birthrate (due to the increase in the resource population) outweighs the increased mortality; this increase may happen before there is much change in C.

Consequences of Adaptive Defense by the Resource in Two-Level Systems

The resource being considered here is a biological population and, as such, is likely to exhibit some form of flexible defense against its consumer. Such defense could be behaviors that make it less vulnerable to the consumer, induction of phenotypes that make it less attractive to the consumer, or evolutionary change in traits that also deter consumption (see, in this volume, Brown and Chivers, ch. 3, Lima and Steury, ch. 8, and Relyea, ch. 9). Can adaptive defense qualitatively change the responses of resource or consumer (or both) to an increase in the mortality rate of either species? It is possible to examine this question with a predator-prey model, as in the previous section. However, the trait, C, is now under the control of the resource species rather than the consumer species.

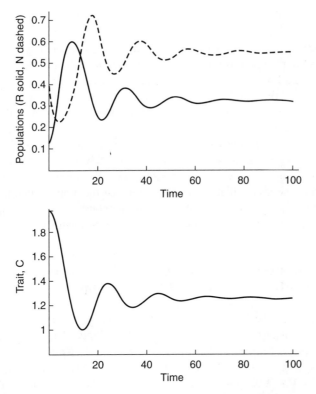

Figure 13.2. The responses of consumer, resource, and foraging trait to an approximate doubling of the consumer mortality in a specific example of the general system given by equations (2). The functional forms of the model components are: $f(R) = r(1 - (R/K))$; $b(CR) = B_0 - B_1 C - D_1/(CR) - d_2$; $\delta = 0$. Parameter values are $r = 1$; $K = 1$; $B_0 = 1$; $B_1 = 0.2$; $D_1 = 0.1$; d_2 (before time zero) $= 0.21$; d_2 (at time zero) $= 0.5$. The adaptive rate constant, V, is 1.

A general model of adaptive change in such a system has the following form:

$$\frac{dR}{dt} = R[f(R,C) - CNg(CR) - d_1]$$

$$\frac{dN}{dt} = N\,[b(CRg(CR)) - \delta\,(N) - d_2]$$

$$\frac{dC}{dt} = V\frac{dW_N}{dC} = V\left[\frac{\partial f}{\partial C} - Ng(CR)\right]$$

(4a–c)

Note that, because decreasing C is assumed to be costly for individuals of the resource population, $\partial f/\partial C < 0$. Here I have again included a satiation function for the consumer, because previous work has shown this function to be an important determinant of the dynamics of evolution or other forms of adaptive change in the defensive traits of the resource species (Matsuda and Abrams 1994, Abrams and Matsuda 1997). The analysis of this model is not presented here, but follows basically the same steps as in the preceding cases. The analysis confirms that the consumer's density can increase with an increase in its own death rate. The mechanism underlying such a response is the increase in resource growth rate that results in both higher consumer and higher resource populations; higher resource population means the consumer is more saturated, so a lower level of defense (higher growth) can be maintained. A specific example of this kind of system is provided by some analyses of one-consumer–two-resource systems (Abrams 1999, Abrams 2002), in which the two resources differ in that one is faster growing and more vulnerable to the consumer. Adaptation of the resource comes about by shifts in the abundances of the two types. Figure 13.3 presents an example of the system described by equations (4), in which adaptation is by means of a change in the growth and vulnerability characteristics of a relatively homogeneous population. This example has the property that increasing the harvest of the consumer increases the equilibrium consumer population size almost up to the point where the consumer abruptly becomes extinct.

Consequences of Adaptive Change in Feeding or Defensive Traits in Three-Level Systems

Three-level systems can be characterized by adaptive change on any or all of the levels. When this kind of system is combined with the possibility of nonlinear functional responses and direct density dependence at some or all levels, a range of complicated results can be generated. I will address only the most frequently studied case (Abrams 1995, Abrams and Vos 2003), in which the consumer species (the middle trophic level) can adjust a trait that determines both its level of exploitation of the resource and its vulnerability to the predator on the top level. To simplify matters further, I will examine only the responses of trophic levels in a system having the following characteristics: (a) all functional responses are linear, (b) the middle species has a potentially nonlinear relationship between its food intake and its per capita growth rate, and (c) only the middle species exhibits adaptive change in foraging/antipredator traits. Moreover, its vulnerability, s, to the predator, and its per capita death rate, D_2, both increase as its capture rate of the resource, C, increases, and (d) the trait C changes at a rate proportional to the fitness gradient with respect to that trait.

All this leads to a model with the following form:

$$\frac{dR}{dt} = R[f(R) - CN - d_1]$$

$$\frac{dN}{dt} = N\,[b(CR) - D_2(C) - d_2 - s(C)P - \gamma(N)]$$

$$\frac{dP}{dt} = P[Es(C)N - d_3 - \alpha(P)]$$

$$\frac{dC}{dt} = V\,[Rb' - D_2' - s'\,P]$$

$$(5a\text{–}d)$$

There are three populations, and each has a per capita death rate that may be varied, giving rise to nine combinations of a population density with a mortality rate, combinations that must be quantified to give a full picture of the responses of the system

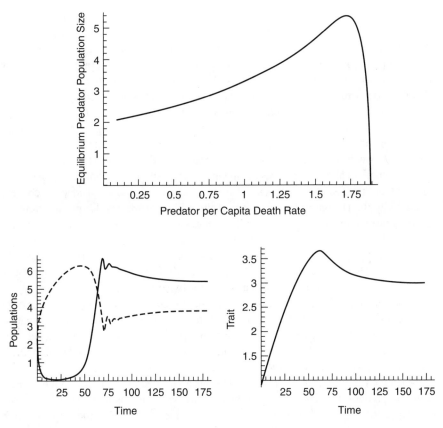

Figure 13.3. The top panel shows equilibrium predator population size as a function of its own death rate in a version of equations (4) in which the component functions are $f(R) = r_0 + r_1C - r_2C^2 - kR$; $g(CR) = 1/(1 + ChR)$; $\delta = 0$; $b(CRg(CR)) = BCR/(1 + hCR)$; the parameter values are $r_0 = 2$; $r_1 = 2$; $r_2 = 1$, $k = 1$; $h = 0.5$, $B = 1$. The bottom two panels show the temporal change in the two populations (predator solid; prey dashed) and the prey's vulnerability trait, C, when the predator per capita mortality is increased from $d_2 = 1$ to $d_2 = 1.7$ at time zero. The adaptive rate constant, V, is 0.05.

to perturbations. These nine responses are measured by the sensitivity of each population to each mortality (i.e., $\partial X/\partial d_i$, where X stands for R, N, or P; and i is 1, 2, or 3). The full details regarding these derivatives may be found in Abrams and Vos (2003). A related model with no interference at either the middle or top levels was analyzed in Abrams (1995). If there is no adaptive change in the trait of the consumer species, each one of these nine responses has a determinate sign. Mortality applied to the resource decreases the abundances of all levels. Mortality of the consumer reduces its population and that of the predator but increases the resource. Finally, mortality applied to the predator increases the consumer level but decreases predator and resource populations. This set of responses of the three populations to the three mortality rates will be referred to as the "traditional" responses.

When adaptive change of the consumer in response to food and predator densities is added to the model, as in equation (5d), there are only two of the nine responses (of one of the three population sizes to one of the three mortality rates) that are characterized by a particular sign. These are that increases in the per capita mortality of either the predator or the resource will decrease the equilibrium abundance of the resource. These two determinate responses have the same sign that is predicted by models with no adaptive change. The remaining seven potential responses of a density to a mortality may all be either positive or negative, depending on the strengths of the interference terms and the shapes of the relationships, b, between food intake and consumer per capita growth rate and between foraging rate, C, and vulnerability to the predator, s. This plethora of alternatives does not mean that the responses are unpredictable. They represent the "traditional" responses that are produced in the absence of adaptive change, modified by the change in foraging trait of the mid-level species. Larger interference terms reduce the magnitudes of the traditional responses and thereby increase the relative size of the changes due to adaptation. As in the two-level model, if the consumer's growth rate saturates strongly as its food intake increases (b'' is negative and large in magnitude), the adaptive response is to reduce costly foraging traits when food abundance increases. This phenomenon leads to a positive effect of reduced resource abundance on the predator, via the increase in consumer foraging. Note that this is the opposite of the density-mediated indirect effect, whereby less resource would have a negative effect on the predator by reducing the consumer population. As in the two-level consumer-resource model, the sign of the quantity $b' + CRb''$ determines the immediate response of C to an increase in R.

When the foraging trait, C, changes, there is an immediate response in the vulnerability to predation, s. The change in the ratio s/C with increasing C determines how the impact of the predator on the consumer changes, relative to the impact of the resource on the consumer as C increases. Thus, the change in this ratio is an important determinant of how effects due to adaptive change in traits are transmitted up or down the food chain. The sign of the change in s/C with C, is given by the sign of the expression $Cs' - s$, and this quantity therefore has an important effect on the sign of indirect effects involving adaptive change in foraging. For example, the resource may decrease when mortality on the consumer is increased, provided $Cs' < s$. This relationship is possible because the increase in foraging results in a less than proportional increase in vulnerability to the predator, which allows the increase in foraging to outweigh the possible decrease in consumer density. The magnitudes of the interference terms, γ and α, influence indirect effects due to changes in the abundance of other trophic levels. A larger interference term essentially creates inertia in the focal population, with a consequent decrease in indirect effects transmitted by changes in density, unaccompanied by any corresponding effect on the trait-transmitted effects.

The possible responses of trophic levels to mortalities under equations (4) include the following outcomes: (a) the abundance of either the consumer or the predator may increase with an increase in their own mortality, (b) the abundance of the predator may increase with an increase in mortality applied to the consumer, and (c) the abundance of the consumer may decrease with an increase in mortality applied to the predator. In particular, it is possible for increased mortality of the predator to increase its own abundance while decreasing the abundance of the other two trophic levels.

This outcome requires that (a) $b' + CRb'' \ll 0$ (i.e., the optimum C decreases rapidly with increased resource), (b) consumer interference be relatively strong (γ' large), and (c) resource density dependence be relatively weak (f' small). The resource always decreases in abundance with increased mortality on the predator (this is one of the two effects with a definite sign). When $b' + CRb'' < 0$, the consumer will increase its foraging in response to decreased resource abundance, but this response automatically makes it more vulnerable to the predator. The resulting increased consumer availability can increase predator numbers and decrease consumer numbers. The strong consumer interference buffers its own numbers, so that the decrease in numbers is not large enough to outweigh the consumers' increased availability to predators. Other types of mortality applied to this food chain typically have similar sorts of interactions between trait-mediated and density-mediated indirect effects (for a more detailed description of some of the other possible responses, see Abrams and Vos 2003).

Allowing the possibility of adaptation of two or all three trophic levels cannot lead to fewer alternative outcomes and is likely to lead to a wider range of potential responses of densities to mortality rates. Abrams (1992b) examined some of the potential responses of a four-level food chain model to changes in parameters, but there is apparently no other analysis of mutual adaptation in two or more levels of a three- or four-species food chain. Loeuille and Loreau (2004) have analyzed the responses of plant and herbivore abundances to nutrient enrichment in a model in which both plant and herbivore traits coevolve. Even though all fitness components of both species in this model are linear functions of densities, the possibility of coevolution implies that every qualitatively different conceivable response of plant and herbivore densities to nutrient enrichment can occur. The results presented here also need to be reexamined in models in which consumers have nonlinear functional responses. Incorporating type-2 functional responses into a three-species chain without adaptation can lead to chaotic dynamics (Hastings and Powell 1991), alternative attractors, and long-lasting transients (McCann and Yodzis 1994). Such complex dynamics may lead to responses of average densities to mortalities at lower or higher trophic levels that differ dramatically from those of the comparable models with linear functional responses and stable equilibria (Abrams and Roth 1994). It will require considerable work just to understand the homogeneous three-level differential equation models of such food chains, once these factors have been added to the model.

Discussion

There are a variety of reasons that it is important to be able to predict or understand the impacts of changes in mortalities at different levels of a food chain on the abundances of all of the levels. Most planning and adaptive management of exploited populations, particularly fisheries, rests on theory that assumes that mortality applied to a population will decrease its abundance. Furthermore, the direction of the initial response of abundance to mortality is often assumed to indicate the final response. These assumptions follow from single-species models but are not generally justified in food chains characterized by adaptive changes in traits related to feeding interactions (Matsuda and

Abrams 2004). Given the rapid pace of global change, many of the changes in density of a focal species of interest are likely to be caused by direct effects of the environment on some other species in the food web. The search for that other species is more likely to be successful if we know what sorts of changes applied to indirectly interacting species might cause an observed change in the focal species.

Because all of the results discussed to this point are based on very simple models, readers are no doubt wondering whether there is any reason to believe that the somewhat counterintuitive predictions discussed here have been, or are likely to ever be, observed. Unfortunately, the evidence bearing on this question is circumstantial, but it is at least strong enough to justify serious research. First, there is a rapidly accumulating body of literature showing that changes in adaptively flexible characteristics of organisms have major consequences for their interactions with other species (Peacor and Werner 2000; recent reviews by Lima 1998, Bolker et al. 2003, Werner and Peacor 2003; see also chapters in this volume). Second, there is an abundance of evidence for responses to manipulations of the top or bottom of food webs or chains that does not fit the traditional predictions. Leibold et al. (1997) carried out a large-scale review, focusing on the correlations between the responses of adjacent trophic levels to manipulations at the top or bottom. Every possible set of pairs of signs of responses was observed in a significant number of studies. A large number of the studies reviewed showed the abundances of herbivore and algal abundances changing in the same direction in response to a fish manipulation, something that should not happen in stable systems that lack adaptive foraging or defense.

Leibold et al. (1997) argue that the responses that are most like the traditional predictions come from studies in which species replacement does not occur. Responses that did not fit the traditional predictions most often came from studies in which species replacement could occur. Because species replacement represents adaptive change of the trophic level, one would expect a wider range of outcomes in such cases. It would be interesting to determine whether exceptions to the Oksanen et al. (1981) predictions were more frequent in systems in which the mid-level species had a greater capacity for adaptive change in foraging traits. Whether some of the counterintuitive predictions of models with adaptive processes are observable in natural systems is, in part, dependent on the magnitude of the shifts in the traits. If foraging or defensive traits change little in response to changes in the abundances of food or enemies, large indirect effects will not be produced. This type of insensitivity of the trait is associated with strong stabilizing selection on the trait (if it is genetically determined) or a strongly concave down relationship between immediate fitness and the trait (if it is behaviorally determined). The calculus-based approach used here determines changes in the variables in response to a small change in a mortality rate. A sufficiently large increase in mortality will often produce a qualitatively different effect. In particular, a large enough mortality will eliminate the species experiencing that mortality. Krivan and Schmitz (2003) have shown that the food web structure itself is likely to change with enrichment when species behave adaptively in some simple three-level models. Thus, the direction of response of a species or level to higher mortality at its own or other levels will often depend on the magnitude of the increase.

It is certainly an open question to what extent the predictions from simple food chain models can be applied to large food webs or ecosystems with many species per level.

However, it is known that many of the consequences of species replacement are identical to those of adaptation within a species (Abrams and Matsuda 1997, Abrams and Vos 2003). The impacts of such factors as stage-structured populations with ontogenetic diet shifts, species feeding on more than one trophic level, or choice behaviors such as predator switching, remain largely unexplored (but see Krivan and Schmitz 2003). However, there is little doubt that adaptive foraging and defense will play an equally important role in determining the indirect effects between species that are parts of larger food webs with a larger range of adaptive traits (Abrams et al. 1996).

One of the key features of the models analyzed here is the important role that direct density dependence ("interference") plays in determining the sign of indirect effects. In the three-species chain described by equations (5), direct density dependence in the consumer reduces the strength of indirect effects transmitted by changes in its population density, while not affecting those transmitted by changes in traits. As a result, seven of the nine responses of abundance of a trophic level to mortality at one level can have either sign. In the absence of direct density dependence, only four of the nine responses have indeterminate signs (Abrams and Vos 2003). This finding adds to the growing list of reasons it is important to determine the strength of such interference terms for species in natural communities. Among the myriad of studies on density dependence performed over the years, there are at best a handful that separate the cause into direct and indirect (food-web-mediated) effects (Ganter 1984 is one notable study that does so). More theoretical work is needed, as well. The work reviewed here has considered only interference affecting mortality, but it is likely to affect food acquisition and conversion, as well. If behavioral or evolutionary changes in foraging and defensive traits prove to be common and large in magnitude, assessing their impact in a food web context will require considerably more knowledge about direct density dependence than we now have.

Another important quantity for determining the responses of food-level abundances to mortality is the relationship between food intake and per capita growth rate. If this relationship increases at a sufficiently decelerating rate (i.e., if $b' + CRb''$ is negative), then the consumer in question is expected to reduce costly foraging in response to greater food abundance. A number of studies (reviewed in Abrams 1991, Werner and Anholt 1993) have demonstrated such responses. In such cases, a variety of counterintuitive food chain responses are possible, such as predators' increasing in response to increased mortality of resources, two levels lower in the food chain. It would certainly be possible to get some information about the shape of the consumer's numerical response to increased food availability (which determines $b' + CRb''$) in many more systems, either by observing short-term responses of foraging to increased food, or by measuring fitness at different food intake rates.

In laboratory systems, it should be possible to directly explore the impacts of mortalities on abundances of different trophic levels. Some microbes have been shown to have phenotypic plasticity in morphology and palatability (Matz et al. 2002). Bohannan and Lenski's (1999) study illustrates that three-level microbial systems can be used to explore some predictions of food web theory. Planktonic systems with rotifers or daphnids at the herbivore level represent other possible laboratory systems where adaptive defenses are known to occur (Persson et al. 2001, Shertzer et al. 2002, Vos et al. 2002).

Most animals act as both predators and prey at different points in time, and these two roles are frequently coupled by trade-offs that relate vulnerability to their predators to the rate at which they themselves exploit prey. The full implications of these trade-offs for the properties of food webs and ecosystems are still not completely understood. There is little doubt that the trade-offs can alter the responses of other trophic levels to factors that affect a particular level. Thus, more theory and empirical work will be needed if we are to predict, understand, or control the ecosystem changes that are going on around us.

Appendix: Stability Analysis and Formulas for Responses of Abundances to Mortality Rates in Two-Level Systems

The local stability of an equilibrium of equations (1) where both species have positive densities can be analyzed by determining the Jacobian matrix and applying the standard Routh-Hurwitz stability criteria (e.g., Yodzis 1989). For equations (1) this yields the conditions

$$- \delta'(f' - C^2Ng') + C^2gb'(g + CRg') > 0$$

and

$$f' - C^2Ng' - \delta' < 0 \tag{A-1a,b}$$

where primes denote derivatives of the functions with respect to their arguments.

The effect of a parameter, such as d_i, on the equilibrium densities can be found by (a) setting the left-hand sides of equations (1) to zero; (b) differentiating the resulting equations with respect to the parameter, taking into account that the population densities, R and N, are now implicit functions of the parameter; and (c) solving the resulting linear equations for the derivative of equilibrium densities with respect to the parameter. In this case,

$$\frac{\partial N_{eq}}{\partial d_2} = \frac{f' - C^2Ng'}{C^2gb'(g + CRg') - \delta'(f' - C^2 Ng')} \tag{A-2}$$

To compare formulas (A-1) and (A-2), one must recognize that the quantity, $g + CRg'$, is the slope of the functional response with respect to resource density, which must be positive. Similarly, the definitions of the symbols imply that $f' < 0$, $g' < 0$, $b' > 0$, and $\delta' \geq 0$. If $\delta' = 0$, which must be true if there is no interference, the condition for N to increase with d_2 is seen to be identical to the condition for local stability. However, if $\delta' > 0$, it becomes possible for the equilibrium N to increase with d_2 in a stable system. This condition requires that the slope of the interference function, δ', be large enough that (A-1b) is satisfied, but not so large that the denominator of (A-2), which is the same as (A-1a), becomes negative.

The analysis of two-level systems with adaptive change follows the same steps. In the case of equations (2), the expression for the change in the consumer population with d_2 is

$$\frac{\partial N_{eq}}{\partial d_2} = \frac{f'(R^2b'' - D_2'') + N(b' + CRb'')}{(b'C^2 - \delta'f')(R^2b'' - D_2'') - \delta'N(b' + C^2CRb'')} \tag{A-3a}$$

Similarly, the change in the resource population with d_2 is

$$\frac{\partial R_{eq}}{\partial d_2} = \frac{C(R^2b'' - D_2'')}{(b'C^2 - \delta'f')(R^2b'' - D_2'') - \delta'N(b' + C^2CRb'')} \tag{A-3b}$$

The denominators of both of these expressions must be positive for the dynamics system, equations (2), to be stable. Fitness maximization at equilibrium guarantees that the numerator of (A-3b) is positive; the sign of (A-3a) is discussed in the text.

Acknowledgments

The work described here was supported by a grant from the Natural Sciences and Engineering Research Council of Canada and a fellowship from the J. S. Guggenheim Foundation. Thanks to M. Vos and H. Matsuda for their roles in some of the work described here. Thanks also to R. Holt, O. Schmitz, P. Barbosa, and two anonymous reviewers for their suggestions on improving the previous version of this chapter.

Literature Cited

Abrams, P. A. 1991. Life history and the relationship between food availability and foraging effort. Ecology 72:1242–1252.

Abrams, P. A. 1992a. Adaptive foraging by predators as a cause of predator-prey cycles. Evol. Ecol. 6:56–72.

Abrams, P. A. 1992b. Predators that benefit prey and prey that harm predators: Unusual effects of interacting foraging adaptations. Am. Nat. 140:573–600.

Abrams, P. A. 1993. Effect of increased productivity on the abundance of trophic levels. Am. Nat. 141:351–371.

Abrams, P. A. 1995. Implications of dynamically variable traits for identifying, classifying, and measuring direct and indirect effects in ecological communities. Am. Nat. 146:112–134.

Abrams, P. A. 1999. Is predator-mediated coexistence possible in unstable systems? Ecology 80:608–621.

Abrams, P. A. 2002. Will declining population sizes warn us of impending extinctions? Am. Nat. 160:293–305.

Abrams, P. A. 2003. The dynamics of adaptive change in traits affecting both resource acquisition and predator vulnerability. Evol. Ecol. Res. 5:553–570.

Abrams, P. A., and Chen, X. 2002. The evolution of traits affecting resource acquisition and predator vulnerability; character displacement under real and apparent competition. Am. Nat. 160:692–704.

Abrams, P. A., and Matsuda, H. 1997. Prey evolution as a cause of predator-prey cycles. Evolution 51:1740–1748.

Abrams, P. A., Matsuda, H., and Harada, Y. 1993. Evolutionarily unstable fitness maxima and stable fitness minima in the evolution of continuous traits. Evol. Ecol. 7:465–487.

Abrams, P. A., Menge, B. A., Mittelbach, G. G., Spiller, D., and Yodzis, P. 1996. The role of indirect effects in food webs. In: Food Webs: Integration of Patterns and Dynamics (Polis, G. A., and Winemiller, K. O., eds.). New York: Chapman & Hall; 371–395.

Abrams, P. A., Namba, T., Mimura, M., and Roth, J. D. 1997. Comment on Abrams and Roth: the relationship between productivity and population densities in cycling predator-prey systems. Evol. Ecol. 11:371–373.

Abrams, P. A., and Roth, J. D. 1994. The effects of enrichment on three-species food chain with nonlinear functional responses. Ecology 75:1118–1130.

Abrams, P. A., and Vos, M. 2003. Adaptation, density dependence, and the responses of trophic level abundances to mortality. Evol. Ecol. Res. 5:1113–1132.

Bender, E. A., Case, T. J., and Gilpin, M. E. 1984. Perturbation experiments in community ecology: theory and practice. Ecology 65:1–13.

Benndorf, J., Böing, W., Koop, J., and Neubauer, I. 2002. Top-down control of phytoplankton: the role of time scale, lake depth and trophic state. Fresh. Biol. 47:2282–2295.

Bohannan, B. J. M., and Lenski, R. E. 1999. Effect of prey heterogeneity on the response of a model food chain to resource enrichment. Am. Nat. 153:73–82.

Bolker, B. M., Holyoak, M., Krivan, V., Rowe, L., and Schmitz, O. J. 2003. Connecting theoretical and empirical studies of trait-mediated interactions. Ecology 84:1101–1115.

Brett, M. T., and Goldman, C. R. 1996. A meta-analysis of the freshwater trophic cascade. Proc. Natl. Acad. Sci. U.S.A. 93:7723–7726.

Carpenter, S. R., and Kitchell, J. F. 1993. The Trophic Cascade in Lakes. Cambridge: Cambridge University Press.

Chase, J. M., Leibold, M. A., Downing, A. L., and Shurin, J. B. 2000. The effects of productivity, herbivory and plant species turnover in grassland food webs. Ecology 81:2485–2497.

DeAngelis, D. L., Persson, L., and Rosemond, A. D. 1996. Interaction of productivity and consumption. In: Food Webs: Integration of Patterns and Dynamics (Polis, G. A., and Winemiller, K. O., eds.). New York: Chapman & Hall; 109–112.

Dyer, L. A., and Letourneau, D. K. 1999. Trophic cascades in a complex terrestrial community. Proc. Natl. Acad. Sci. U.S.A. 96:5072–5076.

Fretwell, S. D. 1977. The regulation of plant communities by food chains exploiting them. Persp. Biol. Med. 20:169–185.

Ganter, P. F. 1984. Effects of crowding on terrestrial isopods. Ecology 65:438–445.

Geritz, S. A. H., Kisdi, E., Meszéna, G., and Metz, J. A. J. 1998. Evolutionarily singular strategies and the adaptive growth and branching of the evolutionary tree. Evol. Ecol. 12:35–57.

Hairston, N. G., Smith, F. E., and Slobodkin, L. B. 1960. Community structure, population control and competition. Am. Nat. 44:421–425.

Halaj, J., and Wise, D. H. 2001. Terrestrial trophic cascades: how much do they trickle? Am. Nat. 157:262–281.

Hastings, A., and Powell, T. 1991. Chaos in a three-species food chain. Ecology 72:896–903.

Iwasa, Y., Pomiankowski, A., and Nee, S. 1991. The evolution of costly mate preferences. II. The "handicap" principle. Evolution 45:1431–1442.

Krivan, V., and Schmitz, O. J. 2003. Adaptive foraging and food web topology. Evol. Ecol. Res. 5:623–652.

Leibold, M. A. 1989. Resource edibility and the effects of predators and productivity on the outcome of trophic interactions. Am. Nat. 134:922–949.

Leibold, M. A. 1996. A graphical model of keystone predators in food webs: trophic regulation of abundance, incidence and diversity patterns in communities. Am. Nat. 147:784–812.

Leibold, M. A., Chase, J. M., Shurin, J. B., and Downing, A. L. 1997. Species turnover and the regulation of trophic structure. Annu. Rev. Ecol. Syst. 28:467–494.

Lima, S. L. 1998. Stress and decision making under the risk of predation: recent developments from behavioral, reproductive, and ecological perspectives. Adv. Stud. Behav. 27:215–290.

Loeuille, N., and Loreau, M. 2004. Nutrient enrichment and food chains: can evolution buffer top-down control? Theor. Pop. Biol. 65:285–298.

Ma, B. O., Abrams, P. A., and Brassil, C. E. 2003. Dynamic vs. instantaneous models of diet choice. Am. Nat. 162:668–684.

Matsuda, H., and Abrams, P. A. 1994. Timid consumers: self-extinction due to adaptive change in foraging and anti-predator effort. Theor. Pop. Biol. 45:76–91.

Matsuda, H., and Abrams, P. A. 2004. Effects of predator-prey interactions and adaptive change on sustainable yield. Can. J. Fish. Aquat. Sci. 61:175–184.

Matz, C., Deines, P., and Jürgens, K. 2002. Phenotypic variation in *Pseudomonas sp. CM10* determines microcolony formation and survival under protozoan grazing. FEMS Microb. Ecol. 39:57–65.

McCann, K., and Yodzis, P. 1994. Nonlinear dynamics and disappearing populations. Am. Nat. 144:873–879.

Menge, B. A. 2000. Top-down and bottom-up community regulation in marine rocky intertidal habitats. J. Exp. Mar. Biol. Ecol. 250:257–289.

Moran, M. D., and Scheidler, A. R. 2002. Effects of nutrients and predators on an old-field food chain: interactions of top-down and bottom-up processes. Oikos 98:116–124.

Oksanen, L., Fretwell, S. D., Arruda, J., and Niemela, P. 1981. Exploitation ecosystems in gradients of primary productivity. Am. Nat. 118:240–261.

Peacor, S. D., and Werner, E. E. 2000. The contribution of trait-mediated indirect effects to the net effects of a predator. Proc. Natl. Acad. Sci. U.S.A. 98:3904–3908.

Persson, A., Hansson, L.-A., Brönmark, C., Lundberg, P., Pettersson, L. B., Greenberg, L., Nilsson, P. A., Nyström, P., Romare, P., and Tranvik, L. 2001. Effects of enrichment on simple aquatic food webs. Am. Nat. 157:654–669.

Persson, L. 1999. Trophic cascades: abiding heterogeneity and the trophic level concept at the end of the road. Oikos 85:385–397.

Puccia, C., and Levins, R. 1986. The Qualitative Analysis of Dynamical Systems. Cambridge, Mass.: Harvard University Press.

Rosenzweig, M. L., and MacArthur, R. H. 1963. Graphical representation and stability conditions of predator-prey interactions. Am. Nat. 97:209–223.

Shertzer, K. W., Ellner, S. P., Fussmann, G. F., and Hairston, N. G., Jr. 2002. Predator-prey cycles in an aquatic microcosm: testing hypotheses of mechanism. J. Anim. Ecol. 71:802–815.

Shurin, J. B., Borer, E. T., Seabloom, E. W., Anderson, K., Blanchette, C. A., Broitman, B., Cooper, S. D., and Halpern, B. S. 2002. A cross-ecosystem comparison of the strength of tropic cascades. Ecol. Lett. 5:785–791.

Taylor, P., and Day, T. 1997. Evolutionary stability under the replicator and the gradient dynamics. Evol. Ecol. 11:579–590.

Vos, M., Flik, B. J. G. Vijverberg, J., Ringelberg, J., and Mooij, W. M. 2002. From inducible defenses to population dynamics: modeling refuge use and life history changes in *Daphnia*. Oikos 99:386–396.

Werner, E. E., and Anholt, B. R. 1993. Ecological consequences of the trade-off between growth and mortality rates mediated by foraging activity. Am. Nat. 142:242–272.

Werner, E. E., and Peacor, S. D. 2003. A review of trait-mediated indirect interactions in ecological communities. Ecology 84:1083–1100.

Yodzis, P. 1988. The indeterminacy of ecological interactions. Ecology 69:508–515.

PART IV

APPLIED CONSEQUENCES OF PREDATOR-PREY INTERACTIONS

The study of the ecology of predator-prey interactions has served as the basis for development of strategies for the management of resources, habitats, and various taxa. Although the theoretical advances in our understanding of the ecology of predator-prey interactions is fundamental to applied concerns, much of the theory does not necessarily translate directly to managed or disturbed habitats that may be shaped and influenced by distinct ecological features. The chapters in this section provide some fundamental exploration into concepts that may have a significant impact on applied subdisciplines and that explore the relevance of theory to practice.

In chapter 14, John Stireman, Lee Dyer, and Robert Matlock review some hotly debated aspects of research on the role of top-down and bottom-up forces, particularly as they relate to differences in predator-prey interactions in managed and unmanaged habitats. The authors use a meta-analysis of existing studies, as well as new data, to address the general question, are top-down forces significantly different in managed and unmanaged habitats? They conclude that both their meta-analysis of recent literature and new data indicate that top-down effects from predators do not differ substantially between managed and unmanaged systems. This lack of difference, despite significant differences in diversity and community complexity, suggests to them that the same features and processes that serve to buffer the strength of top-down effects on herbivores may be present and important in both types of systems, or are insufficient to cause a marked difference in top-down control. Nevertheless, they suggest some critical research that may further identify key differences in managed and unmanaged habitats, and testing for the effects of predators on entire communities.

Differences in diversity between agricultural habitats and unmanaged habitats are also central to discussions in chapter 15 by William Snyder, Gary Chang, and Renée Prasad. They ask whether plant diversity within agroecosystems increases predator diversity and thus improves biological control. Surprisingly, they find little evidence that higher predator density, following plant diversification, leads to improved biological control. Snyder and colleagues argue that increased plant diversity is useful only to the extent that plants are added that supply specific resources that otherwise limit predator densities. They also explore a basic tenet of conservation biological control: that the effectiveness of biological control will increase as predator species diversity increases. Overall, their analysis of available evidence leads them to conclude that the conservation biological control practitioner's positive expectations of the importance of biodiversity may be misplaced. Another often discussed approach to conservation biological control also is evaluated; that is, the prediction that adding alternative prey will lead to higher predator densities and more attacks on pests. They contend that experimental evidence suggests that biological control may not be improved when this tactic is employed, because predators may be satiated by feeding on the abundant alternate prey rather than on the target pest species. A consistent theme, regardless of whether the focus is on how plant or prey diversity (as well as prey abundance and habitat complexity) improve predator performance, is that the key to success is whether the changes made are carefully designed to provide key resources that potentially limit predators.

In contrast to the focus on the structure of the habitat of natural enemies, Pedro Barbosa, Astrid Caldas, and Susan Riechert focus in chapter 16 on the structure of natural enemy assemblages, specifically on their species abundance distribution. Predator assemblages do not comprise equally abundant species but, typically, comprise a few numerically dominant species, whereas most other predators occur in very low abundance, often as singletons. The outcomes of interactions (such as predator-prey interactions) can be affected not only by the type and number of species involved, but also by the relative abundance of each species (i.e., how abundance is distributed across the species in the assemblage). Thus, an understanding of species abundance distributions and the factors that alter those distributions is critical. Barbosa and colleagues suggest that understanding differences in species abundance distribution may help determine the potential effectiveness of natural enemy (e.g., predator) assemblages. The authors provide a new approach (Robbins's curves) for the depiction and statistical comparison of species abundance distributions.

Current conservation biological control tactics target all species in an assemblage. Indeed, a hypothesis such as the species assemblage control hypothesis suggests that targeting the entire assemblage rather than single prominent species is the better approach for effective control of pests. The authors suggest ways in which their new approach might be used to assess hypotheses such as the species assemblage control hypothesis and others in conservation biological control. In addition, they suggest that comparisons of species abundance distributions may be useful in the evaluation of current conservation biological control tactics and strategies.

Predator-prey interactions also are pivotal to other applied disciplines, not the least of which is conservation biology. John Gittleman and Matthew Gompper argue in chapter 17 that the understanding of these interactions and conserving and protecting predators also provide insights that make conservation biology and ecology of predators important models of conservation. After describing global patterns depicting where mammalian predators are threatened, they then highlight factors that influence extinction risk, such as body size, life histories, and trophic level. The consequences of extinction become very vivid when the authors provide interesting illustrations of the role of predators in structuring faunal and floral communities, and thus ecosystem-level changes.

Gittleman and Gompper also ask whether most carnivore global distributions coincide with the distributions of their prey. They conclude that, although predators are obviously reliant on prey and affected by the extent of their distribution, these factors are only partially responsible for extinction risk in predators. Thus, the authors turn to a discussion of the biological correlates of extinction that might help develop a model that predicts why some species are more vulnerable to extinction than others.

The chapters in this part of the book focus on the importance of the ecology of predator-prey interactions to insect pest management, conservation biological control, and conservation biology. Nevertheless, the approach taken in these chapters and the insights presented in them can guide other applied concerns. Further, the analyses undertaken could also be undertaken with regard to other applied issues, such as invasive species, restoration ecology, and the like.

14

Top-Down Forces in Managed versus Unmanaged Habitats

JOHN STIREMAN
LEE A. DYER
ROBERT MATLOCK

The recent surge of terrestrial studies on plant-herbivore-enemy interactions has increased our understanding of top-down forces in complex ecosystems. Concurrently, research in pest management has continued to document the important role of natural enemies in intensively managed terrestrial ecosystems. Although there is considerable theoretical and empirical overlap between basic research on interactions between plant insects and natural enemies and applied research on biological control, it is not yet clear whether similar conclusions can be drawn from studies of top-down trophic interactions in both managed and unmanaged systems. For example, some authors of trophic web studies have argued that top-down cascades will occur only in simpler systems, such as managed monocultures. Enemy and prey physiology, behavior, and population ecology can also vary dramatically between managed and unmanaged systems, creating differing roles for top-down forces. In this chapter, we use a meta-analysis of existing studies, as well as new data, to address the general question, are top-down forces significantly different in managed versus unmanaged habitats? We also summarize other reviews and meta-analyses on top-down forces in terrestrial systems and, as a case study, provide a brief comparison of predation on caterpillars in banana plantations with predation on caterpillars in a tropical forest. We conclude that there are no differences in the strength of predation in natural versus managed systems, despite the obvious physiological, behavioral, population, community, and ecosystem differences. We end this chapter with recommendations for future research on predation along a gradient from the simplest agricultural to the most complex natural communities.

The regulation of prey populations by natural enemies and the indirect effects on other trophic levels are of central interest to community ecologists, population ecologists, wildlife managers, and applied scientists in agriculture and forestry. In terrestrial systems, interest has focused on natural enemies that prey directly on herbivores, thus functioning as plant mutualists (Price et al. 1980). Theoretical and empirical studies in managed agroecosystems have attempted to elucidate criteria for top-down control of herbivore populations by predators. This research has uncovered strong empirical patterns and generated theoretical predictions, but their applicability to "natural" or unmanaged communities is not well understood. Although research on top-down trophic effects in unmanaged terrestrial ecosystems has lagged behind that in managed systems, recent syntheses of the empirical literature (e.g., Schmitz et al. 2000, Halaj and Wise 2001) have begun to establish some general patterns and relationships. Empirical and theoretical research on each of these communities (agricultural and natural) promises to inform the other, but at this point we lack a clear understanding of whether top-down forces differ fundamentally in these systems and, if so, why.

The goal of this chapter is to compare predator-induced top-down trophic forces on prey in managed (agricultural) and unmanaged ("natural") terrestrial communities, assess whether they appear to differ significantly according to community type, and explore the underlying causes for their similarity or dissimilarity. First, we provide an overview of some of the underlying assumptions and ideas concerning trophic interactions and top-down trophic regulation of communities. Next we explore predictions of how and why top-down trophic forces may differ between managed and unmanaged communities, and discuss features of organisms and communities that may shape trophic effects of predators. We test these predictions with a meta-analysis of recent literature addressing top-down effects and an illustrative case study comparing levels of predation in a tropical forest with those in neighboring banana plantations. We conclude with a discussion and synthesis of the managed-unmanaged comparison and an outline of promising future research directions.

We restrict our focus to terrestrial systems. Our meta-analysis and discussion is dominated by arthropod herbivores and predators because of the nature of our comparison (i.e., managed vs. unmanaged communities), the present biases in the experimental literature (e.g., Halaj and Wise 2001), and our own research experiences. However, we address systems involving other organisms and argue that most of the relationships we discuss are independent of the taxa involved. We include parasitoids in our discussion of top-down effects, because they are essentially specialized predators. In general, we focus on the top-down effects of natural enemies on herbivores and do not attempt to assess the relative contribution of bottom-up forces. The goal of this review is not to determine whether communities are structured more by top-down as opposed to bottom-up forces, but to assess how top-down forces due to predators vary across different types of communities.

Top-Down Trophic Control

Predators are ubiquitous throughout the world's ecosystems. The trait that unites these organisms, by definition, is the killing and consumption of other living animals, often (but not exclusively) consumers of primary producers. When predators are capable of regulating prey populations, then they may indirectly influence both the composition and biomass of plant communities by releasing them from herbivory. This simple relationship forms the basis of the *green world hypothesis*, originally formulated by Hairston, Smith, and Slobodkin in their influential 1960 study. Hairston and colleagues argued that terrestrial primary producers are not generally constrained by herbivores, because of top-down control of these herbivores by predators and other natural enemies. Such indirect effects of species in one trophic level on nonadjacent trophic levels have commonly been termed *trophic cascades* (Paine 1980, Carpenter et al. 1985). In other words, the effects of predators on herbivores cascade down indirectly to plants, releasing them from herbivore control. This basic idea has been extended to the *ecosystem exploitation hypothesis* (EEH), which considers the impact of primary productivity and generalizes the hypothesis to low- (few trophic level) and high-productivity (many trophic level) systems, respectively (Fretwell 1977, Fretwell 1987, Oksanen et al. 1981, Oksanen 1991, Hairston and Hairston 1993).

The Hairston et al. and EEH models of multitrophic interactions, which rely on strong top-down interactions, have been criticized by some ecologists who have argued that donor (i.e., bottom-up) control is usually a much more important driving force in community dynamics than predation (e.g., Strong 1992). Ecologists have also argued that most food webs are far too complex for simple chain-like trophic cascades to be very important in determining population dynamics and community structure (Strong 1992, Polis and Strong 1996). Omnivory, intraguild predation, interference competition, spatial heterogeneity, prey refugia, and other factors that should buffer ecological systems from strong top-down effects of predators are found in most terrestrial and aquatic communities (Strong 1992, Polis and Strong 1996, Polis et al. 2000). Both Strong (1992) and Polis (1999) have argued that trophic cascades should be expected only in systems characterized by low within-trophic-level diversity, simple food webs, discrete habitats, and little spatial heterogeneity. In this view, predation can be important in diverse communities for particular imbedded food chains, but trophic cascades are not predicted to be important for an entire complex community. But how simple must a community be to experience strong top-down effects, and can we assume that managed communities are necessarily simpler than natural communities? The incidence of top-down trophic control in managed and unmanaged terrestrial communities and its importance relative to bottom-up effects of plant productivity and defense has become one of the most controversial topics in modern ecology.

Top-Down Control in Terrestrial Systems

There is now a general consensus that predator-driven top-down trophic cascades are commonplace in aquatic ecosystems, including lakes (Carpenter and Kitchell 1988, Persson et al. 1992, Brett and Goldman 1996), rivers and streams (Power 1990, Huryn 1998), and marine communities (Paine and Vadas 1969, Estes and Palmisano 1974, Paine 1980, Shiomoto et al. 1997, Shurin et al. 2002). Consumers tend to monopolize much more of the total productivity in aquatic systems than on land, sometimes resulting in inverted biomass pyramids. Several features of aquatic systems have been proposed to explain the relative prevalence of trophic cascades (Polis 1999). These features include less-developed plant defenses, more rapid cycling and redistribution of nutrients, more rapid life cycles, faster rates of consumption, and less severe abiotic conditions (Strong 1992, Cyr and Pace 1993, Polis 1999). Because of these differences between aquatic and terrestrial systems, it has been argued that all well-documented examples of trophic cascades are aquatic (Polis 1999).

Despite this claim, over the last decade, evidence for trophic cascades in terrestrial systems has mounted (Spiller and Schoener 1994, Dial and Roughgarden 1995, Floyd 1996, Moran and Hurd 1997, Schmitz et al. 1997, Dyer and Letourneau 1999, Pace et al. 1999, Schmitz et al. 2000, Schmitz 2003). Whether many of these examples represent true trophic cascades (Polis et al. 2000), or whether a more restrictive definition of a trophic cascade should be applied (Strong 1992) is still a hotly debated issue. Polis et al. (2000) argue that the vast majority of these studies have demonstrated only "species-level cascades" (i.e., predators affect only one or a few plant species), rather than "community-level cascades" in which the overall plant biomass and species composition of the system is altered by the presence or absence of natural enemies. If the general predictions outlined by Strong (1992) and Polis (1999) are upheld and trophic cascades are likely to be restricted to systems characterized by low diversity and ladder-like food webs, then trophic cascades should have little influence in complex terrestrial ecosystems. Pace et al. (1999) claim, however, that evidence of strong trophic cascades can be found in a variety of systems, irrespective of complexity and ecosystem type, including highly diverse tropical forests (e.g., Dyer and Letourneau 1999, Terborgh et al. 2001). Most recently, Halaj and Wise (2001) concluded from a meta-analysis of studies on terrestrial enemy-herbivore-plant interactions that, although strong effects of predators on their prey were observed, the indirect effects on the primary trophic level were more often a trophic "trickle" than a cascade. That is, the effects of carnivores attenuate rapidly across trophic levels such that their ultimate influence upon primary producers is relatively weak. These conclusions are paralleled by a recent comparison of top-down effects in aquatic and terrestrial systems (Shurin et al. 2002).

The debate about trophic cascades in terrestrial systems, especially arguments concerning high-diversity versus low-diversity systems and whether a species cascade is a true trophic cascade, is fraught with definition problems (Hunter 2001). The original Hairston et al. (1960) hypothesis does not necessarily rely upon "keystone" predators (sensu Paine 1969), focusing instead on the regulatory role of the entire third trophic level. However, most of the examples of top-down trophic cascades in terrestrial systems cited by reviewers such as Pace et al. (1999), Schmitz et al. (2000), and Halaj and Wise (2001) involve the removal or addition of one or a few predators, which in many cases act in keystone fashion, with effects ramifying down to the plant community. If these systems do not represent "true" community-wide trophic cascades, the fact that manipulations of one small component of the predator community (i.e., a single species) often results in potent effects on plant damage and biomass suggests that removal of all predators in the system would likely result in even more dramatic changes in the plant community, qualifying these effects for "community-wide" status.

Top-Down Control in Agricultural Systems

In agricultural systems there is much less argument over whether top-down trophic regulation and trophic cascades are important, in part because they are thought to conform to many of the criteria that Polis (1999) argues are necessary for trophic cascades. Many agricultural systems are dominated by a single plant species with poor intrinsic chemical defense, habitats tend to be discrete, herbivores are unlikely to be food-limited because of resource concentration and periodic population disruption, and trophic chains may be shorter (Polis et al. 2000, Halaj and Wise 2001). These features and their predicted influence on top-down trophic control are outlined in Table 14.1.

It can be seen that agricultural systems appear to posses many or all of the features that should facilitate strong top-down trophic cascades. Indeed, the relative success of predators and parasitoids introduced into agricultural systems in biological control programs (see examples in Huffaker and Messenger 1976, Greathead 1986, Greathead and Greathead 1992) has been attributed to the simple nature of the agricultural communities, with relatively few species occupying each trophic level (Hawkins et al. 1999). In such systems, there may be relatively few ecological connections and reticulations (e.g., omnivores) to dissipate the top-down effect of the natural enemy on the pest species in question and prevent strong indirect effects on plant biomass or productivity (but see Eubanks, ch. 1 in this volume). In more complex systems, such as most natural ecological communities, it is expected that trophic reticulations and dilution of predation effects across multiple herbivore species, and of herbivore effects across multiple plant species, tend to prevent strong top-down trophic control of communities (Polis and Strong 1996, Hawkins et al. 1999). However, this caricature of managed systems and their inherent simplicity may be realistic in only a very few intensively managed systems. Many agricultural systems in

Table 14.1. Predicted Distinctions between Managed and Unmanaged Ecosystems with Respect to the Likelihood and Prevalence of Top-Down Regulation of Communities

Feature	Managed Ecosystem	Unmanaged Ecosystem
Substrate cleared by normal herbivory?	Sometimes	Rarely
Single or few keystone herbivores?	Yes	Rarely
Single or few keystone carnivores?	Yes	Rarely
Consumption	Runaway; unified	Buffered; differentiated
Species number	Relatively low	Moderate to high
Habitat	Discrete	Continuous with adjacent habitats
Trophic architecture	Trophic ladder: levels fairly distinct	Trophic web: levels indistinct
Spatial heterogenetity and environmental grain	Fairly homogenous; fine-grained	Heterogeneous; coarse-grained

Source: Modified from Strong (1992).

which biological control has been successful, such as forest tree plantations, orchards, and certain crops (e.g., alfalfa), are likely to exhibit degrees of trophic complexity and diversity comparable to many natural systems. Despite the apparent pervasiveness of top-down regulation in these systems, few agroecosystems have been thoroughly studied as entire communities (i.e., crops, weeds, attendant herbivores, and natural enemies) from the perspective of community ecology and food web theory (Letourneau and Andow 1999).

Comparing Managed and Unmanaged Communities

Despite the strongly divergent predictions for the strength of top-down forces and trophic cascades in managed and unmanaged ecosystems, relatively few empirical studies have explicitly compared trophic interactions in these systems to test whether the predictions are upheld. Hawkins et al. (1999) made significant progress in this area in their analyses of life table data and biological control introductions in natural and managed habitats. They explicitly sought to determine whether fundamental differences exist between biological control in agroecosystems and the force of natural enemies in noncultivated communities. They concluded that biological control is not a "natural phenomenon," because it overemphasizes the role of parasitoids (i.e., specialized enemies) and because it results more often from a single strong link in simplified food webs. In contrast, natural control is due to multiple links in a more complex food web. One might consider an agricultural landscape a more or less ecologically blank slate, which is colonized by one or a few herbivores against which a single highly specialized enemy is subsequently introduced. Thus, control will more likely be due to that enemy than to whatever remnant generalist natural enemies persist in the highly disturbed, species-poor habitat (but see, in this volume, Barbosa et al., ch. 16, and Snyder et al., ch. 15).

Hawkins et al. (1999) do not conclude that top-down forces are weaker or less important in natural than in managed systems, despite a final statement by the authors that biological control "exaggerates the extent to which natural enemies exert top-down control of insect populations" (p. 502). Their analysis of life table data documents that parasitoids and generalist predators more often comprise key factors in managed and unmanaged systems, respectively. However, it fails to demonstrate a difference in the frequency of top-down control between managed and unmanaged systems. A recent review of experimental studies of generalist predators by Symondson et al. (2002) provides data that seemingly contradict this dominance of parasitoids in agricultural systems. Symondson and colleagues found that 74% of (102) studies demonstrate significant impacts of single generalist predators on herbivore pest abundance, and 95% of cases that measured effects on plants demonstrated reduced crop damage or increased yield. For assemblages of generalist predators, the percentage of studies showing significant reduction of pest densities increased to 79%. However, it may be argued that generalist predators alone are not likely to reduce pest herbivory below economic thresholds as effectively as parasitoids, because much of their prey may consist of other natural enemies or of herbivores and detritivores that are relatively unimportant economically. Further, it is important to recognize that significant effects of predators on herbivores and plants do not always translate to successful biological control or suppression below an economic threshold.

Hawkins et al.'s (1999) analysis of biological control introductions also showed that classical biological control is more successful in managed than in natural systems (especially for parasitoids). This finding is consistent with the hypothesis that trophic interactions are generally more linear in agricultural systems and that there are fewer or weaker intratrophic-level and higher order trophic-level interactions to disrupt herbivore control. Other reviews of biological control programs, which have revealed that control of herbivores is most often accomplished by only a single, often specialist, enemy (68% of successful biological control programs; Myers et al. 1989, Denoth et al. 2002) support this view. However, the increased success of biological control introductions in managed systems does not necessarily indicate that top-down control is more likely to occur, but only that establishment of a foreign enemy will more likely succeed in agricultural or managed than in natural systems.

There is strong evidence that many invasive herbivores are limited by top-down control in their natural native habitats (Hagen et al. 1971, Debach and Rosen 1992, Mack et al. 2000), which provides the motivation for biological control researchers working on exotic pests in natural systems to search for enemies of herbivores in their native ranges (Debach and Rosen 1992). Thus, although biological control programs of exotic pests may be less likely to succeed in unmanaged habitats, these pests are often controlled in their natural, native habitats. In addition, biological control has been effective against many forest pests (see examples in Dahlsten and Mills 1999). Although production forests are managed, they are typically more complex habitats than agricultural systems and support greater arthropod diversity. Despite greater complexity, however, success rates for establishment and control of forest pests by natural enemies are similar to those for more intensively managed crops (Dahlsten and Mills 1999).

Although it may be equivocal whether trophic control of herbivores differs between natural and managed systems, the predicted prevalence of strong top-down trophic cascades in agricultural systems relative to unmanaged systems was upheld by Halaj and Wise's (2001) recent meta-analysis of trophic cascades in terrestrial communities. These authors found significantly stronger effect sizes for herbivory levels in crop communities than in noncrop communities. Predators significantly increased plant biomass in crop systems, whereas the effect of predators on plant biomass in noncrop systems did not differ significantly from zero. Halaj and Wise (2001) observed this difference between ecosystem types, despite their result that overall effects of enemies on herbivore density did not differ between managed and unmanaged systems.

Predation in Natural and Managed Communities

Meta-Analysis

The studies of Hawkins et al. (1999) and Halaj and Wise (2001) are among the few attempts to achieve consensus regarding the influence of top-down forces in managed and unmanaged systems. Here we expand upon the meta-analysis of Halaj and

Wise (2001) in examining effects of natural enemies on herbivores, focusing on the comparison between managed and unmanaged systems and on the most recent ecological literature. Unlike Halaj and Wise (2001), we do not focus exclusively on arthropods, nor do we focus exclusively on the trophic cascade from first-order predators to plants. We analyze the effects of natural enemies on herbivore abundance, or density, to assess whether the strength of this trophic interaction differs among managed and unmanaged systems. We do not explicitly assess the indirect effects of predators on plants, because our primary objective is to compare the direct top-down force due to predators in managed and unmanaged systems. In addition, we restrict the number of studies used from single articles to avoid effect size biases.

Methods

The meta-analysis included data from January 1985 through November 2003. All studies (1999–2003) in the journals *Ecology, Ecological Monographs, Ecological Applications, Ecology Letters*, and *Basic and Applied Ecology*, and all studies (1985–2003) in the journals *Oecologia, Biotropica*, and *Journal of Tropical Ecology* were examined for quantitative measures of the direct effects of predator manipulations (exclusions or additions) on prey response variables (survival, abundance, density, or species richness). A bibliography of the studies examined, together with the effect sizes from each study can be found at Dyer and Gentry's Web site (2002). studies that were actually included in the meta-analysis were those that contained means, measures of dispersion, and sample sizes, or those for which such measures could be reconstructed. We conducted a mixed-model meta-analysis for agricultural versus natural systems to uncover potential differences in the strength of predation.

Equations in Gurevitch and Hedges (2001) were used to calculate combined effect sizes and 95% confidence intervals across all studies. Means and standard deviations were taken directly from tables or text, were calculated from other statistics, or were extracted from figures (with a ruler). We calculated only one effect size per enemy-prey interaction per essay. If more than one effect size was available for an interaction, we randomly selected a value. In this chapter, we report all effect sizes, together with the range of the 95% confidence intervals (after Gurevitch and Hedges 2001). All other measures of dispersion reported here are ±1 standard error. Any effect sizes greater than 1.0 were considered to be large (Gurevitch and Hedges 2001). We compared the strength of specific trophic interactions in natural and agricultural systems by using the between-class heterogeneity statistic, Q_B, which is distributed approximately as χ^2 (Gurevitch and Hedges 2001). We used the same statistic to compare the effect sizes of generalist and specialist predators.

Results

Our meta-analysis included 66 studies in natural systems and 24 studies in agricultural systems. There were strong effects of predator manipulations on prey in natural and agricultural systems, but the magnitude of the effect was not significantly different for agricultural as compared to natural systems ($Q_B = 1$, df = 1, $p > .1$; Figure

14.1). The effect of specialist predators was very large and significantly greater than that of generalists ($Q_B = 5.1$, df = 1, $p < .05$; Figure 14.1).

For the studies in this analysis there was no relationship between effect size and length of study ($F_{(1, 60)} = 1.1$, $p = .3$, $r^2 = .02$), but the longest study included was only 3 years, and much longer studies are often necessary to detect consistently strong top-down effects (Carson and Root 2000). The importance of longer term (5–20 years) experimental studies of trophic interactions has been stressed by ecologists interested in the strength of top-down forces in ecological systems (e.g., Leibold et al. 1997, Carson and Root 2000, Holt 2000, Schmitz et al. 2000, Hunter 2001), because community-wide effects of manipulations may take longer than the short duration of most experiments to be realized, and because strong short-term effects of manipulations may weaken over time as the community adjusts to predator absence or presence (e.g., Sinclair et al. 2000). In either case, this "time effect" may bias comparisons between natural and agricultural systems.

Experiments in managed systems are bimodal. Experimental data are often relatively short term, partly because they take place in systems with very frequent turnover and disturbance, whereas biological control studies usually provide relatively long-term data sets. This difference may result in overestimated degrees of top-down control for short-term experiments. In natural systems, experiments are often of longer duration, but in order to observe community-wide effects, if they exist (especially with regard to the plant communities), even longer time scales (comparable to those of biological control) may be required. For example, it could take decades or longer to observe community consequences of insect herbivory due to release from predation in mature temperate or tropical forest (Carson and Root 2000).

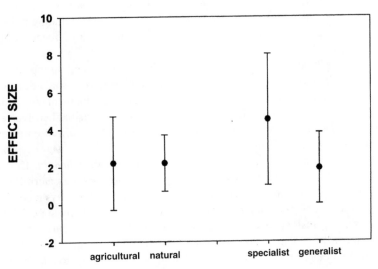

Figure 14.1. Effect sizes and 95% confidence intervals from a meta-analysis on 66 studies of predator-prey interactions in agricultural and 24 in natural systems. The diet breadth of predators is also compared.

An Example from Tropical Forests and Bananas

Banana (*Musa acuminata*) is one of the most important export crops in the Neotropics. During the 1950s through the 1970s, outbreaks by 53 species of lepidopteran defoliators, associated with aerial applications of the insecticides dieldrin and carbaryl, were frequent in Costa Rican banana plantations (Lara 1970, Ostmark 1974, Stephens 1984). In 1973 the aerial insecticide applications ceased, and these Lepidoptera have rarely exceeded their economic thresholds since (Stephens 1984, Thrupp 1990).

Banana is typically grown as a perennial in large monocultural plantations and thus has lower rates of disturbance than most annual crops. Plantations are treated with herbicides (approximately 6 applications per year, typically paraquat and glyphosate), nematicides (2–3 applications per year, carbamates and organophosphates), and fungicides (25–40 applications per year of a variety of compounds, including copper-based inorganics, benzimidazoles, conazoles, triazoles, benzonitriles, carbamates, dithiocarbamates and morpholines) to control weeds, nematodes, and the fungus Sigatoka Negra (*Mycosphaerella fijiensis*), respectively. In addition, plastic bags treated with the organophosphate chlorpyrifos are placed over growing banana stems to control insects that scar the fruit. Here we make comparisons between parasitism and predation rates in banana and in wet tropical forest at the La Selva Biological Station for two of the most important of outbreak Lepidoptera, *Caligo memnon* and *Opsiphanes tamarindi* (both Nymphalidae).

About two-thirds (1611 ha) of the La Selva Biological Station (which is owned and operated by the Organization for Tropical Studies) is composed of primary wet tropical forest, and one-third is composed of regenerating habitats in Cantón Sarapiquí, Heredia Province, Costa Rica. La Selva receives 4200 mm of rain annually, and diurnal temperatures vary less than 3°C annually (24.7°C in January to 27.1°C in August). La Selva's forest reserve hosts a diverse flora and fauna, including more than 1,836 vascular plant species (Wilbur 1994) and an estimated 7,000 species of Lepidoptera (Wagner, D., personal communication).

In November 2002, we determined age-specific parasitism and disappearance of caterpillars on *Heliconia* spp. leaves at La Selva and banana leaves at a nearby (approximately 10 km) banana plantation, throughout larval development. At each site, we randomly selected 40 individual plants and removed all visible herbivores. On 20 plants we placed one early instar (first through third) *O. tamarindi* on an easily reached leaf, and on the remaining 20 plants we added eight early instar *C. memnon*. These densities mimic their natural abundances on banana and *Heliconia*. The plants were marked and then surveyed every 6 hours until all surviving caterpillars had reached their final stadium and were near pupation (3 weeks). Surveys recorded the number of individuals remaining on the plant, their stadia, and any signs of predation, disease, or parasitism events. All predation events witnessed during the survey were also recorded. Upon reaching the final stadium, all remaining individuals were brought to an ambient-temperature laboratory and reared in bags to adulthood.

Levels of predation and parasitism were very high and almost identical in the forest and the plantation (Table 14.2). At both sites, predators included spiders, wasps, and ants, and parasitoids were either tachinids or braconids. Bag-reared individuals of *C. memnon* usually had much higher levels of parasitism at La Selva and at the ba-

Table 14.2. Impact of Natural Enemies on Focal Lepidoptera: La Selva versus Banana (Means of Percent Mortality)

Pest Lepidoptera	Natural Enemy	La Selva	Banana
Caligo memnon	Parasitism	18 (8)	5 (4)
	Predation	74 (12)	60 (9)
Opsiphanes tamarindi	Parasitism	33 (12)	30 (15)
	Predation	60 (18)	50 (17)

Note: Standard errors are in parentheses.

nana plantations, but this was not the case for *O. tamarindi* (33% for *O. tamarindi* and 45% for *C. memnon*; Dyer, L. A., and Gentry, G. L., unpublished data). The lower field parasitism experienced by *C. memnon* suggests that parasitized caterpillars are preyed upon preferentially in both the natural and managed systems (as with the predation observed by Snyder and Ives 2001). Even though the plant and arthropod diversity at La Selva is orders of magnitude greater than that at the banana plantations, the enemy-prey interactions and risk of mortality are the same for *C. memnon* and *O. tamarindi*. It is possible that this pattern could be a general trend, that herbivore species will typically be impacted by an average of a couple specialist and a few generalist enemies, no matter how diverse the community where they are found.

Discussion

Both our meta-analysis of recent literature and the data we present from a comparison of caterpillar mortality due to natural enemies in an agricultural system and a natural system indicate that top-down effects from predators (and parasitoids) to herbivores do not differ substantially between managed and unmanaged systems. This finding is in spite of sizable differences in diversity and community complexity. We do not claim that this result is always the case but that as a whole these systems appear to be comparable. It suggests that either the same features and processes that serve to buffer the strength of top-down effects on herbivores are present and important in both types of systems (e.g., intraguild predation), or, conversely, that these supposed buffering forces are insufficient to cause a marked difference in top-down control between community types. Next we explore some of these features and how they may or may not differ between managed and unmanaged communities.

Predators versus Other Enemies

A common argument in the biological control literature is that specialist natural enemies, such as many hymenopteran parasitoids, are more efficient at controlling prey populations than generalist parasitoids or predators (Huffaker et al. 1971, Beddington et al. 1978, Hassell 1978). That the vast majority of successful biological control programs have employed specialized enemies supports this view (Huffaker et al. 1971, Debach 1974, Horn 1988). This difference may be due to shorter food chain lengths in parasitoid-dominated systems (but see Holyoak 2000), in which the parasitoids themselves do not experience strong top-down regulation and the "unified" runaway consumption (sensu Strong 1992) characterizing these interactions.

However, an increasing number of empirical studies have demonstrated that generalist predators are often quite important in limiting herbivore abundances in agricultural systems (Carter and Rypstra 1995, Losey and Denno 1999, Perfecto and Castiñeiras 1998, Eubanks and Denno 2000). They may occasionally be better at suppressing host populations than specialist parasitoids (Symondson et al. 2002). These and other researchers have identified potential advantages of predators in controlling herbivore species: (a) they may aid in controlling several pests concurrently; (b) because of alternative food resources, they may exhibit more stable population dynamics with lower likelihood of local extinction; (c) they may be able to maintain effective population sizes on alternate prey when primary prey are at low densities and potentially prevent irruptions of pest populations; and (d) conservation (and perhaps augmentation) of naturally occurring predators may be easier and more economical than introduction biological control, with much lower likelihood of dramatic nontarget effects. Whether a predator or parasitoid is a superior natural enemy probably depends on the system and, in particular, on characteristics of the pest and particular enemy (Dyer and Gentry 1999, Dyer and Gentry 2002). The greater average effect size that we identified for specialist natural enemies in our meta-analysis tends to support the traditional view of efficient specialists and the perceived importance of parasitoids as compared to predators in managed communities. However, most studies in this analysis were not conducted over the long time periods often required for detecting the considerable benefits of generalist predators, and most focused on the responses of one or a few herbivore species rather than the total herbivore community.

Effects of Enemy Diversity

Much of the discussion of the differences in top-down regulation in managed and unmanaged systems relies on arguments concerning how differences in the diversity of the natural enemy communities influence the magnitude of their effects on the herbivore community. On one hand, it has been argued that simple enemy complexes made up of one or a few species are likely to have the most dramatic effects on prey suppression, because of unified consumption, lack of trophic reticulations, and intraguild interactions (e.g., parasitoids in agriculture; Hochberg 1996, Polis and Strong 1996, Hawkins et al. 1999, Polis 1999, Denoth et al. 2002).

Conversely, a number of researchers have recently argued that a diverse natural enemy fauna may often result in more effective regulation of prey populations and stronger positive effects on primary producers (Riechert and Lawrence 1997, Losey and Denno 1998, Losey and Denno 1999, Cardinale et al. 2003, Sinclair et al. 2003). These arguments often rely on assumptions of *species complementarity*; that is, that enemies occupy ecological niches that are differentiated so that, as species diversity increases, the total amount of niche space occupied also increases, resulting in more efficient regulation of herbivores (or an "increased ecosystem process rate"; Wilby and Thomas 2002; see also Snyder et al., ch. 15). The "cumulative stress" model of plant biological control, in which the effects of enemies are additive and can control a lower trophic level at some threshold is an analogous argument (Myers 1985). Alternatively, increasing the diversity of enemies in a system may follow a "lottery" model, in which larger predator assemblages are more likely to suppress prey or host populations simply because the probability of sampling the "right" enemy is increased (Myers 1985, Myers et al. 1989). Denoth et al.'s (2002) analysis of biological control programs found no relationship between the number of enemy species introduced and successful regulation of arthropod pests, whereas introductions of herbivores to control invasive plants was consistent with a lottery model. These authors confirmed previous conclusions (Myers et al. 1989) that the success of most biological control programs is due to a single enemy. Again, however, as Hawkins et al. (1999) argue, this finding may not reflect the dominant trophic interactions with herbivores in unmanaged systems.

Whatever the relationship between enemy diversity and prey regulation, a general assumption when comparing managed and unmanaged ecosystems is that the former are characterized by relatively low diversity at all trophic levels (and this low diversity facilitates top-down cascades). However, the validity of this assumption varies widely among different types of managed habitats and management practices. Many managed agricultural communities contain a diverse array of both herbivores and predators (Barbosa 1998). This fact is most evident in forage crops such as alfalfa and tree plantations, where management of weeds and herbivores is typically less intensive. For example, Pimentel and Wheeler (1973) recorded 591 species of arthropods in a single experimental study of less than one acre in New York state alfalfa, and approximately 100 species of arthropods are recorded as causing some degree of injury to alfalfa in the United States (Gyrisco 1958, Armbrust and Gyrisco 1982). The violation of this low-diversity assumption suggests that many of the supposed "buffering" processes that limit top-down trophic control (e.g., intraguild predation and dilution of trophic effects) may also operate in agricultural systems. Thus, even if diversity hinders top-down trophic regulation of communities, which is debatable, managed communities exhibit a wide range of diversity and complexity that may overlap considerably with that of unmanaged communities.

Intraguild and Synergistic Predation

Intraguild predation is a common feature of many natural and managed ecological communities (Rosenheim et al. 1993, Spiller and Schoener 1994, Floyd 1996, Moran et al. 1996, Moran and Hurd 1997, Schoener and Spiller 1999, Rosenheim 2001, Snyder and Ives 2001). Several studies have shown that intraguild or higher order predation can disrupt prey regulation and lead to outbreaks of prey populations (e.g., see review of Rosenheim 1998), though in other cases enemies appear to act additively or synergistically (Losey and Denno 1998, Lang 2003).

Studies of lacewings (*Chrysoperla carnea*) and cotton aphids (*Aphis gossypi*) have demonstrated strong intraguild predation by generalist hemipteran predators, in both enclosures and open fields, that has led to disruption of aphid control (Rosenheim et al. 1993, Rosenheim et al. 1999). A similar disruption in the regulation of the pea aphid (*Acyrthosiphon pisum*) was demonstrated in the interaction between an aphidophagous parasitoid wasp and a predaceous carabid beetle (Snyder and Ives 2001). The carabids in this system, primarily *Pterostichis melanaria*, preferentially preyed upon parasitized aphid mummies and interrupted control of the aphids by the parasitoids. However, in a subsequent experiment, in which these authors manipulated the entire assemblage of predators in the system (including carabid beetles), they showed that, despite intense intraguild predation by generalist predators on parasitoids (50% of parasitoid pupae), the impact of the two enemy groups on pea aphid populations was additive (Snyder and Ives 2003). In a similar system, Losey and Denno (1998, 1999) demonstrated not only that multiple predators can contribute to control of an herbivore, but also that they can act synergistically to control prey densities. In their alfalfa system, escape behaviors of aphid prey (dropping from plants) in response to foliar predators increased their susceptibility to ground-dwelling predators.

Interactions among predators in unmanaged systems have also been shown to both facilitate and retard prey regulation. One of the first and best-documented examples of a terrestrial trophic cascade in a nonagricultural system involved a predator (mantids) that preyed on both herbivores and other predators (Moran et al. 1996, Moran and Hurd 1997). However, as in managed systems, other studies in natural systems have demonstrated deleterious effects of intraguild predation on herbivore control, such as that between *Anolis* lizards and spiders on Caribbean islands (Spiller and Schoener 1994, Schoener and Spiller 1996, Schoener and Spiller 1999, Spiller and Schoener 2001,) and that between leafhoppers and their spider predators in salt marsh communities (Finke and Denno 2002, Denno et al., ch. 10 in this volume). It is notable that several of these studies of intraguild predation demonstrate that trophic cascades from predator to plant may still occur despite potentially complex trophic links that serve to dilute the impact of an enemy (Lang 2003). This finding was demonstrated by our own work in Colorado alfalfa fields, where top-down indirect effects were clearly detected, despite arthropod diversity that rivals that of native Colorado grasslands (Dyer and Stireman 2003, and Pearson, C., unpublished data).

Omnivory

The relative frequency and importance of omnivory on plants and animals in managed versus unmanaged systems is unclear, but the realization of its general importance in terrestrial food webs is growing (Coll and Guershon 2002, Eubanks, ch. 1). Theoretically, omnivory could inhibit trophic cascades, because the indirect benefits that these organisms provide to plants, by killing herbivores, are likely to be countered by their direct consumption of plant tissues (Coll and Guershon 2002). However, because plant feeding may buffer them from erratic population fluctuations of prey, it is possible that omnivores may be better able to control prey populations when the prey are rare. This relationship was demonstrated by Eubanks and Denno (2000) in a lima bean agroecosystem involving the omnivore *Geocoris punctipes* and both lepidopteran and aphid herbivores. Recent simulation studies by Rosenheim and Corbett (2003) indicate that the degree to which omnivores may disrupt or facilitate prey regulation may be related to their foraging strategies. They found that sedentary omnivores may be more likely to disrupt control through intraguild predation, whereas wide-ranging omnivores are less likely to disrupt control and more likely to contribute to prey regulation. Although these studies do not directly bear on the issue of top-down control in managed and unmanaged systems, they do argue that complex trophic interactions such as omnivory do not necessarily negate the possibility of top-down regulation.

Nonlethal Effects of Predators on Herbivores

The realization that strong top-down effects of predators on herbivores can occur without actual prey consumption, and that trophic interactions can be mediated through changes in behavior or other traits of predators or prey, promises to expand our views of community structure and food webs (Lima 2002, Bolker et al. 2003, Werner and Peacor 2003; see other chapters, this volume). Much of the study of these interactions has focused on vertebrates (e.g., Brown et al. 1999), but it has become an increasingly popular and exciting area of study for invertebrate ecologists over the past decade or so (Beckerman et al. 1997, Schmitz et al. 1997, Schmitz 1998, Gastreich 1999). For example, in a recent study, Schmitz (2003) demonstrated the existence of a top-down trophic cascade due to spider predators in an old-field plant community. This cascade, however, was not due primarily to direct consumption of the dominant herbivores (grasshoppers), but to a change in herbivore behavior in the presence of spiders, which increased herbivory on a dominant plant species and indirectly increased diversity of the plant community (see Schmitz, ch. 12 in this volume). Top-down effects can also be thwarted by prey defenses. Denno et al. (2003) have shown that variation in defensive behaviors among planthopper species in a salt marsh community can dramatically influence the strength of top-down control by spiders. These recent studies demonstrate the conditional nature of top-down effects and how they may be mediated by behavior. They promise to modify how we view trophic interactions and thus represent fertile areas for future study.

Conclusions and Recommendations for Future Work

The strength of predator effects on communities and propensity for trophic cascades vary widely among both agricultural and natural systems because of differences in many factors, including habitat heterogeneity and stability; plant, herbivore, and enemy diversity; predator and prey characteristics; and the frequency of anthropogenic control measures. Some of these factors, such as enemy diversity, may increase or decrease the likelihood of top-down control (Cardinale et al. 2003, Dyer and Stireman 2003), depending on the types of interactions between enemies. However, the meta-analysis presented here suggests that differentiating between managed and natural systems will not account for much of the variation in the strength of predation. Our results confirm and strengthen the findings of Halaj and Wise (2001) that top-down effects of enemies on herbivores do not differ significantly between natural and managed systems. This finding suggests that, if a difference in the indirect effects of predators on plants exists as Halaj and Wise (2001) report, it is due to the interaction between herbivores and their host plants, and is perhaps related to the relative lack of intrinsic defenses characteristic of many crop plants. We have also demonstrated that prey defenses are similar in natural and managed systems (Dyer and Gentry 1999), top-down diversity cascades can be detected in natural and managed systems (Dyer and Letourneau 2003, Dyer and Stireman 2003), and that two caterpillar species experience very similar predation risks whether they are in diverse tropical forest or a banana monoculture.

Nevertheless, there are obvious differences between unmanaged and agricultural communities (Table 14.1). For a better understanding of how predator effects might differ on a gradient from simple and complex agricultural systems, to intensively managed natural systems, to large and diverse unmanaged systems, there are many aspects of tritrophic interactions that need to be studied in greater detail. Among other factors, trait-mediated indirect effects, prey defenses, and behaviorally mediated interactions need to be understood more fully so that we may predict where, when, and how top-down forces can structure communities of herbivores and plants in all terrestrial systems (see other chapters, this volume).

It seems clear that predator-initiated trophic cascades occur in all systems, no matter what the complexity or the intensity of management. Our meta-analysis supports the assumption that predation can be equally intense in agricultural and in natural systems. However, the criticisms and interpretive modifications of putative cascade, or trickle, examples from terrestrial systems warrant closer inspection of methods and problems in studying predation in diverse communities. In a full-sized, natural community, it is not possible to experimentally remove all predators and parasitoids, nor is it possible to find a terrestrial community from which all predators have been removed. Thus, there are no direct experimental or correlational tests of trophic cascades in terrestrial communities, which is why the evidence is weak and easy to criticize. There are many solutions to the lack of manipulative or mensurative deletion of predaceous trophic levels, and the existing approaches should be integrated in cohesive research programs aimed at testing the effects of predators on entire communities. Examples of approaches to integrate include modeling (Pimm and Lawton 1977, Hassell and May 1986), mesocosm (Dyer and Letourneau 2003), or island (Terborgh et al. 2001) approaches, and extrapolation from "species cascades" (Hunter 2001).

Finally, it is not difficult to see why trophic cascades are thought to occur more frequently in managed agricultural systems than in unmanaged ecosystems. In these floristically simple systems, there are often relatively few dominant herbivores, so any predator or parasitoid that consumes one or more of these few herbivores has the potential to be a "keystone" species. Contrary to Hawkins et al. (1999), one could have exactly the same essential interactions and relationships occurring in agricultural biological control as in natural systems. However, removing a single predator or parasitoid in the natural system may cause only a very minor effect on plants, because of the diversity of plants, herbivores, and carnivores; whereas the effect may be much more pronounced in the agricultural system. Thus, community-wide trophic cascades may occur in both types of systems; they are just more evident in agricultural systems, because whole trophic levels, or major portions of them, are more easily manipulated.

Acknowledgments

We thank P. Barbosa for inviting us to write this chapter and for his comments on a previous version. We also thank three additional anonymous reviewers for their comments on a previous version. During the writing, data collection, and analysis for this chapter, L.A.D. was supported by grants from the National Science Foundation (DEB 0344250), the U.S. Department of Agriculture (2002-02761), and National Institute for Global Environmental Change; R.M. was supported by a grant from the U.S. Department of Agriculture (2003-03454) during the writing and data analysis. The chemical ecology and tropical entomology lab group provided very useful comments on earlier versions of this chapter. We thank C. Pearson for sharing his data with us from Colorado grasslands and alfalfa.

Literature Cited

Armbrust, E. J., and Gyrisco, G. G. 1982. Forage crops insect pest management. In: Introduction to Insect Pest Management, 2nd ed. (Metcalf, R. L., and Luckmann, W. H., eds.). New York: Wiley-Interscience; 443–463.

Barbosa, P. 1998. Agroecosystems and conservation biological control. In: Conservation Biological Control (Barbosa, P., ed.). San Diego: Academic Press; 39–54.

Beckerman, A. P., Uriarte, M., and Schmitz, O. J. 1997. Experimental evidence for a behavior-mediated trophic cascade in a terrestrial food chain. Proc. Natl. Acad. Sci. U.S.A. 94:10735–10738.

Beddington, J. R., Free, C. A., and Lawton, J. H. 1978. Characteristics of successful natural enemies in models of biological control of insect pests. Nature 273:513–519.

Bolker, B., Holyoak, M., Krivan, V., Rowe, L., and Schmitz, O. J. 2003. Connecting theoretical and empirical studies of trait-mediated interactions. Ecology 84:1101–1114.

Brett, M. T., and Goldman, C. R. 1996. A meta-analysis of the freshwater trophic cascade. Proc. Natl. Acad. Sci. U.S.A. 93:7723–7726.

Brown, J. S., Laundre, J. W., and Gurung, M. 1999. The ecology of fear: optimal foraging, game theory, and trophic interactions. J. Mammal. 80:385–399.

Cardinale, B. J., Harvey, C. T., Gross, K., and Ives, A. R. 2003. Biodiversity and biocontrol: emergent impacts of a multi-enemy assemblage on pest suppression and crop yield in an agroecosystem. Ecol. Lett. 6:857–865.

Carpenter, S. R., and Kitchell, J. F. 1988. Consumer control of lake productivity. BioScience 38:764–769.

Carpenter, S. R., Kitchell, J. F., and Hodgson, J. R. 1985. Cascading trophic interactions and lake productivity. BioScience 35:634–649.

Carson, W. P., and Root, R. B. 2000. Herbivory and plant species coexistence: community regulation by an outbreaking phytophagous insect. Ecol. Mon. 70:73–99.

Carter, P. E., and Rypstra, A. L. 1995. Top-down effects in soybean agroecosystems: spider density affects herbivore damage. Oikos 72:433–439.

Coll, M., and Guershon, M. 2002. Omnivory in terrestrial arthropods: mixing plant and prey diets. Annu. Rev. Entomol. 47:267–297.

Cyr, H., and Pace, M. L. 1993. Magnitude and patterns of herbivory in aquatic and terrestrial ecosystems. Nature 361:148–150.

Dahlsten, D. L., and Mills, S. J. 1999. Biological control of forest insects. In: Handbook of Biological Control (Bellows, T. S., and Fisher, T. W., eds.). New York: Academic Press; 761–788.

Debach, P. 1974. Biological Control by Natural Enemies. Cambridge: Cambridge University Press.

Debach, P., and Rosen, D. 1992. Biological Control of Insect Pests and Weeds. New York: Reinhold.

Denno, R. F., Gratton, C., Dobel, H., and Finke, D. L. 2003. Predation risk affects relative strength of top-down and bottom-up impacts on insect herbivores. Ecology 84:1032–1044.

Denoth, M., Frid, L., and Myers, J. H. 2002. Multiple agents in biological control: improving the odds? Biol. Contr. 24:20–30.

Dial, R., and Roughgarden, J. 1995. Experimental removal of insectivores from rain forest canopy: direct and indirect effects. Ecology 76:1821–1834.

Dyer, L. A., and Gentry, G. L. 1999. Larval defensive mechanisms as predictors of successful biological control. Ecol. Appl. 9:402–408.

Dyer, L. A., and Gentry, G. L. 2002. Caterpillars and parasitoids of a tropical lowland wet forest. http://www.caterpillars.org.

Dyer, L. A., and Letourneau, D. K. 1999. Trophic cascades in a complex, terrestrial community. Proc. Natl. Acad. Sci. U.S.A. 96:5072–5076.

Dyer, L. A., and Letourneau, D. K. 2003. Top-down and bottom-up diversity cascades in detrital vs. living food webs. Ecol. Lett. 6:60–68.

Dyer, L. A., and Stireman, J. O., III. 2003. Trophic control, species diversity, and complex interactions in an agricultural system. Basic Appl. Ecol. 4:423–432.

Estes, J. A., and Palmisano, J. F. 1974. Sea otters—their role in structuring nearshore communities. Science 185:1058–1060.

Eubanks, M. D., and Denno, R. F. 2000. Host plants mediate omnivore-herbivore interactions and influence prey suppression. Ecology 81:936–947.

Finke, D. L., and Denno, R. F. 2002. Intraguild predation diminished in complex-structured vegetation: implications for prey suppression. Ecology 83:643–652.

Floyd, T. 1996. Top-down impacts on creosotebush herbivores in a spatially and temporally complex environment. Ecology 77:1544–1555.

Fretwell, S. D. 1977. The regulation of plant communities by food chains exploiting them. Persp. Biol. Med. 20:169–185.

Fretwell, S. D. 1987. Food chain dynamics: the central theory of ecology? Oikos 50:291–301.

Gastreich, K. R. 1999. Trait-mediated indirect effects of a theridiid spider on an ant-plant mutualism. Ecology 80:1066–1070.

Greathead, D. 1986. Parasitoids in classical biological control. In: Insect Parasitoids, 13th Symposium of the Royal Entomological Society of London (Waage, J., and Greathead, D., eds.). London: Academic Press; 61–83.

Greathead, D., and Greathead, A. H. 1992. Biological control of insect pests by parasitoids and predators: the BIOCAT database. Biocont. News Inform. 13:61–68.

Gurevitch, J., and Hedges, L. V. 2001. Meta-analysis: combining the results of independent experiments. In: Design and Analysis of Ecological Experiments (Scheiner, S. M., and Gurevitch, J., eds.). New York: Chapman & Hall; 346–369.

Gyrisco, G. G. 1958. Forage insects and their control. Annu. Rev. Entomol. 3:421–441.

Hagen, K. S., van den Bosch, R., and Dahlsten, D. A. 1971. The importance of naturally-occurring biological control in the western United States. In: Biological Control (Huffaker, C. B., ed.). New York: Plenum Press; 253–293.

Hairston, N. G., Jr., and Hairston, N. G., Sr. 1993. Cause-effect relationships in energy-flow, trophic structure, and interspecific interactions. Am. Nat. 142:379–411.

Hairston, N. G., Smith, F. E., and Slobodkin, L. B. 1960. Community structure, population control, and competition. Am. Nat. 94:421–425.

Halaj, J., and Wise, D. H. 2001. Terrestrial trophic cascades: how much do they trickle? Am. Nat. 157:262–281.

Hassell, M. P. 1978. The Dynamics of Arthropod Predator-Prey Systems. Princeton, N.J.: Princeton University Press.

Hassell, M. P., and May, R. M. 1986. Generalist and specialist natural enemies in insect predator-prey interactions. J. Anim. Ecol. 55:923–940.

Hawkins, B. A., Mills, N. J., Jervis, M. A., and Price, P. W. 1999. Is the biological control of insects a natural phenomenon? Oikos 86:493–506.

Hochberg, M. E. 1996. Consequences for host population levels of increasing natural enemy species richness in classical biological control. Am. Nat. 147:307–318.

Holt, R. D. 2000. Trophic cascades in terrestrial ecosystems. Reflections on Polis et al. Trends Ecol. Evol. 15:444–445.

Holyoak, M. 2000. Comparing parasitoid-dominated food webs with other food webs: problems and future promises. In: Parasitoid Population Biology (Hochberg, M. E., and Ives, A. R., eds.). Princeton, N.J.: Princeton University Press; 184–197.

Horn, D. J. 1988. Ecological Approach to Pest Management. New York: Guilford Press.

Huffaker, C. B., and Messenger, P. S. 1976. Theory and Practice of Biological Control. New York: Academic Press.

Huffaker, C. B., Messenger, P. S., and Debach, P. 1971. The natural enemy component in natural control and the theory of biological control. In: Biological Control (Huffaker, C. B., ed.). New York: Plenum Press; 16–67.

Hunter, M. D. 2001. Multiple approaches to estimating the relative importance of top-down and bottom-up forces on insect populations: experiments, life tables, and time-series analysis. Basic Appl. Ecol. 2:295–309.

Huryn, A. D. 1998. Ecosystem-level evidence for top-down and bottom-up control of production in a grassland stream system. Oecologia 115:173–183.

Lang, A. 2003. Intraguild interference and biocontrol effects of generalist predators in a winter wheat field. Oecologia 134:144–153.

Lara, E. F., 1970. Problemas y Procedimientos Bananeros en la Zona Atlántica de Costa Rica. San Jose, Costa Rica: Imprenta Trejos Anos.

Leibold, M. A., Chase, J. M., Shurin, J. B., and Downing, A. L. 1997. Species turnover and the regulation of trophic structure. Annu. Rev. Ecol. Syst. 28:467–494.

Letourneau, D. K., and Andow, D. A. 1999. Natural-enemy food webs. Ecol. Appl. 9:363–364.

Lima, S. 2002. Putting predators back into behavioral predator-prey interactions. Trends Ecol. Evol. 17:70–75.

Losey, J. E., and Denno, R. F. 1998. Interspecific variation in the escape responses of aphids: effect on risk of predation from foliar-foraging and ground-foraging predators. Oecologia 115:245–252.

Losey, J. E., and Denno, R. F. 1999. Factors facilitating synergistic predation: the central role of synchrony. Ecol. Appl. 9:378–386.

Mack, R. N., Simberloff, D., Lonsdale, W. M., Evans, H., Clout, M., and Bazzaz, F. A. 2000. Biotic invasions: causes, epidemiology, global consequences, and control. Ecol. Appl. 10:689–710.

Moran, M. D., and Hurd, L. E. 1997. A trophic cascade in a diverse arthropod community caused by a generalist arthropod predator. Oecologia 113:126–132.

Moran, M. D., Rooney, T. P., and Hurd, L. E. 1996. Top-down cascade from a bitrophic predator in an old-field community. Ecology 77:2219–2227.

Myers, J. H. 1985. How many insect species are necessary for successful biocontrol of weeds? In: Proceedings of the 6th International Symposium of the Biological Control of Weeds, Agriculture Canada (Delfoesse, E. S., ed.). Ottawa, Canada: Canadian Government Printing Office; 77–82.

Myers, J. H., Higgins, C., and Kovacs, E. 1989. How many insect species are necessary for the biological control of insects? Environ. Entomol. 18:541–547.

Oksanen, L. 1991. Trophic levels and trophic dynamics—a consensus emerging. Trends Ecol. Evol. 6:58–60.

Oksanen, L., Fretwell, S. D., Arruda, J., and Niemela, P. 1981. Exploitation ecosystems in gradients of primary productivity. Am. Nat. 118:240–261.

Ostmark, E. H. 1974. Economical insect pests of bananas. Annu. Rev. Entomol. 19:161–175.

Pace, M. L., Cole, J. J., Carpenter, S. R., and Kitchell, J. F. 1999. Trophic cascades revealed in diverse ecosystems. Trends Ecol. Evol. 14:483–488.

Paine, R. T. 1969. The *Pisaster-Tegula* interaction: prey patches, predator food preference, and intertidal community structure. Ecology 50:950–961.

Paine, R. T. 1980. Food webs: linkage, interaction strength, and community infrastructure. J. Anim. Ecol. 49:667–685.

Paine, R. T., and Vadas, R. 1969. The effects of grazing by sea urchins, *Strongylocentrotus* spp., on benthic algal populations. Limnol. Oceanogr. 14:710–719.

Perfecto, I., and Castiñeiras, A. 1998. Deployment of the predaceous ants and their conservation in agroecosystems. In: Conservation Biological Control (Barbosa, P., ed.). San Diego: Academic Press; 269–289.

Persson, L., Diehl, S., Johansson, L., Andersson, G., and Hamrin, S. F. 1992. Trophic interactions in temperate lake ecosystems: a test of food chain theory. Am. Nat. 140:59–84.

Pimentel, D., and Wheeler, A. G., Jr. 1973. Species and diversity of arthropods in the alfalfa community. Environ. Entomol. 2:659–668.

Pimm, S. L., and Lawton, J. H. 1977. Number of trophic levels in ecological communities. Nature 268:329–331.

Polis, G. A. 1999. Why are parts of the world green? Multiple factors control productivity and the distribution of biomass. Oikos 86:3–15.

Polis, G. A., Sears, A. L. W., Huxel, G. R., Strong, D. R., and Maron, J. 2000. When is a trophic cascade a trophic cascade? Trends Ecol. Evol. 15:473–475.

Polis, G. A., and Strong, D. R. 1996. Food web complexity and community dynamics. Am. Nat. 147:813–846.

Power, M. E. 1990. Effects of fish in river food webs. Science 250:811–814.

Price, P. W., Bouton, C. E., Gross, P., Mcpheron, B. A., Thompson, J. N., and Weis, A. E. 1980. Interactions among three trophic levels: influence of plants on interactions between insect herbivores and natural enemies. Annu. Rev. Ecol. Syst. 11:41–65.

Riechert, S. E., and Lawrence, K. 1997. Test for predation effects of single versus multiple species of generalist predators: spiders and their insect prey. Entomol. Exp. Appl. 84:147–155.

Rosenheim, J. A. 1998. Higher-order predators and the regulation of insect populations. Annu. Rev. Entomol. 43:421–447.

Rosenheim, J. A. 2001. Source-sink dynamics for a generalist insect predator in habitats with strong higher-order predation. Ecol. Monogr. 71:93–116.

Rosenheim, J. A., and Corbett, A. 2003. Omnivory and the indeterminacy of predator function: can a knowledge of foraging behavior help? Ecology 84:2538–2548.

Rosenheim, J. A., Limburg, D. D., and Colfer, R. G. 1999. Impact of generalist predators on a biological control agent, *Chrysoperla carnea:* direct observations. Ecol. Appl. 9:409–417.

Rosenheim, J. A., Wilhoit, L. R., and Armer, C. A. 1993. Influence of intraguild predation among generalist insect predators on the suppression of an herbivore population. Oecologia 96:439–449.

Schmitz, O. J. 1998. Direct and indirect effects of predation and predation risk in old-field interaction webs. Am. Nat. 151:327–342.

Schmitz, O. J. 2003. Top predator control of plant biodiversity and productivity in an old-field ecosystem. Ecol. Lett. 6:156–163.

Schmitz, O. J., Beckerman, A. P., and O'Brien, K. M. 1997. Behaviorally mediated trophic cascades: predation risk on food-web interactions. Ecology 78:1388–1399.

Schmitz, O. J., Hambäck, P. A., and Beckerman, A. P. 2000. Trophic cascades in terrestrial systems: a review of the effects of carnivore removals on plants. Am. Nat. 155:141–153.

Schoener, T. W., and Spiller, D. A. 1996. Devastation of prey diversity by experimentally introduced predators in the field. Nature 381:691–694.

Schoener, T. W., and Spiller, D. A. 1999. Indirect effects in an experimentally staged invasion by a major predator. Am. Nat. 153:347–358.

Shiomoto, A., Tadokoro, K., Nagasawa, K., and Ishida, Y. 1997. Trophic relations in the subarctic North Pacific ecosystem: possible feeding effect from pink salmon. Mar. Ecol. Prog. Ser. 150:75–85.

Shurin, J. B., Borer, E. T., Seabloom, E. W., Anderson, K., Blanchette, C. A., Broitman, B., Cooper, S. D., and Halpern, B. S. 2002. A cross-ecosystem comparison of the strength of trophic cascades. Ecol. Lett. 5:785–791.

Sinclair, A. R. E., Krebs, C. J., Fryxell, J. M., Turkington, R., Boutin, S., Boonstra, R., Seccombe-Hett, P., Lundberg, P., and Oksanen, L. 2000. Testing hypotheses of trophic level interactions: a boreal forest ecosystem. Oikos 89:313–328.

Sinclair, A. R. E., Mduma, S., and Brashares, J. S. 2003. Patterns of predation in a diverse predator-prey system. Nature 425:288–290.

Snyder, W. E., and Ives, A. R. 2001. Generalist predators disrupt biological control by a specialist parasitoid. Ecology 82:705–716.

Snyder, W. E., and Ives, A. R. 2003. Interactions between specialist and generalist natural enemies: parasitoids, predators and pea aphid biocontrol. Ecology 84:91–107.

Spiller, D. A., and Schoener, T. W. 1994. Effects of top and intermediate predators in a terrestrial food web. Ecology 75:182–196.

Spiller, D. A., and Schoener, T. W. 2001. An experimental test for predator-mediated interactions among spider species. Ecology 82:1560–1570.

Stephens, C. S. 1984. Ecological upset and recuperation of natural control of insect pests in some Costa Rican banana plantations. Turrialba 34:101–105.

Strong, D. R. 1992. Are trophic cascades all wet? Differentiation and donor-control in speciose ecosystems. Ecology 73:747–754.

Symondson, W. O. C., Suderland, K. D., and Greenstone, M. H. 2002. Can generalist predators be effective biocontrol agents? Annu. Rev. Entomol. 47:561–594.

Terborgh, J., Lopez, L., Nunez, P., Rao, M., Shahabuddin, G., Orihuela, G., Riveros, M., Ascanio, R., Adler, G. H., Lambert, T. D., and Balbas, L. 2001. Ecological meltdown in predator-free forest fragments. Science 294:1923–1926.

Thrupp, L. A. 1990. Entrapment and escape from fruitless insecticide use: lessons from the banana sector of Costa Rica. Intern. J. Environ. Stud. 36:173–189.

Werner, E. E., and Peacor, S. D. 2003. A review of trait-mediated indirect interactions in ecological communities. Ecology 84:1083–1100.

Wilbur, R. L. 1994. Vascular plants: an interim checklist. In: La Selva: Ecology and Natural History of a Neotropical Rain Forest (McDade, L., Bawa, K. S., Hespenheide, H. A., and G. S. Hartshorn., eds.). Chicago: Chicago University Press; 350–378.

Wilby, A., and Thomas, M. B. 2002. Natural enemy diversity and pest control: patterns of pest emergence with agricultural intensification. Ecol. Lett. 5:353–360.

15

Conservation Biological Control

*Biodiversity Influences the
Effectiveness of Predators*

WILLIAM E. SNYDER
GARY C. CHANG
RENÉE P. PRASAD

In conservation biological control, we seek to increase the density and diversity of natural enemies by making agroecosystems more hospitable to them, in part under the assumption that greater predator diversity will lead to more effective pest control. However, much recent work in the trophic ecology literature suggests that the top-down impact of predators actually weakens with increasing biodiversity. Work in ecosystem ecology, in contrast, suggests that "ecosystem functions" such as pest control should become more effective with increasing biodiversity. Case studies in a variety of agroecosystems reveal mechanisms that either improve or disrupt biological control as predator biodiversity increases. Pest control likely will improve if predators facilitate prey capture by one another, if, for example, one predator chases prey from a refuge where it is then eaten by a second predator. However, predators can interfere with one another if intraguild predation is common. Similarly, supplying alternative (nonpest) prey can improve biological control, if these prey supplement predator nutrition and thereby increase overall predator densities or disrupt pest control, if nonpest prey distract predators from feeding on target pests. It has long been suggested that increasing plant diversity within agroecosystems might be an effective way to increase predator diversity and thus improve biological control. However, the literature suggests that this strategy is effective only when plants are added that supply specific resources that otherwise limit predator densities. In summary, recent work in ecology now supplies the conservation biological control practitioner with a catalog of species interactions that lead to either the improvement or disruption of pest suppression. The challenge is to gain a better understanding of how common is each mechanism and to develop strategies to promote those mechanisms that lead to improved pest control.

Conservation biological control (CBC) is the improvement of farming practices to make agroecosystems more hospitable to natural enemies. Conservation is one of the three primary approaches to biological control (see Barbosa et al., ch. 16 in this volume), along with natural enemy augmentation and introduction, but has not been the focus of much rigorous research. However, recent trends in agriculture, including the loss of broad-spectrum pesticides because of changing federal regulations, increasing availability of selective pesticides, growing pesticide resistance among pests, concerns about the risk of introducing exotic biological control agents, and the expansion of organics and other new approaches under the larger heading of *sustainable agriculture*, have revitalized interest in CBC. However, despite this renewed interest, growers have adopted few CBC strategies (Ehler 1998, Gurr et al. 2000). Here we critically examine the central assumption of many CBC efforts: the effectiveness of biological control will increase as predator species diversity increases. We suggest that a better understanding of the relationship between biodiversity and biological control will be necessary to developing effective CBC strategies.

With the growth in interest in CBC, there has been a flurry of books (Barbosa 1998, Pickett and Bugg 1998) and reviews of CBC (van Emden 1965, Altieri and Whitcomb 1979, Russell 1989, van Emden 1990, Altieri 1991, Andow 1991, Trenbath 1993, Chang and Kareiva 1999, Landis et al. 2000, Bommarco and Banks 2003) that are excellent and far more comprehensive than treatment that we could provide in this one chapter. Therefore, we focus our attention on recent advances in understanding the relationship between biodiversity and the effectiveness of CBC. First, we examine divergent views among community and ecosystem ecologists, and ecologists interested in biological control, regarding the likely relationship between species diversity and the strength of herbivore suppression. It is interesting that basic and applied ecologists have envisioned widely varying views on the relationship between predator diversity and predator impact.

Ultimately, the diversity–biological control relationship hinges on the summed impact of community members and the myriad of species interactions that provide the mechanisms underlying CBC success or failure. So, we next look at the range of both positive and negative interactions within communities of natural enemies, and between multiple natural enemies and their prey. We demonstrate that complex interactions abound within communities of arthropods in agroecosystems, sometimes to the benefit of biological control but often to its detriment. Similarly, plant diversity can influence predator diversity, so we examine attempts to increase predator density and diversity (and thus improve biological control) through the addition of plant species to agroecosystems. Surprisingly, we find little evidence for higher predator densities' leading to improved biological control following plant diversification. The exceptions to this generalization were cases in which plants added to agroecosystems were targeted to provide limiting resources for a particular natural enemy. We conclude with recommendations of areas that need more study.

Biodiversity and the Strength of Herbivore Suppression

Different hypotheses make wildly opposing predictions about the relationship between community diversity and the ability of predators to regulate herbivores, including no relationship, a positive relationship, and a negative relationship. Ecosystem ecologists have generally proposed an initial increase in function as species are added, but an eventual plateau with increasing functional redundancy. Research we discuss below leads us to suggest that an idiosyncratic relationship between increasing diversity and pest control is most likely at the lower range of biodiversity typical of many agroecosystems.

Trophic Ecology Views of Predator Diversity and Biological Control

Community ecologists interested in herbivore regulation by natural enemies have proposed three hypotheses concerning the relationship between predator diversity and predator impact, each making conflicting predictions: the green world hypothesis, the trophic-level omnivory hypothesis, and the enemies hypothesis. Most of these hypotheses were not created with pest control in mind, but we take some liberty in extrapolating the implications of each for biological control. The first hypothesis was introduced in a classic study by Hairston, Smith, and Slobodkin (1960), wherein they argued that terrestrial communities are made up of distinct, strongly connected trophic levels. These authors posit that natural enemies can be grouped into a cohesive third trophic level acting in concert to regulate herbivore densities (Hairston and Hairston 1993). This *green world hypothesis* predicts that predators regulate herbivore populations, freeing plants from limitation by herbivores (i.e., a *trophic cascade*). The original description of the green world hypothesis (Hairston et al. 1960) does not explicitly discuss the relationship between predator diversity and the strength of trophic cascades (see Stireman et al., ch. 14 in this volume); however, recently the hypothesis has been expanded to predict consistently strong trophic cascades across a range of communities, regardless of species diversity (Hairston and Hairston 1993, Hairston and Hairston 1997).

Some ecologists have questioned the validity of the green world hypothesis (Polis 1991, Strong 1992, Polis and Strong 1996), noting that terrestrial communities often include generalist predators. Generalist predators will feed on species from several trophic levels, often including herbivores, detritivores, and other predators, leading these authors to propose the *trophic-level omnivory* hypothesis. These ecologists argue that trophic-level omnivory weakens the impact of predators on other trophic levels—if, indeed, distinct trophic levels can be identified. So, the impact of predators will not cascade through only three trophic levels (i.e., from predators to herbivore to plants), but also through four or more trophic levels (i.e., from large predators to small predators to herbivores to plants). Simultaneous trophic cascades through three and four trophic levels can counteract one another, so that predators cause no net change in herbivore densities or plant biomass (e.g., Snyder and Wise 2001). However, even these opponents of the green world hypothesis accept that in some simplified terrestrial systems (e.g., small islands or agricultural monocultures) predators can have a strong impact on herbivores. The prediction here, then, is that strong trophic cascades are unlikely to be found in species-diverse systems.

Theory in biological control posits a third hypothesis on the relationship between community complexity and predator impact. Biological control practitioners have predicted that diversifying agroecosystems, for example by intercropping or leaving less-disturbed refuges near crop fields, should lead to a diversified prey base and thus a more abundant and diverse community of natural enemies (Pimentel 1961, Root 1973, van Emden and Williams 1974, Wilby and Thomas 2002a, Wilby and Thomas 2002b, Kean et al. 2003). For example, Root's *enemies hypothesis* predicts that predator impact should increase with increasing species diversity (Root 1973).

Thus, these three viewpoints predict unchanging, decreasing, or increasing strength of predator impact with increasing predator diversity (Figure 15.1A), so it is unclear whether managing agroecosystems for increased predator diversity is likely to improve biological control. Two recent meta-analyses have summarized the experimental evidence for trophic cascades in terrestrial communities (Schmitz et al. 2000, Halaj and Wise 2001). The authors reach somewhat different conclusions, perhaps related to the view of trophic cascades each has advocated in the past (e.g., Schmitz 1994, Wagner and Wise 1996, Beckerman et al. 1997, Wise and Chen 1999). However, a close look at the results of the two meta-analyses reveals a similar pattern in both. Trophic cascades occur consistently across a range of systems. In both analyses (i.e., Schmitz et al. 2000, Halaj and Wise 2001), it appears that the strength of trophic cascades ameliorates in more complex communities. That is, predators may strongly

A.

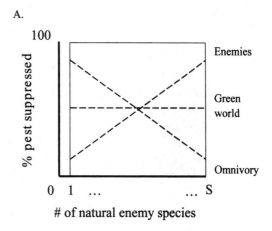

Figure 15.1. Pest control represented as a hypothetical function of the number of natural enemy species based on perspectives from community and ecosystem ecology. S represents the total number of natural enemy species in the species pool (Wootton and Downing 2003). When no natural enemies are present, no biological control occurs. The shaded area represents a broad range of trajectories that may be possible if species identity is important to biological control (Wilby and Thomas 2002a). Trajectories may be nonlinear and may have maximum values anywhere between 1 and S (Naeem et al. 1995). (A) Interactions may cause pest suppression to increase, decrease, or remain unchanged as the number of natural enemy species increases. (B) Selected hypotheses drawn from the debate over the function of biodiversity: (a) "equivalent species" hypothesis, (b) "redundant species" hypothesis, and (c) "idiosyncratic response" hypothesis.

B.

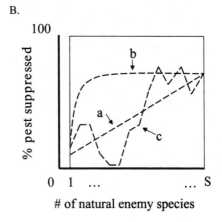

impact herbivore densities, but this reduction in herbivores only slightly increases plant biomass. Overall, these results suggest that real-world trophic interactions fall somewhere between the green world and trophic-level omnivory hypotheses. There is no evidence for the view common in CBC that the strength of herbivore suppression will increase with increasing trophic complexity. Indeed, Halaj and Wise (2001) found that the strongest cascading effects of predators were seen in agricultural systems, suggesting that the CBC practitioner's biodiversity envy may be misplaced (but see Stireman et al., ch. 14).

Pest Control as an Ecosystem Function

Trophic ecologists typically concentrate on the role of predators in suppressing herbivore densities, but pay less explicit attention to how increasing biodiversity might change the strength of herbivore suppression. In contrast, ecosystem ecologists concentrate on how biodiversity impacts a variety of "ecosystem services," but they have rarely explicitly considered herbivore suppression by predators as one of those services (but see Ehrlich and Wilson 1991, Naylor and Ehrlich 1997). The most commonly studied ecosystem service, primary production, generally increases as plant diversity increases (Mulder et al. 1999, Naeem et al. 1999, Kinzig et al. 2002). Ecosystem ecologists present at least four relevant hypotheses (Naeem et al.1995, Yachi and Loreau 1999, Wootton and Downing 2003). We extrapolate the likely implications of each hypothesis for biological control.

1. Biodiversity has no effect on pest control. This hypothesis would predict that having one or having 100 natural enemy species provides the same level of control ("Green world" line, Figure 15.1A).
2. Species are equivalent. Each natural enemy species contributes uniquely and equally to pest control. For simplicity, we illustrate a linear relationship, but nonlinearity is possible if the effect of each species varies with diversity but not identity (Wootton and Downing 2003; line a, Figure 15.1B).
3. Species are functionally redundant (i.e., several species occupy the same niche). A minimal number of natural enemies provide control, and additional species do not increase the amount of control (line b, Figure 15.1B).
4. Species are idiosyncratic. That is, species interact both additively and nonadditively (as will be further discussed). Natural enemy species are unique and unequal in their effect on the level of pest control. Under this hypothesis, pest control can increase, decrease, or remain unchanged as natural enemy diversity increases (line c, Figure 15.1B).

We suggest that in agroecosystems at higher levels of natural enemy diversity some natural enemies will perform redundant functional roles (but see Chalcraft and Resetarits 2003) so that control efficiency will eventually plateau. However, at the lower range of biodiversity, the occurrence of both positive and negative biotic interactions will create an idiosyncratic relationship between pest control and the number of natural enemy species present. That is, addition of some predators will actually disrupt pest control, through negative predator-predator interactions, whereas others will improve control through positive predator-predator interactions. We emphasize, however, that *idiosyncratic* does not mean *unpredictable*. Rather, to succeed in the presence of an idiosyncratic relationship between biodiversity and biological control, the CBC practitioner needs an intimate understanding of the multispecies interactions leading to increased or decreased pest suppression with increasing species diversity.

Arthropod Biodiversity and Biological Control

With increasing interest in CBC, there has been an explosion of research on interactions between multiple predator species, and between predators and multiple prey species, within agricultural food webs. Although few of these studies directly consider CBC, all have relevance to the field. It is important for CBC practitioners to understand the complex trophic interactions that can occur in agroecosystems, because these interactions are the mechanisms that lead to improved, or disrupted, biological control as biodiversity increases. For increasing diversity to improve biological control, natural enemies must complement one another so that a general strengthening of pest control occurs as species are added. Similarly, adding alternative, nonpest prey must increase predator densities and their impact on pests. That is, we need to identify and foster what CBC practitioners have called the "right" kind of diversity (Landis et al. 2000, p. 177), mirroring similar arguments in the ecosystem ecology literature (Schoener and Spiller 2003). Unfortunately, the literature suggests that negative predator-predator interactions, such as predators feeding on one another rather than on pests, are at least as likely as positive ones. Similarly, alternative prey can disrupt biological control by distracting natural enemies from attacking target pests. We first review the diversity of interactions among predators, and then review the diversity of interactions among predators and multiple prey. Alas, most systems demonstrate convoluted trophic connections that do not reliably improve pest suppression.

Additive and Nonadditive Interactions:
The Consequences of Predator Diversity

Broadly, from the standpoint of herbivore regulation, interactions among predators fall within two categories: additive and nonadditive (Wootton 1994, Ferguson and Stiling 1996, Sih et al. 1998). When interactions are additive, the impacts of predator A and predator B alone can be summed to predict the combined impact of the two when together. Nonadditive interactions can be of two types: synergistic or negatively nonadditive. When synergism occurs, the impact of the two predators together is greater than the summed impacts of both alone. For example, one predator may chase prey from a refuge, making the prey more susceptible to attack by a second predator outside the refuge (e.g., Soluk and Collins 1988, Huang and Sih 1991). Negatively nonadditive interactions occur, for example, when intraguild predation between predators leads to overall lower rates of pest suppression (Polis et al. 1989). Indeed, predators can interfere with one another even in the absence of actual intraguild predation, if the threat of predation leads one predator to leave a patch to avoid being eaten by another (Moran and Hurd 1994). With increasing predator diversity, there is greater opportunity for nonadditive interactions among natural enemies, making combined predator impacts less predictable. So, conservation efforts will likely produce indirect effects that both improve and detract from overall biological control effectiveness.

Many of the interactions between natural enemies we have outlined here have been observed in alfalfa. Losey and Denno (1998) provided an example of synergism between two taxa of predatory Coleoptera. In this case, ladybirds feeding in foliage caused pea aphids to drop to the ground, where they became prey for carabid beetles foraging on the soil surface, such that the combined impact of the two predators was greater than would be predicted by adding the individual impacts of each. Snyder and Ives (2001) found that predatory ground beetles disrupted biological control of pea aphids by the parasitoid *Aphidius ervi*. This is an example of negative nonadditivity, mediated in this case by the ground beetles' ease in capturing the immobile parasitoid pupae but relative inability to capture mobile aphids (Snyder and Ives 2001). However, these same authors followed up with a more comprehensive study, this time separately manipulating parasitoids and the entire predator community, which in addition to ground beetles includes several species of predatory bugs, spiders, and ladybird beetles (Snyder and Ives 2003). There was evidence of interference, with parasitoid densities halved in the presence of the predator community. Still, the impact of predators and parasitoid was additive, with the most effective aphid control where both classes of natural enemy were present. The results apparently differed from the earlier study, because other predators in the system preferentially fed on aphids, rather than on parasitoid pupae, counteracting any negative impact of ground beetles. Thus, different components of the alfalfa system act synergistically, negatively nonadditively, and strictly additively, depending on the specifics of community composition within a site or year.

Alfalfa is not the only crop in which a complex mix of additive and nonadditive predator impacts has been found. Other well-studied cropping systems have displayed a similarly complex web of trophic connections, including grains in Europe (Sunderland 1975, Edwards et al. 1979, Sunderland and Vickerman 1980, Helenius 1990, Lang 2003), cotton in California (Rosenheim et al. 1993), vegetables in the eastern United States (Walsh and Riley 1868, Riechert and Bishop 1990, Eubanks and Denno 2000a, Eubanks and Denno 2000b, Snyder and Wise 2001), and Asian rice (Settle et al. 1996, Fagan et al. 1998). Intraguild predation among biological control agents has been extensively reviewed (Polis et al. 1989, Polis and Holt 1992, Rosenheim et al. 1995, Rosenheim 1998, Brodeur and Rosenheim 2000), but there is far less information on positive predator-predator interactions. It is unclear whether positive predator-predator interactions are actually uncommon within agroecosystems; or rather, ecologists are fascinated by and preoccupied with searching for negative interactions. Regardless, the studies we have listed suggest no reason to automatically assume that more convoluted trophic connections will lead to improved pest suppression.

Interactions between Predators and Multiple (Alternative) Prey

Like predator-predator interactions, interactions between predators and multiple prey contribute to the mechanistic underpinnings of the biodiversity–biological control relationship. Conservation biological control practitioners have long realized that alternative, nonpest prey can be managed to increase densities of beneficial organisms (van Emden 1965). However, ecologists have considered a range of outcomes from adding multiple prey species to food webs, some of which would improve biological control and some of which would detract from it. Positive prey-prey interactions arise when predators concentrate their attacks on one prey species, freeing another, less preferred or less accessible prey from predation (Holt 1977). Negative prey-prey interactions occur when one prey supplements a predator's diet, increasing predator density and thus increasing predation pressure on a second prey species (so-called *apparent competition*; Holt 1977). This is the result that CBC practitioners pray for: that adding alternative prey will lead to higher predator densities and more attacks on pests. Theoreticians have long been interested in predator–multiple prey interactions (Noy-Meir 1981, Holt and Kotler 1987, Abrams and Matsuda 1996, Abrams et al. 1998, van Baalen et al. 2001), although empirical studies have been fewer (Settle and Wilson 1990, Holt and Lawton 1994, Chaneton and Bonsall 2000).

An elucidating case study in alfalfa provides an example of a negative prey-prey interaction. Evans and Englund (1996) uncovered a complex web of interactions linking a predator (the ladybird beetle *Coccinella septempunctata*) and a parasitoid (*Bathyplectes curculionis*), both of which attack larvae of the alfalfa weevil (*Hypera postica*). Here the two natural enemies interfered with one another, but this interaction was mediated by shared alternative prey rather than through intraguild predation. The weevil parasitoid benefited from the presence of pea aphids, because adult wasps fed on the sugary honeydew produced by aphids. Honeydew thus increased the longevity and fecundity of adult wasps, a positive effect from the standpoint of weevil control. However, when an artificial sugar solution was applied to plant foliage in an attempt to further supplement food for the parasitoid, ladybird beetles were attracted and then ate pea aphids. With reduced honeydew from pea aphids, parasitism actually declined. Thus, the natural web of interactions in the absence of ladybirds is an example of a negative herbivore-herbivore interaction, as the presence of pea aphids leads to more parasitoids and fewer weevils. Disruption of this relationship by adding predatory ladybird beetles to the system, in essence by increasing natural enemy diversity, weakened biological control of the target weevil pest.

Also working in the well-studied alfalfa system, Cardinale et al. (2003) established experimental communities in field cages that contained no natural enemies, a single natural enemy (either just ladybird beetles, just damsel bugs, or just the parasitoid *A. ervi*), or all three natural enemies together. All other members of the predator community (e.g., ground beetles, spiders) were excluded from all cages. After a month, at approximately the time of cutting, pea aphid densities were found not to differ from controls in the single natural enemy treatments, but were greatly reduced when all natural enemies were present together. Thus, synergism among natural enemies was recorded but mediated by interactions among multiple shared prey. The authors asserted that a second herbivore, the cowpea aphid (*Aphis craccivora*), was a preferred prey for the parasitoid. Where predators were present, cowpea aphid densities were reduced, and parasitoids redirected their attacks to pea aphids. Here, in the absence of predators, a positive prey-prey interaction existed, with cowpea aphids protecting the more injurious pea aphids from parasitoids. Under these circumstances, increasing natural enemy diversity, by adding predators, improved biological control.

Although the study of the importance of alternative herbivore prey perhaps has received the most attention, there is growing interest in the flow of energy from decomposition food webs to generalist predators that also feed on herbivorous target pests (Wise et al. 1999). In the context of CBC, addition of alternative prey that do not feed on the crop may be a more desirable means of increasing predator densities. Predator populations that are subsidized by energy from the detrital food web may reach higher densities than if they were supported by herbivores alone, leading to stronger predator impact on herbivores (Polis et al. 1997). Polis and colleagues, working on small islands in the Gulf of California, found that predator populations were subsidized by marine-derived food such as dead marine mammals washed up on the shore (Polis and Hurd 1995, Polis et al. 1998). These dead animals provided food for saprophytes, which in turn were additional prey for predators. Elevated predator populations supplemented by this marine subsidy then strongly suppressed herbivore populations, because predators fed both on saprophytes and on herbivores. Thus, islands with marine subsidies had higher predator densities and fewer herbivores (Polis and Hurd 1995, Polis et al. 1998). Similarly, Settle et al. (1996) found strong control of tropical rice pests by generalists. In their system, populations of generalist predators increased before planting, when herbivores were not yet abundant, because predators fed on early-season detritivores. As detritivore densities declined later in the season, the generalists switched to feeding on herbivores and thus prevented later season outbreaks of leafhopper pests (Settle et al. 1996).

In both the island and rice systems, the temporal separation or spatial separation (or both) of detritivore and herbivore prey may have kept the alternative prey from disrupting herbivore regulation. This case is strengthened by the results of Halaj and Wise (2002). In their system of cucumber and squash gardens, detritivores and herbivores overlapped both spatially and temporally. Detritivore densities increased after the experimental addition of detritus to plots, leading to a dramatic increase in predator densities. However, biological control was not improved, apparently because predators were gorging themselves on the abundant detritivores rather than attacking herbivorous pests (McNabb et al. 2001, Halaj and Wise 2002). Thus, alternative prey that enter the system early in the season and then decline do the most to improve biological control, by augmenting early-season predator populations that then switch to the attack of herbivores. In contrast, alternative prey that occur in synchrony with target pests can distract predators from attacking herbivores.

Increasing Plant Biodiversity to Benefit Predators

One factor influencing predator diversity is plant diversity; thus, plants can also influence the biodiversity–biological control relationship. Increasing plant diversity is one strategy that can be used to indirectly increase predator diversity and abundance, by providing alternative prey or other resources natural enemies might lack in a strict monoculture. These manipulations have been considered on both field and landscape scales, and we consider each in turn. In studies at both scales, it was the targeted provision of limiting resources to particular natural enemies that improved biological control, rather than an emergent property of diversity itself.

Diversifying Fields

Intercropping is the increase of plant diversity at the field scale by intermingled plantings of two or more plant species within a field, rather than the conventional monoculture. Intercropping has been one of the most intensively studied and frequently reviewed areas of CBC (van Emden and Williams 1974, Altieri and Whitcomb 1979, Russell 1989, Andow 1991, Trenbath 1993, Bommarco and Banks 2003), but there remains little convincing evidence that intercropping leads to stronger top-down effects of predators. Most studies confound the enemies and resource-concentration hypotheses as explanations for observed pest declines, and the limited spatial scale of most studies further confounds their interpretation (Andow 1991, Bommarco and Banks 2003). We suggest that intercropping may be most effective when pairs of plant species are chosen to provide complementary resources to one another's natural enemies.

Root's seminal article on intercropped collards (Root 1973) established two hypotheses to explain the lower pest densities he observed in intercropped, compared to monocropped, plots. The first is the *resource-concentration hypothesis*, proposing that per plant herbivore densities increase when crops are planted in monoculture. The second hypothesis, and most relevant to our chapter, is the *enemies hypothesis*. This hypothesis suggests that more diverse cropping systems support larger and more diverse communities of natural enemies, strengthening predator impact on pests. In his original study, Root (1973) demonstrated that lower pest densities in intercropped collards were primarily due to the disruption of host location by herbivores, with natural enemies playing little role. This conclusion has set a trend yet to change in the intervening 30 years.

Andow (1991) provides a particularly complete review of the literature (at least, through the 1980s). Andow tabulated the results of over 200 intercropping studies and found that pest densities were lower in intercropped plots more than 50% of the time. However, a problem common to nearly all intercropping work was that the enemies and resource-concentration hypotheses could not be distinguished as the cause of pest suppression, because natural enemies and host location by herbivores were not examined independently (Andow 1991). Andow (1991) does cleverly provide indirect evidence for the supremacy of the resource-concentration hypothesis, realizing that polyphagous herbivores should perceive the polycultures essentially as a monoculture, so they would have immunity from resource concentration. However, these generalist herbivores should be no less susceptible to natural enemies. Indeed, when studies examining polyphagous herbivores were singled out, these herbivores actually tended to have higher densities in plots containing multiple plant species.

Attempted CBC in California wine grapes through the addition of flowering cover crops provides a good example of how herbivore decline can be incorrectly attributed to the impact of natural enemies. Nicholls et al. (2000, 2001) examined intercropping rows of flowering cover crops in vineyards to conserve natural enemies of the grape leafhopper, *Erythroneura elegantula*. They found that vineyards undersown with buckwheat and sunflower contained fewer pests, fewer leafhopper parasitoids (*Anagrus* spp.), and more spiders and other generalist predators. Nicholls et al. (2000,

2001) concluded that cover cropping enhanced biological control by providing alternative prey to support a higher density of generalist predators, which then preyed more heavily on leafhoppers. However, Costello and Daane (2003) found another possible cause of herbivore suppression in this wine grape system when they compared arthropod densities and vine vigor among three treatments: monocultures, plots with a cover crop of purple vetch and barley, and another set of plots with the same cover crops but with predators excluded. Monoculture vines were the most vigorous and had the highest leafhopper densities, whereas the cover crop plots had lower vine vigor (probably because cover crops competed with grape vines for water) and lower leafhopper densities, regardless of whether predators were present or absent. Thus, Costello and Daane (2003) concluded that cover crops primarily suppress leafhoppers directly by weakening the grape vines and making them less suitable hosts, with changes in predator density being merely a red herring. Costello and Daane included in their study the experimental manipulation nearly all intercropping studies lack: in essence, a direct manipulation of predator densities. Thus, they were able to reject the enemies hypothesis as an explanation for lower pest densities.

Where intercropping is effective at improving biological control, it is through the addition of a plant that attracts or retains a specific, particularly effective natural enemy. A good example of this situation is found in the effective use of cover crops to increase densities of natural enemies and improve pest control in African maize and sorghum crops. Khan et al. (1997, 2000) assessed several plant species for their potential to minimize lepidopteran stem-borer pests in maize and sorghum. After preliminary screening, molasses grass was selected for further study as an intercrop plant, because this plant seemed to increase parasitoid and reduce pest densities (Khan et al. 1997). In experimental plots comparing molasses grass intercrops to maize monoculture, percentage of parasitism was higher (20.7% vs. 5.4%) and stem-borer damage was lower (4.6% vs. 39.2%) in the intercrop (Khan et al. 1997, Khan et al. 2000), and these differences translated into increased yields in on-farm trials (Khan et al. 2000). As with many other studies, the relative strengths of enemies versus resource-concentration effects have not been experimentally determined. However, the chemical ecology of the system suggests that molasses grass affects both stem-borers and parasitoids to increase pest suppression in maize polycultures. Volatile chemicals from molasses grass both repelled gravid stem-borers and attracted the dominant parasitoid *Cotesia sesamiae*, in Y-tube olfactometer assays (Khan et al. 1997). Thus, intercropping was effective because the added plants were specifically attractive to an important natural enemy and repellent to an important pest.

Diversifying Landscapes

Habitat diversification at scales greater than single fields has not been as extensively studied as intercropping as a means of increasing natural enemy abundance and diversity. However, a few recent studies suggest that extrafield refuges can be effective in conserving natural enemies. At the landscape scale, work in corn and rapeseed has yielded little evidence that landscape diversity per se improves CBC. Rather, shorter distances between crops containing target pests and refuges containing some resource-limiting key natural enemies led to higher percentage of parasitism in the diverse landscapes. This is a promising result for the CBC practitioner, suggesting that adding patches of a limiting resource to fields, or within relatively small-scale farm landscapes, could provide the benefits of diverse landscape structure. This possibility would eliminate the need to reengineer landscapes on a large scale, which would involve the cooperation of multiple landowners, and thus untold political and economic impediments to implementation.

Landis and coworkers have examined the relationship between landscape structure and parasitism of the armyworm (*Pseudaletia unipuncta*) in midwestern U.S. cornfields (Marino and Landis 1996, Menalled et al. 1999, Menalled et al. 2003). These authors compared percentage of parasitism in cornfields embedded within landscape matrices of two types: diverse landscapes, characterized by smaller fields and so greater perimeter area, and simple landscapes with larger fields and proportionally less edge habitat. In an initial study within a single pair of diverse and simple landscapes, Marino and Landis (1996) found that parasitism was much higher (13.1% vs. 2.4%) in fields embedded in the more diverse landscape. This difference was not due to more parasitoid species attacking the larvae, as the overall number of species was nearly identical in both landscapes. Rather, a single dominant braconid parasitoid, *Meteoris communis*, had a higher attack rate in the diverse landscape.

Unfortunately, this first study was pseudoreplicated, because all replicate "simple" fields were within the same area, and separate from the area housing all "complex" fields. The authors recognized this problem and expanded their study (Menalled et al. 1999), this time locating cornfields in three different geographical areas, with each region containing both simple and complex landscapes. Here, the results were surprisingly different from those in the original pilot study. The area used in the first study again showed higher percentage parasitism in the complex landscape than in its paired simple landscape, but this pattern did not hold in the two added regions. It appeared that parasitoid abundance was primarily related to field proximity to black cherry (*Prunus serotina*) trees, which house several alternative hosts of the dominant, generalist parasitoid *M. communis*. Only the region where black cherry was common showed an effect of increasing landscape complexity. It was proximity to this single limiting resource, cherry trees and their associated alternative hosts, rather than any inherent property of heightened plant diversity, that improved CBC.

Thies and Tscharntke (1999) conducted a similar study and found remarkably similar results. Here, the study crop was oilseed rape (*Brassica napus*); the focal pest, the rape pollen beetle (*Meligethes aeneus*); and the natural enemies, a complex of larval parasitoids. Thies and Tscharntke measured percentage parasitism in rapeseed fields embedded in simple versus complex landscapes, defined much as Landis and coworkers' landscapes were. Fields also differed in their proximity to less-disturbed refuge strips, which were either planted in an annual crop (wildflowers, wheat, or weedy), or had been undisturbed for over 6 years and so had developed more mature natural vegetation. In diverse landscapes, percentage parasitism was high both at field edges and in field interiors, whereas in simple landscapes parasitoid attack rates were elevated only in field interiors when the field was near an old refuge. Presumably, it was only the older refuges that were suitable for parasitoid overwintering. Thus, parasitism rates were consistently high in diverse landscapes because, with smaller fields, most field centers were within a relatively short distance of an old refuge. Again, it was proximity to a single limiting resource, overwintering sites, that was the mechanism driving improved CBC with increasing landscape diversity.

Conclusions

In CBC we seek to increase biodiversity within agroecosystems, assuming that this increase will lead to more effective pest control. Theories of herbivore population regulation in community and ecosystem ecology predict that increasing predator diversity can improve biological control, but recent meta-analyses of experimental data show that the strength of top-down effects tends to weaken as complexity increases. Much is now known about the variety of pairwise interactions likely to lead to positive versus negative predator-predator interactions. Understanding species interactions appears to be the key to predicting the impact of multiple enemies on a pest population, because the three potential types of predator-predator interactions (i.e., synergism, negative nonadditivity, and additivity) can have very different consequences for pest control. The co-occurrence of both positive and negative interactions among predators likely will lead to an idiosyncratic relationship between biodiversity and biological control at lower levels of diversity, with the addition of some predators improving but others disrupting biological control.

However, more work is needed in order to understand interactions among complex (greater than two species) guilds of natural enemies and to allow us to engineer communities to encourage positive interactions and reduce negative ones. A consistent finding is that diversity of alternative prey or plants has been effective in subsequently improving predator performance when that diversity is carefully targeted (e.g., alternative prey that disappear before pest colonization, or plant species targeted to provide key limiting resources). The challenge lies in identifying strategies to manipulate alternative prey phenology, and in identifying important natural enemies and the resources that limit them. Additionally, we need more information on predator movement from refuges to crop fields, and back again. More comprehensive CBC case studies, from mechanism through implementation, will be invaluable. In most systems, we still are far from understanding just what constitutes "the right kind of diversity" (sensu Landis et al. 2000).

Future Directions

We suggest three research areas in which good experimental data are particularly lacking. Progress in these areas would assist in the design of effective CBC programs.

Experimental Manipulations of More Than Two Natural Enemies

Few conservation efforts can be narrowly targeted. Usually, many species representing many trophic levels will respond (e.g., Letourneau and Goldstein 2001). Because habitat diversification often increases the abundance of many natural enemy species, experiments that manipulate two or more natural enemies in the style of Cardinale et al. (2003) may be particularly valuable. Conditions that promote positive interactions and minimize negative interactions among natural enemies will improve our mechanistic understanding of CBC. It will also be important to assess the functional similarity and potential redundancy of natural enemies (e.g., Chalcraft and Resetarits 2003), to determine if efforts to diversify natural enemy communities will have diminishing returns for pest control.

Better Knowledge of Limiting Resources and Insect Movement

To develop effective CBC strategies, we need to understand what resources limit natural enemy populations, patterns of natural enemy movement, and the interaction between the two. Successful CBC through addition of extrafield resources will depend on predator movement between resource and crop (Wissinger 1997). Thus, it is crucial to know typical foraging distances for important natural enemies (Corbett 1998, Bommarco and Fagan 2002). Conservation biologists have frequently studied animal movement in relation to reserve geometry and connectivity (Schultz 1998, Gurd et al. 2001, Berggren et al. 2002, Steffan-Dewenter 2003), and interactions between reserves and surrounding habitats (Cantrell et al. 2001, Vandermeer and Carvajal 2001, Cronin 2003). Thus, CBC might be able to draw on the conservation literature to design refuges to foster conservation and movement of beneficial arthropods (Letourneau 1998, Shea et al. 1998). The challenge lies in adapting this literature for the needs of CBC. Conservation biologists generally want to contain biodiversity within refuges, whereas the CBC practitioner requires that natural enemies freely move in and out of the refuge.

More Case Studies

Conservation biological control research has four components: enhancement, efficacy, mechanism, and implementation. Numerous studies and reviews examine different strategies for enhancing natural enemies, but there is a paucity of studies that examine any of the other three components. The efficacy of conserved enemies needs to be demonstrated not only at a small scale (field cage or small plots), but also at the larger scale of real agricultural fields. Understanding the variety of predator–predator and predator–multiple prey interactions, and which of these interactions leads to better pest control, would help improve the success rate of CBC. Further, when the mechanistic causes of failure are known (or strongly suspected), they can enlighten as much, perhaps more so, than successes. Implementation of any successful CBC tactic will require demonstration of pest suppression below the relevant economic threshold, and monitoring of secondary pest insurgence (reviewed in Landis et al. 2000) as a consequence of the conservation tactic. Implementation of CBC remains a challenge. Even the very effective pesticide-reduction-based CBC schemes in rice have been adopted by only 1–10% of the farmers who could practice them (Matteson 2000, Berg 2002).

Acknowledgments

G. Clevenger, A. Jorgensen, D. Prischmann, C. Straub, T. Ealom, and D. Bessey provided valuable suggestions that improved this chapter. M. Costello, J. Banks, and J. Rosenheim shared unpublished data with us. Grant funds from the U.S. Department of Agriculture National Research Initiative and Western Sustainable Research and Education programs, the Organic Farming Research Foundation, the Washington State Potato Commission, and the Washington State Commission on Pesticide Registration supported us during the preparation of this chapter.

Literature Cited

Abrams, P. A., Holt, R. D., and Roth, J. D. 1998. Apparent competition or apparent mutualism? Shared predation when populations cycle. Ecology 79:201–212.

Abrams, P. A., and Matsuda, H. 1996. Positive indirect effects between prey species that share predators. Ecology 77:610–616.

Altieri, M. A. 1991. Increasing biodiversity to improve insect management in agro-ecosytems. In: The Biodiversity of Microorganisms and Invertebrates: Its Role in Sustainable Agriculture (Hawksworth, D. L., ed.). Wallingford, U.K.: CAB International; 165–182.

Altieri, M. A., and Whitcomb, W. H. 1979. The potential use of weeds in the manipulation of beneficial insects. HortScience 14:12–18.

Andow, D. A. 1991. Vegetational diversity and arthropod population response. Annu. Rev. Entomol. 36:561–586.

Barbosa, P. 1998. Conservation Biological Control. New York: Academic Press.

Beckerman, A. P., Uriarte, M., and Schmitz, O. J. 1997. Experimental evidence for a behavior-mediated trophic cascade in a terrestrial food chain. Proc. Natl. Acad. Sci. U.S.A. 94:10735–10738.

Berg, H. 2002. Rice monoculture and integrated rice-fish farming in the Mekong Delta, Vietnam—economic and ecological considerations. Ecol. Econ. 41:95–107.

Berggren, A., Birath, B., and Kindvall, O. 2002. Effect of corridors and habitat edges on dispersal behavior, movement rates, and movement angles in Roesel's bush-cricket (*Metrioptera roeseli*). Cons. Biol. 16:1562–1569.

Bommarco, R., and Banks, J. E. 2003. Scale as a modifier in vegetation diversity experiments: effects on herbivores and predators. Oikos 102:440–448.

Bommarco, R., and Fagan, W. F. 2002. Influence of crop edges on movement of generalist predators: a diffusion approach. Agric. For. Entomol. 4:21–30.

Brodeur, J., and Rosenheim, J. A. 2000. Intraguild interactions in aphid parasitoids. Entomol. Exp. Appl. 97:93–108.

Cantrell, R. S., Cosner, C., and Fagan, W. F. 2001. How predator incursions affect critical patch size: the role of the functional response. Am. Nat. 158:368–375.

Cardinale, B. J., Harvey, C. T., Gross, K., and Ives, A. R. 2003. Biodiversity and biocontrol: emergent impacts of a multi-enemy assemblage on pest suppression and crop yield in an agroecosystem. Ecol. Lett. 6:857–865.

Chalcraft, D. R., and Resetarits, W. J. 2003. Predator identity and ecological impacts: functional redundancy or functional diversity? Ecology 84:2407–2418.

Chaneton, E. J., and Bonsall, M. B. 2000. Enemy-mediated apparent competition: empirical patterns and the evidence. Oikos 88:380–394.

Chang, G. C., and Kareiva, P. 1999. The case for indigenous generalists in biological control. In: Theoretical Approaches to Biological Control (Hawkins, B. A., and Cornell, H. V., eds.). New York: Cambridge University Press; 103–115.

Corbett, A. 1998. The importance of movement in the response of natural enemies to habitat manipulation. In: Enhancing Biological Control: Habitat Management to Promote Natural Enemies of Agricultural Pests (Pickett, C. H., and Bugg, R. L., eds.). Berkeley: University of California Press; 25–47.

Costello, M. J., and Daane, K. M. 2003. Spider and leafhopper (*Erythroneura* spp.) response to vineyard ground cover. Environ. Entomol. 32:1085–1098.

Cronin, J. T. 2003. Matrix heterogeneity and host-parasitoid interactions in space. Ecology 84:1506–1516.

Edwards, C. A., Sunderland, K. D., and George, K. S. 1979. Studies on polyphagous predators of cereal aphids. J. Appl. Ecol. 16:811–823.

Ehler, L. E. 1998. Conservation biological control: past, present, and future. In: Conservation Biological Control (Barbosa, P., ed.). San Diego: Academic Press; 1–8.

Ehrlich, P. R., and Wilson, E. O. 1991. Biodiversity studies: science and policy. Science 253:758–762.

Eubanks, M. D., and Denno, R. F. 2000a. Health food versus fast food: the effects of prey quality and mobility on prey selection by a generalist predator and indirect interactions among prey species. Ecol. Entomol. 25:140–146.

Eubanks, M. D., and Denno, R. F. 2000b. Host plants mediate ominivore-herbivore interactions and influence prey suppression. Ecology 81:936–947.

Evans, E. W., and England, S. 1996. Indirect interactions in biological control of insects: pests and natural enemies in alfalfa. Ecol. Appl. 6:920–930.

Fagan, W. F., Hakim, A. L., Ariawan, H., and Yuliyantiningsih, S. 1998. Interactions between biological control efforts and insecticide applications in tropical rice agroecosystems: the potential role of intraguild predation. Biol. Contr. 13:121–126.

Ferguson, K. I., and Stiling, P. 1996. Nonadditive effects of multiple natural enemies on aphid populations. Oecologia 108:375–379.

Gurd, D. B., Nudds, T. D., and Rivard, D. H. 2001. Conservation of mammals in eastern North American wildlife reserves: how small is too small? Cons. Biol. 15:1355–1363.

Gurr, G. M., Wratten, S. D., and Barbosa, P. 2000. Success in conservation biological control of arthropods. In: Biological Control: Measures of Success (Gurr, G., and Wratten, S., eds.). Dordrecht, the Netherlands: Kluwer Academic; 105–132.

Hairston, N. G., Jr., and Hairston, N. G., Sr. 1993. Cause-effect relationships in energy flow, trophic structure, and interspecies interactions. Am. Nat. 142:379–411.

Hairston, N. G., Jr., and Hairston, N. G., Sr. 1997. Does food-web complexity eliminate trophic-level dynamics? Am. Nat. 149:1001–1007.

Hairston, N. G., Smith, F. E., and Slobodkin, L. B. 1960. Community structure, population control and competition. Am. Nat. 94:421–425.

Halaj, J., and Wise, D. H. 2001. Terrestrial trophic cascades: how much do they trickle? Am. Nat. 157:262–281.

Halaj, J., and Wise, D. H. 2002. Impact of a detrital subsidy on trophic cascades in a terrestrial grazing food web. Ecology 83:3141–3151.

Helenius, J. 1990. Effect of epigeal predators on infestation by the aphid *Rhopalosiphum padi* and on grain yield of oats in monocrops and mixed intercrops. Entomol. Exp. Appl. 54:225–236.

Holt, R. D. 1977. Predation, apparent competition, and the structure of prey communities. Theor. Pop. Biol. 12:197–229.

Holt, R. D., and Kotler, B. P. 1987. Short-term apparent competition. Am. Nat. 130:412–430.

Holt, R. D., and Lawton, J. H. 1993. Apparent competition and enemy-free space in insect host–parasitoid communities. Am. Nat. 142:623–645.

Huang, C., and Sih, A. 1991. Experimental studies on direct and indirect interactions in a three trophic–level stream system. Oecologia 85:530–536.

Kean, J., Wratten, S., Tylianakis, J., and Barlow, N. 2003. The population consequences of natural enemy enhancement, and implications for conservation biological control. Ecol. Lett. 6:604–612.

Khan, Z. R., Ampong-Nyarko, K., Chiliswa, P., Hassanali, A., Kimani, S, Lwande, W., Overholt, W. A., Pickett, J. A., Smart, L. E., Wadhams, L. J., et al. 1997. Intercropping increases parasitism of pests. Nature 388:631–632.

Khan, Z. R., Pickett, J. A., Van den Berg, J., Wadhams, L. J., and Woodcock, C. M. 2000. Exploiting chemical ecology and species diversity: stem borer and striga control for maize and sorghum in Africa. Pest Manag. Sci. 56:957–962.

Kinzig, A. P., Pacala, S., and Tilman, G. D. 2002. The Functional Consequences of Biodiversity: Empirical Progress and Theoretical Extensions. Princeton, N.J.: Princeton University Press.

Landis, D. A., Wratten, S. D., and Gurr, G. M. 2000. Habitat management to conserve natural enemies of arthropod pests in agriculture. Annu. Rev. Entomol. 45:175–201.

Lang, A. 2003. Intraguild interference and biocontrol effects of generalist predators in a winter wheat field. Oecologia 134:144–153.

Letourneau, D. K. 1998. Conservation biology: lessons for conserving natural enemies. In: Conservation Biological Control (Barbosa, P., ed.). San Diego: Academic Press; 9–38.

Letourneau, D. K., and Goldstein, B. 2001. Pest damage and arthropod community structure in organic vs. conventional tomato production in California. J. Appl. Ecol. 38:557–570.

Losey, J. E., and Denno, R. F. 1998. Positive predator-predator interactions: enhanced predation rates and synergistic suppression of aphid populations. Ecology 79:2143–2152.

Marino, P. C., and Landis, D. A. 1996. Effect of landscape structure on parasitoid diversity and parasitism in agroecosystems. Ecol. Appl. 6:276–284.

Matteson, P. C. 2000. Insect pest management in tropical Asian irrigated rice. Annu. Rev. Entomol. 45:549–574.

McNabb, D. M., Halaj, J., and Wise, D. H. 2001. Inferring trophic position of generalist predators and their linkage to the detrital food web in agroecosystems: a stable isotope analysis. Pedobiologia 45:289–297.

Menalled, F. D., Costamagna, A. C., Marino, P. C., and Landis, D. A. 2003. Temporal variation in the response of parasitoids to agricultural landscape structure. Agric. Ecosyst. Environ. 96:29–35.

Menalled, F. D., Marino, P. C., Gage, S. H., and Landis, D. A. 1999. Does agricultural landscape structure affect parasitism and parasitoid diversity? Ecol. Appl. 9:634–641.

Moran, M. D., and Hurd, L. E. 1994. Short-term responses to elevated predator densities: noncompetitive intraguild interactions and behaviour. Oecologia 98:269–273.

Mulder, C. P. H., Koricheva, J., Huss-Danell, K., Hogberg, P., and Joshi, J. 1999. Insects affect relationships between plant species richness and ecosystem processes. Ecol. Lett. 2:237–246.

Naeem, S., Chapin, F. S., III, Costanza, R., Ehrlich, P. R., Golley, F. B., Hooper, D. U., Lawton, J. H., O'Neill, R. V., Mooney, H. A., Sala, O. E., et al. 1999. Biodiversity and Ecosystem Functioning: Maintaining Natural Life Support Processes. Issues in Ecology no. 4. Washington, D.C.: Ecological Society of America.

Naeem, S., Thompson, L. J., Lawler, S. P., Lawton, J. H., and Woodfin, R. M. 1995. Empirical evidence that declining species diversity may alter the performance of terrestrial ecosystems. Phil. Trans. R. Soc. Lond. B 347:249–262.

Naylor, R. L., and Ehrlich, P. R. 1997. Natural pest control services and agriculture. In: Nature's Services: Societal Dependence on Natural Ecosystems (Daily, G. C., ed.). Washington, D.C.: Island Press; 151–174.

Nicholls, C. I., Parella, M., and Altieri, M. A. 2001. The effects of a vegetational corridor on the abundance and dispersal of insect biodiversity within a northern California organic vineyard. Landsc. Ecol. 16:133–146.

Nicholls, C. I., Parella, M. P., and Altieri, M. A. 2000. Reducing the abundance of leafhoppers and thrips in a northern California organic vineyard through maintenance of full season floral diversity with summer cover crops. Agric. For. Entomol. 2:107–113.

Noy-Meir, I. 1981. Theoretical dynamics of competitors under predation. Oecologia 50:277–284.

Pickett, C. H., and Bugg, R. L. 1998. Enhancing Biological Control: Habitat Management to Promote Natural Enemies of Agricultural Pests. Berkeley: University of California Press.

Pimentel, D. 1961. Species diversity and insect population outbreaks. Ann. Entomol. Soc. Am. 54:76–86.

Polis, G. A. 1991. Complex trophic interactions in deserts: an empirical critique of food-web theory. Am. Nat. 138:123–155.

Polis, G. A., Anderson, W. B., and Holt, R. D. 1997. Toward an integration of landscape and food web ecology: the dynamics of spatially subsidized food webs. Annu. Rev. Ecol. Syst. 28:289–316.

Polis, G. A., and Holt, R. D. 1992. Intraguild predation: the dynamics of complex trophic interactions. Trends Ecol. Evol. 7:151–154.

Polis, G. A., and Hurd, S. D. 1995. Extraordinarily high spider densities on islands: flow of energy from the marine to terrestrial food webs and the absence of predation. Proc. Natl. Acad. Sci. U.S.A. 92:4382–4386.

Polis, G. A., Hurd, S. D., Jackson, C. T., and Sanchez-Pinero, F. 1998. Multifactor population limitation: variable spatial and temporal control of spiders on Gulf of California islands. Ecology 79:490–502.

Polis, G. A., Myers, C. A., and Holt, R. D. 1989. The ecology and evolution of intraguild predation: potential competitors that eat each other. Annu. Rev. Ecol. Syst. 20:297–330.

Polis, G. A., and Strong, D. R. 1996. Food web complexity and community dynamics. Am. Nat. 147:813–846.

Riechert, S. E., and Bishop, L. 1990. Prey control by an assemblage of generalist predators: spiders in garden test systems. Ecology 71:1441–1450.

Root, R. B. 1973. Organization of a plant-arthropod association in simple and diverse habitats: the fauna of collards. Ecol. Monogr. 43:95–124.

Rosenheim, J. A. 1998. Higher-order predators and the regulation of insect herbivore populations. Annu. Rev. Entomol. 43:421–447.

Rosenheim, J. A., Kaya, H. K., Ehler, L. E., Marois, J. J., and Jaffee, B. A. 1995. Intraguild predation among biological-control agents: theory and practice. Biol. Contr. 5:303–335.

Rosenheim, J. A., Wilhoit, L. R., and Armer, C. A. 1993. Influence of intraguild predation among generalist insect predators on the suppression of an herbivore population. Oecologia 96:439–449.

Russell, E. P. 1989. Enemies hypothesis: a review of the effect of vegetational diversity on predatory insects and parasitoids. Environ. Entomol. 18:590–599.

Schmitz, O. J. 1994. Resource edibility and trophic exploitation in an old-field food web. Proc. Natl. Acad. Sci. U.S.A. 91:5364–5367.

Schmitz, O. J., Hamback, P. A., and Beckerman, A. P. 2000. Trophic cascades in terrestrial systems: a review of the effect of predator removals on plants. Am. Nat. 155:141–153.

Schoener, T. W., and Spiller, D. A. 2003. Effects of removing a vertebrate versus an invertebrate predator on a food web, and what is their relative importance? In: The Importance of Species: Perspectives on Expendability and Triage (Kareiva, P., and Levin, S. A., eds.). Princeton, N.J.: Princeton University Press; 69–84.

Schultz, C. B. 1998. Dispersal behavior and its implications for reserve design in a rare Oregon butterfly. Cons. Biol. 12:284–292.

Settle, W. H., Ariawan, H., Astuti, E. T., Cahyana, W., Hakim, A. L., Hindayana, D., Lestari, A. S., and Sartanto, P. 1996. Managing tropical rice pests through conservation of generalist natural enemies and alternative prey. Ecology 77:1975–1988.

Settle, W. H., and Wilson, L. T. 1990. Invasion by the variegated leafhopper and biotic interactions: parasitism, competition, and apparent competition. Ecology 71:1461–1470.

Shea, K., and NCEAS Working Group on Population Management. 1998. Management of populations in conservation, harvesting and control. Trends Ecol. Evol. 13:371–375.

Sih, A., England, G., and Wooster, D. 1998. Emergent impacts of multiple predators on multiple prey. Trends Ecol. Evol. 13:350–355.

Snyder, W. E., and Ives, A. R. 2001. Generalist predators disrupt biological control by a specialist parasitoid. Ecology 82:705–716.

Snyder, W. E., and Ives, A. R. 2003. Interactions between specialist and generalist natural enemies: parasitoids, predators, and pea aphid biocontrol. Ecology 84:91–107.

Snyder, W. E., and Wise, D. H. 2001. Contrasting trophic cascades generated by a community of generalist predators. Ecology 82:1571–1583.

Soluk, D. A., and Collins, N. C. 1988. Synergistic interactions between fish and stoneflies: facilitation and interference among predators. Oikos 52:94–100.

Steffan-Dewenter, I. 2003. Importance of habitat area and landscape context for species richness of bees and wasps in fragmented orchard meadows. Cons. Biol. 17:1036–1044.

Strong, D. R. 1992. Are trophic cascades all wet? Differentiation and donor-control in speciose ecosystems. Ecology 73:747–754.

Sunderland, K. D. 1975. The diets of some predatory arthropods in cereal crops. J. Appl. Ecol. 12:507–515.

Sunderland, K. D., and Vickerman, P. 1980. Aphid feeding by some polyphagous predators in relation to aphid density in cereal fields. J. Appl. Ecol. 17:389–396.

Thies, C., and Tscharntke, T. 1999. Landscape structure and biological control in agroecosystems. Science 285:893–895.

Trenbath, B. R. 1993. Intercropping and the management of pests and diseases. Field Crops Res. 34:381–405.

van Baalen, M., Krivan, V., van Rijn, P. C. J., and Sabelis, M. W. 2001. Alternative food, switching predators, and the persistence of predator-prey systems. Am. Nat. 157:512–524.

Vandermeer, J., and Carvajal, R. 2001. Metapopulation dynamics and the quality of the matrix. Am. Nat. 158:211–220.

van Emden, H. F. 1965. The role of uncultivated land in the biology of crop pests and beneficial insects. Sci. Hortic. 17:121–136.

van Emden, H. F. 1990. Plant diversity and natural enemy efficiency in agroecosytems. In: Critical Issues in Biological Control (Mackauer, M., Ehler, L. E., and Roland, J., eds.). Andover, U.K.: Intercept; 63–80.

van Emden, H. F., and Williams, G. F. 1974. Insect stability and diversity in agro-ecosystems. Annu. Rev. Entomol. 19:455–475.

Wagner, J. D., and Wise, D. H. 1996. Cannibalism regulates densities of young wolf spiders: evidence from field and laboratory experiments. Ecology 77:639–652.

Walsh, B. D., and Riley, C. V. 1868. Potato bugs. Amer. Entomol. 1:21–49.

Wilby, A., and Thomas, M. B. 2002a. Are the ecological concepts of assembly and function of biodiversity useful frameworks for understanding natural pest control? Agric. For. Entomol. 4:237–243.

Wilby, A., and Thomas, M. B. 2002b. Natural enemy diversity and pest control: patterns of pest emergence with agricultural intensification. Ecol. Lett. 5:353–360.

Wise, D. H., and Chen, B. R. 1999. Impact of intraguild predators on survival of a forest-floor wolf spider. Oecologia 121:129–137.

Wise, D. H., Snyder, W. E., Tuntibunpakul, P., and Halaj, J. 1999. Spiders in decomposition food webs of agroecosystems: theory and evidence. J. Arachnol. 27:363–370.

Wissinger, S. A. 1997. Cyclic colonization in predictably ephemeral habitats: a template for biological control in annual crop systems. Biol. Contr. 10:4–15.

Wootton, J. T. 1994. Putting the pieces together: testing the independence of interactions among organisms. Ecology 75:1544–1551.

Wootton, J. T., and Downing, A. L. 2003. Understanding the effects of reduced biodiversity: a comparison of two approaches. In: The Importance of Species: Perspectives on Expendability and Triage (Kareiva, P., and Levin, S. A., eds.). Princeton, N.J.: Princeton University Press; 85–104.

Yachi, S., and Loreau, M. 1999. Biodiversity and ecosystem productivity in a fluctuating environment: the insurance hypothesis. Proc. Natl. Acad. Sci. U.S.A. 96:1463–1468.

16

Species Abundance Distribution and Predator-Prey Interactions

Theoretical and Applied Consequences

PEDRO BARBOSA

ASTRID CALDAS

SUSAN E. RIECHERT

In this chapter, we focus on interactions between species in assemblages of predators and their prey, and the importance of these interactions to biological control, in particular, conservation biological control (CBC). Conservation biological control is the approach to biological control that depends on a thorough understanding of interactions among species in natural enemy assemblages, and between natural enemy assemblages and their prey or hosts. In general, assemblages or communities of both predators and herbivore prey are characterized by similar patterns of species abundance distribution. This pattern is one in which a relative few species account for the vast majority of total abundance. However, these patterns do differ in the proportion of the species in each assemblage that are numerically dominant.

Species abundance distributions have been depicted with various mathematical models which, although useful, cannot be readily compared statistically in an ecologically meaningful fashion. We present a new method that can be used to compare species abundance distributions. In addition, we illustrate how the new approach might be used with data from available studies to generate curves depicting the abundance distributions of various assemblages or communities, and compare them to speculate on the influences of differences in habitat types, taxa, and changes over seasons and years. We then provide insights into the potential use of species abundance distribution comparisons in CBC research. Further, we suggest that comparisons of distributions may be helpful in assessing and refining current CBC hypotheses like the species assemblage control hypothesis.

Biological control is defined as the intentional use or manipulation of predators, parasitoids, pathogens, antagonists or competitors (Van Driesche and Bellows

1996) to control insects and other animals, weeds, or diseases. This approach to the management of pest species has been conducted with three basic strategies. In *classical biological control*, natural enemies of a target pest are selected from the country or area of origin of an introduced pest. Introduced natural enemies are released and expected to establish viable populations capable of regulating pest populations. *Augmentation*, the second strategy, aims to increase the population of an introduced or native natural enemy of a target pest. Typically, this strategy is accomplished by mass-rearing and releasing one or more natural enemies to maintain needed levels of control. The third strategy, and the focus of this chapter, is *conservation biological control* (CBC), which generally is defined as the manipulation of the environment to favor natural enemies by either removing or mitigating adverse factors, or providing requisites that are lacking in natural enemies' habitat (DeBach 1974, Van Driesche and Bellows 1996).

Classical biological control and augmentation typically focus on interactions involving single predator species and single pest species (or, at best, a small handful of species). In contrast, CBC typically involves the maintenance and enhancement of a variety of natural enemy species (i.e., multiple predator species, or predators and parasitoids) in different classes (e.g., spiders, insects, and centipedes), orders (e.g., Coleoptera and Hymenoptera), or families (e.g., Trichogrammatidae and Braconidae). The success of CBC relies heavily on an understanding of the ecology and life histories of the many different taxa living together as a community of natural enemies. Nevertheless, relatively little biological control research has taken a broad taxonomic or community perspective. Specifically, few studies have identified the aspects of natural enemy assemblages or communities that enhance biological control. This is, at least in part, due to the difficulty of obtaining insights into the ecology of the diverse taxa that make up many natural enemy assemblages in arthropod communities. Thus, CBC is perhaps the least-studied branch of biological control.

Most agroecosystems have been altered to such an extent that often there is little of the habitat left to conserve; indeed, *restoration biological control* might be a better term than *conservation biological control*. Nevertheless, the study and (for the most part) the practice of CBC has been guided by four general principles. First and foremost is the principle that enhanced biodiversity enhances natural and biological control (Altieri 1991, Andow 1991, Hawksworth 1991). Enhanced biodiversity can be achieved with changes outside the crop (e.g., minimeadows or beetle banks; Boller 1992, Collins et al. 2002), inside the crop (e.g., cover crops, mulch, or within-crop strips; Riechert and Bishop 1990, Bugg and Waddington 1994, Platt et al. 1999, Berndt et al. 2002), or both. A variety of factors and requisites that enhance and maintain natural enemies are presumed to be made available as a result of enhanced biodiversity (Pickett and Bugg 1998; but see Snyder et al., ch. 15 in this volume). These factors and requisites include food (e.g., pollen and nectar), alternate host and prey, shelter, and favorable microclimates or overwintering sites (Corbett and Rosenheim 1996, Barbosa 1998b, Pickett and Bugg 1998).

A second principle is that mitigation of habitat traits or agronomic conditions unfavorable to natural enemies enhances natural enemies and biological control (Brust et al. 1985, Barbosa 1998b, Pickett and Bugg 1998, Riechert 1999). This principle typically involves protection of biological control agents from insecticides but may also involve mitigating the effects of other conditions or practices detrimental to their survival and effectiveness. These disruptive forces can include soil disruption (tillage), excessive dust, flooding (or irrigation), or other cultural practices that limit their population numbers and/or ability to forage maximally on pest species.

A third principle is that mitigation of potentially detrimental interactions among natural enemies, such as intraguild predation and hyperparasitism (so-called negative nonadditive interactions; sensu Snyder et al., ch. 15), enhances biological control (Rosenheim et al. 1993, Rosenheim et al. 1999, Sullivan and Völkl 1999, Rosenheim 2001).

The fourth principle is that host-plant traits that enhance biological control should be encouraged or manipulated and detrimental host-plant traits should be minimized or eliminated (van den Bosch and Telford 1964, Rabb et al. 1976, Bottrell et al. 1998, De Moraes et al. 1998).

Clearly, the principles that guide CBC, implicitly or explicitly, focus on a community or assemblage of species rather than on one or two species. The CBC research, guided by these four principles, has attempted to identify favorable traits or the traits or conditions that limit natural enemies, the circumstances under which they are detrimental, and how their detrimental impact might be mitigated. However, little effort has been devoted to determining if, and how, changes in the structure of pest and natural enemy communities influence their ability to regulate pest populations (but see Heong et al. 1991, Barrion et al. 1994, Provencher and Riechert 1994, Hawkins and Mills 1996, Schoenly et al. 1996, Colunga-Garcia et al. 1997, Schoenly et al. 1998, Marino and Landis 2000). Further, we know little about how tactics or strategies meant to conserve natural enemies change the structure of their community.

For CBC to be more effective and readily adopted, we must enhance our understanding of natural enemy assemblages and how they limit the abundance of pest species. Specifically, we need research that will address questions such as, how is abundance distributed across the species in assemblages, and what are the consequences of changes in species abundance distribution (SAD)? Are the numerically dominant species in an assemblage always the same species, and, if not, why not? What environmental factors alter SAD predictably? With regard to the regulation of pest species, are all species that comprise assemblages important? What are the critical characteristics of a predator assemblage that permit a group of predators to limit prey better than any single species can? In this chapter, we argue for the importance of research focusing on SAD to answer these types of questions and provide a new approach for rigorous comparisons of species abundance distribution.

Why do we focus on SAD? Over the last several decades the importance of direct and indirect higher order interactions has been repeatedly emphasized (Abrams et al. 1996, Schmitz et al. 1997, Schmitz 1998). However, in this chapter, we pay particular attention to interactions among species in assemblages. We consider assemblages to be multispecies collections of organisms usually associated with some (food) resource (sensu Claridge 1987). Clearly, within herbivore assemblages, higher order interactions are influenced by a variety of factors. The outcomes of interactions (such as predator-prey interactions) are affected not only by the type and number of species involved, but also by the relative abundance of each species (i.e., how abundance is distributed across the species in the assemblage; Cushman and Whitman

1991, Letourneau and Dyer 1998a, Letourneau and Dyer 1998b). For example, in theory, two predator assemblages may have the same species and be of similar size in terms of total number of individuals; however, the extent and outcome of interactions among the species in each assemblage or between each species and its prey could differ if abundance were distributed differently across predator species (i.e., if the numerically dominant species differed in each assemblage).

The concept that the community or assemblage of natural enemies in an agroecosystem is important in regulating pests has been a central tenet of CBC since at least the 1970s, yet it has been only relatively recently that hypotheses have emerged formally proposing that effective biological control can be achieved by a community or assemblage of natural enemies rather than single species. The *species assemblage control hypothesis* (SACH) suggests that interactions between predator assemblages and a prey species suppress increases of the prey species (Provencher and Riechert 1994, Riechert and Lawrence 1997). Riechert (1999) noted that "all evidence indicates that successful pest suppression by spiders will best be achieved through the maintenance of high spider densities, and in many cases also high species diversities." "Maximization of spider densities and species richness," Riechert elaborates, "are steps that logically must be taken in agricultural systems to increase the beneficial functioning of spiders in them" (p. 391). The same may be true for most predators. Whether assemblages do indeed provide better control of pests needs to be further evaluated experimentally. We argue that the use of SAD may be helpful in research assessing the SACH and other hypotheses, as well as the conditions under which assemblages are likely to be more effective than single species.

However, in this chapter, we argue that the ways in which SADs have been depicted (i.e., with use of log-series, log-normal, and other distributions) are less than useful in an experimental context. As an alternative, we suggest an approach (based on cumulative distribution functions) that allows for the depiction and statistical comparison of the SADs of natural enemy and herbivore host or prey assemblages and communities. We use data from a variety of published studies to illustrate how this new approach might be used to compare and contrast the SADs of a variety of phytophagous herbivores and various taxa of natural enemies, focusing primarily on predators. Finally, we focus on, and speculate about, the potential impact that an understanding and comparison of SADs may have on CBC, in general, and the SACH, in particular.

Patterns of SAD

Theoretical models have been useful for both describing and comparing the structure of assemblages and communities. Depictions of SAD with models such as the log-normal, log-series, or broken-stick distributions often have been used to compare and contrast communities and assemblages (Fisher et al. 1943, Preston 1948, Preston 1962a, Preston 1962b, Taylor et al. 1976) of a variety of invertebrate, vertebrate, and plant taxa (Preston 1948, Preston 1962a, Preston 1962b, Taylor 1978, Sugihara 1980, Ludwig and Reynolds 1988, DeVries et al. 1997). These models are implausible approximations of reality (Routledge 1980), because natural populations rarely meet assumptions associated with these and other mathematical models. Nevertheless, the log-series and the log-normal distributions have been among the most commonly used theoretical distributions (but see Dewdney 2000). They are often used to make inferences regarding species diversity, richness, and other characteristics of assemblages and communities (Taylor et al. 1976).

Constraints in Comparing Differences in SAD

Although distribution models have been useful, fitting the SAD of a particular assemblage to a particular model does not describe the structure of the assemblage in a straightforward fashion. Depicting a community or assemblage using a particular model does not uniquely characterize the community or assemblage. That is, if one finds that two SADs fit the log-series distribution and that their parameters differ, one cannot identify where critical differences lie. In addition, SAD data sets often simultaneously fit more than one distribution model (May 1975, Hughes 1986, Magurran 1988, Krebs 1989). Assemblage or community parameter estimates derived from the use of these distribution models do not readily lend themselves to statistical comparison and thus are not particularly informative. Although approaches that allow comparisons of community or assemblage traits are available (Gotelli and Colwell 2001), no current approach provides a uniform or standardized comparison of SAD, be it a comparison of predator or of prey SAD.

An Approach for Depicting and Comparing SADs

In this chapter, we provide an alternative approach (suggested by Robbins, R. K., personal communication) that we will refer to as *Robbins's curves*. Robbins's curves describe SADs through cumulative distribution functions, which then can be statistically compared with a standard test such as the Kolmogorov-Smirnov two-group test (Sokal and Rohlf 1981). This approach is not biased by sample size or total numbers of individuals in the assemblages or communities being compared. In our approach, SADs are tabulated with incremental percentage classes (i.e., tabulations of the number of individuals within percentage species classes, such as increments of 10% of all species) rather than abundance classes. In other words, one would determine the number of individuals that make up the rarest 10% of the species, then the number that make up the rarest 20%, then those that make up the rarest 30%, and so forth. Alternatively, one could tabulate the number that make up the most abundant 10%, then the most abundant 20%, and so forth. One can choose any number of classes (i.e., percentage increments), so long as they are the same for both assemblages. We use increments from 10%, up to 100% of all species. The resulting graphic representation provides a relatively straightforward depiction of the SAD of an assemblage. The Kolmogorov-Smirnov two-group test is applied to statistically compare SADs derived from the Robbins's curves technique. In applying it, one is comparing the rare species in one assemblage to the rare species in the other and, similarly, common species to common species, independent of the numbers collected of each particular species or the number of species present.

An Illustrative Example

For a working example, we used data on the abundance of horseflies and deerflies (Table 16.1) collected with Malaise traps and netting (Tallamy et al. 1976) to generate Robbins's curves and compare them statistically. The Malaise trap data consisted of 44 species (see Malaise, Table 16.2), and 10% of the species (starting from the rarest) was 4.4 species. The 4.4 rarest species were part of the first group of (5) species (i.e., those for which only one individual was collected). The abundance of the first (10%) class was 4.4 × 1, or 4.4 individuals. The next class was 20% of 44, which corresponds to 8.8 species. That would include the 5 species (from the first abundance category) with one individual per species, and 3.8 species of the next abundance category (i.e., the species for which only 2 individuals were collected). Thus, the abundance of the 20% class (the 8.8 species) was (5 × 1) + (3.8 × 2), or 12.6. The next class was 30% of 44 species, which corresponds to 13.2 species. This class includes the first 5 species for which only one individual was collected, plus the 6 species of which 2 individuals were collected, plus 2.2 of the species in the third abundance category (i.e., species for which 3 individuals were collected). The abundance for this class was 23.6: (5 × 1) + (6 × 2) + (2.2 × 3). The 40% class comprises 17.6 species, which include the 5 species for which only one individual was collected, 6 from the 2-individual category, 3 from the 3-individual category, and 3.6 from the 4-individual category, for a total of 40.4. The process is repeated for all data obtained by Malaise sampling and then for the netting data (Table 16.1). Because absolute abundance values vary from assemblage to assemblage, the totals for each abundance category (4.4 for 10%, 12.6 for 20%, 23.6 for 30%, etc., in the example) are graphed as percentage of the total (i.e., cumulative number of individuals; Figure 16.1).

We then apply the Kolmogorov-Smirnov two-group test to the increment in abundance from one class to the next (see Table 16.2). For instance, the 10% species class has 4.4 individuals, and the 20% species class has 12.6 individuals; so the increment is 8.2 individuals. The 30% species class has 23.6 individuals, so the increment from the 20% class is 11 individuals, and so on. The Kolmogorov-Smirnov test compares each of the incremental increases of SADs being compared, and the maximum difference (Dmax) serves as the basis on which one distribution is designated as statistically different from another. In our example, the SAD of the community sampled by Malaise traps and the distribution sampled by netting were significantly different from each other ($\chi^2 = 633.43$, df = 2, $p < .0001$).

We have described an approach that can be useful in a variety of community ecology studies for comparisons and assessments of natural enemy or herbivore assemblages in different agroecosystems or in unmanaged habitats. The use of the approach allows for more robust, statistically contrasted, and thus unequivocal comparisons that are more informative to theoretical and applied ecologists. Next, we use abundance data from various studies to compare SADs of assemblages affected by different variables. Further, we show how comparisons of SADs may be important in CBC, in general, and more specifically in the evaluation of important CBC hypotheses, such as the SACH.

Table 16.1. Raw Data on Species Abundance Distributions of Horsefly and Deerfly Community Sampled with Malaise Traps and Netting

Abundance Category	Malaise	Netting
1	5	7
2	6	1
3	3	3
4	5	1
5	0	2
6	3	1
8	0	1
9	2	1
11	1	1
13	3	0
14	2	0
15	1	1
16	0	2
17	0	1
18	0	3
21	0	1
22	1	0
25	0	1
29	0	2
35	0	1
37	0	1
39	0	1
40	1	0
44	0	1
49	1	0
60	1	0
82	0	1
84	1	0
90	0	1
96	1	0
103	1	0
112	1	0
130	1	0
174	1	0
236	1	0
327	0	1
366	1	0
429	0	1
437	1	0
527	0	1
855	0	1
7594	0	1

Source: Data from Tallamy et al. (1976).

Table 16.2. Calculated Class Increments (CI) Used for Test of Statistical Significance (Kolmogorov-Smirnov Two-Group Test) and Cumulative Abundances (CA) Used in the SAD Graphs

Percentage of Species	Malaise		Netting	
	CI	CA	CI	CA
10	4.4	4.4	4	4
20	8.2	12.6	5	9
30	11.0	23.6	11	20
40	16.8	40.4	24	44
50	23.6	64.0	51	95
60	47.2	111.2	69	164
70	60.8	172.0	93	257
80	190.8	362.8	140	397
90	456.2	819.0	543	940
100	1,265.0	2,084.0	9,407	10,347

Comparisons of Communities and Assemblages

What is most remarkable when one begins to generate SADs for communities and assemblages is the broad similarity of the observed patterns across different assemblages, regardless of taxa, habitat type, or species niche. In general, in most assemblages and communities, relatively few species account for most of the abundance of the assemblage or community. This widespread pattern has been found repeatedly in studies using models like the log-series and log-normal distributions.

However, although SAD patterns are similar, there are nevertheless differences in the number of numerically dominant species in each assemblage. For example,

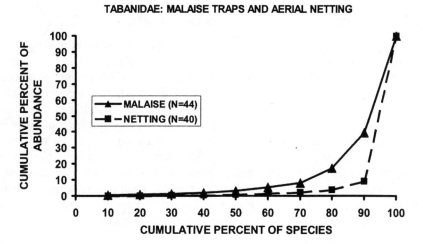

Figure 16.1. Comparison of the SADs of tabanids sampled from Malaise traps or aerial netting.

the number of species that represent the 10% most abundant species in an assemblage may be two species or six. It is in the detection of these important differences that our method is particularly useful. Given that a straightforward method for the statistical comparison of SADs has been unavailable until now, it is difficult to assert with direct evidence that our method is useful for understanding predator-prey interactions and CBC. However, in the absence of direct evidence, we use data from various studies to generate Robbins's curves and compare SADs, in order to illustrate the circumstances under which SADs differ and the factors that may cause differences in SADs. Further, we highlight the potential of Robbins's curve comparisons in research on CBC.

There are at least three major applications of SAD comparisons. The first assumes that the number of abundant species has a major influence on the nature, frequency, and outcome of predator-prey interactions, so SAD comparisons that give quantitative insights into the identity of numerically dominant species are useful. For example, SAD comparisons can identify the species that represent the 10% most abundant species in an assemblage, and with this information one can determine whether these species persist over time as numerically dominant species in any given habitat and whether effective biological control persists. A second application of SAD comparisons could be in the determination of the ecological forces that may alter the number and identity of, for example, the 10% most abundant species in the assemblage. Knowing the circumstances under which SADs differ may help identify potential traits that can alter the nature, frequency, and outcome of predator-prey interactions. The third application of SAD comparisons could be in the assessment of the impact of CBC tactics on natural enemy assemblages. The implementation of CBC tactics should alter the SAD of natural enemies like predators, or the SAD of the prey on which they feed. We need to know whether, and in what way, CBC tactics enhance the number of numerically abundant predators (or of all predators) and whether this enhancement actually enhances biological control.

The Influence of Ecological Factors on SADs

There are, of course, many unanswered questions about the SADs of predators. One can ask if temporal and spatial forces, assemblage taxa, habitat (or prey's host plant), or even sampling method significantly alter patterns of SAD. The ultimate determination of the relative importance of these factors must be based on rigorous experiments in which one or more factors are simultaneously evaluated, with a minimum of confounding factors. However, until such research is conducted, use of data from available studies to generate and compare SADs is a good first step. For example, we can use Robbins's curve analyses to identify temporal and spatial differences in predator and prey assemblages, and perhaps correlate these with potential causal factors. Further, we might determine how changing plant or host herbivore availability, habitat characteristics, or shifts in environmental parameters alter SAD. Last, using this new approach, we may be able to examine the relative stability of assemblages or communities in time and space.

Does the SAD of an assemblage vary over time? A comparison of the SAD of predators of aphids in soybean (Rutledge et al. 2004) indicates that patterns of SAD in certain predator assemblages may change from year to year. Indeed, the SAD of predators may change significantly within a growing season. The SAD of aphid predators changes significantly from early season to mid-season in each of two years, 2001 and 2002 (χ^2 = 32.82 and 53.79, respectively; p < .0001, Figure 16.2A,B). Other seasonal changes also may be highlighted with SAD comparisons. For example, although the most abundant 10% of the species accounted for a higher proportion of the total abundance in mid-season (compared to early season) in 2002, the reverse

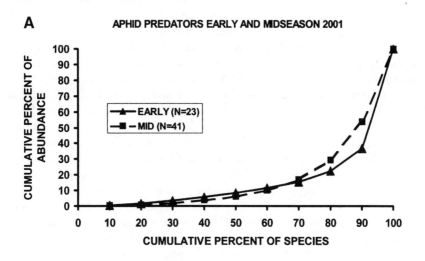

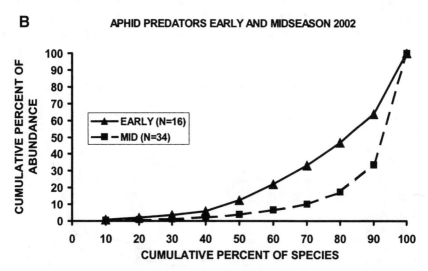

Figure 16.2. (A) Comparison of the SADs of early- and mid-season of the predator assemblage associated with the soybean aphid on soybean in 2001 and 2002. (B) Comparison of the SADs of early-season predators of the soybean aphid collected in 2001 to early-season predators collected in 2002, and of mid-season predators of the soybean aphid collected in 2001 to mid-season predators collected in 2002.

was true in 2001. In addition, there was a significant difference between early-season 2001 and early-season 2002, and between mid-season 2001 and mid-season 2002, SADs (χ^2 = 34.83 and 86.54, respectively; $p < .0001$, Figure 16.2A,B; Rutledge et al. 2004). In different agroecosystems, however, patterns of variation in the SADs of assemblages, over time, may differ. In some agroecosystems, there may be little difference in SAD across years. For example, there was no significant difference in the SAD of all arthropods on hops (*Humulus lupulus*) cultivated in 2000 and in 2001 (χ^2 = 4.14, $p = .126$, Figure 16.3; Gardiner et al. 2003), perhaps because taxa on this crop, microclimate, or crop plant quality changed little from year to year.

When comparing the SADs of assemblages in distant habitats, one might expect them to differ. For example, the SADs of Lepidoptera (collected in light traps) in two distant geographical areas such as Malaya and Rothamstead, England, are significantly different (χ^2 = 1245.05, $p < .0001$, Figure 16.4; Williams 1964). However, this difference may not always occur. For example, the SADs of sphingids in Ghana and Nigeria are not significantly different (χ^2 = 0.48, $p = .7853$, Figure 16.5; Williams 1964). Although significant geographic distances may not produce differences in SAD, the use by taxonomically similar herbivores of different plant species in the same habitat may result in different SADs. Thus, the SADs of lepidopteran assemblages on oak and blueberry differ—in essence, abundance is spread across many more species on oak than among the species in the assemblage feeding on the understory shrub, blueberry (χ^2 = 133.97, $p < .0001$, Figure 16.6; Wagner et al. 1995). There also may be a significant difference in the SADs of assemblages of taxonomically related species on the same host plant. For example, the SADs of micro- and macrolepidoptera on blueberry plants (two lepidopteran assemblages with different species that differ most preeminently in size and feeding mode) are significantly different (χ^2 = 133.97, $p < .0001$, Figure 16.7; Wagner et al. 1995). Even the ways in which data used to generate SAD are collected may influence the ultimate depiction

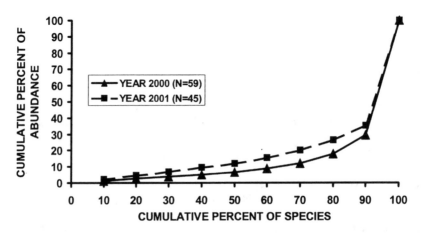

Figure 16.3. Comparison of the SADs of arthropods on hops cultivated in 2000 and in 2001.

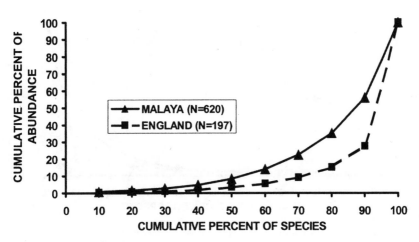

Figure 16.4. Comparison of the SADs of Lepidoptera (collected in light traps) in Malaya and Rothamstead, England.

of SAD. As we noted in our illustrative example, the depiction of the SAD of the same dipteran (tabanid) community can be significantly different ($\chi^2 = 633.43$, $p <$ .0001, Figure 16.1; Tallamy et al. 1976), depending on whether the data used to generate Robbins's curves are based on samples acquired with aerial netting or Malaise traps.

In general, a variety of factors may alter the SAD of species in any given assemblage or community. These changes in SAD have the potential to alter the number and ultimate outcome of critical interactions between predators and their prey. From

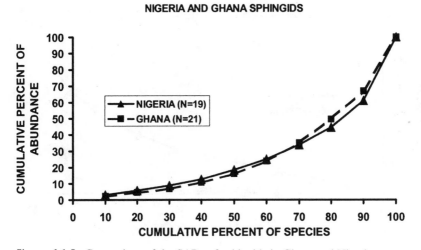

Figure 16.5. Comparison of the SADs of sphingids in Ghana and Nigeria.

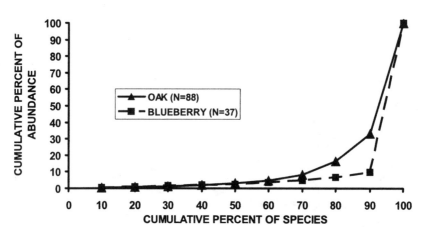

Figure 16.6. Comparison of the SADs of Lepidoptera on oak and blueberry.

an applied perspective, changes in the nature of interactions between predators and prey may determine the success of biological control efforts directed at pest species. The ability to compare the SAD of predator assemblages provides a useful tool for evaluating important hypotheses that attempt to predict the outcome of predator-prey interactions or predict the mechanisms that regulate pest populations. Thus, changes in SADs may provide useful comparative criteria when hypotheses such as the SACH are tested, because they focus on the importance of interactions between multiple predators and target pest species. Rigorous comparisons of SADs of predator assemblages may be useful in the refinement or evaluation of not only the SACH, but also other CBC hypotheses, as they arise.

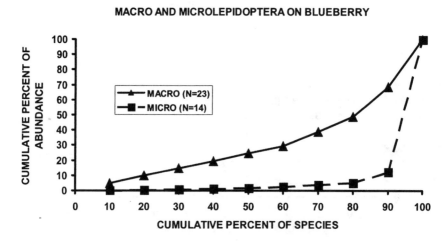

Figure 16.7. Comparison of the SADs of micro- and macrolepidoptera occurring on blueberry.

SAD and the SACH

Predators can be dominant and diverse members of many agroecosystems (Barbosa 1998a). More important, generalist predators can be effective regulators of herbivore abundance (Carter and Rypstra 1995, Settle et al. 1996, Perfecto and Castiñeiras 1998, Eubanks and Denno 2000). Riechert and Lawrence (1997), Cardinale et al. (2003), and Sinclair et al. (2003) have argued or provided support for the SACH, which specifically predicts that predator species assemblages can limit prey species abundance to a greater extent than single predator species.

Symondson et al. (2002) found that in 74% of the 102 studies that they examined, single generalist predators had a significant impact on herbivore pest abundance. In their analysis, the number of studies demonstrating reduced pest densities due to predators increased to 79% if studies assessing the impact of assemblages of generalist predators were included. Further, in those studies that assessed the impact on crops, 95% found decreased crop damage, or yield increases. Although the Symondson et al. (2002) review might be viewed as support for the value of the use of single predator species rather than reliance on assemblages of predators (although not by the authors, personal communication), it is not necessarily the conclusion that should be reached. The results of most of the studies of single predators, included in the review, may be subject to misinterpretation. That is, in most of the studies, assessment of the agent responsible for control focused solely on single target predator species, and assessment was made of other predators in the habitat, predators that may have had a significant synergistic impact on the pest mortality. Further, most studies on predation in agroecosystems rarely, if ever, evaluate the indirect impacts of predators (but see Schmitz, ch. 12 in this volume).

Some studies have reported that biological control of herbivores is most often accomplished by only a single, often specialist, enemy (Myers 1985, Myers et al. 1989, Denoth et al. 2002). However, these and other studies have focused primarily on parasitoids and address whether it is advantageous to release one or several natural enemy species, not whether a community of natural enemies is more or less effective than a single species. Further, nonbiological factors, including political and economic factors, also may favor and perpetuate the use of single species in insect pest management. For example, there is the tendency of implementation agencies to duplicate the success of single (successful) releases rather than release additional species, as well as the tendency to rely on single species because of the logistic and economic constraints of rearing and release of additional biological control agents. These and other factors have reinforced the assumed merits of reliance on single species instead of assemblage control (i.e., reliance on an array of species).

The SACH is a hypothesis that provides an alternative perspective to single species control, but it depends on the validity of several assumptions that remain to be tested. Further, the SACH emphasizes the importance of an assemblage of species but fails to predict whether all natural enemies in the assemblage, or a select subset of natural enemies in the assemblage, are necessary for effective biological control. Development and analyses of SAD curves can tell us what proportion of predator assemblages account for most predator abundance, how and under what circumstances the proportion changes, and the identity of the predators involved.

An alternative, but untested, application of the Robbins's curves might be to consider predator size-class categories rather than species. Predator size-class abundance categories can be compared to those of the prey assemblage, as these categories vary over space (i.e., within a field or across habitats) and time (throughout a season). One could determine whether the size-class abundance category better correlates with an assemblage's effect on prey than would the classical SAD based on predator taxa (see Riechert et al. 1999, in which they found that changes in spider size through the growing season had strong influences on predator effectiveness).

Assumptions of the SACH

A key assumption of the SACH is that predation by the species in an assemblage is synergistic (Losey and Denno 1998, Lang 2003), or at least additive (Snyder and Ives 2003). However, a number of researchers have reported the existence and potentially negative impact of intraguild predation on herbivore regulation (Rosenheim et al. 1993, Spiller and Schoener 1994, Ferguson and Stiling 1996, Moran et al. 1996, Rosenheim et al. 1999, Schoener and Spiller 1999, Wise and Chen 1999, Rosenheim 2001, Finke and Denno 2002). Yet there is abundant evidence of significant prey reduction by predator assemblages, despite the potential for intraguild predation. A variety of factors may preclude potential intraguild predation from ever being a significant factor in the field, even in a highly speciose assemblage. Differences in structural complexity, ranging from relatively small changes in microhabitat (Roda et al. 2000, Norton et al. 2001), to landscape-scale habitat differences (Gunnarsson 1990, Langellotto and Denno 2004), can mitigate any potential detrimental impacts of intraguild predation (see Denno et al., ch. 10 in this volume). In fact, the role of these mitigating factors could be confirmed by comparison of SADs of predator assemblages in the presence or absence of structural complexity or other factors that mitigate intraguild predation. These types of comparisons will show whether the presence of mitigation factors supports higher abundance of many more predators than when those factors are absent.

The Use of SAD Curves

As we have noted, the pattern for most assemblages and communities is that a few predator species account for most of the abundance of natural enemies. Therefore, maximizing the density of all predators may not be necessary. Understanding patterns of SAD may be useful in the identification of the subset of numerically dominant predator species that might be most responsible for pest regulation, and SAD comparisons can show if the number or the identity (or both) of this subset of dominant predators change under varying conditions. Of course, SAD determinations and comparisons must be complemented by studies that determine whether numerically dominant species are also the key predators responsible for the regulation of the target pest herbivore in the system. Further, the life histories of the predators must be such that they exert a significant influence on the targeted prey throughout the life cycle of the prey and its growing season.

If the numerically dominant species also are the key predators, then, rather than promoting CBC aimed at all natural enemies in the assemblage, CBC could be aimed only at the species that are most abundant and likely to be effective. This possibility assumes a relationship that has not been empirically demonstrated across different agroecosystems. That is, it is unclear whether in most agroecosystems the most abundant predator is the most effective against the pest species. There is a further assumption: that the pest is the most abundant, or among the most abundant, of the herbivores; otherwise, one would be confronting the unlikely scenario of a highly abundant predator feeding on something other than the most abundant prey. This unlikely scenario might exist in circumstances in which predators exhibit a high degree of specificity in prey choice, a preference uninfluenced by the density of alternative potential prey. Further, pest status is not always based on density. Clearly, there are pest species that incur significant economic damage at relatively low abundance, because of the nature of their damage, their ability to transmit disease, or consumer demands for unblemished produce. Nevertheless, this type of pest is very much in the minority among agricultural pests.

Targeting a particular subset of predators with specific tactics may be increasingly important, given recent findings that important resources (e.g., pollen and nectar) vary in quality and quantity (Broufas and Koveosa 2000, Wäckers 2001) and in their availability (Pickett and Bugg 1998, Baggen et al. 1999, Landis et al. 2000). Many current tactics assume that these and other required resources are equally advantageous to all natural enemies. However, they may not be of universal or equivalent benefit to all natural enemies. If a subset of predators is to be targeted, it may be more practical, feasible, and effective to develop specific conservation tactics and strategies that provide the type and quantity of resources needed by these targeted species. These conceptual changes in CBC have the potential of making CBC much more practical and readily acceptable to growers. Given the current view of CBC as an approach that aims to conserve all species in natural enemy assemblages, these recommendations for tactics that target a select set of natural enemies represent a marked departure from the current aims of CBC.

SAD, Predator-Prey Interactions, CBC, and Future Directions

Relatively few studies have thoroughly determined how critical components of natural enemy communities influence CBC. Thus, there are many unanswered questions. It is only through rigorous temporal and spatial comparisons of SADs that we can know whether species that are numerically dominant continue to be so over time and space, and whether they are critical in the regulation of pests. There are some suggestions that high abundances of species do not appear to be maintained at particular sites and latitudinal positions, and that climatic conditions and other environmental factors may determine whether species' abundances remain constant in space and time (Coull and Fleeger 1977, Jarvinen 1979, Taylor and Taylor 1979, Grubb et al. 1982, Grubb 1986, Bohning-Gaese et al. 1994, Wilson et al. 1996). Brown et al. (1995)

noted that species are found in low abundance at most sites throughout their respective geographical ranges and achieve high abundances at only a small number of sites. A detailed assessment of SADs across a geographical range of habitats encompassing changing climatic conditions could go a long way toward resolving many of these unanswered questions. The SADs might also be used to examine how communities assemble, particularly predator communities that result from the recolonization of disturbed systems (e.g., tilled fields) each season. However, given the inability, in the past, to statistically contrast SADs, these remain open questions.

Although experiments that specifically address the role of one or more factors on SAD are required, comparisons of SADs based on data taken from available studies suggest that a variety of factors have the potential to cause qualitative and quantitative changes in SAD, predator-prey interactions, and CBC. Next, we provide illustrations of how the SADs of assemblages or communities change as a consequence of habitat type, predator type, species niche, and pest management tactics. These examples in no way represent a synthesis or consensus using all available data on the subject but are provided merely to illustrate the type of result that may emerge from appropriate experimental research.

The impact of management practices on predators may be reflected in differences in the SAD of predators in agroecosystems and unmanaged habitats. The SAD of spiders in a soybean agroecosystem does not differ significantly from that of spiders in a rural (conifer-forest-dominated) habitat ($\chi^2 = 3.09$, $p = .213$, Figure 16.8A; Alaruikka et al. 2002). However, the SAD of spiders may differ if other agroecosystems or unmanaged habitats are compared. Might differences in SAD, or a lack thereof, be a taxon-specific phenomenon? For spiders in urban and rural habitats there were not significant differences in SAD ($\chi^2 = 1.64$, $p = .4396$, Figure 16.8B; Alaruikka et al. 2002). Yet the SAD of carabids in urban and rural habitats were significantly different ($\chi^2 = 10.72$, $p = .005$, Figure 16.8C; Alaruikka et al. 2002). The SADs of different types of predators may differ even in the same habitat. For example, the SAD of carabids were significantly different from that of spiders in the rural habitat, as were they in the urban habitat ($\chi^2 = 45.63$, $p < .0001$, and $\chi^2 = 8.88$, $p = .012$, respectively; Figure 16.8B,C; Alaruikka et al. 2002). Interestingly, about 33% of the carabids collected occurred in one but not both types of habitats. In contrast, about 55% of the spider species collected occurred in one but not both types of habitats.

The effectiveness of biological control may be reflected in differences in the SAD of predators and that of their prey. Could it be that the likely success of control by an assemblage is influenced by the number of species that are numerically dominant, relative to the number of numerically dominant pest species? The SAD of alfalfa herbivores can be significantly different from that of predators in alfalfa ($\chi^2 = 59.05$, $p < .0001$, Figure 16.9; Wheeler 1971). The difference may reflect the relatively larger number of abundant predators in alfalfa, compared to the relatively limited number of numerically dominant herbivore species. Still another use for SAD comparisons may be to depict the identity and relative abundance of numerically dominant species of predators and herbivores, in their respective assemblages when associated with a plant species either cultivated or growing as a wild plant. In hops, it makes a difference in terms of the SAD of potential prey. Abundance was spread across many more species

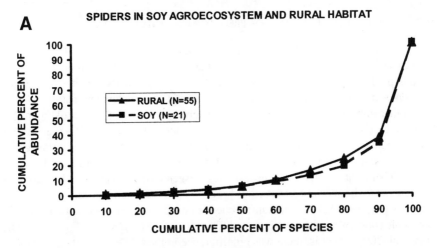

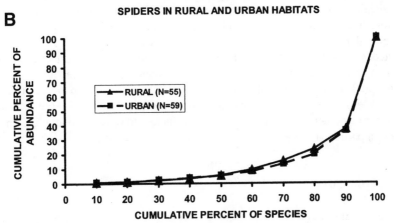

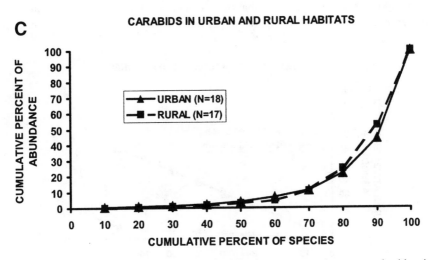

Figure 16.8. (A) Comparison of the SADs of spiders in a soy agroecosystem and spiders in a rural habitat. (B) Comparison of the SAD of spiders in urban and rural habitats. (C) Comparison of the SAD of carabids in urban and rural habitats.

of arthropods on cultivated hops compared to feral hops ($\chi^2 = 48.75, p < .0001$, Figure 16.10; Gardiner et al. 2003). That is, there were more abundant arthropods on cultivated hops. This type of difference may result in quantitative and qualitative changes in predator-prey interactions and in the effectiveness of predators and parasitoids.

Another important use of SAD is in the determination of the impact of "management strategies" on natural enemy and pest communities. For example, the SAD of pests associated with transgenic *Bt* cotton and that of the pests on non-*Bt* transgenic cotton was significantly different ($\chi^2 = 121.93, p < .0001$, Figure 16.11A; Men et al. 2003). This might be expected, because the transgenic *Bt* cotton targets Lepidoptera and can cause the mortality of various species. However, perhaps somewhat unexpected is that the SADs of the natural enemies on transgenic *Bt* cotton and non-*Bt* transgenic cotton were not significantly different ($\chi^2 = 0.313, p = .855$, Figure 16.11B; Men et al. 2003). That is, there was no difference in the SADs, even though the use of transgenic cotton results in the elimination of the eggs, larvae, pupae, and adults of various lepidopteran species used by predators and parasitoids as hosts or prey (Huang et al.1999). In the same fashion, tactics or approaches to CBC might be evaluated with SAD comparisons. That is, what, if any, are the CBC tactics (or approaches) that alter SAD, and do they result in changes, in the predator-prey and predator-predator interactions, that enhance biological control. Comparisons of SADs can help determine whether specific CBC tactics significantly alter SAD (i.e., whether the number of numerically dominant species is increased or reduced after the application of a particular tactic).

Epilogue

We have provided a tool (Robbins's curves) for depicting and comparing SADs, and examples of important questions that may be addressed, in the hope of stimulating further research. The SAD comparisons provided in the chapter can give some guid-

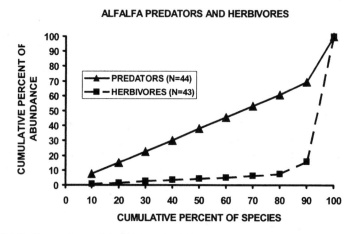

Figure 16.9. Comparison of the SADs of alfalfa herbivores and predators.

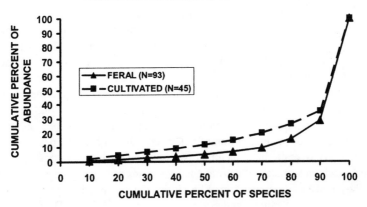

Figure 16.10. Comparison of the SADs of feral and cultivated hops.

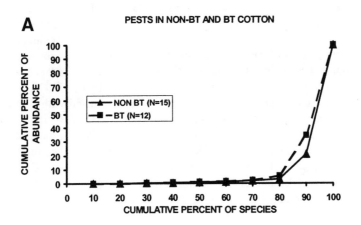

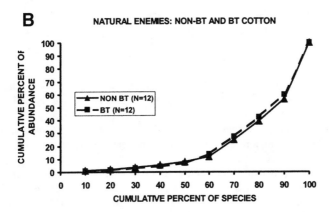

Figure 16.11. (A) Comparison of the SADs of pest species on transgenic *Bt* cotton and non-*Bt* transgenic cotton. (B) Comparison of the SADs of natural enemies on transgenic *Bt* cotton and non-*Bt* transgenic cotton.

ance as to the types of comparisons that might be useful in an experimental context. The experimental evaluation of CBC tactics and strategies is a major area of future research. Specifically, we need to generate and compare the SADs of assemblages of predators and herbivores in plots with and without flowers (and thus nectar and pollen), refuges (e.g., beetle banks), and ground covers, or any other CBC tactic, to determine if they influence the proportion or identity of numerically dominant predators. Such changes have potentially significant impacts on the number, type, and outcome of predator-prey interactions and thus on biological control.

Research on the SAD of predator assemblages can help determine if abundance is spread across a larger or smaller number of species than it is in the pest assemblage, the proportion of overall abundance that is accounted for by major pest species, and the extent to which some or all natural enemy species of target pests need to be conserved. Research in agroecosystems like alfalfa (as discussed) may provide an opportunity for us to determine whether differences between the SAD of pests and the SAD of their natural enemies can predict the likelihood of species assemblage control. If the difference between predator and prey SADs is confirmed (e.g., that there is a greater number of species represented in the most abundant 10% of predators than in most abundant 10% of the prey), one might determine whether all or some of the numerically dominant predators are important regulators of pest abundance. This type of assessment can help determine if CBC tactics should be directed to the entire suite of numerically dominant predators or to a smaller subset.

Conservation biological control tactics also can be evaluated with comparisons of the SADs of the assemblages of key predators exposed to the CBC tactic and control plots. Ideally, the SAD of the predator assemblage in CBC tactic plots should show a greater proportion of total abundance represented by the key predator species than that in control plots. These are but a few examples of the type of research that is needed to broaden our understanding of the ecology of predator-prey interactions and increase the potential for the successful use of CBC.

Acknowledgments

We thank R. K. Robbins for the original idea leading to the method described here, and H. Obrecht, biologist of the Patuxent Wildlife Refuge, for his assistance and support.

Literature Cited

Abrams, P. A., Menge, B. A., Mittelbach, G. G., Spiller, D., and Yodzis, P. 1996. The role of indirect effects in food webs. In: Food Webs: Integration of Patterns and Dynamics (Polis, G., and Winemiller, K., eds.). New York: Chapman & Hall; 371–395.

Alaruikka, D., Johan Kotze, D., Matveinen, K., and Niemela, J. 2002. Carabid beetle and spider assemblages along a forested urban-rural gradient in southern Finland. J. Insect Cons. 6:195–206.

Altieri, M. A. 1991. Increasing biodiversity to improve insect pest management in agroecosystems. In: Biodiversity of Microorganisms and Invertebrates: Its Role in Sustainable Agriculture (Hawksworth, D. L., ed.). Wallingford, U.K.: CABI; 165–182.

Andow, D. 1991. Vegetational diversity and arthropod population response. Annu. Rev. Entomol. 36:561–586.

Baggen, L. R., Gurr, G. M., and Meats, A. 1999. Flowers in tri-trophic systems: mechanisms allowing selective exploitation by insect natural enemies for conservation biological control. Entomol. Exp. Appl. 91:155–161.

Barbosa, P. 1998a. Agroecosystems and conservation biological control. In: Conservation Biological Control (Barbosa, P., ed.). San Diego: Academic Press; 39–54.

Barbosa, P. 1998b. Conservation Biological Control. San Diego: Academic Press.

Barrion, A. T., Aquino, G. B., and Heong, K. L. 1994. Community structures and population dynamics of rice arthropods in irrigated rice fields in the Philippines. Philipp. J. Crop Sci. 9:73–85.

Berndt, L. A., Wratten, S. D., and Hassan, P. G. 2002. Effects of buckwheat flowers on leafroller (Lepidoptera: Tortricidae) parasitoids in a New Zealand vineyard. Agric. For. Entomol. 4:39–45.

Bohning-Gaese, K., Taper, M. L., and Brown, J. H. 1994. Avian community dynamics are discordant in space and time. Oikos 70:121–126.

Boller, E. F. 1992. The role of integrated pest management in integrated production of viticulture in Europe. In: Proceedings of the British Crop Protection Conference, Pests and Diseases, Brighton. Brighton, U.K.: British Crop Protection Council; 499–506.

Bottrell, D., Barbosa, P., and Gould, F. 1998. Manipulating natural enemies by plant variety selection and modification: a realistic strategy? Annu. Rev. Entomol. 43:347–367.

Broufas, G. D., and Koveosa, D. S. 2000. Effect of different pollens on development, survivorship and reproduction of *Euseius finlandicus* (Acari: Phytoseiidae). Environ. Entomol. 29:743–749.

Brown, J. H., Mehlman, D. W., and Stevens, G. C. 1995. Spatial variation in abundance. Ecology 76:2028–2043.

Brust, G. E., Stinner, B. R., and McCartney, D. A. 1985. Tillage and soil insecticide effects on predator–black cutworm (Lepidoptera: Noctuidae) interactions in corn agroecosystems. J. Econ. Entomol. 78:1389–1392.

Bugg, R. L., and Waddington, C. 1994. Using cover crops to manage arthropod pests of orchards: a review. Agric. Ecosyst. Environ. 50:11–28.

Cardinale, B. J., Harvey, C. T., Gross, K., and Ives, A. R. 2003. Biodiversity and biocontrol: emergent impacts of a multi-enemy assemblage on pest suppression and crop yield in an agroecosystem. Ecol. Lett. 6:857–865.

Carter, P. E., and Rypstra, A. L. 1995. Top-down effects in soybean agroecosystems: spider density affects herbivore damage. Oikos 72:433–439.

Claridge, M. F. 1987. Insect assemblages—diversity, organization, and evolution. In: Organization of Communities. Past and Present. 27th Symposium British Ecological Society (Gee, J. H. R., and Giller, P. S., eds.). Oxford: Blackwell Scientific Publications; 141–162.

Collins, K. L., Boatman, N. D., Wilcox, A., and Holland, J. M. 2002. The influence of beetlebanks on cereal aphid population predation in winter wheat. Agric. Ecosyst. Environ. 93:337–350.

Colunga-Garcia, M., Gage, S. H., Landis, D. A. 1997. Response of an assemblage of Coccinellidae (Coleoptera) to a diverse agricultural landscape. Environ. Entomol. 26:797–804.

Corbett, A., and Rosenheim, J. A. 1996. Impact of a natural enemy overwintering refuge and its interaction with the surrounding landscape. Ecol. Entomol. 21:155–164.

Coull, B. C., and Fleeger, J. W. 1977. Long-term temporal variation and community dynamics of meiobenthic copepods. Ecology 58:1136–1143.

Cushman, J. H., and Whitman, T. G. 1991. Competition mediating the outcome of a mutualism: protective services of ants as a limiting resource for membracids. Am. Nat. 138:851–865.

DeBach, P. 1974. Biological Control of Pests and Weeds. New York: Reinhold.

De Moraes, C. M., Lewis, W. J., Pare, P. W., Alborn, H. T., and Tumlinson, J. H. 1998. Herbivore-infested plants selectively attract parasitoids. Nature 393:570–573.

Denoth, M., Frid, L., and Myers, J. H. 2002. Multiple agents in biological control: improving the odds? Biol. Contr. 24:20–30.

DeVries, P. J., Murray, D., and Lande, R. 1997. Species diversity in vertical, horizontal, and temporal dimensions of a fruit feeding butterfly community in an Ecuadorian rainforest. Biol. J. Linn. Soc. 62:343–364.

Dewdney, A. K. 2000. A dynamical model of communities and a new species-abundance distribution. Biol. Bull. 198:152–163.

Eubanks, M. D., and Denno, R. F. 2000. Host plants mediate omnivore-herbivore interactions and influence prey suppression. Ecology 81:936–947.

Ferguson, K. I., and Stiling, P. 1996. Non-additive effects of multiple natural enemies on aphid populations. Oecologia 108:375–379.

Finke, D. L., and Denno, R. F. 2002. Intraguild predation diminished in complex-structured vegetation: implications for prey suppression. Ecology 83:643–652.

Fisher, R. A., Corbet, A. S., and Williams, C. B. 1943. The relation between the number of species and the number of individuals in a random sample of an animal population. J. Anim. Ecol. 12:42–58.

Gardiner, M. M., Barbour, J. D., and Johnson, J. B. 2003. Arthropod diversity and abundance on feral and cultivated *Humulus lupulus* (Urticales: Cannabaceae) in Idaho. Environ. Entomol. 32:564–574.

Gotelli, N. J., and Colwell, R. K. 2001. Quantifying biodiversity: procedures and pitfalls in the measurement and comparison of species richness. Ecol. Lett. 4:379–391.

Grubb, P. J. 1986. Problems posed by sparse and patchily distributed species in species-rich plant communities. In: Community Ecology (Diamond, J. M., and Case, T. J., eds.). New York: Harper & Row; 207–225.

Grubb, P. J., Kelly, D., and Mitchley, J. 1982. The control of relative abundance in communities of herbaceous plants. In: The Plant Community as a Working Mechanism. Special Publications Series of the British Ecological Society (Newman, E. I., ed.). Oxford: Blackwell Scientific Publications; 79–97.

Gunnarsson, B. 1990. Vegetation structure and the abundance and size distribution of spruce-living spiders. J. Anim. Ecol. 59:743–752.

Hawkins, B. A., and Mills, N. J. 1996. Variability in parasitoid community structure. J. Anim. Ecol. 65:501–516.

Hawksworth, D. L. 1991. Biodiversity of Microorganisms and Invertebrates: Its Role in Sustainable Agriculture. Wallingford, U.K.: CABI.

Heong, K. L., Aquino, G. B., and Barrion, A. T. 1991. Arthropod community structures of rice ecosystems in the Philippines. Bull. Entomol. Res. 81:407–416.

Huang, F. N., Higgins, R. A., and Buschman, L. L. 1999. Transgenic *Bt*-plants: successes, challenges and strategies. Pestology 2:2–29.

Hughes, R. G. 1986. Theories and models of species abundance. Am. Nat. 128:879–899.

Jarvinen, O. 1979. Geographical gradients of stability in European land bird communities. Oecologia 38:51–69.

Krebs, C. J. 1989. Ecological Methodology. New York: Harper & Row.

Landis, D. A., Wratten, S. W., and Gurr, G. M. 2000. Habitat management to conserve natural enemies of arthropod pests in agriculture. Annu. Rev. Entomol. 45:175–201.

Lang, A. 2003. Intraguild interference and biocontrol effects of generalist predators in a winter wheat field. Oecologia 134:144–153.

Langellotto, G. A., and Denno, R. F. 2004. Responses of invertebrate natural enemies to complex-structured habitats: a meta-analytical synthesis. Oecologia 139:1–10.

Letourneau, D. K., and Dyer, L. A. 1998a. Density patterns of *Piper* ant-plants and associated arthropods: top predator cascades in a terrestrial system? Biotropica 30:162–169.

Letourneau, D. K., and Dyer, L. A. 1998b. Experimental test in lowland tropical forest shows top-down effects through four trophic levels. Ecology 79:1678–1687.

Losey, J. E., and Denno, R. F. 1998. Positive predator-predator interactions: enhanced predation rates and synergistic suppression of aphid populations. Ecology 79:2143–2152.

Ludwig, J. A., and Reynolds, J. F. 1988. Statistical Ecology. New York: Wiley.

Magurran, A. E. 1988. Ecological Diversity and Its Measurement. Princeton, N.J.: Princeton University Press.

Marino, P. C., and Landis, D. A. 2000. Parasitoid community structure: implications for biological control in agricultural landscapes. In: Interchanges of Insects between Agricultural and Surrounding Habitats (Ekbom, B., ed.). Dordrecht, the Netherlands: Kluwer Academic; 183–194.

May, R. M. 1975. Patterns of species abundance and diversity. In: Ecology and Evolution of Communities (Cody, M. L., and Diamond, J. M., eds.). Cambridge, Mass.: Belknap/Harvard University Press; 81–120.

Men, X., Ge, F., Liu, X., and Yardim, E. N. 2003. Diversity of arthropod communities in transgenic *Bt* cotton and nontransgenic cotton agroecosystems. Environ. Entomol. 32:270–275.

Moran, M. D., Rooney, T. P., and Hurd, L. E. 1996. Top-down cascade from a bitrophic predator in an old-field community. Ecology 77:2219–2227.

Myers, J. H. 1985. How many insect species are necessary for successful biocontrol of weeds? In: Proceedings of the 6th International Symposium of the Biological Control of Weeds, Agriculture Canada (Delfosse, E. S., ed.). Ottawa, Canada: Canadian Government Printing Office; 77–82.

Myers, J. H., Higgins, C., and Kovacs, E. 1989. How many insect species are necessary for the biological control of insects? Environ. Entomol. 18:541–547.

Norton, A. P., English-Loeb, G., and Belden, E. 2001. Host plant manipulation of natural enemies: leaf domatia protect beneficial mites from insect predators. Oecologia 126: 535–542.

Perfecto, I., and Castiñeiras, A. 1998. Deployment of the predaceous ants and their conservation in agroecosystems. In: Conservation Biological Control (Barbosa, P., ed.). San Diego: Academic Press; 269–289.

Pickett, C. H., and Bugg, R. L. 1998. Enhancing Biological Control: Habitat Management to Promote Natural Enemies of Agricultural Pests. Berkeley: University of California Press.

Platt, J. O., Caldwell, J. S., and Kok, L. T. 1999. Effect of buckwheat as a flowering border on populations of cucumber beetles and their natural enemies in cucumber and squash. Crop Prot. 18:305–313.

Preston, F. W. 1948. The commonness, and rarity, of species. Ecology 29:254–283.

Preston, F. W. 1962a. The canonical distribution of commonness and rarity: part I. Ecology 43:185–215.

Preston, F. W. 1962b. The canonical distribution of commonness and rarity: part II. Ecology 43:410–432.

Provencher, L., and Riechert, S. E. 1994. Model and field test of prey control effects by spider assemblages. Environ. Entomol. 23:1–17.

Rabb, R. L., Stinner, R. E., and van den Bosch, R. 1976. Conservation and augmentation of natural enemies. In: Theory and Practice of Biological Control (Huffaker, C. B., and Messenger, P. S., eds.). New York: Academic Press; 233–254.

Riechert, S. E. 1999. The hows and whys of successful pest suppression by spiders: insights from case studies. J. Arachnol. 27:387–396.

Riechert, S. E., and Bishop, L. 1990. Prey control by an assemblage of generalist predators in a garden test system. Ecology 71:1441–1450.

Riechert, S. E., and Lawrence, K. 1997. Test for predation effects of single versus multiple species of generalist predators: spiders and their insect prey. Entomol. Exp. Appl. 84:147–155.

Riechert, S. E., Provencher, L., and Lawrence, K. 1999. The potential of spiders to exhibit stable equilibrium point control of prey: tests of two criteria. Ecol. Appl. 9:365–377.

Roda, A., Nyrop, J., Dicke, M., and English-Loeb, G. 2000. Trichomes and spider-mite webbing protect predatory mite eggs from intraguild predation. Oecologia 125:428–435.

Rosenheim, J. A. 2001. Source-sink dynamics for a generalist insect predator in habitats with strong higher-order predation. Ecol. Monogr. 71:93–116.

Rosenheim, J. A., Limburg, D. D., and Colfer, R. G. 1999. Impact of generalist predators on a biological control agent, *Chrysoperla carnea*: direct observations. Ecol. Appl. 9:409–417.

Rosenheim, J. A., Wilhoit, L. R., and Armer, C. A. 1993. Influence of intraguild predation among generalist insect predators on the suppression of an herbivore population. Oecologia 96:439–449.

Routledge, R. D. 1980. The form of species-abundance distributions. J. Theor. Biol. 82:547–558.

Rutledge, C. E., O'Neil, R. J., Fox, T. B., and Landis, D. A. 2004. Soybean aphid predators and their use in IPM. Ann. Entomol. Soc. Am. 97:240–248.

Schmitz, O. J. 1998. Direct and indirect effects of predation and predation risk in old-field interaction webs. Am. Nat. 151:327–342.

Schmitz, O. J., Beckerman, A. P., and O'Brien, K. M. 1997. Behaviorally-mediated trophic cascades: the effects of predation risk on food web interactions. Ecology 78:1388–1399.

Schoener, T. W., and Spiller, D. A. 1999. Indirect effects in an experimentally staged invasion by a major predator. Am. Nat. 153:347–358.

Schoenly, K. G., Cohen, J. E., Heong, K. L., Litsinger, J. A., Aquino, G. B., Barrion, A. T., and Arida, G. 1996. Foodweb dynamics of irrigated rice fields at five elevations in Luzon, Philippines. Bull. Entomol. Res. 86:451–466.

Schoenly, K. G., Justo, H. D., Jr., Barrion, A. T., Harris, M. K., and Bottrell, D. G. 1998. Analysis of invertebrate agrobiodiversity in a Philippine farmer's irrigated rice field. Environ. Entomol. 27:1125–1136.

Settle, W. H., Ariawan, H., Astuti, E. T., Cahyana, W., Hakim, A. L., Hindayana, D., Lestari, A. S., and Sartanto, P. 1996. Managing tropical rice pests through conservation of generalist natural enemies and alternative prey. Ecology 77:1975–1988.

Sinclair, A. R. E., Mduma, S., and Brashares, J. S. 2003. Patterns of predation in a diverse predator-prey system. Nature 425:288–290.

Snyder, W. E., and Ives, A. R. 2003. Interactions between specialist and generalist natural enemies: parasitoids, predators, and pea aphid biocontrol. Ecology 84:91–107.

Sokal, R. R., and Rohlf, F. J. 1981. Biometry. The Principles and Practice of Statistics in Biological Research. San Francisco: Freeman.

Spiller, D. A., and Schoener, T. W. 1994. Effects of top and intermediate predators in a terrestrial food web. Ecology 75:182–196.

Sugihara, G. 1980. Minimal community structure: an explanation of species-abundance patterns. Am. Nat. 116:770–787.

Sullivan, D. J., and Völkl, W. 1999. Hyperparasitism: multitrophic ecology and behavior. Annu. Rev. Entomol. 44:291–315.

Symondson, W. O. C., Suderland, K. D., and Greenstone, M. H. 2002. Can generalist predators be effective biocontrol agents? Annu. Rev. Entomol. 47:561–594.

Tallamy, D. W., Hansens, E. J., and Denno, R. F. 1976. A comparison of Malaise trapping and aerial netting for sampling a horsefly and deerfly community. Environ. Entomol. 5:788–792.

Taylor, L. R. 1978. Bates, Williams, Hutchinson—a variety of diversities. In: Diversity of Insect Faunas. Ninth Symposium of the Royal Entomological Society (Mound, L. A., and Waloff, N., eds.). Oxford: Blackwell Scientific Publications; 1–18.

Taylor, L. R., Kempton, R. A., and Woiwod, I. P. 1976. Diversity statistics and the log-series model. J. Anim. Ecol. 45:255–272.

Taylor, R. A. J., and Taylor, L. R. 1979. A behavioural model for the evolution of spatial dynamics. In: Population Dynamics (Anderson, R. M., Turner, B. D., and Taylor, L. R., eds.). Oxford: Blackwell Scientific Publications; 1–27.

van den Bosch, R., and Telford, A. D. 1964. Environmental modification and biological control. In: Biological Control of Pests and Weeds (DeBach, P., ed.). New York: Reinhold; 459–488.

Van Driesche, R. G., and Bellows, T. S., Jr. 1996. Biological Control. New York: Chapman & Hall.

Wäckers, F. L. 2001. A comparison of nectar- and honeydew sugars with respect to their utilization by the hymenopteran parasitoid *Cotesia glomerata*. J. Insect Physiol. 47:1077–1084.

Wagner, D. L., Peacock, J. W., Carter, J. L., and Talley, S. E. 1995. Spring caterpillar fauna of oak and blueberry in a Virginia deciduous forest. Ann. Entomol. Soc. Am. 88:416–426.

Wheeler, A. G. 1971. A study of the arthropod fauna of alfalfa (Ph.D. dissertation, Cornell University, Ithaca, N.Y.).

Williams, C. B. 1964. Patterns in the Balance of Nature—and Related Problems in Quantitative Ecology. New York: Academic Press.

Wilson, J. B., Wells, T. C. E., Trueman, I. C., Jones, G., Atkinson, M. D., Crawley, M. J., Dodds, M. E., and Silvertown, J. 1996. Are there assembly rules for plant species abundance? An investigation in relation to soil resources and successional trends. J. Ecol. 84:527–538.

Wise, D. H., and Chen, B. 1999. Impact of intraguild predators on survival of a forest-floor wolf spider. Oecologia 121:129–137.

17

Plight of Predators

The Importance of Carnivores for Understanding Patterns of Biodiversity and Extinction Risk

JOHN L. GITTLEMAN
MATTHEW E. GOMPPER

Predator species such as tigers and wolves are often symbols of conservation science, frequently serving as umbrella or flagship taxa for species-based conservation efforts. Predators also serve as important examples for how and why biodiversity is facing an unparalleled global extinction crisis. Here, we focus on the mammalian order Carnivora to assess arguments for the importance of a predator-based agenda in conservation, to observe global patterns of where mammalian predators are threatened, and to highlight factors influencing extinction risk, such as body size, life histories, and trophic level. On a local scale, evidence suggests that large predators may play a particularly important role in structuring ecological processes, and the loss of a few top predators may dramatically alter communities or even entire ecosystems. The role and relative importance of mid-sized or small predators, however, remains unclear, as does the role of predators in complex or highly biodiverse landscapes. Yet, given the potential importance of predators, it is essential to understand causes of their decline on a global framework. Toward this end, we show that there are particular global regions that are likely to suffer unusually high rates of extinction and describe the intricate interplay between anthropogenic effects and species' biology.

Predators receive considerable attention in the conservation sciences. Much of this attention is undoubtedly due to a charismatic image that draws attention from diverse segments of society. Nevertheless, there are good biological reasons for directing conservation resources to predators. Many of the biological and ecological characteristics of predators, such as rarity (Gaston 1994), reliance on scarce and fluctuating resources (Fuller and Sievert 2001), and susceptibility to hunting pressures, raise the chances of extinction. Lessons learned from protecting predators also indicate that they sometimes legitimately serve as important models of conservation (Gittleman et al. 2002). For this reason, we do not ask in this chapter why predators often represent focal points for conservation agendas. Rather, we consider whether these species as a collective guild *deserve* such extensive attention. We do so by assessing the effects of the loss or gain of predators in natural systems. Because most predator-based conservation agendas focus on large, terrestrial mammalian carnivores, such as tigers, wolves, jaguars, and grizzly bears, we use examples mainly from the mammalian order Carnivora. Further, we ask if it is just these large predators that deserve special conservation emphasis because of their perceived ecological importance, or whether smaller or more omnivorous members of the order (e.g., weasels, jackals, foxes, small cats, and mongooses) also deserve such attention. How robust is the evidence that large and mid-sized Carnivora always play an especially important role where they are present? And, if this guild of species is indeed especially important, what and where are the threats to their continued persistence?

We begin by highlighting several cases showing the importance of top predators in structuring ecological communities. The top (or apex) predators in particular, rather than predators in general, are emphasized for two reasons. First, these animals exist at the very top of the food pyramid and are therefore relatively rare in terms of proportional biomass of the broader community and in terms of the absolute numbers of individuals found within any particular study site. As a result, where the influence of top predators is found to be important in structuring the ecological community, the per capita impact of each predator is especially great. Thus, although the population size of a top carnivore in any given region may be quite small, the behavior of any given individual can have tremendous importance for the rest of the community. This seems to be a point that even researchers actively involved in studies of large carnivores have failed to appreciate. It is in the study of apex carnivores that ethology interfaces with community and ecosystem ecology: the behavior of just a handful of individuals can fundamentally influence entire ecosystems.

Second, apex predators are themselves rarely the subjects of interspecific predation. Although this is not to say that predation on these animals never occurs, their populations are rarely limited by top-down effects, but rather are generally limited by bottom-up resource availability. This situation is in contrast to that of mid-sized carnivores, which are often limited by top-down effects and may therefore have their ecological impact on the broader community mediated by the presence of other predator species.

If indeed carnivores are deemed especially important biologically, then we need to conserve them. A first step toward this goal is to identify global patterns of carnivore distribution, diversity, and threat. By revealing these patterns, we attempt to apply lessons learned from local and regional studies to develop a more comprehensive and predictive science-based framework for identifying potentially vulnerable species before they decline.

Wolves, Otters, and Orcas

Two "textbook" cases reveal the importance of apex predators as top-down controls of communities and even ecosystems: wolves (*Canis lupus*) in central North America, and sea otters (*Enhydra lutris*) in northern Pacific near-shore environments. Although wolves have been studied intensively in North America and Eurasia, most relevant here is the long-term work from Isle Royale, Michigan. This 544 sq. km national park lacked wolves until the late 1940s, when Lake Michigan froze over, allowing a pair of wolves to colonize the site (Mech 1966, Wayne et al. 1991). Over the past 40 years, wolf numbers have fluctuated between 10 and 50 individuals in two to three packs, and these individuals have had a significant effect on the island's community. Wolves regulate moose (*Alces alces*) that in turn regulate the growth rates of balsam fir (*Abies balsamea*) in a tightly linked three-trophic-level system (Figure 17.1A; McLauren and Peterson 1994). Balsam fir is a dominant tree species on Isle Royale, and browsing on fir by moose determines its relative abundance in the over-story, seedling establishment, sapling recruitment, forest litter production and, on an ecosystem-level edaphic nutrient control (McInnes et al. 1992, Pastor et al. 1993,

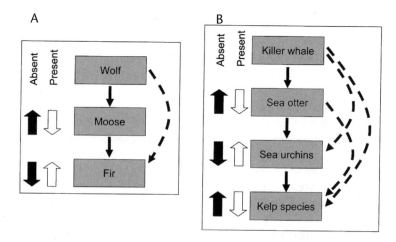

Figure 17.1. Conceptualized trophic interactions involving large top carnivores. (A) Three trophic-level systems in which wolves limit moose and beaver and thereby indirectly facilitate fir tree growth. Thin solid and thin dashed arrows represent direct and indirect trophic interactions, respectively. Thick arrows represent the relative influence of top predator presence (open arrow) or absence (filled arrow) on the relative biomass of lower trophic levels (upward arrows = increased biomass; downward arrows = decreased biomass). (B) Four-trophic-level interaction in which killer whales limit sea otters, thereby indirectly facilitating increased sea urchin biomass and decreased kelp biomass. (Adapted with permission from the following sources: Post, E., Peterson, R. O., Stenseth, N. C., and McLaren, B. E. 1999. Ecosystem consequences of wolf behavioural response to climate. Nature 401:905–907. © 1999 Nature Publishing Group. Estes, J. A., Tinker, M. T., Williams, T. M., and Doak, D. F. 1998. Killer whale predation on sea otters linking oceanic and nearshore ecosystems. Science 282:473–476. © 1998 AAAS. McLaren, B. E., and Peterson, R. O. 1994. Wolves, moose, and tree rings on Isle Rale. Science 266:1555–1558. © 1994 AAAS.)

Post et al. 1999). Although these top-down effects do not indicate that bottom-up or abiotic factors are unimportant (Vucetich and Peterson 2004), they do indicate that through indirect effects a small number of wolves can dramatically influence the structure of an entire community (see Schmitz, ch. 12 in this volume).

The role of wolves as top-down regulators of community structure is also observed in other areas. For instance wolves were reintroduced in the Greater Yellowstone Ecosystem (Idaho, Wyoming, and Montana) in 1995, and results of studies assessing the impacts of this restoration on other components of the ecosystem are now emerging (Smith et al. 2003). Most intriguing are the altered behavioral and foraging ecologies of prey species and mesocarnivores, and the direct and indirect impact of these changes. For instance, coyote numbers have declined by about 50%, but those that remain are using a novel resource, the carcasses of ungulates killed by wolves (Smith et al. 2003). Altered behavior of elk, moose, and other ungulates has resulted in changes to the vegetation community in Yellowstone and may be facilitating the return of beavers, despite this species' also being a prey of wolves (Ripple et al. 2001, Ripple and Beschta 2003, Smith et al. 2003, Soulé et al. 2003). If the return of wolves truly facilitates the return of beaver to the landscape, the ecosystem-level effects are likely vast, because beavers themselves are ecosystem engineers (Wright et al. 2002). If these shifts in the foraging pressures of ungulates continue, one might expect broad changes to riparian vegetation structure and an associated increase in faunal richness, as already observed in cross-site comparisons that showed differences in moose browsing pressures (Berger et al. 2001).

Like wolves, sea otters also are the top predators in a three-trophic-level community (Paine 1966, Leibold 1996, Steiner 2003). Sea otters limit sea urchin size and densities. In the absence of otters, grazing pressures from urchins devastate kelp forests and result in increased habitat homogeneity and decreased faunal richness (Figure 17.1B). Conversely, colonization of an area by sea otters promotes the growth and maintenance of kelp forests and its associated communities. This top-down cause-and-effect relationship has been observed at multiple sites, from northern California to Alaska's Aleutian Islands (Estes and Duggins 1995). How many otters are necessary to derive the interactive effects of the presence of a top predator? The question is difficult to assess. Recent work (Doroff et al. 2003, Soulé et al. 2003) has shown that the functional response relationships between otters, urchins, and kelp are nonlinear and differ not only between sites, but also within sites. The key is whether the site starts as urchin dominated (without otters) or kelp dominated (with otters). Nonetheless, the proportional biomass of sea otters is likely minor relative to other components of the community.

An intriguing facet of the sea otter case study has been the impact of a new top carnivore, the killer whale (*Orcinus orca*), which has changed the three-trophic-level community to a four-trophic-level community, with relatively few otters, high urchin numbers, and low kelp density (Figure 17.1B; Estes et al. 1998). Recent evidence suggests that the arrival of orcas in this community is a function of prey switching. The traditional prey of killer whales are great whales, and the decline of the great whales may have resulted in orcas' switching to progressively smaller prey before devastating sea otter populations (Springer et al. 2003). Nonetheless, the generality that a top predator is especially important because a few individuals can have such a great effect is upheld. The decline in sea otters and kelp forests over 3300 km of Alaskan Aleutian Island near-shore environments may have been caused by changes in the foraging behavior of as few as *four* whales (Estes et al. 1998). The ability of non-Carnivora such as killer whales to determine community structure in systems that also include large Carnivora again emphasizes that it is not taxonomic membership per se that defines whether a species is an apex predator. Species such as large sharks, eagles, snakes, or crocodiles all represent likely candidate organisms that may drive the shape of communities and perhaps ecosystems. A similar phenomenon, called *keystone predators*, has been described among invertebrate communities.

Mesocarnivores

These studies clearly indicate that some predators have the potential to strongly structure communities. However, these predators represent large top carnivores in relatively simple systems. Most carnivores are neither large nor at the apex of trophic pyramids. Especially relevant in the context of predator conservation is the ecological role of *mesocarnivores* (mid-sized carnivores) in structuring communities. Mesocarnivores make up the majority of species in the order Carnivora, but our understanding of their ecology is superficial. Indeed, although the term *mesocarnivore* itself has long been used in the ecological literature, we are not aware of a clear quantitative definition. Buskirk (1999) simply defines mesocarnivores as being 1–15 kg, but the term is often construed in a general sense to include all Carnivora that are not large top predators. Consequently, more than 90% of terrestrial Carnivora are mesocarnivores (Gittleman 1985).

The importance of small and mid-sized carnivores can be assessed at two levels: the role assumed by these predators when they are de facto top carnivores and the importance of these predators within communities that also contain large top carnivores. The former is a topic that has received a fair amount of attention under the heading of *mesocarnivore* (or *mesopredator*) *release* (Terborgh and Winter 1980, Soulé et al. 1988, Crooks and Soulé 1999). Because carnivore communities are strongly influenced by intraguild predation and competition, the loss of a top predator may result in a rise and fall of some secondary and tertiary carnivores, respectively. For instance, the loss of wolves from a community can result in an increase in coyotes that, in turn, can result in a decrease in foxes (Johnson et al. 1996). Conversely, a loss of coyotes can result in increases in foxes and other mesopredators (Sovada et al. 1995, Henke and Bryant 1999). The consequences of these interactions, which fundamentally involve a body-size-based shift in trophic status of mesocarnivores from formerly secondary or tertiary predator to apex predator, can be seen throughout the food web and in fundamental measures of biodiversity (Crooks and Soulé 1999).

Scenarios in which mesocarnivores are apex carnivores do not occur solely in disturbed systems where top carnivores have been lost. There are numerous regions without larger mammalian carnivores in which mesocarnivores such as procyonids, mustelids, and small canids (<8 kg) act as top carnivores (e.g., island systems; Roemer et al. 2002, Cuarón et al. 2004). There also are habitats within areas that contain larger species that mesocarnivores species do not use. For example, otters (e.g., river otter, *Lutra canadensis*; range, 6–14 kg) dominate many freshwater habitats. Might river otters control aquatic systems in the same way sea otters and wolves may control marine near-shore and boreal forest communities?

In Missouri, river otters were virtually extirpated but were restored through intensive reintroduction efforts in the early 1980s and currently number about 10,000–18,000 (Missouri Department of Conservation, unpublished data). For this region, a stream system food web conceptualized from otter, fisheries, and macroinvertebrate studies suggests a four- or five-trophic-level system (Figure 17.2) in which river otters feed on centrarchids, carps, suckers, and the large crayfish that these fish, in turn, usually avoid (Rabeni 1992, Roberts 2003). Crayfish are known to strongly influence lower trophic levels (Whitledge and Rabeni 1997), so we expect that as an indirect effect of otter reintroduction, invertebrate and algal biomass should increase and decrease, respectively. A possible river-otter-driven trophic cascade, illustrates that even mid-sized carnivores can be important drivers (sensu Soulé et al. 2003) of communities and ecosystems. Indeed, such a finding would be in congruence with results of other aquatic community studies that have stressed the importance of top-down interactions (Wooton et al. 1996, Finlay et al. 2002). In freshwater systems where communities are often simple and there is a tight link between primary and secondary productivity, the top carnivores may greatly reduce prey populations, thereby causing trophic cascade (i.e., a series of strong indirect effects on populations two or more links down the food chain; Vanni et al. 1990, Strong 1992, Carpenter and Kitchell 1993, Polis and Strong 1996).

The role of mesocarnivores is less clear in systems where opportunities exist to interact with larger carnivores, because few researchers have closely and explicitly studied trophic cascades involving the loss or gain of mammalian mesocarnivores in systems (a) that contain top mammalian predators and (b) in which the loss or gain of the mesocarnivore is not due to the presence or absence of a larger top carnivore. Instead, studies of mesocarnivores focus more on the direct interactions with one or a series of prey species, assessing to what extent prey numbers are a function of top-down or bottom-up processes (e.g., Krebs et al. 1995, Jedrzejewska and Jedrzejewski 1998). This is not to suggest that indirect effects are unimportant but rather that the trophic cascades associated with these events have not been well documented. In the rare system where mesocarnivore-prey dynamics have been well studied, in a two-trophic-level (or more) framework, results are complex. Long-term time series analyses and large-scale experimental studies of lynx (*Lynx canadensis*) and hare (*Lepus americanus*) interactions (Stenseth et al. 1997, Krebs, Boonstra, et al. 2001, Krebs, Boutin, et al. 2001) indicate that prey population dynamics are caused not by top-down or bottom-up factors acting alone or in the absence of other predators, but by the combination of all of these processes.

If indeed species such as river otter and lynx are fundamental drivers of community structure even in systems that have larger carnivores, are these examples typical of the importance of mesocarnivores? There is currently no clear answer, although several issues are certainly relevant. Most mesocarnivores are omnivores, feeding on multiple prey species and on multiple trophic levels. For example, the Procyonidae

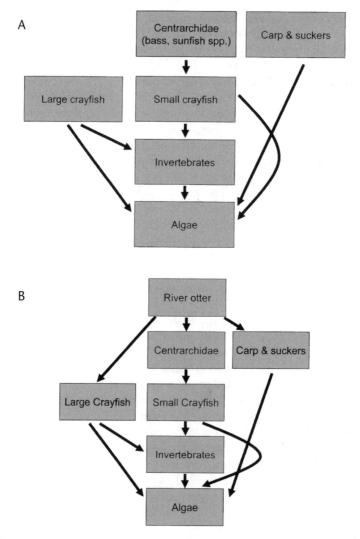

Figure 17.2. Hypothesized primary trophic interactions in a Missouri stream with and without river otters. (A) In systems without river otters, stream fish communities are dominated by Centrarchidae, carps, and suckers. Carps and suckers feed primarily on algae. Centrarchids feed primarily on small crayfish, which in turn feed on invertebrates. While crayfish and some fish also feed on algae, biomass of this trophic level is primarily limited by invertebrate biomass. (B) River otters feed primarily on large fish and on large crayfish (thereby also reducing small crayfish biomass). A decrease in fish and crayfish foraging pressures results in increased invertebrate biomass and an associated decrease in algal biomass. All arrows represent known, ecologically robust linkages (Rabeni 1992, Whitledge and Rabeni 1997, Roberts 2003).

(e.g., raccoons, coatis, and ringtails), Mephitidae (e.g., skunks), and small Canidae (e.g., foxes) all feed on plants, invertebrates, and vertebrates. Diverse feeding habits might act to buffer predator populations against fluctuations in nutrient availability of particular prey species (Eubanks and Denno 1999, and see Eubanks, ch. 1 in this volume). However, most vertebrate prey species of mesocarnivores are more strongly influenced by resource availability than by predation (Desy and Batzli 1989, Krebs et al. 1995, Jedrzejewska and Jedrzejewski 1998). In addition, invertebrates and plant material, principally fruit or nuts, dominate the diet of many mesocarnivore species. Yet, from the perspective of primary producers, mast or fruit output (and perhaps invertebrate population dynamics) is more a function of abiotic factors, and meso-carnivores rarely limit seed-to-seedling transition rates.

Carnivores in Hyperdiverse Environments

Most of the world's biodiversity (including the Carnivora) exists in the tropics. Yet our knowledge about the role of carnivores is from simple, species-poor communities. In such communities, trophic cascades driven by the presence of top predators mimic theoretical expectations (Hairston et al. 1960, Oksanen et al. 1981). In high-diversity tropical environments, however, food webs are far more complex, and have increased capacities for buffering and compensation among species (Strong 1992, Polis and Strong 1996). What is the empirical support for top vertebrate predators' controlling community structure in biodiverse terrestrial systems?

This issue has been studied mainly in Central and South America, where the apex predators are puma (*Felis concolor*) and jaguar (*Panthera onca*). If a single species of carnivore can drive community structure in biodiverse regions of the neotropics, evidence should be most apparent from studies of these species. Loss of these predators would result in increased population densities of the mesocarnivores, browsing mammals (e.g., deer), and seed predators (e.g., large rodent species) that are generally limited by large felids. In a scenario suggested by Terborgh and colleagues, outbreaks of these later species would form part of a trophic cascade that ultimately results in severely reduced seed and seedling survival, high rates of avian nest predation, and overall forest degradation (Terborgh and Winter 1980, Terborgh 1990, Terborgh 1992, Terborgh et al. 1997, Terborgh et al. 2001).

Problematically, much of the evidence for the importance of predators in neotropical forests is based on observations from extremely degraded environments or very small habitat remnants that have lost not one or two species, but rather about 75% of their vertebrate fauna (Terborgh et al. 1997, Terborgh et al. 2001, Lambert et al. 2003). Studies of more intact systems, or systems that have lost only a single predator, are rare. Such studies that have implicated the importance of large cats have often focused on Barro Colorado Island (BCI), Panama, a site where large felids sometimes visit but are not resident, so predation pressures from large felids is considered weak. For instance, comparisons of BCI and Manu, Peru, where large felids are common and predation risk is probably high, suggest that densities of many mid-sized mammals on BCI are inordinately intense because of loss of large predators (Terborgh and Winter 1980, Terborgh 1990). Yet some (Wright et al. 1994) question both the notion that mammalian densities on BCI are extreme and the conclusions taken from the BCI-Manu comparison regarding the importance of top-down processes in structuring neotropical communities. Studies of seed and seedling herbivory by mid-sized mammals at both sites also identified no differences in herbivore pressures (Terborgh and Wright 1994), although comparisons between BCI and an adjacent mainland site where large felids persist did identify greater levels of seed loss (Asquith et al. 1997). These equivocal results indicate the need for additional research to buttress any suggestion for the importance of large predators as community or ecosystem drivers in regions with high biodiversity.

Synopsis: Are Carnivora Important?

From the evidence discussed we conclude that in some systems mammalian carnivores do play a fundamental role in structuring faunal and floral communities, and it is likely that this influence is of such magnitude as to drive ecosystem-level changes. We also observe that the influence of carnivores derives not from the size of the animal per se, but rather from its ability to limit the biomass or resource use of an entire lower trophic level or functional guild. Given relatively few long-term studies of the indirect effects of Carnivora outside temperate systems, it is too early to determine whether the disproportional importance of individual predators is a general rule of thumb for more biodiverse regions. There has also been little study of the importance of the presence or absence of Carnivora that are true omnivores.

If carnivores are potentially very important in structuring communities, it follows that we should be especially mindful of the persistence of these species. Carnivores not only may be disproportionately important, but also may suffer disproportionately high risks of local extirpation and global extinction (Gittleman et al. 2002). We therefore now address where and why there is an interesting disparity between some carnivores that are at a high risk of extinction, such as the giant otter and black-footed ferret, and others, like the small Indian mongoose and red fox, whose populations are thriving.

Global Range and Species Richness Patterns

Where species live and what factors influence the processes of distributional decline are paramount for understanding changes in global biodiversity (Gaston and Spicer 1998). Thus, for our discussion of how mammalian carnivores can be used as models for understanding predators, it is important to know where carnivores live around the planet as a first step toward global conservation. Admittedly, worldwide distributions of species biodiversity are poorly described; most species have not been mapped even in the most rudimentary fashion (Gaston 2003). This situation is frustrating, because, of all the rubrics of conservation, we know that small geographic range size is the single greatest biological factor contributing to extinction. Thus, the fact that over half of all mammal species live in a land area smaller than the country of Spain, or that, among single species, the black-footed ferret has an extremely small range compared to the tens of millions of sq. km of the red fox, the widest of all globally distributed mammals, is both biologically meaningful and importantly symbolic for conservation science. Our lack of geographic knowledge is perhaps not surprising, given that the alpha taxonomy for most species is not known (Wilson 2002). For mammals, a complete geographic range database has only recently been placed in any sort of geographic information system (GIS) platform so that global distributions of species richness and extinction risk can be assessed (Brown et al. 1996). Here, we summarize results from Sechrest's (2003) new geographic range database for all mammals, results based on the digitizing of more than 1,700 source maps and on using them in a GIS to assess general patterns of carnivore extinction risk.

In terms of global mammal distributions, most mammal species have fairly small geographic ranges (Figure 17.3), the extent of occurrences being less than 250,000 sq. km (Sechrest 2003). Across all organisms studied so far and across different ecological scales, species range size distributions show that the smallest range size class is the most common (Gaston 2003). That is, there are many more species with narrow global ranges and few with large ranges, suggesting that most species are relatively rare, at least spatially. For mammals, this skew is observed in both predatory carnivores (including canids, felids, and mustelids) and clades that usually comprise their prey (artiodactyls, lagomorphs, and rodents).

The geographic range size of terrestrial carnivores is generally large and statistically greater than species in other mammalian orders, with an average of 6 373 986 sq. km for Carnivora, compared to an average for all mammals of over 1.5 million sq. km, areas roughly comparable to Australia and Alaska, respectively. Carnivores characteristically are able to move over large areas because of their relatively large body size, extensive home ranges, flexibility in diet and physiology, habitat breadth, and overall need to move in response to fluctuations in prey. Simultaneously, some carnivores are also relatively immune to anthropogenic effects, and, as their habitats are shrinking, the range declines are not as great. For example, across a select group of mammals, the spotted hyaena has lost only 14% of its global range (Ceballos and Ehrlich 2002). Of the 28 orders of mammals (Wilson and Reeder 1993), only the Carnivora, Artiodactyla, and Didelphimorphia have a small proportion (<20%) that are endemic with restricted ranges of less than 50 000 sq. km, a threshold for generally defining endemism (Terborgh and Winter 1983). The only truly endemic carnivores are island forms such as the Malagasy clade of herpestids and viverrids and the California Channel island fox (*Urocyon littoralis*). Overlaying all range distributions across mammals reveals that the greatest species richness is around the equator, with 34% (1,621 of 4,740 terrestrial mammals) ranging in and around the tropics (Figure 17.4A). A similar pattern is observed across genera and families, indicating that geographic ranges are to some extent phylogenetically correlated (Jones et al. 2004).

Do most carnivore global distributions coincide with the distributions of their prey? Surprisingly, this question has never been addressed quantitatively on a global scale. The reason is a combination of lack of accurate maps for most individual species, lack of a complete database of maps across large taxa, and lack of analytical tools for overlaying hundreds of the maps onto one another and simultaneously assessing

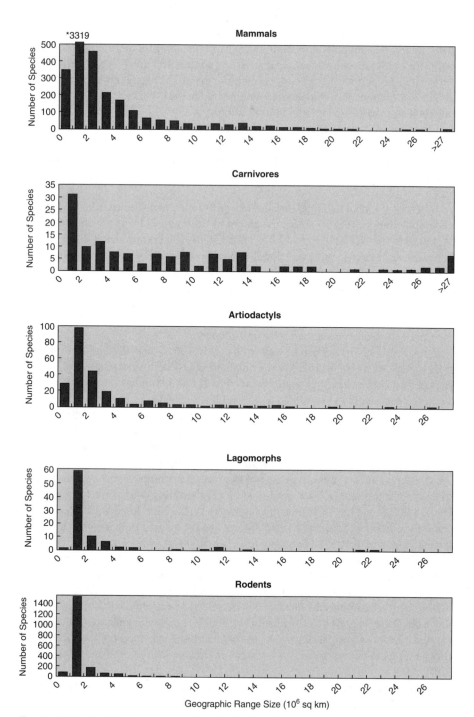

Figure 17.3. Global species range size distributions for mammals, carnivores, and associated taxa (artiodactyls, lagomorphs, rodents) representing prey. Range size × one million sq. km. Based on data compiled from Sechrest (2003).

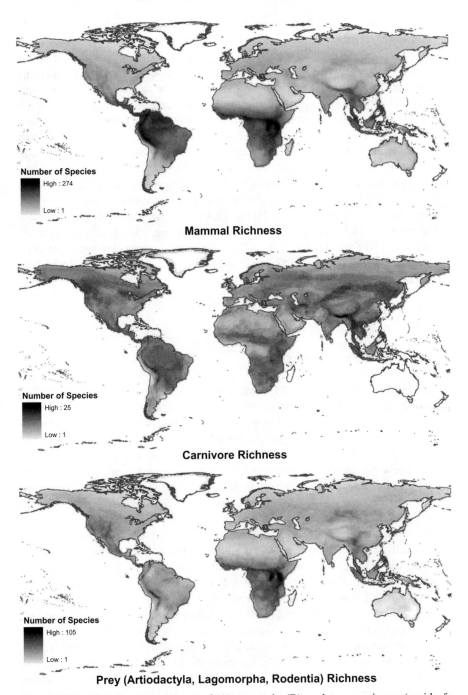

Mammal Richness

Number of Species
High : 274
Low : 1

Carnivore Richness

Number of Species
High : 25
Low : 1

Prey (Artiodactyla, Lagomorpha, Rodentia) Richness

Number of Species
High : 105
Low : 1

Figure 17.4. Global species richness of (A) mammals, (B) predatory carnivores (canids, felids, mustelids), and (C) associated prey (artiodactyls, lagomorphs, rodents). Equal area map projection, divided into grid cells of equal area (~111 sq. km). Levels of species richness are represented by color-scale intensities (see scale). Based on data compiled from Sechrest (2003).

degree of overlap. Recently, a program, Biodiversity Research Analysis Tool (BRAT; Smith, E., et al. unpublished), was developed that can sort through species range maps digitized into a GIS platform, quantitatively show the extent of occurrence across hundreds of species, isolate range distributions at any critical level of resolution (e.g., designations for endemics), and use grid overlays to produce species richness maps.

Using the global range database of Sechrest (2003) and calculating richness maps with a gridded one-tenth degree geographic projection, we can compare species richness among carnivores and their prey. Species richness for the carnivore families that comprise species whose diets are primarily predatory (Canidae, Felidae, and Mustelidae) is highest across sub-Saharan tropical Africa and Southeast Asia. Highest species richness for prey is primarily in Africa (again, mainly sub-Saharan), though with a generally more restricted range boundary of species richness, as shown in distributions of different clades in Figure 17.4. The relationship between predator species richness and prey species richness is significant, though only because the sample size is so great (Figure 17.5). Although clearly predators are reliant on prey, the factors that influence extinction risk in predators are only partly related to the characteristics, or at least distribution, of their prey. We now turn to what comparative analyses tell us about the general processes causing species declines of carnivores worldwide.

Processes of Extinction

McKinney (1997) listed at least 24 factors that have been used to explain patterns of extinction. Surprisingly, even though there is an extensive literature on the subject (Lawton and May 1995, Purvis, Jones, et al. 2000, Reynolds 2003), relatively few studies have statistically linked biological traits with known extinction or associated risks. This is a crucial omission in conservation and biodiversity studies, because we need to know which factors currently influence threatened species to predict the species that are likely to be threatened in the future. That is, a science that will serve to prevent species extinctions must develop ways to anticipate which species are likely to be vulnerable.

One of the problems, of course, in developing a method for studying extinction is that it is difficult to assess which species are vulnerable before they become extinct. Currently, the only global classification for all extant species is the IUCN Red List (Hilton-Taylor et al. 2000). The Red List is a system based on demographic and distributional data that are composed by working groups of taxonomic and biological experts. Initially, species classifications were subjective and sporadically updated. Now, the expert groups are more worldwide in their acquisition of data, and the time frame for revising species' status is modernized in step with the length of time it takes to upload information onto the Web. For example, the Red List for mammals was revised every few years, whereas now some species are checked and shifted in status on monthly intervals, with the entire class reevaluated approximately every year. Although Red List classification is not perfect, the database is useful for assessing biological reasons underlying the variance in species extinction risk (Mace 1995, Mace and Balmford 2000).

Carnivore vs. Prey Density

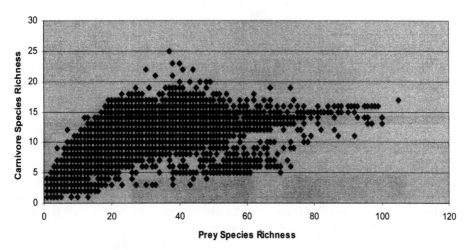

Figure 17.5. Global species richness of predatory carnivores and associated prey taxa.

Using the IUCN Red List as a measure of extinction risk, we can ask two kinds of process questions: Are there biological (intrinsic) traits that increase the likelihood of extinction, and how do these traits interact with the human (extrinsic) impacts? When we code the IUCN Red List into quantitative values (0 = least concerned, 1 = near threatened, 2 = vulnerable, etc.) representative of extinction risk, various ecological and life history traits significantly correlate with increasing levels of threat. In carnivores, over a third of variation is explained by small geographic range size, slow gestation length, late age at sexual maturity, and foraging on high-trophic-level foods (Purvis, Gittleman, et al. 2000). Representative species having such characteristics are many felids, wild dog (*Lycaon pictus*), and black-footed ferret (*Mustela nigripes*). To date, geographic range size is the sole factor that consistently and negatively correlates with threat status in other mammal groups, such as bats, primates, and marsupials (Purvis, Gittleman, et al. 2000, Jones et al. 2003). Other factors are taxon-unique in how they influence extinction risk. For example, in primates, large body size seems to be important, because it enhances conspicuousness and perhaps reduces an ability to adapt to habitat fragmentation. Size in predatory carnivores (i.e., mesocarnivore as opposed to large carnivore) is less important than life histories for increasing vulnerability, whereas in bats wing aspect ratio that represents dispersal ability is the most salient factor. Therefore, small geographic range distribution is the first step toward increasing extinction risk, with other taxon-specific traits becoming significant as they reflect uniquely adaptive characteristics for adjusting to fluctuating environmental conditions.

Biological correlates of extinction thus help in developing a model that may predict why some species are more vulnerable to extinction while others are seemingly immune. We can now address the second part of our question: Do we find that a coupling of anthropogenic factors with these biological correlates increases explanatory power of determining taxonomic distribution of risk? It does, but in an interesting way. In carnivores, contrary to expectation, human population density alone does not correlate with extinction risk, at least as represented by the IUCN classification (Cardillo et al. 2004). But when combined with the significant biological traits just discussed, human population density increases explanatory power in the model up to 80%, revealing that there are strong interactive effects between exposure to human populations and a species' biology. It will be insightful to find out whether other taxonomic groups, as well as other anthropogenic qualities outside human density (e.g., global climate change, exotic species introductions, habitat fragmentation), show similar comparative patterns. Perhaps the interactive effects show something that we have intuitively known. The observation that African wild dogs are going extinct while black bears are proliferating in regions with high human impact (Beckmann and Berger 2003) is clearly not the result of exposure to humans per se. Rather, it is the biological differences and the way each species' evolutionary history and intrinsic characteristics allow the species to adapt to environmental change that is critical. For predators, it is necessary to find out how the distributional change of their food and habitat will impact their extinction risk. As we previously noted, many carnivore populations are influenced by relatively few variables. Identifying these critical variables must increasingly demand our attention.

Saving Predators: A Cruel Bind

In this chapter, we have summarized many factors that are relevant to understanding problems of predator conservation. This is not a thorough list, yet it reveals the complexity of what is involved in making decisions at various scales, from local to regional to global issues, all confronting which predator species are threatened, where the threat is happening, and why the threat is occurring in some species and not in others. Clearly, from the body of evidence reviewed here and described throughout this book, we are amassing effective data to adequately answer many scientific questions of conservation, and developing the means for establishing a predictive science in which we can anticipate what species and geographic areas require close vigilance to prevent extinction.

But, in the end, this is only part of the problem. There remain exceptionally hard decisions that will force us to select which predators to save when we know that the scientific answer is a stalemate. An example is provided by the island fox (*Urocyon littoralis*) endemic to the California Channel islands. Colonization of the islands by golden eagles (*Aquila chrysaetos*) has all but eliminated foxes on three of seven islands (Roemer et al. 2002). What happens when the primary food source (feral pigs), the culprit that initially attracted eagles to cause declines in the foxes, is eradicated? Simulation studies show that the problem will not be solved: eagles will simply shift their attention to foxes, thus increasing the foxes' chances of extinction. The real solution will rest with our decision about whether to protect an endangered predator (the island fox) or a protected predator (the golden eagle). This example should serve as a warning. We will need not only solid scientific data, but also an awareness that it is our decision-making values, in the end, that must conserve the ecology of the world's predators.

Acknowledgments

Thanks to J. Estes for insights on sea otters and near-shore communities and to C. Rabini for insights on river-otter-dominated communities. The work on global geographic ranges was support by a grant from the National Science Foundation (DEB/0129009).

Literature Cited

Asquith, N. M., Wright, S. J., and Claus, M. J. 1997. Does mammal community composition control seedling recruitment in neotropical forests? Evidence from islands in central Panama. Ecology 78:941–946.

Beckmann, J. P., and Berger J. 2003. Rapid ecological and behavioural changes in carnivores: the responses of black bears (*Ursus americanus*) to altered food. J. Zool. 261:207–212.

Berger, J., Stacey, P. B., Bellis, L., and Johnson, M. P. 2001. A mammalian predator-prey imbalance: grizzly bears and wolf extinction affect avian neotropical migrants. Ecol. Applic. 11:947–960

Brown, J. H., Stevens, G. C., and Kaufmann, D. M. 1996. The geographic range: size, shape, boundaries and internal structure. Annu. Rev. Ecol. Syst. 27:597–623.

Buskirk, S. W. 1999. Mesocarnivores of Yellowstone. In: Carnivores in Ecosystems: The Yellowstone Experience (Clark, T. W., Curlee, P. M., Minta, S. C., and Kareiva, P. M., eds.). New Haven, Conn.: Yale University Press; 165–187.

Cardillo, M., Purvis, A., Sechrest, W., Gittleman, J. L., Bielby, J., and Mace, G. M. 2004. Human Population Density and Extinction Risk in the World's Carnivores. Public Library of Science.

Carpenter, S. R., and Kitchell, J. F. 1993. The Trophic Cascade in Lakes. Cambridge: Cambridge University Press.

Ceballos, G., and Ehrlich, P. R. 2002. Mammal population losses and the extinction crisis. Science 296:904–907.

Crooks K. R., and Soulé, M. E. 1999. Mesopredator release and avifaunal extinctions in a fragmented system. Nature 400:563–566.

Cuarón, A., Morales-Martinez, M., McFadden, K. W., Valenzuela, D., and Gompper, M. E. 2004. The status of dwarf carnivores on Cozumel Island, Mexico. Biodiver. Conserv. 13:317–331.

Desy, E. A., and Batzli, G. O. 1989. Effects of food availability and predation on prairie vole demography: a field experiment. Ecology 70:411–421.

Doroff, A. M., Estes, J. A., Tinker, M. T., Burn, D. M., and Evans, J. A. 2003. Sea otter population declines in the Aleutian archipelago. J. Mammal. 84:55–64.

Estes, J. A., and Duggins, D. O. 1995. Sea otters and kelp forests in Alaska: generality and variation in a community ecological paradigm. Ecol. Monogr. 65:75–100.

Estes, J. A., Tinker, M. T., Williams, T. M., and Doak, D. F. 1998. Killer whale predation on sea otters linking oceanic and nearshore ecosystems. Science 282:473–476.

Eubanks, M. D., and Denno, R. F. 1999. The ecological consequences of variation in plants and prey for an omnivorous insect. Ecology 80:1253–1266.

Finlay, J. C., Khandwala, S., and Power, M. E. 2002. Spatial scales of energy flow in food webs of the South Fork Eel River. Ecology 83:1845–1859.

Fuller, T. K., and Sievert, P. R. 2001. Carnivore demography and the consequences of changes in prey availability. In: Carnivore Conservation (Gittleman, J. L., Funk, S., Macdonald, D., and Wayne, R. K., eds.). Cambridge: Cambridge University Press; 163–178.

Gaston, K. J. 1994. Rarity. London: Chapman & Hall.

Gaston, K. J. 2003. The Structure and Dynamics of Geographic Ranges. Oxford: Oxford University Press.

Gaston, K. H., and Spicer, J. I. 1998. Biodiversity: An Introduction. Oxford: Blackwell.

Gittleman, J. L. 1985. Carnivore body size: ecological and taxonomic correlates. Oecologia 67:540–554.

Gittleman, J. L., Funk, S. M., Macdonald, D., and Wayne, R. K. 2002. Carnivore Conservation. Cambridge: Cambridge University Press.

Hairston, N. G., Smith, F. E., and Slobodkin, L. B. 1960. Community structure, population control, and competition. Am. Nat. 94:421–425.

Henke, S. E., and Bryant, F. C. 1999. Effect of coyote removal on the faunal community in western Texas. J. Wildlife Manag. 63:1066–1081.

Hilton-Taylor, C. 2000. 2000 IUCN Red List of Threatened Species. Gland, Switzerland: IUCN.

Jedrzejewska, B., and Jedrzejewski, W. 1998. Predation in Vertebrate Communities: The Bialowieza Primeval Forest as a Case Study. New York: Springer-Verlag.

Johnson, W. E., Fuller, T. K., and Franklin, W. L. 1996. Sympatry in canids: a review and assessment. In: Carnivore Behavior, Ecology, and Evolution (Gittleman, J. L., ed.). Ithaca, N.Y.: Cornell University Press; 189–218.

Jones, K. E., Purvis, A., and Gittleman, J. L. 2003. Biological correlates of extinction risk in bats. Am. Nat. 161:601–614.

Jones, K. E., Sechrest, W., and Gittleman, J. L. 2004. Age and area revisited: identifying global patterns and implications for conservation. In: Phylogeny and Conservation (Purvis, A., Gittleman, J. L., and Brooks, T. M., eds.). Cambridge: Cambridge University Press.

Krebs, C. J., Boonstra, R., Boutin, S., and Sinclair, A. R. E. 2001. What drives the 10-year cycle of snowshoe hares? Bioscience 51:25–35.

Krebs, C. J., Boutin, S., and Boonstra, R. 2001. Ecosystem Dynamics of The Boreal Forest: The Kluane Project. New York: Oxford University Press.

Krebs, C. J., Boutin, S., Boonstra, R., Sinclair, A. R. E., Smith, J. N. M., Dale, M. R. T., Martin, K., and Turkington, R. 1995. Impact of food and predation on the snowshoe hare cycle. Science 269:1112–1115.

Lambert, T. D., Adler, G. H., Riveros, C. M., Lopez, L., Ascanio, R., and Terborgh, J. 2003. Rodents on tropical land-bridge islands. J. Zool. 260:179–187.

Lawton, J., and May, R. M. 1995. Extinction Rates. Oxford: Oxford University Press.

Leibold, M. A. 1996. A graphical model of keystone predators in food webs: trophic regulation of abundance, incidence, and diversity patterns in communities. Am. Nat. 147:784–812.

Mace, G. M. 1995. Classification of threatened species and its role in conservation planning. In: Extinction Rates (Lawton, J. H., and May, R. M., eds.). Oxford: Oxford University Press; 197–213.

Mace, G. M., and Balmford, A. 2000. Patterns and processes in contemporary mammalian extinction. In: Priorities for the Conservation of Mammalian Diversity (Entwhistle, A., and Dunstone, N., eds.). Cambridge: Cambridge University Press; 27–52.

McInnes, P. F., Naiman, R. J., Pastor, J., and Cohen, Y. 1992. Effects of moose browsing on vegetation and litter of the boreal forest, Isle Rale, Michigan, USA. Ecology 73:2059–2075.

McKinney, M. L. 1997. Extinction vulnerability and selectivity: combining ecological and paleontological views. Annu. Rev. Ecol. Syst. 28:495–516.

McLaren, B. E., and Peterson, R. O. 1994. Wolves, moose, and tree rings on Isle Rale. Science 266:1555–1558.

Mech, L. D. 1966. The wolves of Isle Rale. Fauna Series no. 7. Washington, D.C.: U.S. National Park Service.

Oksanen, L., Fretwell, S. D., Arruda, J., and Niemela, P. 1981. Exploitation ecosystems in gradients of primary productivity. Am. Nat. 118:240–261.

Paine, R. T. 1966. Food web complexity and species diversity. Am. Nat. 100:65–75.

Pastor, J., Dewey, B., Naiman, R. J., MacInnes, P. F., and Cohen, Y. 1993. Moose browsing and soil fertility in the boreal forests of Isle Rale National Park. Ecology 74:467–480.

Polis, G. A., and Strong, D. R. 1996. Food web complexity and community dynamics. Am. Nat. 147:813–846.

Post, E., Peterson, R. O., Stenseth, N. C., and McLaren, B. E. 1999. Ecosystem consequences of wolf behavioural response to climate. Nature 401:905–907.

Purvis, A., Gittleman, J. L., Cowlishaw, G., and Mace, G. M. 2000. Predicting extinction risk in declining species. Proc. R. Soc. Lond. 267:1947–1952.

Purvis, A., Jones, K. E., and Mace, G. M. 2000. Extinction. BioEssays 22:1123–1133.

Rabeni, C. F. 1992. Trophic linkage between stream centrarchids and their crayfish prey. Can. J. Fish. Aquat. Sci. 49:1714–1721.

Reynolds, J. D. 2003. Life histories and extinction risk. In: Macroecology (Blackburn, T. M., and Gaston, K. J., eds.). Oxford: Blackwell; 195–217.

Ripple, W. J., and Beschta, R. J. 2003. Wolf reintroduction, predation risk, and cottonwood recovery in Yellowstone National Park. For. Ecol. Manag. 184:299–313.

Ripple, W. J., Larsen, E. J., Renkin, R. A., and Smith, D. W. 2001. Trophic cascades among wolves, elk and aspen on Yellowstone National Park's northern range. Biol. Conser. 102:227–234.

Roberts, N. 2003. River otter food habits in Missouri (M.S. thesis, University of Missouri, Columbia, Mo.).

Roemer, G. W., Donlan, C. J., and Courchamp, F. 2002. Golden eagles, feral pigs and insular carnivores: how exotic species turn native predators into prey. Proc. Nat. Acad. Sci. U.S.A. 99:791–796.

Sechrest, W. W. 2003. Global diversity, endemism, and the conservation of mammals (Ph.D. dissertation, University of Virginia, Charlottesville, Va.).

Smith, D. W., Peterson, R. O., and Houston, D. B. 2003. Yellowstone after wolves. Bioscience 53:330–340.

Soulé, M. E., Bolger, D. T., Alberts, A. C., Wright, J., Sorice, M., and Hill, S. 1988. Reconstructed dynamics of rapid extinctions of chaparral-requiring birds in urban habitat islands. Conser. Biol. 2:75–92

Soulé, M. E., Estes, J. A., Berger, J., and Martinez del Rio, C. 2003. Ecological effectiveness: conservation goals for interactive species. Conser. Biol. 17:1238–1250.

Sovada, M. A., Sargeant, A. B., and Grier, J. W. 1995. Differential effects of coyotes and red foxes on duck nest success. J. Wild. Manag. 59:1–9.

Springer, A. M., Estes, J. A., van Vliet, G. B., Williams, T. M., Doak, D. F., Danner, E. M., Forney, K. A., and Pfister, B. 2003. Sequential megafaunal collapse in the North Pacific Ocean: an ongoing legacy of industrial whaling? Proc. Nat. Acad. Sci. U.S.A. 100:12223–12228.

Steiner, C. F. 2003. Keystone predator effects and grazer control of planktonic primary production. Oikos 101:569–577.

Stenseth, N. C., Falck, W., Bjørnstad, O. N., and Krebs, C. J. 1997. Population regulation in snowshoe hare and Canadian lynx: asymmetric food web configurations between lynx and hare. Proc. Natl. Acad. Sci. U.S.A. 94:5147–5152.

Strong, D. R. 1992. Are trophic cascades all wet? Differentiation and donor-control in speciose ecosystems. Ecology 73:747–754.

Terborgh, J. 1990. The role of felid predators in neotropical forests. Vida Silvest. Neotrop. 2:3–5.

Terborgh, J. 1992. Maintenance of diversity in tropical forests. Biotropica 24:283–292.

Terborgh, J., Lopez, L., Nunez, P., Rao, M., Shahabuddin, G., Orihuela, G., Riveros, M., Ascanio, R., Adler, G. H., Lambert, T. D., et al. 2001. Ecological meltdown in predator-free forest fragments. Science 294:1923–1926.

Terborgh, J., Lopez, L., and Tellos, J. S. 1997. Bird communities in transition: the Lago Guri Islands. Ecology 78:1494–1501.

Terborgh, J., and Winter, B. 1980. Some causes of extinction. In: Conservation Biology: An Evolutionary Ecological Perspective (Soule, M. E., and Wilcox, B. A., eds.). Sunderland, Mass.: Sinauer; 119–133.

Terborgh, J., and Winter, B. 1983. A method for siting parks and reserves with special reference to Columbia and Ecuador. Biol. Conser. 27:45–58.

Terborgh, J., and Wright, S. J. 1994. Effects of mammalian herbivores on plant recruitment in two neotropical forests. Ecology 75:1829–1833.

Vanni, M. J., Luecke, C., Kitchell, J. F., Allen, Y., Temte, J., and Magnuson, J. J. 1990. Effects on lower trophic levels of massive fish mortality. Nature 344:333–335.

Vucetich, J. A., and Peterson, R. O. 2004. The influence of top-down, bottom-up, and abiotic factors of the moose (*Alces alces*) population of Isle Rale. Proc. R. Soc. Lond. B 271:183–189.

Wayne, R. K., Gilbert, D. A., Lehman, N., Hansen, K., Eisenhawer, A., Girman, D., Peterson, R. O., Mech, L. D., Gogan, P. J. P., Seal, U. S., and Krumenaker, R. J. 1991. Conservation genetics of the endangered Isle Rale gray wolf. Conser. Biol. 5:41–51.

Whitledge, G. W., and Rabeni, C. F. 1997. Energy sources and ecological role of crayfishes in an Ozark stream: insights from stable isotopes and gut analyses. Can. J. Fish. Aquat. Sci. 54:2555–2563.

Wilson, D. S., and Reeder, D.-A. 1993. Mammals Species of the World. Washington, D.C.: Smithsonian Institution Press.

Wilson, E. O. 2002. The Future of Life. New York: Knopf.

Wootton, J. T., Parker, M. S., and Power, M. E. 1996. The effect of disturbance on river food webs. Science 273:1558–1560

Wright, J. P., Jones, C. G., and Flecker, A. S. 2002. An ecosystem engineer, the beaver, increases species richness at the landscape scale. Oecologia 132:96–101.

Wright, S. J., Gompper, M. E., and DeLeon, B. 1994. Are large predators keystone species in neotropical forests? The evidence from Barro Colorado Island. Oikos 71:279–294.

Index